Signal Processing and Radioengineering

Signal Processing and Radioengineering

Edited by **George Pilato**

NY RESEARCH PRESS
New York

Published by NY Research Press,
23 West, 55th Street, Suite 816,
New York, NY 10019, USA
www.nyresearchpress.com

Signal Processing and Radioengineering
Edited by George Pilato

International Standard Book Number: 978-1-63238-517-8 (Hardback)

Printed in the United States of America.

Contents

Preface

Signal processing is the technology used in transferring information contained in different formats called signals. It uses different tools for acquisition, extraction, recovery, sensing, learning and modeling. Radio engineering is mainly concerned with control, coverage and generation of signals of transmission systems. The important topics covered under this discipline include antennas, receivers, amplifiers, tuners, modulators and demodulators. This book discusses the fundamentals as well as modern approaches of signal processing and radioengineering. It includes some of the vital pieces of work being conducted across the globe on various topics related to signal processing and radioengineering. This book attempts to understand the multiple branches that fall under these fields and how such concepts have practical applications. It will serve as a valuable source of reference for graduate and post graduate students.

After months of intensive research and writing, this book is the end result of all who devoted their time and efforts in the initiation and progress of this book. It will surely be a source of reference in enhancing the required knowledge of the new developments in the area. During the course of developing this book, certain measures such as accuracy, authenticity and research focused analytical studies were given preference in order to produce a comprehensive book in the area of study.

This book would not have been possible without the efforts of the authors and the publisher. I extend my sincere thanks to them. Secondly, I express my gratitude to my family and well-wishers. And most importantly, I thank my students for constantly expressing their willingness and curiosity in enhancing their knowledge in the field, which encourages me to take up further research projects for the advancement of the area.

Editor

1

Visible Light Communications towards 5G

Stanislav ZVANOVEC [1], *Petr CHVOJKA* [1], *Paul Anthony HAIGH* [2], *Zabih GHASSEMLOOY* [3]

[1] Dept. of Electromagnetic Field, Czech Technical University in Prague, Technicka 2, 166 27 Prague, Czech Republic
[2] Faculty of Engineering, University of Bristol, Bristol, BS8 1TR, UK
[3] Optical Communications Research Group, Faculty of Engineering and Environment, Northumbria University, Newcastle-upon-Tyne NE1 8ST, UK

xzvanove@fel.cvut.cz, petr.chvojka@fel.cvut.cz, paul.anthony.haigh@bristol.ac.uk, z.ghassemlooy@northumbria.ac.uk

Abstract. *5G networks have to offer extremely high capacity for novel streaming applications. One of the most promising approaches is to embed large numbers of co-operating small cells into the macro-cell coverage area. Alternatively, optical wireless based technologies can be adopted as an alternative physical layer offering higher data rates. Visible light communications (VLC) is an emerging technology for future high capacity communication links (it has been accepted to 5GPP) in the visible range of the electromagnetic spectrum (~370–780 nm) utilizing light-emitting diodes (LEDs) simultaneously provide data transmission and room illumination. A major challenge in VLC is the LED modulation bandwidths, which are limited to a few MHz. However, myriad gigabit speed transmission links have already been demonstrated. Non line-of-sight (NLOS) optical wireless is resistant to blocking by people and obstacles and is capable of adapting its' throughput according to the current channel state information. Concurrently, organic polymer LEDs (PLEDs) have become the focus of enormous attention for solid-state lighting applications due to their advantages over conventional white LEDs such as ultra-low costs, low heating temperature, mechanical flexibility and large photoactive areas when produced with wet processing methods. This paper discusses development of such VLC links with a view to implementing ubiquitous broadcasting networks featuring advanced modulation formats such as orthogonal frequency division multiplexing (OFDM) or carrier-less amplitude and phase modulation (CAP) in conjunction with equalization techniques. Finally, this paper will also summarize the results of the European project ICT COST IC1101 OPTICWISE (Optical Wireless Communications - An Emerging Technology) dealing VLC and OLEDs towards 5G networks.*

Keywords

5G networks, light emitting diodes, visible light communications

1. Introduction

In recent years, the worldwide growth in mobile data traffic has led to the development of new technologies for future high capacity communication systems. Every year the number of wireless devices such smartphones, laptops and tablets increases, thus multimedia content becomes the main part of the overall mobile data transferred. This fact results in an increasing throughput requirement from the next generation of communication networks (5G), which are expected to be deployed beyond 2020. Network designers face several critical challenges, all of which need to be addressed, such as optimal spectra allocation, high capacity broadband links, power consumption, quality of services (QoS) and mobility. For instance, approximately one exabyte (EB) of data was transferred across the entire global internet in 2000 (0.083 EB/month) [1]. In contrast, ~30 times more data was carried by the mobile networks per month in 2014, which corresponds to ~2.5 EB/month (refer to Fig. 1). Moreover, the latest projections from Cisco predict that the overall mobile data traffic will reach approximately 24 EB/month by 2019, which is approximately one order of magnitude larger than 2014 (Fig. 1). This corresponds to a compound annual growth rate (CAGR) of 57% for the 2014-2019 period. Following the projection Fig. 1 also shows the fit for this data, which estimates that > 30 EB/month will be transmitted beyond 2020. Most of this mobile data traffic (up to 69%) is expected to consist of video and media by the end of 2018 [1].

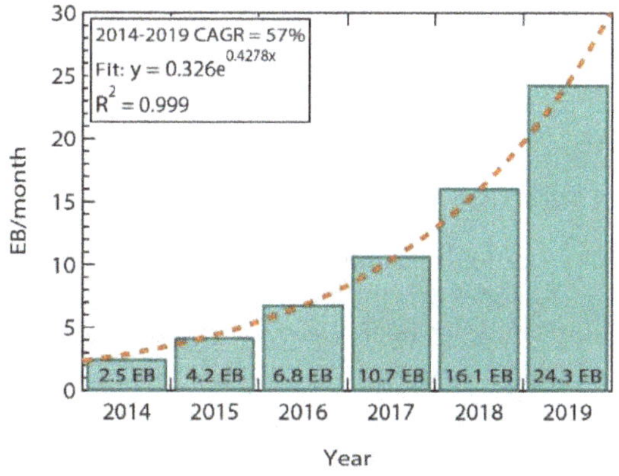

Fig. 1. A prediction of the mobile data traffic per month.

5G networks are expected to meet all the mentioned requirements [2], [3]. The network architecture is expected to be changed dramatically and the limited frequency bands must be used more efficiently. Future systems will be based on heterogeneous networks (HetNets) and advanced radio access technologies (RATs). HetNets include several small cells featuring low transmission power and small coverage area, thus enabling high cell density. Such a system configuration allows spectral reuse, hence improving the capacity of the wireless channel [2], [3], [4]. The cellular architecture should be designed to separate indoor and outdoor scenarios and support massive multiple-input multiple-output (MIMO) technologies with distributed antenna systems [2], [3].

As the number of communication devices increases and the demand for connections grows, power efficiency becomes one of the most important issues for 5G networks. Around 2% of all carbon-dioxide emissions worldwide are produced by communication technologies; this will increase significantly with the aforementioned increasing mobile data traffic demand [5]. Moreover, ~57% of overall wireless network energy consumption is dissipated in radio access nodes [6]. Thus, 'green' and energy efficient systems should reduce the CO_2 emissions and decrease the operating costs.

Indoor communication systems can offer a solution to all the above issues, such as mm-wave systems (3 to 300 GHz) or optical wireless communications such as VLC, which is carried at 400–490 THz. 5G-VLC offers a number of small cells, also referred to as attocells in the literature, in the indoor environment, thus enabling many advantages such as high capacity data transmission, excellent mobility and energy efficient management. Recent experiments demonstrated up to 1.6 Tbit/s for optical wireless backhaul links at 1550 nm [7] and up to 3.4 Gbit/s [8] in VLC based access networks. Moreover, two functionalities are provided with VLC, i.e. the combining of data transmission with illumination of the room using light-emitting diodes (LEDs), which is not available with other network technologies. This paper provides a wide overview of the VLC technology and summarizes its development and the state-of-the-art. The rest of the paper is organized as follows: Section 2 describes the fundamental VLC principles with emphasis on inorganic and organic LEDs, Section 3 is focused on the modulation formats and finally in Section 4, perspective VLC applications are discussed.

2. Visible Light Communications Technology

License-free spectrum, practically unlimited channel bandwidth, high capacity links and energy efficiency are the main features of light fidelity (Li-Fi) networks. Li-Fi is a subset of the VLC domain, which refers to broadcasting access networks with multiple users. The transmitter consists of LEDs either singularly or in an array that are intensity modulated at a rate above which it is not perceivable by the human eye. Since full room illumination is required, the visible range of the electromagnetic spectrum (~370 to 780 nm) is utilized in VLC. Moreover, VLC has been proposed for future 5G networks standards.

There are two types of LEDs used to produce white light: (*i*) inorganic metal alloy semiconductor LEDs (usually a blue emitting gallium nitride (GaN) interface with a cerium doped yttrium aluminum garnet (Ce:YAG) color converting phosphor) and (*ii*) organic LEDs (OLEDs) made from either small molecules or polymers, using epitaxial or wet processing methods, respectively [9], [10]. Inorganic LEDs are commonly used in VLC, while generic OLEDs are attracting significant attention for future VLC networks. When dissolved into solvents and processed with wet methods (i.e. inkjet printing, spray coating) polymer based OLEDs offer several significant advantages over inorganic devices, namely; large, arbitrarily shaped photoactive areas limited only by the size of the printing apparatus, mechanical flexibility, low temperature and ultralow costs.

2.1 Inorganic LEDs

The most common type of inorganic LEDs producing white light are Ce:YAG converted GaN diodes, as mentioned, which are known as white phosphor LEDs (WPLEDs). Alternatively, white light can be produced using a single LED package with on-board red, green and blue (RGB) chips. Considering WPLEDs, the main impediment in achieving high data rates is the modulation bandwidth. GaN diodes can offer modulation bandwidths up to several hundred MHz [11], however the Ce:YAG phosphor layer has a slow transient response, reducing the bandwidth down to the low MHz region, which common values around 4-5 MHz [12]. Thus, increasing the transmission capacity is the key challenge undertaken by researchers. However, the concept of micro-LEDs was introduced in [13], where the photoactive area of the device is reduced significantly to the μm scale, thus relieving the plate capacitance of the device and increasing the bandwidth. Depending on the photoactive area diameter, bandwidths exceeding 400 MHz were demonstrated. Such a device can provide a 3.22 Gb/s throughput when using an adaptive bit- and power-loading technique [14]. On the other hand, reducing the photoactive area means a reduction in optical power, which in turn severely limits the transmission distance. In [13], the optical power ranged from ~0.5 mW to ~5 mW, which is significantly smaller than standard GaN LEDs.

VLC links at Gb/s data rates have already been demonstrated, despite the very limited LED modulation bandwidths. A popular method to increase the capacity of VLC links is to use spectrally efficiency modulation formats such as discrete multi-tone (DMT) [8], [15]. A 3.4 Gb/s transmission speed was achieved in [8] using a single RGB LED and DMT modulation at a distance < 30 cm. Wavelength division multiplexing (WDM) was utilized to transmit independent streams of information on each wavelength. Nevertheless, DMT requires complex signal processing and a feedback channel for bit- and power-loading. A similar approach was adopted in [15] resulting in a transmission speed of 1 Gb/s at a 10 cm distance.

Another possibility for increasing the channel capacity is the implementation of an equalizer [12]. On-off keying (OOK) is the most common modulation format due to simplicity of implementation and compatibility with equalizers. In [12] several equalizers were tested on a commercially available digital signal processing (DSP) board. The LED had low modulation bandwidth (4.5 MHz) and transmission speeds up to 170 Mb/s were recorded using a highly complex artificial neural network (ANN) based equalizer. Real time data processing on field programmable gate arrays (FPGAs) is the next step towards building a fully real time system, as currently most reports in the literature feature offline processing.

2.2 Organic LEDs

There are two types of OLEDs, the first produced using small molecules (SMOLEDs) and the second with polymers (PLEDs). Due to the previously mentioned advantages, organic materials have been focus of the research to be implemented in communication systems. According to [16], the global organic electronic market will approach $80 billion by 2020, thus resulting in (CAGR) of 29.5% between 2014 – 2020.

A possible and generic PLED structure is depicted in Fig. 2. It consists of thin films (total thickness 1-200 nm) of organic emissive and charge transport layers. Polymers such as:

(i) red: poly[2-methoxy-5-(3′,7′-dimethyloctyloxy)-1,4-phenylenevinylene] (MDMO-PPV)

(ii) green: poly[(9,9-di-n-octylfluorenyl-2,7-diyl)-alt-(benzo[2,1,3] thiadiazol-4,8-diyl)] (F8BT) and

(iii) blue: poly(9,9-dioctylfluorene), (poly(9,9-dioctylfluorene-alt-N-(4-butylphenyl)diphenylamine) and poly(9,90-dioctylfluorene-alt-bis-N,N0-(4-butylphenyl)-bis-N,N0-phenyl-1,4-phenylenediamine) (F8:TFB:PFB)

are evaluated as the emissive layer in [17] while a poly(3,4-ethylenedioxythiophene) (PEDOT) doped with poly(styrenesulfonic acid) (PSS) interlayer is used consistently for the charge transport layer in order to minimize the energy gap between the electrode and emissive layer. In Fig. 2 the PLED is a bottom emitter and hence the anode must be transparent (generally indium tin oxide (ITO) is used). When positive and negative charges recombine, the charges combine for a fraction of a second as Frenkel excitons before the energy is emitted as photons. The material can be placed on various substrates including plastic, enabling flexibility.

However, the most limiting factor for OLEDs is very low modulation bandwidth, which is much lower than the bandwidth of conventional inorganic diodes. The reason for this is because the charge transport characteristics of organic materials are approximately three orders of magnitude lower than those of amorphous silicon (a-Si). OLEDs act as a low pass filter with a certain cut-off frequency. The -3 dB frequency is given by:

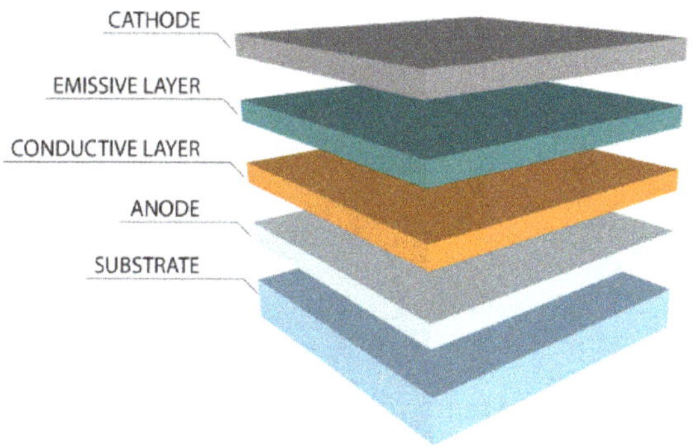

Fig. 2. The structure of OLED.

$$f_c = \frac{1}{2\pi RC} \tag{1}$$

where R is the effective resistance and C is the plate capacitance, given as:

$$C = A\varepsilon_0\varepsilon_r / d \tag{2}$$

where A is the photoactive area and d is thickness of OLED, ε_0 and ε_r are the relative permittivity of free space and the emissive layer, respectively. As d is generally very small (1-200 nm), this tends to a large C, hence a very high capacitance and low bandwidth, which is typically in the order of several hundred kHz [18]. On the other hand, OLEDs with bandwidths up to ~60 MHz were reported in [19], which was achieved by reducing the photoactive area to 0.018 mm^2. Using commercial and custom OLEDs, promising experimental results have been reported in the literature. A SMOLED with a bandwidth 93 kHz is modulated by DMT signal to demonstrate to data rate up to 1.4 Mb/s in [20]. A two-fold improvement of this transmission speed was published in [21] using the same SMOLED as in [20]; a 2.7 Mb/s link is shown using a multi-layer perception (MLP) ANN equalizer offline in MATLAB.

A breakthrough in organic VLC was made in [22], where PLEDs were used to demonstrate transmission speeds in excess of that required for Ethernet connectivity. The data rate of 20 Mb/s was reported in [22], [23] using OOK format and an MLP equalizer. More recently in [17], an aggregated transmission speed of 55 Mb/s was experimentally demonstrated and was achieved using an RGB PLED and WDM, which is a promising improvement for future research activities.

2.3 Other Organic Components

Organic photodetectors (OPD) are also a promising technology and can substitute silicon photodiodes in specific applications. OPDs used in the authors' previous work are based on the bulk heterojunction principle [24]; an interpenetrated and disordered blend of an electron donor and electron acceptor and were fabricated using spray coating [25]. The material costs for the poly(3-hexylthiophene): [6,6]-phenyl C61-butyric acid methylester (P3HT:PCBM) blend are around ~€0.20/cm^2. Furthermore, OPDs can offer superior responsivity in comparison to Si photodetectors in the visible spectrum as shown in Fig. 3.

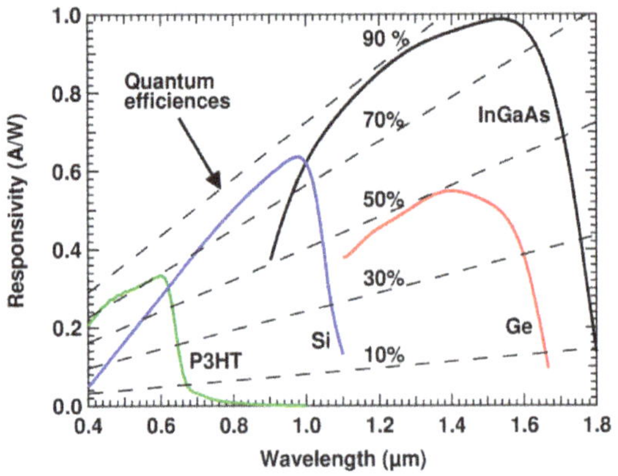

Fig. 3. Photodetector material responsivities.

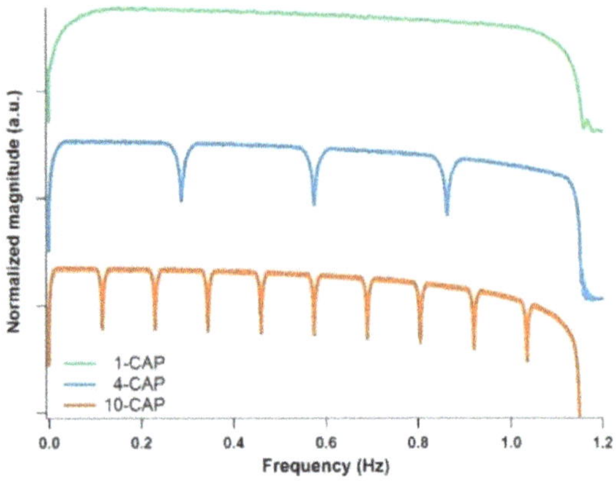

Fig. 4. The concept of m-CAP modulation, where the available system bandwidth is split into 1, 4 and 10 subbands.

The OPD with raw 3 dB bandwidth of 30 kHz is tested in [26] with the resulting data rate of 750 kb/s, which is ~20 times increment representing huge potential of such devices. An OPD with ~160 kHz modulation bandwidth is utilized in [27]. Fourth-order pulse position modulation (4-PPM) and an ANN equalizer were used to achieve a transmission speed of 3.75 Mb/s. Fully organic VLC link was introduced in [28], where the low bandwidth components (up to 135 kHz) with OOK modulation and ANN based filtering were used to achieve data rate exceeding 1 Mb/s.

3. Modulation Formats

Besides equalization techniques, spectrally efficient modulation formats such as orthogonal frequency division multiplexing (OFDM) are a popular way to increase the VLC link capacity. For instance, Gb/s transmissions were reported in [8] and [29] by implementing DMT. Power efficient schemes such as low peak-to-average power ratio (PAPR) single carrier frequency-domain equalization (SC-FDE) was proposed in [30]. Here the IFFT block was moved to the receiver side to avoid the generation of complicated waveforms and consequently reduced PAPR. Further spectral efficiency improvements have been achieved by adopting novel OFDM SC-FDE signal formats - polar OFDM and polar SC-FDE in [31].

Recently, carrier-less amplitude and phase modulation (CAP) has appeared as a candidate for optical systems, which outperforms OFDM channel using the same experimental setup [32]. CAP systems have several advantages over OFDM including no Fourier transform as in OFDM or local oscillator, which is utilized in a single carrier modulations such as quadrature amplitude modulation (QAM). The carrier frequencies in CAP are generated by either analogue or digital finite impulse response filters (FIR).

Unlike OFDM, a flat frequency response is required for CAP meaning the low bandwidths available are the major problem once more, due to the 20 dB/decade attenuation for frequencies outside the bandwidth. A possible solution for this was introduced in [33], where the available bandwidth is divided into a multiband (m-CAP) format for an optical fiber channel. The transmission bandwidth was split into 6 sub-bands (subcarriers) and the performance was compared with the traditional CAP (1-CAP) system. The fiber system transmission speed was 102.4 Gb/s and 100 Gb/s for m-CAP and 1-CAP, respectively, which is not a significant gain, however a gain in transmission speed was not the focus of the article. The m-CAP system noted a significant improvement in dispersion tolerance. Nevertheless, for 1-CAP transmission, two FIR filters are required for signal generation. As m increases, the number of FIR filters grows by $2m$. Thus, one must consider an increment in the system complexity resulting from a multiband approach. On the other hand, such a concept allows optimization of the modulation format used in each subband according to measured signal-to-noise ratio (SNR), i.e. bit- and power-loading in OFDM. The principle of m-CAP is illustrated in Fig. 4. The available system bandwidth is split into m sub-bands. The more sub-bands are utilized, the less bandwidth is occupied by a single subcarrier, thus resulting in a reduced attenuation caused by the LED low-pass frequency response.

A detailed description of the m-CAP modulation format and signal generation can be found in [34], [35]. The first VLC experiment utilizing m-CAP was reported in [35], where a data rate of 31.5 Mb/s was demonstrated using 10-CAP; resulting in a spectral efficiency of 4.85 b/s/Hz, which offers huge potential for a future research.

4. Applications

As the technology has evolved over the last decade and with rapid increase of mobile data requirements, there are several challenging areas for specific deployment of VLC systems. Alongside classic indoor communication schemes (i.e., broadcasting networks), VLC can also be utilized for localization or car-to-car and car-to-infrastructure communications. The following sections highlight the main approaches and principles.

4.1 Indoor Communications

Alongside illumination and data communications, VLC systems have been proposed for indoor positioning with a very high accuracy (a few cm). The IEEE 802.15.7 task group [36] has been forming the new standard for VLC, orienting their efforts towards the PHY and MAC standards since 2009. The main focus from the first releases had been given on slow VLC for indoor positioning by means of optical camera communication. More recently high bit rate VLC transmission systems are under standardization, especially based on results produced by the EU COST IC1101 project OPTICWISE consortium.

For indoor applications with a static environment, VLC can offer high QoS even when the movement of people and shadowing by obstacles perturb the beam (so called non-line of sight (NLOS)). This has been proved by several theoretical and experimental investigations [37], [38]. Typically, such NLOS systems use diffuse reflections, where the transmitter illuminates the ceiling, and/or wide field-of-view or receiving apertures instead of direct links. Data rates up to 400 Mbit/s have been reported for a NLOS link in [39]. The first mobile VLC system was reported for typical indoor distances between 2 m and 20 m with data rates decreasing with distance from approx. 500 Mbit/s to 100 Mbit/s in [40].

The typical model of diffuse reflection was described in [41]. The power efficiency for the diffuse signal can be derived as [41]:

$$\mu_{\mathrm{diff}} = \frac{P_{\mathrm{diff}}}{P_{\mathrm{T}}} = \frac{A_{\mathrm{R}} \sin^2(FOV/2)}{A_{\mathrm{room}}} \frac{\langle \rho \rangle}{1-\langle \rho \rangle} \tag{3}$$

where $\langle \rho \rangle$ is mean reflectivity, FOV stands for field of view of detector and A_{R} and A_{room} are areas of detector and room, respectively. The decay time then can be given as [41]:

$$\tau = -\frac{\langle t \rangle}{\ln \langle \rho \rangle}. \tag{4}$$

The average time $\langle t \rangle$ between two diffuse reflections for a rectangular room with dimensions $l \times w \times h$ (length × width × height) is expressed by [41]:

$$\langle t \rangle = \frac{2lwh}{c(lw + lh + wh)}. \tag{5}$$

The DC gain of the reflected path can afterwards be determined as [10]:

$$H_{\mathrm{ref}}(O) = \begin{cases} \frac{(m+1)}{2(\pi d_1 d_2)^2} \rho A_{\mathrm{det}} dA_{\mathrm{w}} \cos^m(\theta_r) \cos(\alpha) & \\ \cdot \cos(\beta) T_{\mathrm{S}}(\vartheta) g(\vartheta) \cos(\vartheta) & \text{for } 0 \le \vartheta \le \vartheta_{\mathrm{FOV}} \\ 0 & \text{for } \vartheta > \vartheta_{\mathrm{FOV}} \end{cases} \tag{6}$$

where d_1 is the distance between the transmitter and the reflective point, d_2 is the distance between the reflective point and the receiver, ρ is the reflectance coefficient, dA_{w} is a small reflective area on the wall, α is the angle of incidence from the transmitter and β is the angle of irradiance from a reflected point.

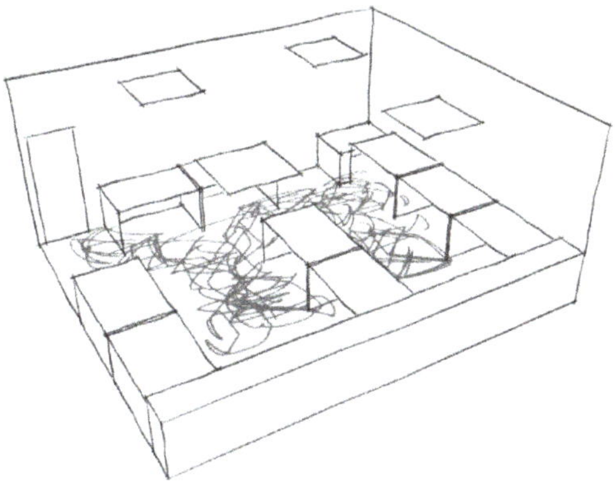

Fig. 5. The simulation of people's movement within an office using VLC with 4-LED.

Fig. 5 illustrates people's movement within a typical office environment utilizing VLC by employing 4 LEDs mounted on the ceiling.

In such a case sometimes the direct paths from transmitter to receiver are blocked or temporally shadowed by a human. Several studies derived the percentage of shadowing. For instance [42] reported shadowing with a probability of < 2% for a multiple-input multiple-output (MIMO) system covering a typical office. Similar results were derived in [43] where time division multiple access was investigated. Higher order reflections induce significant influence on the temporal dispersion according to [37], [38]. The reflection component always appears respective to

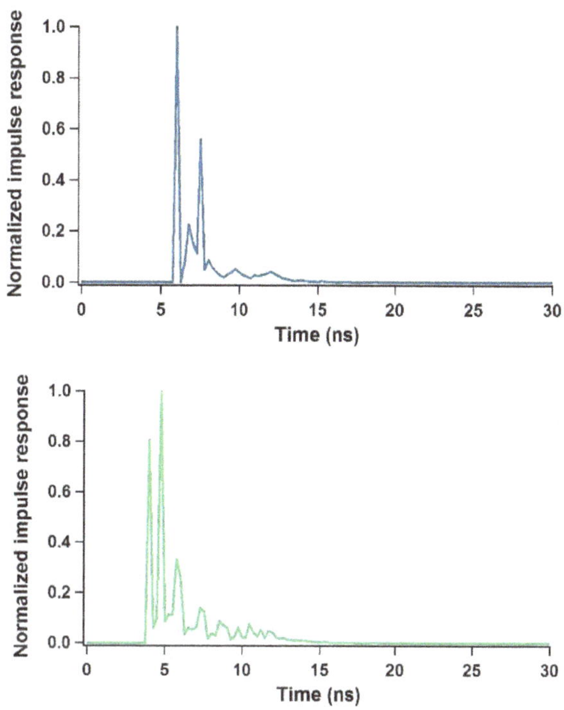

Fig. 6. Examples of the normalized channel impulse response for LOS/NLOS scenario considering blocking signal path by people in an office with: a) 4 LEDs, b) 18 LEDs.

LOS incident paths [38]. This influence can be easily observed in the transmission bandwidth. Two examples of a normalized channel response for the scenario shown in Fig. 5 are illustrated in Fig. 6.

The measured probability density function of the normalized received optical power considering people's movement as well as the analysis of the RMS delay spread for different indoor scenarios and people densities in rooms was carried out in [44]. Based on the measurement campaign the normalized received power showed a Rayleigh distribution with the scale parameter varying from 0.98 to 1.79 for an empty to a crowded room. The RMS delay spread statistics have been derived for three different indoor scenarios. For the case of furnished office environment (people density > 0.16 people/m^2), the cumulative distribution function (CDF) of the received power differs in the worst case by up to 7% contrary to an RMS delay of 2% that was experienced under the same people density in the corridor [44].

4.2 Positioning and Localization

Several positioning systems have been tested over the last few years. In [45] a digital camera was used as a receiver to capture a sequence of images of the LED positioning beacon transmitter. By using image-processing algorithms, the system was able to decode the location information encoded in the visual patterns transmitted by LEDs. The system demonstrated that improved performance can be attained even at low values of SNR. A system for localization of vehicles using the global positioning system (GPS) together with a light beacon device mounted on the vehicle to receive information from transmitter positioned at the road intersections is developed by Honda motors Co., Ltd. in 2010 [46].

Another challenge represents the utilization of the localization within the indoor scenario, where standard GPS signals cannot be received (see illustrative deployment in Fig. 7). In recent years, we have seen research and development in optical based indoor positioning schemes (IPS) offers multitude of advantageous including smaller transceiver size, immunity to electromagnetic interference, and inherent security [47]. VLC based IPS offers the advantage of LED and VLC technologies such as ubiquitous coverage, static channel, multiple lighting elements etc. [48]. Using VLC with synchronization between the transmitter

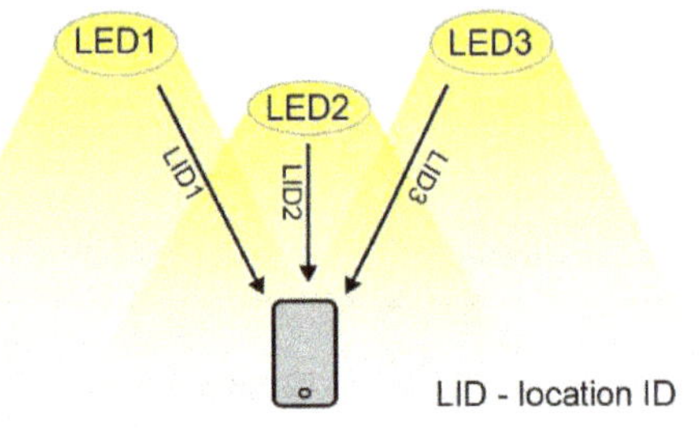

Fig. 7. Concept of VLC positioning system (multi-access mechanism among single terminal and multiple lights + position estimation).

and receiver, the bounds on position estimation accuracy are typically in the order of millimeters or centimeters depending on the geometry of the room, the frequency and power of the transmitted signal and the properties of the LED and the photoreceiver [49]. A VLC-based IPS employing 3-LED, the dual-tone multi-frequency technique and a dedicated algorithm was reported in [48]. Unlike the time-division multiplexing based VLC-IPS, this scheme does not require synchronization between the transmitter and receiver, thus makes it simple, robust and cost effective. VLC-IPS is highly accurate offering an average positioning error of about 1.6 cm, which is much less than many existing IPS.

Typically several transmitters serve as beacons and the RMS delay spread can be used to quantify the amount of multipath distortion that can occur at a particular point within a room [50]. Parameter $D_{\mathrm{rms\_Max}}$ corresponds to the largest multipath distortion, which limits the maximum transmission data rate $R_{\max}$ of the system. $R_{\max}$ for the indoor VLC channel can be calculated following [51], which is given by:

$$R_{\max} \le \frac{1}{10 D_{\mathrm{rms}_{\mathrm{Max}}}} . \tag{7}$$

In order to evaluate the positioning accuracy, we have to consider the Cramer-Rao bound (CRB) [52] as a performance reference, which is the lower bound on the mean square estimation error in the set of unbiased estimates. Typical CRB ranges within the rooms for 4-, 6- and 9-cell configurations were derived for an optimized Lambertian order (OLO) LED case for an indoor cellular optical wireless communication system in [53]. Simulations of particular scenario revealed reached values of CRB from 12.8 cm, from 8.6 cm and from 5.8 cm, respectively, for above mentioned cells' deployment in for rooms of $5 \times 5 \times 3$ m^3, $4 \times 6 \times 3$ m^3, and $5 \times 5 \times 3$ m^3 [53]. This has shown VLC as very useful tool for developers of indoor positioning systems.

4.3 Car-to-Car Communications

VLC can also be utilized for outdoor applications such as the public transport. Note how the infrastructure of public lights has changed over the last 5 years. Typical incandescent lamps have been replaced by LED lighting across whole cities. For example the Los Angeles LED Streetlight Replacement Program has replaced over 140,000 existing streetlights in the city with LED units which has brought energy saving 68 GWh/year and money saving $ 10M/year [54].

Several test use-cases and experimental results have been published for a vehicular VLC network consisting of on-board units, vehicles, and road side units, i.e., traffic lights, street lamps, digital signage etc. Cars fitted with LED-based front and back lights can communicate with each other and with the road side units (RSUs) through the VLC technology. Furthermore, LED-based RSUs can be used for both signaling and broadcasting safety-related

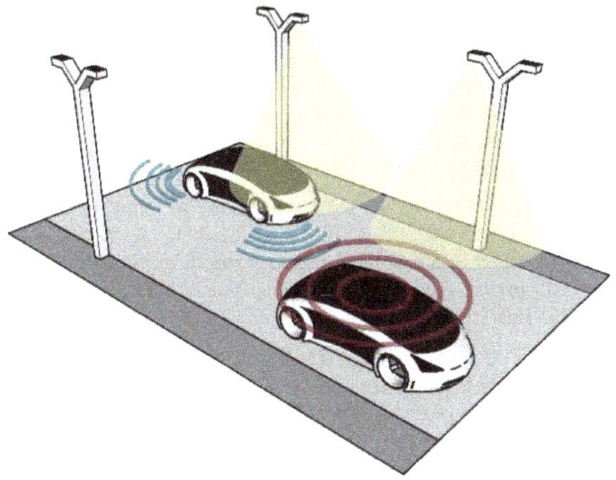

Fig. 8. Concept of VLC for car to car and car to infrastructure.

information to vehicles on the road. Such a network is illustrated in Fig. 8.

An analytical performance analysis of VLC based car-to-car communications for a range of communication geometries consider both the LOS and NLOS paths over a link span of 20 m at a data rate of 2 Mbps was outlined in [55]. Optical wireless communications systems based on an LED transmitter and camera receiver was proposed for automotive applications in [56]. The signal reception experiment has been performed for static and moving camera receivers. Up to 15 Mb/s error-free throughput under fixed conditions was sustained. This represents very good performance for optical wireless communication systems, since the experiment did not involve further correction methods like coding and equalization. In [57] it was shown that the receiver in the driving situation can detect and accurately track an LED transmitter array with error-free communication over distances of 25–80 m. Further tests and experiments are however needed to prove these concepts.

5. Conclusion

The research and development in VLC at a global level has increased more than ten times for last two years. This paper gave an overview of VLC and its use in a number of applications. In outdoor environment VLC can provide internet hot spots using street lighting and mobile access as part of the 5G technology in highly congested areas and within indoor environment it can be used for localization and small cells coverage networks. VLC has been accepted as part of the 802.15.7 task group and is being proposed as a supplement technology in 5G networks. We assume further increase of research efforts within selected applications.

Acknowledgements

This joint research is supported by the EU COST ICT Action IC1101 Optical Wireless Communications: An Emerging Technology (OPTICWISE) and by the Grant Agency of the Czech Technical University in Prague, grant no. SGS14/190/OHK3/3T/13.

References

[1] C. V. N. Index, *Global Mobile Data Traffic Forecast Update*, 2014-2019, White paper, ed. 2013.

[2] BANGERTER, B., TALWAR, S., AREFI, R., STEWART, K. Networks and devices for the 5G era. *IEEE Communications Magazine,* 2014, vol. 52, p. 90–96. DOI: 10.1109/MCOM.2014.6736748

[3] WANG CHENG-XIANG, HAIDER, F., XIQI GAO, XIAO-HU YOU, YANG YANG, DONGFENG YUAN, et al. Cellular architecture and key technologies for 5G wireless communication networks. *IEEE Communications Magazine,* 2014, vol. 52, p. 122 to 130. DOI: 10.1109/MCOM.2014.6736752

[4] LI, Q. C., HUANING NIU, PAPATHANASSIOU, A., GENG WU. 5G network capacity: Key elements and technologies. *IEEE Vehicular Technology Magazine,* 2014, vol. 9, no. 1, p. 71–78. DOI: 10.1109/MVT.2013.2295070

[5] YONG SHENG SOH, QUEK, T. Q. S., KOUNTOURIS, M., HYUNDONG SHIN. Energy efficient heterogeneous cellular networks. *IEEE Journal on Selected Areas in Communications,* 2013, vol. 31, p. 840–850. DOI: 10.1109/JSAC.2013.130503 .

[6] HU, R. Q., YI QUIAN. An energy efficient and spectrum efficient wireless heterogeneous network framework for 5G systems. *IEEE Communications Magazine,* 2014, vol. 52, p. 94–101. DOI: 10.1109/MCOM.2014.6815898

[7] PARCA, G., SHAHPARI, A., CARROZZO, V., TOSI BELEFFI, G. M., TEIXEIRA, A. L. J. Optical wireless transmission at 1.6-Tbit/s (16×100 Gbit/s) for next-generation convergent urban infrastructures. *Optical Engineering,* 2013, vol. 52, no. 11, p. 116102. DOI: 10.1117/1.OE.52.11.116102

[8] COSSU, G., KHALID, A. M., CHOUDHURY, P., CORSINI, R., CIARAMELLA, E. 3.4 Gbit/s visible optical wireless transmission based on RGB LED. *Optics Express,* 2012, vol. 20, no. 26, p. B501–B506. DOI: 10.1364/OE.20.00B501

[9] BURROUGHES, J. H., BRADLEY, D. D. C., BROWN, A. R., MARKS, R. N., MACKAY, K., FRIEND, R. H., et al. Light-emitting diodes based on conjugated polymers. *Nature,* 1990, vol. 347, p. 539–541. DOI:10.1038/347539a0

[10] TANG, C. W., VANSLYKE, S. A. Organic electroluminescent diodes. *Applied Physics Letters,* 1987, vol. 51, p. 913–915. DOI: 10.1063/1.98799

[11] MCKENDRY, J. J. D., GREEN, R. P., KELLY, A. E., ZHENG, G., GUILHABERT, B., MASSOUBRE, D., et al. High-speed visible light communications using individual pixels in a micro light-emitting diode array. *IEEE Photonics Technology Lett.,* 2010, vol. 22, no. 18, p. 1346–1348. DOI: 10.1109/LPT.2010.2056360

[12] HAIGH, P. A., GHASSEMLOOY, Z., RAJBHANDARI, S., PAPAKONSTANTINOU, I., POPOOLA, W. Visible light communications: 170 Mb/s using an artificial neural network equalizer in a low bandwidth white light configuration. *Journal of Lightwave Technology,* 2014, vol. 32, no. 9, p. 1807–1813. DOI: 10.1109/JLT.2014.2314635

[13] MCKENDRY, J. J. D., MASSOUBRE, D., ZHANG, S., RAE, B. R., GREEN, R. P., GU, E., et al. Visible-light communications using a CMOS-controlled micro-light-emitting-diode array. *Journal of Lightwave Technology,* 2012, vol. 30, no. 1, p. 61–67. DOI: 10.1109/JLT.2011.2175090

[14] TSONEV, D., HYUNCHAE, C., RAJBHANDARI, S., MCKENDRY, J. J. D., VIDEV, S., GU, E., et al. A 3-Gb/s single-LED

OFDM-based wireless VLC link using a gallium nitride μLED. *IEEE Photonics Technology Letters,* 2014, vol. 26, p. 637–640. DOI: 10.1109/LPT.2013.2297621

[15] KHALID, A. M., COSSU, G., CORSINI, R., CHOUDHURY, P., CIARAMELLA, E. 1-Gb/s transmission over a phosphorescent white LED by using rate-adaptive discrete multitone modulation. *IEEE Photonics Journal,* 2012, vol. 4, p. 1465–1473. DOI: 10.1109/JPHOT.2012.2210397

[16] SINGH, R. *Global Organic Electronics Market (Application and Geography) - Size, Share, Global Trends, Company Profiles, Demand, Insights, Analysis, Research, Report, Opportunities, Segmentation and Forecast, 2013 – 2020.* 2014.

[17] HAIGH, P. A., BAUSI, F., LE MINH, H., PAPAKONSTANTINOU, I., POPOOLA, W., BURTON, A., et al. Wavelength-multiplexed polymer LEDs: Towards 55 Mb/s organic visible light communications. *IEEE Journal on Selected Areas in Communications.* Accepted, 2014.

[18] HAIGH, P. A., GHASSEMLOOY, Z., RAJBHANDARI, S., PAPAKONSTANTINOU, I. Visible light communications using organic light emitting diodes. *IEEE Communications Magazine,* 2013, vol. 51, p. 148–154. DOI: 10.1109/MCOM.2013.6576353

[19] BARLOW, I. A., KREOUZIS, T., LIDZEY, D. G. High-speed electroluminescence modulation of a conjugated-polymer light emitting diode. *Applied Physics Letters,* 2009, vol. 94, p. 243301–3. DOI: 10.1063/1.3147208

[20] HAIGH, P. A., GHASSEMLOOY, Z., PAPAKONSTANTINOU, I. 1.4-Mb/s white organic LED transmission system using discrete multitone modulation. *IEEE Photonics Technology Letters,* 2013, vol. 25, no. 6, p. 615–618. DOI: 10.1109/LPT.2013.2244879

[21] HAIGH, P. A., GHASSEMLOOY, Z., PAPAKONSTANTINOU, I. HOA LE MINH. 2.7 Mb/s with a 93-kHz white organic light emitting diode and real time ANN equalizer. *IEEE Photonics Technology Letters,* 2013, vol. 25, no. 17, p. 1687–1690. DOI: 10.1109/LPT.2013.2273850

[22] HAIGH, P. A., BAUSI, F., KANESAN, T., LE, S. T., RAJBHANDARI, S., GHASSEMLOOY, Z., et al. A 20-Mb/s VLC link with a polymer LED and a multilayer perceptron equalizer. *IEEE Photonics Technology Letters,* 2014, vol. 26, p. 1975–1978. DOI: 10.1109/LPT.2014.2343692

[23] LE, S. T., KANESAN, T., BAUSI, F., HAIGH, P. A., RAJBHANDARI, S., GHASSEMLOOY, Z., et al. 10 Mb/s visible light transmission system using a polymer light-emitting diode with orthogonal frequency division multiplexing. *Optics Letters,* 2014, vol. 39, p. 3876–3879. DOI: 10.1364/OL.39.003876

[24] BRABEC, C. J., SARICIFTCI, N. S., HUMMELEN, J. C. Plastic solar cells. *Advanced Functional Materials,* 2001, vol. 11, no. 1, p. 15–26.

[25] TEDDE, S. F., KERN, J., STERZL, T., FURST, J., LUGLI, P., HAYDEN, O. Fully spray coated organic photodiodes. *Nano Letters,* 2009, vol. 9, p. 980-3. DOI: 10.1021/nl803386y

[26] HAIGH, P. A., GHASSEMLOOY, Z., HOA LE MINH, RAJBHANDARI, S., ARCA, F., TEDDE, S. F., et al. Exploiting equalization techniques for improving data rates in organic optoelectronic devices for visible light communications. *Journal of Lightwave Technology,* 2012, vol. 30, no. 19, p. 3081–3088. DOI: 10.1109/JLT.2012.2210028

[27] GHASSEMLOOY, Z., HAIGH, P. A., ARCA, F., TEDDE, S. F., HAYDEN, O., PAPAKONSTANTINOU, I., et al. Visible light communications: 3.75 Mbits/s data rate with a 160 kHz bandwidth organic photodetector and artificial neural network equalization. [Invited] *Photonics Research,* 2013, vol. 1, no. 2, p. 65–68. DOI: 10.1364/PRJ.1.000065

[28] HAIGH, P. A., GHASSEMLOOY, Z., PAPAKONSTANTINOU, I., ARCA, F., TEDDE, S. F., HAYDEN, O., et al. A 1-Mb/s visible light communications link with low bandwidth organic components. *IEEE Photonics Technology Letters,* 2014, vol. 26, no. 13, p. 1295–1298. DOI: 10.1109/LPT.2014.2321412

[29] AZHAR, A. H., TRAN, T., O'BRIEN, D. A Gigabit/s indoor wireless transmission using MIMO-OFDM visible-light communications. *IEEE Photonics Technology Letters,* 2013, vol. 25, no. 2, p. 171–174. DOI: 10.1109/LPT.2012.2231857

[30] TEICHMANN, V. S. C., BARRETO, A. N., PHAM, T. T., RODES, R., MONROY, I. T., MELLO, D. A. A. SC-FDE for MMF short reach optical interconnects using directly modulated 850 nm VCSELs. *Optics Express,* Nov 5 2012, vol. 20, no. 23, p. 25369–25377. DOI: 10.1364/OE.20.025369

[31] ELGALA, H., LITTLE, T. D. C. Polar-based OFDM and SC-FDE links toward energy-efficient Gbps transmission under IM-DD optical system constraints [Invited]. *Journal of Optical Communications and Networking,* 2015/02/01, vol. 7, no. 2, p. A277–A284. DOI: 10.1364/JOCN.7.00A277

[32] WU, F. M., LIN, C. T., WEI, C. C., CHEN, C. W., CHEN, Z. Y., HUANG, H. T., et al. Performance comparison of OFDM signal and CAP signal over high capacity RGB-LED-based WDM visible light communication. *IEEE Photonics Journal,* 2013, vol. 5, article no. 7901507. DOI: 10.1109/JPHOT.2013.2271637

[33] OLMEDO, M. I., TIANJIAN ZUO, JENSEN, J. B., QIWEN ZHONG, XIAOGENG XU, POPOV, S., et al. Multiband carrierless amplitude phase modulation for high capacity optical data links. *Journal of Lightwave Technology,* 2014, vol. 32, no. 4, p. 798–804. DOI: 10.1109/JLT.2013.2284926

[34] HAIGH, P. A., LE, S. T., ZVANOVEC, S., GHASSEMLOOY, Z., LUO, P., XU, T., et al. Multi-band carrier-less amplitude and phase modulation for bandlimited visible light communications systems. *IEEE Wireless Communication Magazine.* In print, 2015.

[35] HAIGH, P. A., BURTON, A., WERFLI, K., HOA LE MINH, BENTLEY, E., CHVOJKA, P., et al. A multi-CAP visible light communications system with 4.85 b/s/Hz spectral efficiency. *IEEE Journal on Selected Areas in Communications.* Accepted, 2015.

[36] *IEEE 802.15 WPAN™ Task Group 7 (TG7) Visible Light Communication.* [Online] Cited 2015-03-05. Available at: http://www.ieee802.org/15/pub/TG7.html

[37] LEE, K., PARK, H., BARRY, J. R. Indoor channel characteristics for visible light communications. *IEEE Communications Letters,* 2011, vol. 15, no. 2, p. 217–219. DOI: 10.1109/LCOMM.2011.010411.101945

[38] BARRY, J. R., KAHN, J. M., KRAUSE, W. J., LEE, E. A., MESSERSCHMITT, D. G. Simulation of multipath impulse response for indoor wireless optical channels. *IEEE Journal on Selected Areas in Communications,* 1993, vol. 11, p. 367–379. DOI: 10.1109/49.219552

[39] LANGER, K.-D., HILT, J., SHULZ, D., LASSAK, F., HARTLIEB, F., KOTTKE, C., et al. Rate-adaptive visible light communication at 500Mb/s arrives at plug and play. *Optoelectronics & Communications, SPIE Newsroom,* 2013. DOI: 10.1117/2.1201311.005196

[40] GROBE, L., PARASKEVOPOULOS, A., HILT, J., SCHULZ, D., LASSAK, F., HARTLIEB, F., et al. High-speed visible light communication systems. *IEEE Communications Magazine,* 2013, vol. 51, no. 12, p. 60–66. DOI: 10.1109/MCOM.2013.6685758

[41] JUNGNICKEL, V., POHL, V., NONNIG, S., VON HELMOLT, C. A physical model of the wireless infrared communication channel. *IEEE Journal on Selected Areas in Communications,* 2002, vol. 20, p. 631–640. DOI: 10.1109/49.995522

[42] JIVKOVA, S., KAVEHRAD, M. Shadowing and blockage in indoor optical wireless communications. In *IEEE Global Telecommunications Conference GLOBECOM '03.* 2003, vol. 6, p. 3269–3273. DOI: 10.1109/GLOCOM.2003.1258840

[43] KOMINE, T., NAKAGAWA, M. A study of shadowing on indoor visible-light wireless communication utilizing plural white LED

lightings. In *1st International Symposium on Wireless Communication Systems.* 2004, p. 36–40. DOI: 10.1109/ISWCS.2004.1407204

[44] CHVOJKA, P., ZVANOVEC, S., HAIGH, P. A., GHASSEMLOOY, Z. Channel characteristics of visible light communications within dynamic indoor environment. *Journal of Lightwave Technology,* 2015, vol. 33, no. 9, p. 1719–1725. DOI: 10.1109/JLT.2015.2398894

[45] LIU, H. S., PANG, G. Positioning beacon system using digital camera and LEDs. *IEEE Transactions on Vehicular Technology,* 2003, vol. 52, no. 2, p. 406–419. DOI: 10.1109/TVT.2002.808800

[46] KATAYAMA, MUTSUMI, M. KAZUYUKI, K. KAZUMITSU, *Vehicle position detection system.* JP Patent, 2010.

[47] ARAFA, A., XIAN JIN, KLUKAS, R. Wireless indoor optical positioning with a differential photosensor. *IEEE Photonics Technology Letters,* 2012, vol. 24, no. 12, p. 1027–1029. DOI: 10.1109/LPT.2012.2194140

[48] PENGFEI LUO, GHASSEMLOOY, Z., HOA LE MINH, KHALIGHI, A., XIANG ZHANG, MIN ZHANG, et al. Experimental demonstration of an indoor visible light communication positioning system using dual-tone multi-frequency technique. In *2014 3rd International Workshop in Optical Wireless Communications (IWOW).* 2014, p. 55–59. DOI: 10.1109/IWOW.2014.6950776

[49] WANG, T. Q., SEKERCIOGLU, Y. A., NEILD, A., ARMSTRONG, J. Position accuracy of time-of-arrival based ranging using visible light with application in indoor localization systems. *Journal of Lightwave Technology,* 2013, vol. 31, no. 20, p. 3302–3308. DOI: 10.1109/JLT.2013.2281592

[50] CARRUTHERS, J. B., CAROLL, S. M., KANNAN, P. Propagation modelling for indoor optical wireless communications using fast multi-receiver channel estimation. *IEE Proceedings-Optoelectronics,* 2003, vol. 150, no. 5, p. 473–481. DOI: 10.1049/ip-opt:20030527

[51] RAPPAPORT, T. S. *Wireless Communications.* Prentice-Hall, 2002.

[52] MCDONOUGH, R. N., WHALEN, A. D. *Detection of Signals in Noise.* San Diego (CA, USA): Wiley, 1995.

[53] WU, D., GHASSEMLOOY, Z., ZHONG, W.-D., KHALIGHI, M.-A., HOA LE MINH, CHEN, C., et al. Effect of optimal Lambertian order on the performance of cellular indoor optical wireless communications and positioning. *Journal of Lightwave Technologies,* 2015, submitted.

[54] *The LED Streetlight Replacement Program.* [Online] Cited 2014-02-10 Available at: http://bsl.lacity.org/led.html

[55] PENGFEI LUO, GHASSEMLOOY, Z., HOA LE MINH, BENTLEY, E., BURTON, A., TANG, X. Performance analysis of a car-to-car visible light communication system. *Applied Optics,* 2015, vol. 54, no. 7, p. 1696–1706. DOI: 10.1364/AO.54.001696

[56] TAKAI, I., ITO, S., YASUTOMI, K., KAGAWA, K., ANDOH, M., KAWAHITO, S. LED and CMOS image sensor based optical wireless communication system for automotive applications. *IEEE Photonics Journal,* 2013, vol. 5, article no. 6801418. DOI: 10.1109/JPHOT.2013.2277881

[57] NAGURA, T., YAMAZATO, T., KATAYAMA, M., YENDO, T., FUJII, T., OKADA, H. Tracking an LED array transmitter for visible light communications in the driving situation. In *2010 7th International Symposium on Wireless Communication Systems (ISWCS).* 2010, p. 765–769. DOI: 10.1109/ISWCS.2010.5624361

About the Authors ...

Stanislav ZVANOVEC received his M.Sc. and Ph.D. from the Czech Technical University in Prague, in 2002 and 2006, respectively. Now he is a full professor and vice-head of the Department of Electromagnetic Field at the Faculty of Electrical Engineering, Czech Technical University in Prague. He leads a Free-space and Fiber Optics team from the Faculty of Electrical Engineering, CTU and several research projects. His current research interests include wireless optical communications, visible light communications, remote sensing and optical fiber sensors.

Petr CHVOJKA was born in 1987. He received his M.Sc. from the Czech Technical University in Prague in 2013. Now he is a postgraduate student and a researcher at the Department of Electromagnetic Fields, Czech Technical University in Prague, where he is a member of a Free-space and Fiber Optics team. His research area includes visible light communications, OLED technologies and wireless optical communications.

Paul Anthony HAIGH received the PhD degree in Visible Light Communications from Northumbria University in 2014, publishing 13 articles in high ranking journals. Between 2010 and 2011 he held the prestigious Marie Curie Fellowship at CERN where he worked on optoelectronic links for large hadron collider experiments. In 2010 Paul received the BEng (Hons) degree in Communications Engineering from Northumbria University. Currently, he is a Research Associate within the High Performance Networks group at the University of Bristol working on the EPSRC TOUCAN project. His research is focused on real time seamless, transparent and adaptive and programmable interfaces between wireless and wired multi-technology networks.

Zabih GHASSEMLOOY, CEng, Fellow of IET, Senior Member of IEEE received his BSc (Hons) from the Manchester Metropolitan University in 1981, and MSc and PhD from the University of Manchester, Institute of Science and Technology (UMIST), in 1984 and 1987, respectively. 1986-87 worked in UMIST and from 1987 to 1988 was a Post-doctoral Research Fellow at the City University, London. 1988 joined Sheffield Hallam University as a Lecturer, becoming a Professor in Optical Communications in 1997. 2004-2012 was an Associate Dean for Research in the School of Computing, Engineering and from 2012-2014 Associate Dean for Research and Innovation in the Faculty of Engineering and Environment, Northumbria University at Newcastle, UK. He currently heads the Northumbria Communications Research Laboratories within the Faculty. He is the Editor-in-Chief of the International Journal of Optics and Applications, and British Journal of Applied Science & Technology. His researches interests are on optical wireless communications, visible light communications and radio over fiber/free space optics. He has supervised over 48 PhD students and published more than 550 papers (195 in journals + 4 books). He is a co-author of a CRC book on "Optical Wireless Communications – Systems and Channel Modelling with Matlab (2012)". From 2004-06 he was the IEEE UK/IR Communications Chapter Secretary, the Vice-Chairman (2004-2008), the Chairman (2008-2011), and Chairman of the IET Northumbria Network (Oct 2011-).

RF MEMS Based Tunable Bowtie Shaped Substrate Integrated Waveguide Filter

Muhammad Zaka Ur REHMAN [1, 2], Zuhairi BAHARUDIN [1], Mohd Azman ZAKARIYA [1], Mohd Haris Md. KHIR [1], Muhammad Taha JILANI [1]

[1] Dept. of Electrical and Electronics Engineering, Universiti Teknologi Petronas, Tronoh, Perak, Malaysia
[2] Dept. of Physics, COMSATS Institute of Information Technology, Park Road, Islamabad, Pakistan

zaka_g01951@utp.edu.my, zuhairb@petronas.com.my

Abstract. *A tunable bandpass filter based on a technique that utilizes substrate integrated waveguide (SIW) and double coupling is presented. The SIW based bandpass filter is implemented using a bowtie shaped resonator structure. The bowtie shaped filter exhibits similar performance as found in rectangular and circular shaped SIW based bandpass filters. This concept reduces the circuit footprint of SIW; along with miniaturization high quality factor is maintained by the structure. The design methodology for single-pole triangular resonator structure is presented. Two different inter-resonator couplings of the resonators are incorporated in the design of the two-pole bowtie shaped SIW bandpass filter, and switching between the two couplings using a packaged RF MEMS switch delivers the tunable filter. A tuning of 1 GHz is achieved for two frequency states of 6.3 and 7.3 GHz. The total size of the circuit is 70 mm x 36 mm x 0.787 mm (L x W x H).*

Keywords

Substrate integrated waveguide (SIW), tunable filter, bowtie filter, RF MEMS, double coupling

1. Introduction

Filters have received a particular attention with the advent of various wireless systems, this interest has dramatically increased with the introduction and development of new millimeter waves applications over the past decade. Various applications have been recently proposed including wireless local area networks [1], radars [2], intelligent transportation systems [3] and imaging sensors [4]. Efficient filters demand has also increased with the development of chipsets operating at 60 GHz or higher frequencies by a number of semiconductor industries [5].

Filters based on Substrate Integrated Waveguide (SIW) structures are achieved through incorporating the rectangular waveguide structure into the microstrip substrate [6]. SIWs are dielectric filled and are formed from the substrate material utilizing two rows of conducting vias connecting bottom and top metal plates, these vias are embedded in dielectric filled substrate; hence providing easy combination with other planar circuits and a reduction in size. The size reduction along with involving dielectric filled substrate instead of air-filled reduces the quality factor (Q), but the entire circuitry including waveguide and microstrip transitions can be realized by using printed circuit board (PCB) technology or other techniques, like LCP [7] and LTCC [8].

The design of an SIW bandpass filter can either utilize a design methodology based on coupling matrix method [9], or it can also follow a methodology used for designing air filled waveguide filters. The design of an SIW filter based on the methodology adopted in a rectangular waveguide, a shunt inductive coupling realization is adopted. Vias of irregular diameters placed in the center of the cavity may possibly occur in an inductive post filter; which is based on a requirement of control couplings. Large couplings might occur in the use of a small diameter. The utilization of shunt inductive vias at the couplings of the filter realizes a shunt inductive coupling filter as depicted in Fig. 1 (a) or an iris (aperture) coupling post as shown in Fig. 1 (b). A detailed literature on the development of SIW filters has been reported in [10].

A three pole structure of a SIW bandpass filter based on shunt inductive vias is shown in Fig. 1 (a). It utilizes four coupling vias centered in the guide and two microstrip to SIW transitions at the input and output. The two vias at the transitions facilitate in input and output couplings, while the centered vias are for coupling between the resonators.

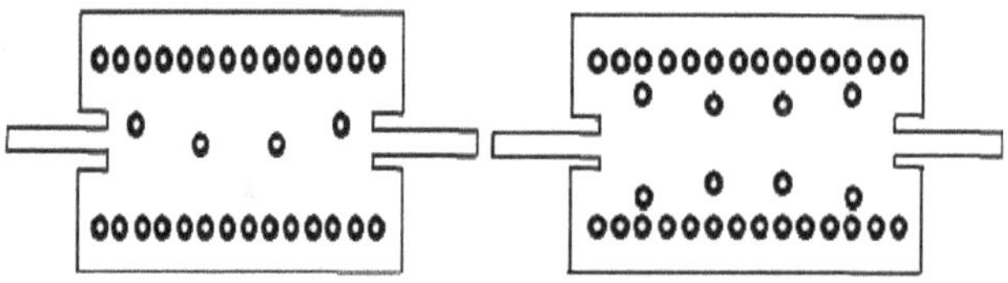

Fig. 1. (a) Shunt inductive coupling post filter. (b) Iris coupling based filter.

An SIW bandpass filter based on iris coupling posts is shown Fig. 1 (b); the apertures form three resonators. The filter's structure is such that the three cavities of half wavelength are formed in the center while SIW to microstrip transition are on the two edges of the filter; such a filter operational at 60 GHz has been presented in [11].

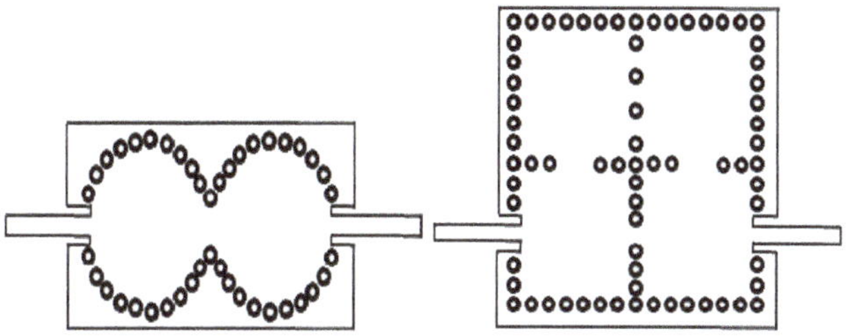

Fig. 2. Cavity filters with (a) circular cavities; (b) rectangular cavities.

Cavity filters with circular cavities[12] as presented in Fig. 2(a) and rectangular cavities [13] as depicted in Fig. 2(b) has been observed in literature. These variants of SIW allow more design variations and transmission zeros are also introduced due to cross coupling, better selectivity is also presented by these designs. Various SIW filters structures have been proposed in the literature; however there still exists a need to further miniaturize the structure. Furthermore the cavities are only either in circular or rectangular shape; therefore a triangular shaped cavity would reduce the circuit footprint.

In this paper, a tunable bandpass filter based on triangular cavity is presented. The tunable filter utilizes packaged RF MEMS switches which can be directly soldered on the filter circuit to switch the filter at two distinct frequencies. The tunable filter is designed through additional incorporation of a switchable extended coupling mechanism. As a validation of the proposed tunable filter, the design, fabrication and measured response of the two pole bowtie shaped bandpass filter are presented. The proposed bandpass filter configuration is suitable for integration with planar devices and its small footprint area allows other devices to be easily integrated on a single board.

2. Triangular SIW Cavity Filter

Design and implementation of SIW filters are being performed through defined practical methods so far. The most common technique is to form the SIW cavity through metallic sidewalls [14], a resemblance of which is displayed in Fig. 3. A dielectric substrate having thickness of H forms the cavity and it is of length L and width W. The bottom and top of the cavity are constructed through placing metallic plates and conducting posts/vias going through the substrate connecting the top and bottom plates; hence forming the sidewalls of the cavity. The vias are of diameter d and the separation between two neighboring vias is given as s. The choice of diameter and separation between the two vias forms the basis of the SIW filters; therefore these should be selected in a manner that minimum radiation loss is exhibited.

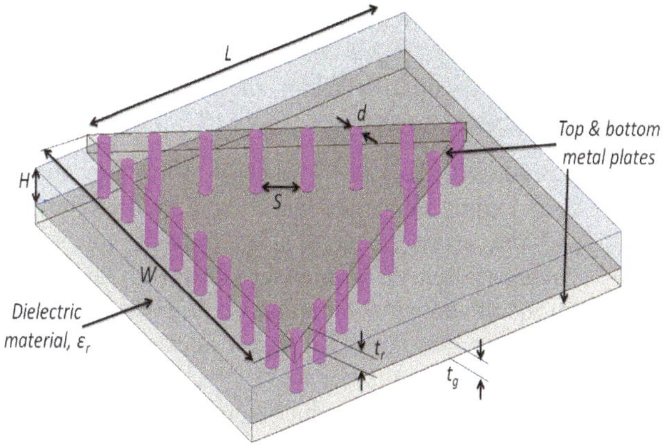

Fig. 3. Substrate Integrated Waveguide structure geometry.

The Deslandes and Wu [14] study reveals two primary design rules for SIW structures as given in (1) and (2); these rules are followed in order to ensure the same design and modeling methodology adopted for rectangular waveguides. These rules pertain to the diameter d of the via posts and the via post spacing s:

$$d < \frac{\lambda_g}{5}, \quad (1)$$

$$s \le 4d \quad (2)$$

where λ_g is the guided wavelength. In the design d and s are chosen to be 0.8 mm and 2 mm respectively, these values ensure less radiation losses and the SIW cavity acts closely to a triangular waveguide. For the TE_{101} mode, the dimensions of the SIW resonator structure are calculated by using the relation in (3) [14]

$$f_{TE_{101}} = \frac{c}{2\sqrt{\mu_r \varepsilon_r}} \sqrt{\left(\frac{1}{W_{eff}}\right)^2 + \left(\frac{1}{L_{eff}}\right)^2}. \quad (3)$$

W_{eff} and L_{eff} denote the effective width and length of the SIW resonator, respectively, and are given as:

$$W_{eff} = W - \frac{d^2}{0.95s}, \quad L_{eff} = L - \frac{d^2}{0.95s} \quad (4)$$

where W and L are the real width and length of the SIW resonator, μ_r and ε_r are the relative permeability and permittivity of the substrate respectively, and c is the velocity of light in free space. In this design the width and length of the triangular resonator structure is computed using (3) and (4) as displayed in Fig. 3. Utilizing this method the cavity is designed for the specifications laid out in Tab. 1.

Parameter	Value
Tunable center frequency	7.3 GHz
Chebyshev response filter order	2
Passband ripples (dB)	0.01
Passband bandwidth at -3 dB	> 300 MHz

Tab. 1. Design specifications of the tunable bandpass filter.

The printed circuit of triangular resonator and its subsequent bowtie shaped two pole filter is etched over using Rogers RT/Duriod 5880 material substrate having dielectric constant $\varepsilon_r = 2.2$ and substrate height H of 787 µm, and a dissipation factor of $\tan\delta = 9 \times 10^{-4}$. The copper thickness denoted as t_r and t_g is 17.5 µm.

The proposed triangular SIW cavity filter's resonance frequency is mainly dependent upon the dimensions of the cavity and the arrangement of vias forming the cavity walls. Theoretically, the resonance frequency does not depend on the thickness of the substrate H. However, it has been observed in literature [15] that it does play a role on the loss (mainly on radiation loss). The thicker the substrate the lower is the loss or higher is the Quality factor. Therefore, keeping in view the fabrication limitations a relatively thicker substrate is utilized for the SIW cavity.

2.1 Single Inter-resonator and I/O Coupling

In order to accomplish two pole bandpass SIW filter, once the triangular resonator is created for a specific resonant mode, the design methodology closely resembles conventional simulation-based microstrip filter design [14]. Two single resonators are coupled through inter-resonator coupling dimensions. In this coupling, a resonator length and coupling openings in both the resonators are used to couple the two resonators.

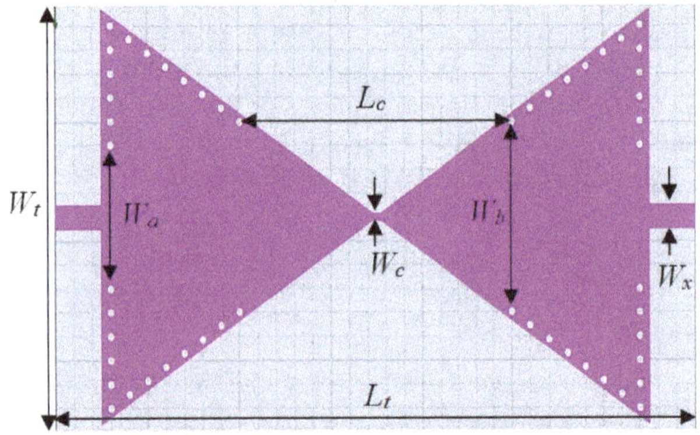

Fig. 4. Bowtie shaped two pole bandpass filter dimensions.

In Fig. 4 the dimensions are labeled and the values are given in Tab. 2, while the fabricated two pole bandpass filter with single coupling is depicted in Fig. 7.

External quality factor and coupling coefficients are calculated from derived expressions based on lowpass prototype parameters [14].

$$K_{1,2} = \frac{FBW}{\sqrt{g_1 g_2}}, \quad Q_{e1} = \frac{g_0 g_1}{FBW}, \quad Q_{e2} = \frac{g_2 g_3}{FBW} \tag{5}$$

where FBW is the fractional bandwidth, and g_0, g_1, g_2, g_3 are the Chebyshev lowpass prototype values which are used to calculate the external quality factors Q_{e1} and Q_{e2}. The coupling coefficient $K_{1,2}$ computed based on second order Chebyshev response prototype parameters is compared with the simulated coupling coefficient obtained through weak coupling. The coupling coefficient's value is dependent of two parameters; the coupling opening width denoted as W_b and inter-resonator combining width denoted as W_c. For 0.01 dB passband ripple, (5) results in values of $K_{1,2} = 0.1402$ and $Q_{e1} = Q_{e2} = 7.48$, then widths of W_c and W_b are selected from a closely matched coupling coefficients values. The values obtained through various combinations of the dimensions are depicted in Fig. 5.

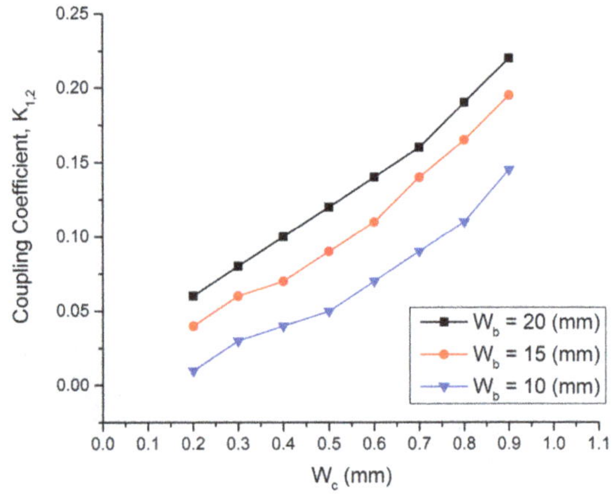

Fig. 5. Variation of inter-resonator dimensions against coupling coefficient.

The microstrip to SIW transitions at the input and output ports are of width W_x, the dimensions of the feed lines are computed through transmission line calculator. However the input and output coupling openings denoted as W_a are selected as a result of comparison of the simulated extracted external quality factors computed through (6) and external quality factors calculated on basis of theoretical LPF prototype parameters as given in (5)

$$Q_{ext} = \frac{f_0}{\Delta f_{-3dB}}. \tag{6}$$

A selection of dimension resulting in a close match of theoretical and simulated external quality factors dictates the I/O coupling as depicted in Fig. 6.

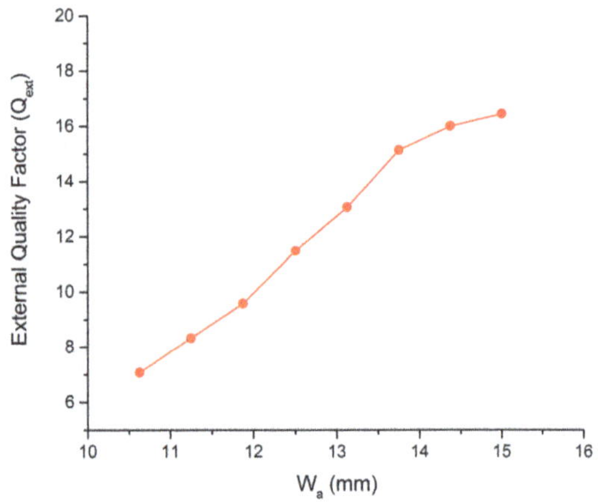

Fig. 6. External Q-factor for input/output coupling dimension.

The coupling is a result of iterations and adjustments to the dimensions of the coupling areas of the filter through full-wave simulations until desired filter and response is achieved.

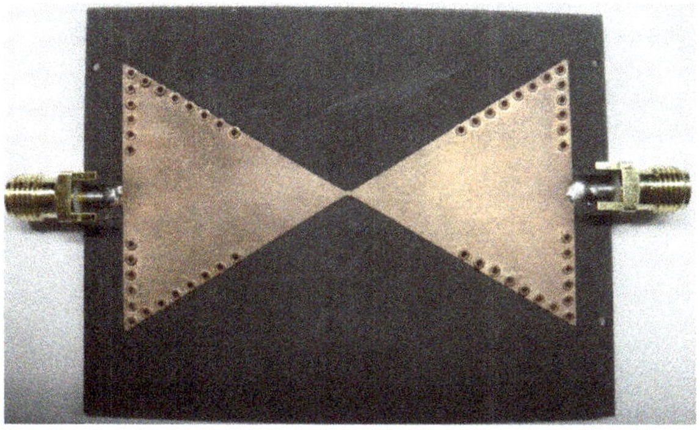

Fig. 7. Fabricated two pole fixed bandpass filter.

2.2 Double Coupling and Tunable Structure

An extended coupling overlaying the single coupling builds the basis of frequency shifting. The switching between the extended coupling and the single coupling forms tunable bandpass filter mechanism. The extended coupling is placed on top and bottom of the single coupling along with vias placed in the center of each length. The overlaying double coupling is shown in Fig. 8(a), while the notation values are displayed in Tab. 2.

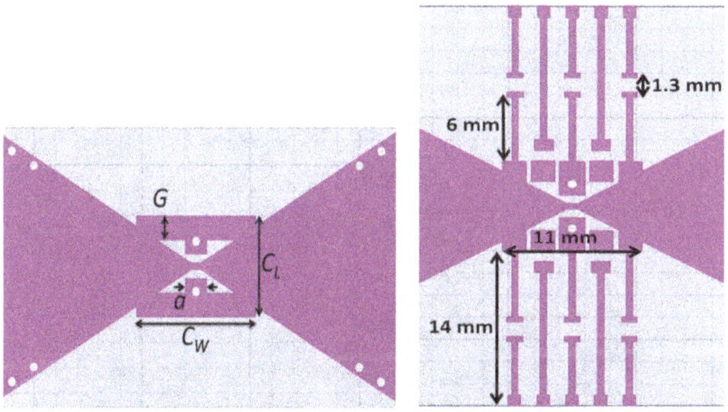

(a) (b)

Fig. 8. (a) Overlaying double coupling, (b) Dimensions of bias circuitry.

Notation	Value (mm)	Notation	Value (mm)
W_t	36.0	W_c	0.5
L_t	70.0	W_x	2.0
L	35.0	L_c	30.0
W	30.0	C_w	11
W_a	12.0	C_L	8
W_b	15.9	G, a	2

Tab. 2. Tunable filter parameters dimensions.

The tunable filter is constructed based on packaged RF MEMS switches, a total of four packaged RMSW201 RADANT MEMS [16]. The filters are actuated with a voltage of 90 V, and an onboard bias circuitry is built to place the switches. The bias lines of 0.5 mm thickness with spacing of 1.5 mm are placed on the board to actuate the RF MEMS switches, the bias circuitry and its corresponding dimensions are shown in Fig. 8(b).

Fig. 9 shows the operational diagram of the RF MEMS switches along with its bias circuitry, a voltage of 90 V is applied at the gate of the switch, while GND is connected through 100 kΩ resistors to the drain and source of the switch. The switches are placed on pads created on top of the board in between the coupling area, an on state of the switch refers to forming the double coupling whereas the off state refers to single coupling. The pads are wire-bonded to the gate, drain and source of the RF MEMS switches.

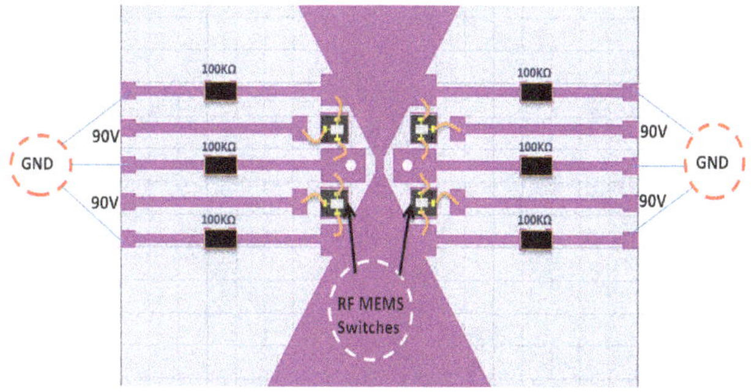

Fig. 9. The functionality and placement of RF MEMS and bias resistors (the figure is rotated).

3. Results and Discussion

The simulations to obtain the filter responses from the designed structures are conducted using ANSYS High Frequency Structure Simulator (HFSS). In addition Agilent Vector Network Analyzer (VNA) is utilized for the measurements of the fabricated filters.

The fixed filter described in the specifications in Tab. 1 and its corresponding design structure shown in Fig. 4 is realized with the responses shown in Fig. 10. The simulated S_{21} and S_{11} response of the bowtie shaped bandpass filter reveals that the S_{11} value at the center frequency of 7.3 GHz is less than -30 dB, whereas the S_{21} response is greater than -0.5 dB and the passband bandwidth at -3 dB is greater than 400 MHz. The measured S_{21} response at the center frequency of 7.3 GHz is better than -1.7 dB and its corresponding S_{11} response at the center frequency is less than -20 dB. The passband bandwidth at -3 dB is greater than 350 MHz.

The desired tunable filter with the double coupling shown in Fig. 8 and Fig. 9 is realized with the responses shown in Fig. 11 and Fig. 12. The details of the passband are depicted in Fig. 11(a) and Fig. 12(a) for the simulated and measured S_{21} responses, respectively.

Fig. 11 shows the response of the simulated two pole tunable filter designed for two resonant frequencies of 6.3 and 7.3 GHz. Response of the filter is obtained through realizing the resonator structure designed using the equations presented in Sec. 2.

The simulated S_{21} and S_{11} response of the bowtie shaped tunable bandpass filter reveals that when the switch is disconnected a two pole bandpass response is achieved at 6.3 GHz, whereas when the RF MEMS switch connects

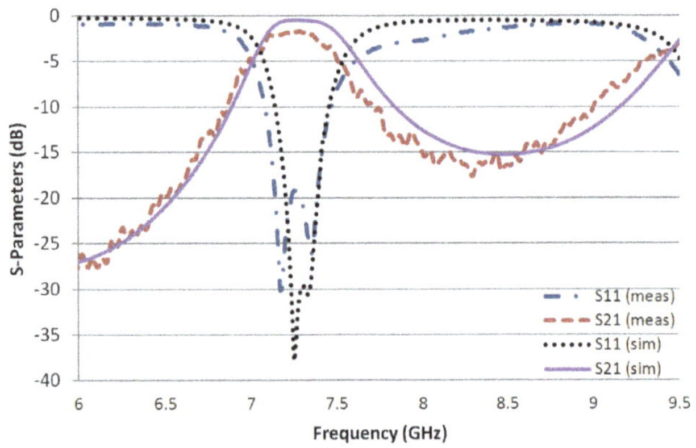

Fig. 10. Response of the fixed bowtie bandpass filter.

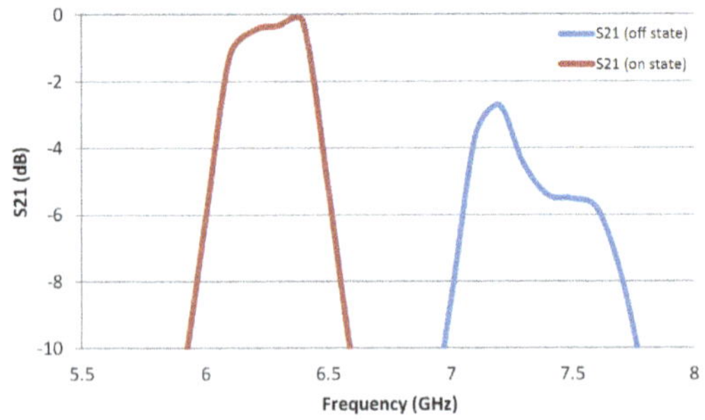

Fig. 11. (a) Passband details for the simulated S_{21} response.

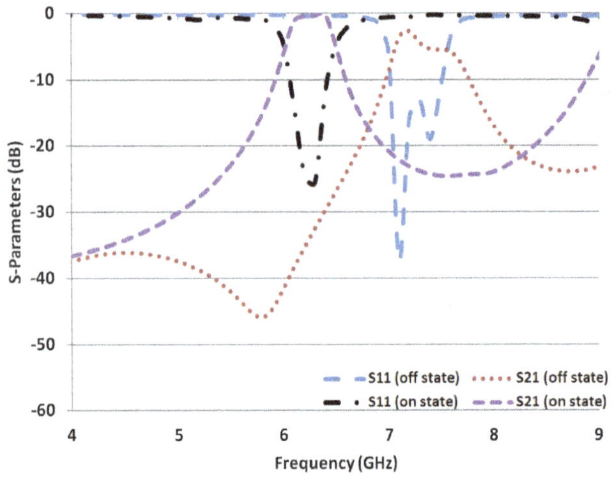

Fig. 11. (b) Simulated response of the tunable bandpass filter.

the double layered coupling a resonant frequency of 7.3 GHz is achieved. The passband bandwidth at -3 dB for both the resonant frequencies is observed to be greater than 400 MHz.

Measured S_{21} and S_{11} response of the two pole bandpass filter is presented in Fig. 12. The measured responses show similar trends in terms of frequency tuning between the two resonant modes as observed in the simulated responses. The S_{21} responses at the center frequencies of 6.44 and 7.4 GHz are better than -7 dB and their corresponding S_{11} responses are less than -14 dB and -18 dB respectively. The passband bandwidth at -3 dB is greater than 400 MHz, whereas the lower and upper stopband rejections are better than -25 dB. These responses are summarized in Tab. 3.

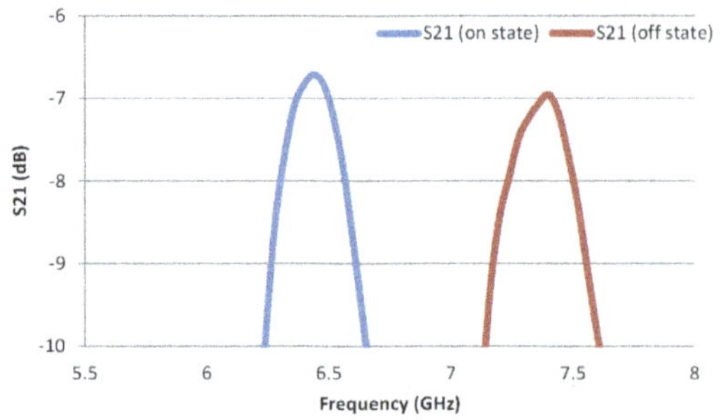

Fig. 12. (a) Passband details for the measured S_{21} response.

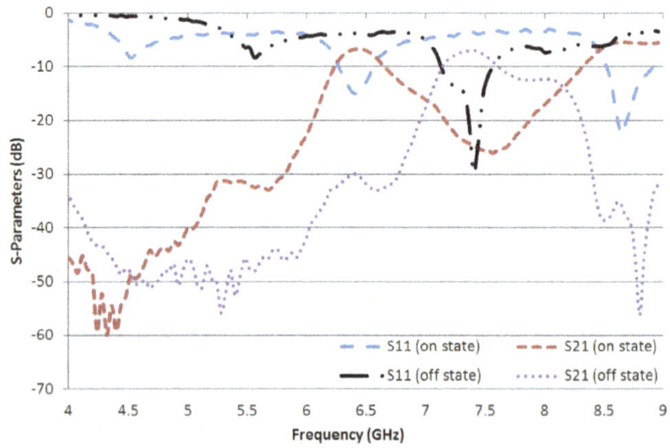

Fig. 12. (b) Measured response of the two pole tunable filter.

Fig. 13. Fabricated bowtie shaped tunable two pole filter.

Key Parameters	Simulated	Measured
Passband center frequency (GHz)	6.3, 7.3	6.44, 7.4
Passband S_{11} (dB)	< 25,20	< 14, 18
Passband S_{21} (dB)	> -0.5,-3.5	> -7,-7.5
Stopband rejection (dB)	> 25	> 25
Passband bandwidth at -3 dB level	0.4 GHz	0.435 GHz

Tab. 3. Summary of the tunable bandpass filter performance.

The measured insertion loss response includes the losses of the SMA connectors at the input and output of the tunable bandpass filter. The upper passband produced due to cancellation of the TE_{101} and TE_{201} modes is distanced enough from the filter passband, hence a pure Chebyshev response is observed as the design was based on Chebyshev LPF prototype parameters. However high order resonant modes presents spurious at 10 GHz. Consequently suppression of spurious is achievable by employing lowpass filters at the input and output of the resonators. Therefore the cut-off for the lowpass filters has to be matched with the upper passband of the bandpass filter.

4. Conclusion

A tunable bowtie shaped bandpass filter based on RF MEMS switches utilizing triangular structure SIW is proposed in this paper, the filter exhibits good performance and a miniaturized version of the SIW structure is exploited in the design process. The tunable filter is unique in terms of the tuning mechanism exploiting the coupling area. The RF MEMS two states switchable bandpass filter is designed, simulated, fabricated and the measured filter response shows two distinct frequency states at 6.4 and 7.4 GHz with a nearly constant bandwidth of 400 MHz. The simple structure of the filter allows readily integration with planar circuits and devices.

Acknowledgements

This work was supported by the Fundamental Research Grant Scheme (FRGS) of the Department of Higher Education, Malaysia. The authors would like to thank Dr. John Ojur Dennis and his student Mr. Farooq Ahmed of Nano fabrication facility in Universiti Teknologi PETRONAS, Malaysia for providing their guidance and wire-bonding facility.

References

[1] JAMES, J., PENGBO SHEN, NKANSAH, A., XING LIANG, GOMES, N. J. Millimeter-wave wireless local area network over multimode fiber system demonstration. *IEEE Photonics Technology Letters,* 2010, vol. 22, no. 9, p. 601–603. DOI: 10.1109/LPT.2010.2043249

[2] LI WANG, GLISIC, S., BORNGRAEBER, J., WINKLER, W., SCHEYTT, J. C. A single-ended fully integrated SiGe 77/79 GHz receiver for automotive radar. *IEEE Journal of Solid-State Circuits,* 2008, vol. 43, no. 9, p. 1897–1908. DOI: 10.1109/JSSC.2008.2003994

[3] YAN, X., ZHANG, H., WU, C. Research and development of intelligent transportation systems. In *Proceedings of the 11th International Symposium on Distributed Computing and Applications to Business Engineering & Science.* Guilin (China), 2012, p. 321–327.

[4] WILSON, J. P., SCHUETZ, C. A., MARTIN, R., DILLON, T. E., YAO, P., PRATHER, D. W. Polarization sensitive millimeter-wave imaging sensor based on optical up-conversion scaled to a distributed aperture. In *Proceedings of the 37th International conference on Infrared, Millimeter, and Terahertz Waves.* Wollongong (Australia), 2012, vol. 1, p. 23–28.

[5] NIKNEJAD, A. M., HASHEMI, H. *Millimetre-wave Silicon Technology: 60 GHz and beyond.* Springer, 2008.

[6] HIROKAWA, J., ANDO, M. Single-layer feed waveguide consisting of posts for plane TEM wave excitation in parallel plates. *IEEE Transactions on Antennas and Propagation,* 1998, vol. 46, no. 5, p. 625–630. DOI: 10.1109/8.668903

[7] KI SEOK YANG, PINEL, S., IL KWON KIM, LASKAR, J. Low-loss integrated-waveguide passive circuits using liquid-crystal polymer System-on-Package (SOP) technology for millimeter-wave applications. *IEEE Transactions on Microwave Theory and Techniques,* 2006, vol. 54, no. 12, p. 4572–4579. DOI: 10.1109/TMTT.2006.886004

[8] JUNFENG XU, ZHI NING CHEN, XIANMING QING, WEI HONG. 140-GHz planar broadband LTCC SIW slot antenna array. *IEEE Transactions on Antennas and Propagation,* 2012, vol. 60, no. 6, p. 3025–3028. DOI: 10.1109/TAP.2012.2194673

[9] MOKHTAARI, M., BORNEMANN, J., RAMBABU, K., AMARI, S. Coupling-matrix design of dual and triple passband filters. *IEEE Transactions on Microwave Theory and Techniques,* 2006, vol. 54, no. 11, p. 3940–3946. DOI: 10.1109/TMTT.2006.884687

[10] REHMAN, M. Z. U., BAHARUDIN, Z., ZAKARIYA, M., KHIR, M., KHAN, M., WENG, P. W. Recent advances in miniaturization of substrate integrated waveguide bandpass filters and its applications in tunable filters. In *Proceedings of Business Engineering and Industrial Applications Colloquium.* (Malaysia), 2013, p. 109–114. DOI: 10.1109/BEIAC.2013.6560093

[11] SUNG TAE CHOI, KI SEOK YANG, TOKUDA, K., YONG HOON KIM. A V-band planar narrow bandpass filter using a new type integrated waveguide transition. *IEEE Microwave and Wireless Components Letters,* 2004, vol. 14, no. 12, p. 545–547. DOI: 10.1109/LMWC.2004.837386

[12] TANG, H. J., HONG, W., HAO, Z. C., CHEN, J. X., WU, K. Optimal design of compact millimetre-wave SIW circular cavity filters. *Electronics Letters,* 2006, vol. 41, no. 19, p. 1068–1069. DOI: 10.1049/el:20052251

[13] XIAO-PING CHEN, KE WU. Substrate integrated waveguide cross-coupled filter with negative coupling structure. *IEEE Transactions on Microwave Theory and Techniques,* 2008, vol. 56, no. 1, p. 142–149. DOI: 10.1109/TMTT.2007.912222

[14] DESLANDES, D., WU, K. Accurate modeling, wave mechanisms, and design considerations of a substrate integrated waveguide. *IEEE Transactions on Microwave Theory and Techniques,* 2006, vol. 54, no. 6, p. 2516–2526. DOI: 10.1109/TMTT.2006.875807

[15] ALI KHAN, A., MANDAL, M. K., SANYAL, S. Unloaded quality factor of a substrate integrated waveguide resonator and its variation with the substrate parameters. In *Proceedings of International Conference on Microwave and Photonics.* Dhanbad (India), 2013, p. 1–4. DOI: 10.1109/ICMAP.2013.6733496

[16] RADANT MEMS, MA, USA. *RMSW201 SPST RF MEMS switch (datasheet).* 2 pages. [Online] Cited 2014-05-10. Available at: http://www.radantmems.com/radantmems.data/Library/RMSW201.pdf

About the Authors ...

Muhammad Zaka Ur REHMAN was born in Lahore, Pakistan in 1986. He is serving as a Lecturer in COMSATS Inst. of Information Technology (CIIT), Islamabad, Pakistan. He received his B.S. degree in Electronics from CIIT in 2007, the MSc Degree in Digital Signal Processing in Communication Systems from Lancaster University, UK, in 2010, and is currently working towards the Ph.D. degree in Electrical Engineering (with an emphasis on RF and Microwave Circuits) at the Universiti Teknologi PETRONAS, Perak, Malaysia. His research interests include RF MEMS for microwave applications, substrate integrated waveguide structures and reconfigurable filters design.

Zuhairi BAHARUDIN is a faculty member at Universiti Teknologi PETRONAS, Malaysia. He obtained his Diploma and B.Eng. Hons. Electrical from the University Technology MARA, Shah Alam, Malaysia and subsequently his M.Eng in Electrical Power Engineering from the University of South Australia, Australia. He received his Ph.D. degree in Electrical Engineering from the Universiti Teknologi PETRONAS in 2010. His research interests are in effects of harmonics on power systems, applications of artificial intelligence of load forecasting, and power economics operation and control.

Mohd Azman Bin ZAKARIYA received bachelors in Electrical Engineering from Universiti Teknologi Malaysia, and Master of Science in Communications and Signal Processing from University of Newcastle upon Tyne, UK. He is a lecturer in Universiti Teknologi PETRONAS, Malaysia. He is also working towards his Ph.D. from Universiti Sains Malaysia. His research interests include dielectric resonator antennas, defected ground structure.

Mohd Haris Md KHIR received the B. Eng. degree in Electrical and Electronic Engineering from Universiti Teknologi MARA, Selangor, Malaysia, in 1999, the Masters of Science degree in Computer & System Engineering from Rensselaer Polytechnic Institute, New York, USA, in 2001, and the Ph.D. degree in Systems Engineering from Oakland University, Michigan, USA, in 2010. Since 2002, he has been with Universiti Teknologi PETRONAS, Perak, Malaysia, where he is currently a Senior Lecturer in Electrical & Electronic Engineering Department. His research interests include Micro-Electro-Mechanical Systems (MEMS) sensors/actuators design and fabrication based on CMOS and MUMPS technologies. He has successfully fabricated a number of MEMS devices such as accelerometers, micro-mirror, micro switches, energy harvester, electromagnetic sensors, gas sensors, and thermal electric generator (TEG) system.

Muhammad Taha JILANI received bachelor's degree in Electrical Technology followed by masters in Telecommunication, in 2007 & 2009 respectively. He is currently working toward Ph.D. in the area of dielectric material characterization using RF and microwave frequencies at Universiti Teknologi PETRONAS, Malaysia.

3

Dielectric Properties Determination of a Stratified Medium

Paiboon YOIYOD, Monai KRAIRIKSH

Faculty of Engineering, King Mongkut's Institute of Technology Ladkrabang, Bangkok 10520, Thailand

S36101 16@kmitl.ac.th, kkmonai@kmitl.ac.th

Abstract. *The method of detection of variation in dielectric properties of a material covered with another material, which requires nondestructive measurement, has numerous applications and the accurate measurement system is desirable. This paper presents a dielectric properties determination technique whereby the dielectric constant and loss factor are extracted from the measured reflection coefficient. The high frequency reflection coefficient shows the effect of the upper layer, while the dielectric properties of the lower layer can be determined at the lower frequency. The proposed technique is illustrated in 1-11 GHz band using 5 mm-thick water and 5% saline solution. The fluctuation of the dielectric properties between the high frequency and the low frequency, results from the edge diffraction in the material and the multiple reflections at the boundary of the two media, are invalid results. With the proposed technique, the dielectric properties of the lower layer can be accurately determined. The system is validated by measurement and good agreement is obtained at the frequency below 3.5 GHz. It can be applied for justifying variation of the material in the lower layer which is important in industrial process.*

Keywords

Dielectric properties, stratified medium, reflection coefficient, diffraction coefficient

1. Introduction

The dielectric properties determination of a covered material or stratified medium by using a microwave technique is essential and has numerous applications. One of the determination techniques of dielectric properties is the free space technique whose main advantages include distinguishing capability of inhomogeneous materials, nondestructiveness, non-contact, and no machinery that fits the sample required despite sophisticated procedure to obtain accurate results [1], [2]. The techniques need an insertion of a perfectly conducting plate behind the plate of unknown material. The comprehensive reviews of dielectric properties measurement techniques and nondestructive testing using both millimeter and microwave are respectively shown in [3]. Such measurement techniques together with nondestructive testing are employed in a number of applications, such as moisture content detection [4–6], determination of liquids [7], skin cancer detection [8], surface crack detection [9], corrosion detection [10], [11], in which some sensors require contact [7–11] while the other are contactless [5], [6]. Apart from the contactless measurement, certain techniques require embedding of a scatterer in the medium under test [12], [13] which inevitably results in limitation in some applications as in agricultural applications [14], [15].

The work in [16] utilized a free space technique to characterize ripeness of mango but sample of mango must be prepared in a planar sample holder. Therefore, it is unsuitable in practice. A number of the free space techniques widely employed in the past, the sole magnitude measurement technique is less complex and thus inexpensive; nevertheless, it requires transmission measurement [17]. In addition, the use of millimeter wave reflectometer [18] is one of the attractive solutions. Fruit testing nevertheless must be carried out by measuring through the peel nondestructively. The methods in [19], [20] estimate the dielectric properties and thickness of multilayer object but they are contact measurement. The contactless measurement can however be accomplished by the technique of through-wall measurement [21]. From the aforementioned statement, it is desirable to estimate dielectric properties of a stratified object, consisting of the upper layer, lower layer, and the thickness of the upper layer. The issue is that the low variation of dielectric properties of the lower layer has little effect on the total variation of the measured results. Therefore, it is necessary to propose an accurate measurement technique to estimate the dielectric properties and thickness of the upper layer, which leads to the accurate estimation of the dielectric properties of the lower layer.

To develop a measurement system, a measurement technique for determination of not merely the dielectric properties of upper and lower layers but also the thickness of upper layer must be well established. To this end, in this paper the reflection measurement in wideband to determine the dielectric properties of the upper layer with high frequencies in which depth of penetration is shorter than the depth of the upper layer is introduced. The measured material is supposed to be in a data base for comparing dielectric properties of the upper layer when measured at high frequency, and then the values at the lower frequency can be obtained. The application of interest is measurement of material in a container in industrial process. The relevant work is the reflection measurement for estimating thickness of snow on the road [22]. Using the full band dielectric properties, it is possible to calculate depth of penetration of wave through the upper layer and thus velocity of wave in the upper layer. As a result, the thickness of the upper layer is determined. The suitable frequency at the lower frequency band is selected from the frequency response of the dielectric properties. Finally, the dielectric properties of the lower layer can be determined. To obtain a tangible insight into the aforesaid technique, the planar structure is investigated in this paper.

The organization of this paper begins with the introduction in Sec. 1. Section 2 discusses the principles of the proposed technique. The calculation results of the measurement technique are illustrated in Sec. 3. Validation of the calculation results, together with the discussions, is carried out by measurements shown in Sec. 4. Section 5 discusses the limitation of the system and addresses the factors affecting the required bandwidth. This paper is then ended with a conclusion in Sec. 6.

2. Dielectric Properties Determination of a Stratified Medium

Let us consider a stratified medium of planar structure with upper layer thickness d as shown in Fig. 1. The upper medium has dielectric properties of μ_1, ε_1 and σ_1. The lower layer is infinite extent in thickness and possesses dielectric properties of μ_2, ε_2 and σ_2. The width and length of the structure are finite size with the width of W. This structure is illuminated by a uniform plane wave in free space in which dielectric properties are μ_0, ε_0 and σ_0, where σ_0 is zero. Fig. 1 depicts the various components of waves for derivation of the reflection coefficient. The transmitting and receiving antennas are at $O(x',y',z')$ and $P(x,y,z)$, respectively, where position of y' is same as y at $W/2$, $O(x',y',z')$ and $P(x,y,z)$ are almost identical position to act as a monostatic radar. The transmitter transmits incident electric wave E_i on the upper surface at $z = -d$. With perfect isolation between the two antennas, the receiving antenna receives reflected and diffracted waves from four edges (D_1, D_2, D_3 and D_4). The reflected wave is the sum of the reflection at the upper surface and the multiple reflections between the upper surface and the interface between dielectric 1 and dielectric 2. The reflection coefficient at $z = -d$ ($\Gamma_{in}(z = -d)$) can be expressed as (1).

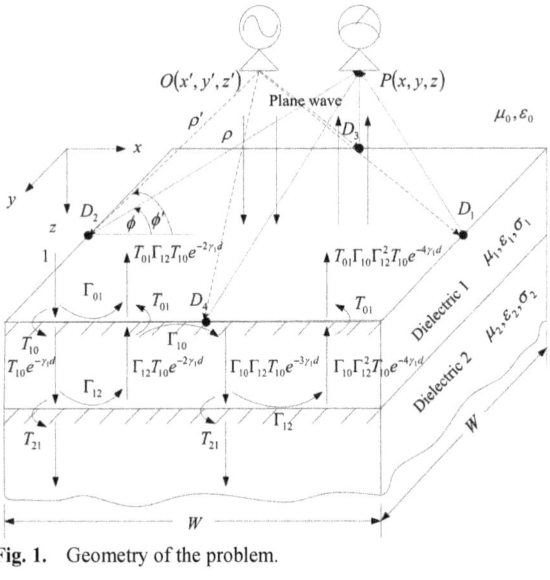

Fig. 1. Geometry of the problem.

$$\Gamma_{in}(z=-d)=\Gamma_{01}+T_{01}T_{10}\sum_{i=1}^{n}\Gamma_{12}^{i}\Gamma_{10}^{i-1}e^{-2i\gamma_1 d}+D \qquad (1)$$

where Γ_{01} and Γ_{10} are the reflection coefficients between the free space and the upper layer for propagation in upward direction and downward direction, respectively, T_{01} and T_{10} are the transmission coefficients between the free space and the upper layer for propagation in upward direction and downward direction, respectively, Γ_{12} is the reflection coefficient at the boundary of the two media, γ_1 is propagation constant in dielectric 1 ($\gamma_1 = \alpha_1 + j\beta_1$), i represents the ith reflection, n is the number of multiple reflections in the material, and D is diffraction coefficient as shown in (2).

$$D=\frac{E_{D_1}+E_{D_2}+E_{D_3}+E_{D_4}}{E_i}. \qquad (2)$$

Diffracted electric field intensity on each side from [23] is expressed in rectangular coordinate as follows

$$\begin{aligned}E_{D_k}(x,y,z)=\pm E_i\Bigg\{&T\left(\sqrt{\varepsilon_{r1}}\sin\left(\tan^{-1}\left(\frac{z}{x}\right)\right)-\sqrt{\varepsilon_{r1}-\cos^2\left(\tan^{-1}\left(\frac{z'}{x'}\right)\right)}\right)\\&\times\frac{e^{-j\frac{\pi}{4}}}{2\sqrt{2\pi k_d}}\frac{F_t\left(2k_d\left(\sqrt{(x^2+z^2)}\right)\cos^2\left(\dfrac{\tan^{-1}\left(\frac{z}{x}\right)-\cos^{-1}\left(\dfrac{\cos\left(\tan^{-1}\left(\frac{z'}{x'}\right)\right)}{\sqrt{\varepsilon_{r1}}}\right)}{2}\right)\right)}{\sqrt{\varepsilon_{r1}}\cos\left(\tan^{-1}\left(\frac{z}{x}\right)\right)+\cos\left(\tan^{-1}\left(\frac{z'}{x'}\right)\right)}\\&+T\left((1-R)\cos\left(\tan^{-1}\left(\frac{z'}{x'}\right)\right)-(1+R)\sqrt{\varepsilon_{r1}}\cos\left(\tan^{-1}\left(\frac{z}{x}\right)\right)\right)\frac{e^{-j\frac{\pi}{4}}}{2\sqrt{2\pi k_d}}\\&\times\frac{F_t\left(2k_d\left(\sqrt{(x^2+z^2)}\right)\cos^2\left(\dfrac{\cos^{-1}\left(\sqrt{1-\frac{\cos^2(\tan^{-1}(z'/x'))}{\varepsilon_{r1}}}\right)-\left(\tan^{-1}\left(\frac{z}{x}\right)-\frac{\pi}{2}\right)}{2}\right)\right)}{\sqrt{\varepsilon_{r1}}\cos\left(\tan^{-1}\left(\frac{z}{x}\right)+\frac{\pi}{2}\right)-\sqrt{\varepsilon_{r1}-\cos^2\left(\tan^{-1}\left(\frac{z'}{x'}\right)\right)}}\Bigg\}\end{aligned} \qquad (3)$$

where E_i is the incident electric field intensity, T and R are Fresnel transmission coefficient and reflection coefficient, respectively, k = 1, 2, 3 and 4 for electric field intensity on each side, (+) is for perpendicular polarized and (-) is for parallel polarized diffraction coefficients. ρ is a distance from the diffraction point to the observation point $\rho = (x^2 + z^2)^{1/2}$, ϕ is an angle between the line from the diffraction point to the observation point with respect to x-axis $\phi = \tan^{-1}(z/x)$, ρ' is a distance from source point to diffraction point $\rho' = (x'^2 + z'^2)^{1/2}$, ϕ' is an angle between the line from the source point to the diffraction point with respect to x-axis $\phi' = \tan^{-1}(z'/x')$, and ε_{r1} is relative permittivity of dielectric 1.

Since every reflection and transmission term in the summation has a very low magnitude, therefore the summation shrinks very rapidly. Hence, the summation in (1) can be truncated as shown in (4a).

$$\Gamma_{in} = \Gamma_{01} + \frac{T_{01}T_{10}\Gamma_{12}e^{-2\gamma_1 d}}{1 - \Gamma_{10}\Gamma_{12}e^{-2\gamma_1 d}} + D \tag{4a}$$

where $\Gamma_{10} = -\Gamma_{01}$, $T_{10} = 1 + \Gamma_{01}$ and $T_{01} = 1 + \Gamma_{10} = 1 - \Gamma_{01}$. Hence

$$\Gamma_{in} = \frac{\Gamma_{01} + \Gamma_{12}e^{-2\gamma_1 d}}{1 + \Gamma_{01}\Gamma_{12}e^{-2\gamma_1 d}} + D. \tag{4b}$$

Rewriting (4b), one can solve for Γ_{12} as shown in (5). In order to get d, we use the reflection coefficient in time domain. The detail is explained in Sec. 3.1.

$$\Gamma_{12} = \frac{\Gamma_{in} - \Gamma_{01} - D}{\left(1 - \Gamma_{01}\left(\Gamma_{in} - D\right)\right)e^{-2\alpha_1 d}e^{-j2\beta_1 d}}. \tag{5}$$

α_1 and β_1 are respectively the attenuation constant and phase constant in the upper medium. For very high frequencies, attenuation through the upper layer is great enough such that $\Gamma_{in} \approx \Gamma_{01}$. The intrinsic impedance of the upper layer, η_1, can be directly calculated and then curve fitted to the database of the material to estimate η_1 values for lower frequencies. Using η_1 to characterize other parameters including η_2, the intrinsic impedance of the lower layer derived from (5), can be solved.

$$\eta_2 = \frac{\eta_1[(\Gamma_{in} - \Gamma_{01}) + (1 - \Gamma_{in}\Gamma_{01})e^{-2\alpha_1 d}e^{-j2\beta_1 d}]}{(1 - \Gamma_{in}\Gamma_{01})e^{-2\alpha_1 d}e^{-j2\beta_1 d} - (\Gamma_{in} - \Gamma_{01})} \tag{6}$$

or

$$\frac{\sigma_2 + j\omega\varepsilon_2}{j\omega\mu_2} = 1 \Bigg/ \left[\frac{\eta_1\left[\left(\Gamma_{in} - \Gamma_{01}\right) + \left(1 - \Gamma_{in}\Gamma_{01}\right)e^{-2\alpha_1 d}e^{-j2\beta_1 d}\right]}{\left(1 - \Gamma_{in}\Gamma_{01}\right)e^{-2\alpha_1 d}e^{-j2\beta_1 d} - \left(\Gamma_{in} - \Gamma_{01}\right)}\right]^2 \tag{7}$$

where $\eta_2 = x + jy$. Hence

$$\frac{\varepsilon_0\varepsilon''_{r2}}{j\mu_0\mu_{r2}} + \frac{\varepsilon_0\left(\varepsilon'_{r2} - j\varepsilon''_{r2}\right)}{\mu_0\mu_{r2}} = \frac{1}{\left(x + jy\right)^2} \tag{8}$$

and

$$\frac{\varepsilon_0\varepsilon'_{r2}}{\mu_0\mu_{r2}} - j\frac{2\varepsilon_0\varepsilon''_{r2}}{\mu_0\mu_{r2}} = \frac{x^2 - y^2 - j2xy}{\left(x^2 - y^2\right)^2 + \left(2xy\right)^2}. \tag{9}$$

By equating the real part of the left hand side to that of the right hand side of (9), and the imaginary part is also equated in the same manner, the dielectric properties can be derived from the real part x and imaginary part y of η_2

$$\frac{\varepsilon_0\varepsilon'_{r2}}{\mu_0\mu_{r2}} = \frac{x^2 - y^2}{\left(x^2 - y^2\right)^2 + \left(2xy\right)^2}, \tag{10}$$

$$\frac{2\varepsilon_0\varepsilon''_{r2}}{\mu_0\mu_{r2}} = \frac{2xy}{\left(x^2 - y^2\right)^2 + \left(2xy\right)^2}. \tag{11}$$

The dielectric constant and loss factor of the lower layer are expressed in a closed form as shown in (12) and (13), respectively.

$$\varepsilon'_{r2} = \frac{\mu_0\mu_{r2}(x^2 - y^2)}{\varepsilon_0((x^2 - y^2)^2 + (2xy)^2)}, \tag{12}$$

$$\varepsilon''_{r2} = \frac{\mu_0\mu_{r2}(2xy)}{2\varepsilon_0((x^2 - y^2)^2 + (2xy)^2)}. \tag{13}$$

From the above expressions, it can be seen that by measuring the reflection coefficient (Γ_{in}) at the upper surface in a wideband, the reflection coefficient between the free space and the upper layer (Γ_{01}) can be found from the high frequency which is unable to penetrate to the lower layer. Then, the dielectric properties of the upper layer at low frequency can be determined from the database. The above procedure is shown in a block diagram in Fig. 2.

Using the full band dielectric properties, it is possible to calculate the depth of penetration through the upper layer and consequently velocity of wave in the upper layer. From the result of reflection coefficient, the thickness d of the upper layer is determined by inverse Fourier transform

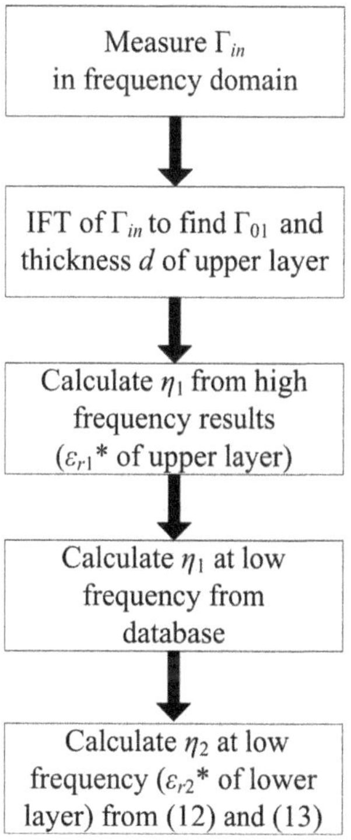
Measure Γin in frequency domain
IFT of Γin to find Γ01 and thickness d of upper layer
Calculate η1 from high frequency results (εr1* of upper layer)
Calculate η1 at low frequency from database
Calculate η2 at low frequency (εr2* of lower layer) from (12) and (13)

Fig. 2. Block diagram of the proposed technique.

(IFT). Using the high frequency for determining $\varepsilon_{r1}^* = \varepsilon_{r1}' - j\varepsilon_{r1}''$), then η_1 at low frequency is determined from the database of the material. Finally, the dielectric properties $\varepsilon_{r2}^* = \varepsilon_{r2}' - j\varepsilon_{r2}''$ of the lower layer η_2 are determined. This principle can be illustrated in Sec. 3.

3. Calculation Results

3.1 Determination of Thickness of the Upper Layer

At high frequency, the intrinsic impedance of the upper layer is calculated from (14)

$$\eta_1 = \eta_0 \left(\frac{1+\Gamma_{in}}{1-\Gamma_{in}} \right). \tag{14}$$

From the obtained dielectric properties at the high frequency band, the full band one can be obtained from the database of the material. For illustration, the material size of 18λ for the frequency of 6 GHz, the upper and lower layers are water and 5% saline solution [24], respectively. The resultant dielectric properties and depth of penetration of water and 5% saline solution are listed in Tab. 1.

Fig. 3(a) depicts the magnitude of reflection coefficient in frequency domain which was calculated using (4). The magnitude of reflection coefficient varies in a similar manner to damped sinusoidal below 5.5 GHz due to the effect of reflection at the upper surface and at the interface of the materials. Then, the steady response can be observed between 5.5 and 8.5 GHz since reflection takes place only at the upper surface. Note that the fluctuation at frequency higher than 8.5 GHz is caused by diffraction at the four edges of the material.

An inverse Fourier transforms of Γ_{in} provides the time-domain reflectometry as in Fig. 3(b). The first peak at 0 ns represents the reflection at the upper surface whereas the other peaks of the response represent the reflection between the upper and lower layers. Observing the time t, these peaks occur and accounting for the velocity of wave in the upper layer, a thickness of the upper layer can be determined from

Freq. (GHz)	Water	dp (mm)	5% saline	dp (mm)
1	77.960-j3.969	106	63.164-j141.956	3.51
2	77.230-j7.738	27	62.611-j75.443	2.83
3	76.144-j11.323	12	61.789-j55.085	2.46
4	74.741-j14.963	7.05	60.727-j46.141	2.14
5	73.066-j17.820	4.61	59.455-j41.464	1.86
6	71.164-j20.682	3.28	58.008-j39.268	1.62
7	69.082-j23.267	2.47	56.419-j38.006	1.42
8	66.866-j25.569	1.94	54.722-j37.358	1.24
9	64.588-j27.592	1.58	52.949-j37.051	1.1
10	62.196-j29.344	1.32	51.128-j36.923	0.98
11	59.814-j30.838	1.12	49.285-j36.879	0.88

Tab. 1. Dielectric properties and depth of penetration of water and 5% saline solution.

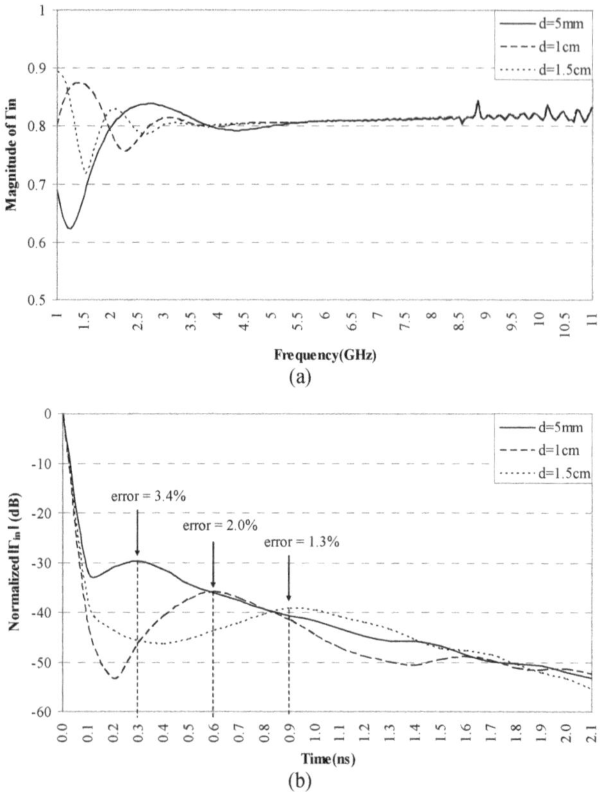

Fig. 3. Magnitude of reflection coefficient: (a) Frequency domain, (b) Time domain.

$$d = v_1 \cdot \frac{t}{2} \tag{15}$$

where v_1 is velocity of wave in the upper layer calculated from $v_1 = 1/(\mu\varepsilon_1)^{1/2}$ and t is the total time wave traveling forth and back. This velocity is obtained from the dielectric properties of water at the frequency corresponding to the time the peaks occur (at $f = 1/t$; dielectric properties of water are 75.754 – j12.358, 77.488 – j6.625 and 77.895 – j4.417). v_1 at these frequencies are 0.344×10^8 m/s, 0.341×10^8 m/s and 0.340×10^8 m/s.

The estimated thickness for 5 mm, 1 cm and 1.5 cm layers are 5.15 mm (error 3.4%), 1.02 cm (error 2.0%) and 1.52 cm (error 1.3%), respectively. From Tab. 1, for the thickness of the upper layer of 5 mm, the frequencies lower than 5 GHz can penetrate to the lower layer. One can hence determine the dielectric properties of the lower layer. From the above results, it can be concluded that for the unknown upper and lower media, one can not only determine the dielectric properties and thickness of the upper layer by wideband measurement with the resulting thickness taken from inversed Fourier transform to time domain response, but also determine dielectric properties of the lower layer from the lower frequency.

With the above procedure, the different materials were investigated. It was found that when the upper and lower layers were respectively 5% saline solution and water, the frequency range for wideband measurement decreased since saline solution is lossier than water.

3.2 Dielectric Properties of Material with Different Dimensions

In practice, the size of the material is finite. Therefore, the effect of size contributes to the reflection coefficient and dielectric properties. In this regard, the size of the material was varied; the upper and lower layers are water ($d = 5$ mm) and 5% saline solution, respectively. Fig. 4 shows the determined dielectric properties of the material of interest with the size of 18λ and 42λ for the frequency of 6 GHz. Note that the fluctuations of the determined dielectric properties between 4.5-10.5 GHz are due to the edge diffraction in the material and the multiple reflection at the boundary of the two media that affect calculation of dielectric properties. The dielectric properties in this frequency band are invalid results. Those above 10.5 GHz are the dielectric properties of the upper layer whereas those below 4.5 GHz are for the lower layer. For the size of the material of 18λ and 42λ, the similar determined dielectric properties are obtained. The dielectric properties of material are not significantly affected by the size of the material. For dielectric constant in Fig. 4(a), the dielectric constant below 4.5 GHz approaches the dielectric constant of 5% saline solution and the one above 10.5 GHz approaches the dielectric constant of water. For loss factor, the frequency below 5 GHz approaches that of 5% saline solution and the one above 10.5 GHz approaches that of water as seen in Fig. 4(b).

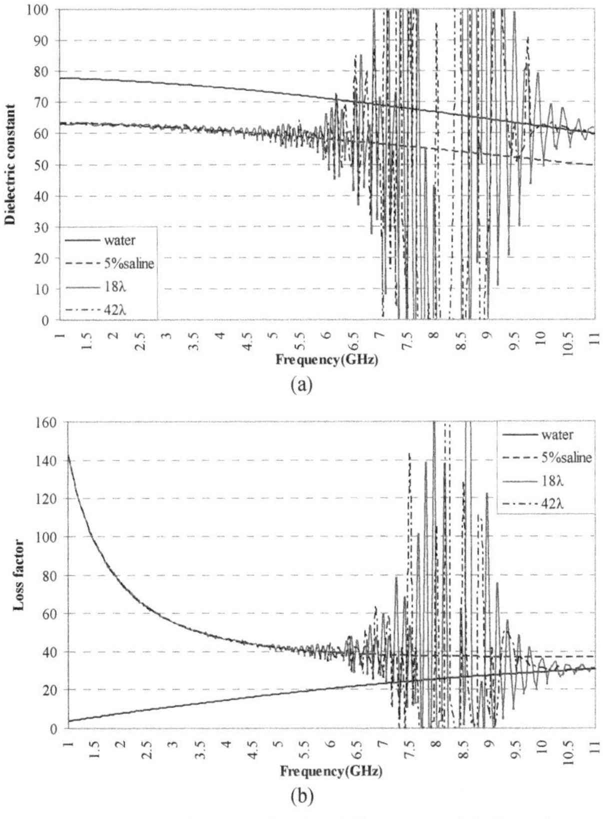

Fig. 4. Dielectric properties for different material dimensions (d = 5 mm, W = 18λ, 42λ at 6 GHz): (a) Dielectric constant, (b) Loss factor.

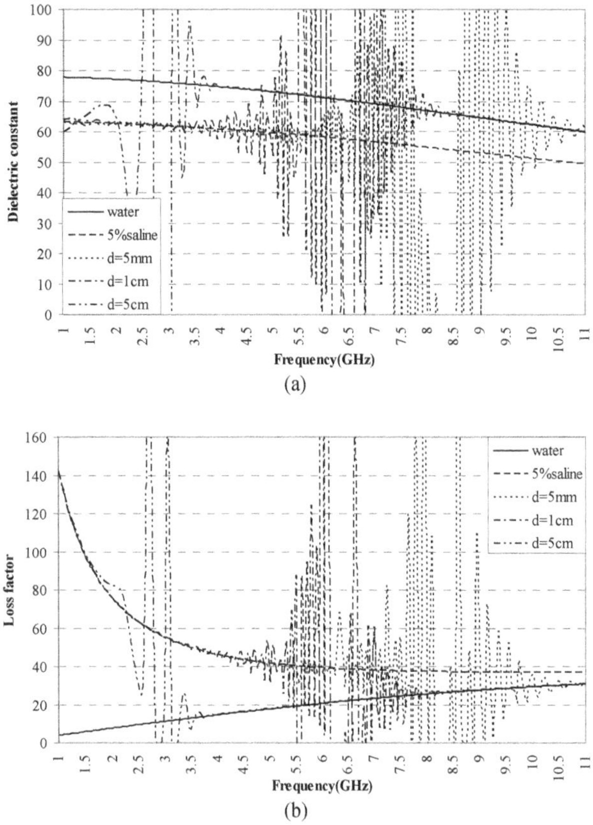

Fig. 5. Dielectric properties for different thickness of the upper layer (W = 18λ at 6 GHz): (a) Dielectric constant, (b) Loss factor.

Let us consider the variation of dielectric properties for various thicknesses of the upper layer of water, lower layer of 5% saline solution and material size of 18λ at 6 GHz, respectively. Fig. 5(a) shows variation of dielectric constant whereas Fig. 5 (b) shows variation of loss factor; for d equal to 5 mm, 1 cm and 5 cm. Obviously, the obtained dielectric properties of material depends on the thickness of the upper layer. The thin layer has a wide range of low frequency of the lower layer. On the other hand, the thick layer has a wide range of high frequency of the upper layer. The transition between the low and the high frequency responses has fluctuation which is related to thickness of the upper layer. Consider Fig. 5(a) the values of dielectric constant of 5% saline solution are obtained below 5 GHz (d = 5 mm), 3.5 GHz (d = 1 cm) and 1.25 GHz (d = 5 cm). On the other hand, the values of water are obtained above 10.5 GHz (d = 5 mm), 8 GHz (d = 1 cm) and 4 GHz (d = 5 cm), respectively. In Fig. 5(b), the values of loss factor of 5% saline solution are obtained below 5.5 GHz (d = 5 mm), 4.25 GHz (d = 1 cm) and 1.75 GHz (d = 5 cm). The values of water are obtained above 10.5 GHz (d = 5 mm), 8 GHz (d = 1 cm) and 3.75 GHz (d = 5 cm), respectively. It should be pointed out that the thickness of the upper layer has significant effect on frequency range of determined dielectric properties.

3.3 Variation of Dielectric Properties of the Lower Layer

In this illustration, the upper layer is water ($d = 5$ mm) and the material size is 18λ for the frequency of 6 GHz. When the material of the lower layer is changed from 5% to 15% saline solution, the frequency response is in the same fashion. From Fig. 6, the saline solution with higher concentration possesses lower dielectric constant and higher loss factor. Clearly, we can determine the variation of dielectric properties of the lower layer from the proposed technique.

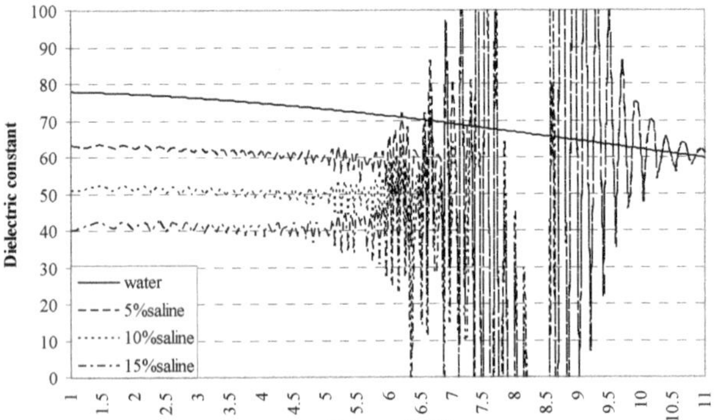

(a)

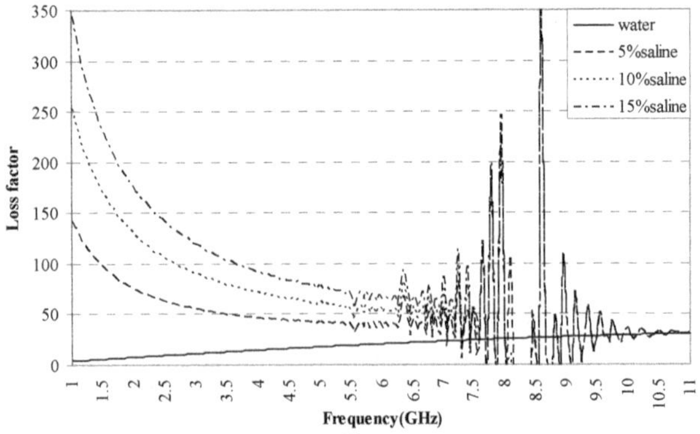

(b)

Fig. 6. Variation of dielectric properties of lower layer ($d = 5$ mm, $W = 18\lambda$ at 6 GHz): (a) Dielectric constant, (b) Loss factor.

4. Experimental Results

4.1 Experimental Setup

A plastic container of 60 cm × 90 cm × 52 cm in size, which the aperture area is 0.54 m^2, was first filled with 5% saline solution until reaching 40 cm in depth from the bottom. A large plastic sheet (thickness of 0.1 mm) was laid over the surface of saline solution. Its edges are wrapped on the edges of the plastic container. A plastic sheet was pressed to delete air bubbles. This plastic sheet can be used as a flat separator between saline solution and water before water of thickness of 5 mm was filled on top of the plastic sheet. The temperatures of water and the 5% saline solution were 25 °C. This container was surrounded by wave absorbers, and two conical log antennas were used to transmit and receive microwave signal. The photograph of the measurement setup is depicted in Fig. 7. The frequency was varied from 1 GHz to 11 GHz, with the transmitting power of 10 mW using a vector network analyzer. The polarization of the transmitting and receiving antennas was respectively right-hand and left-hand circular polarization as these are the available wideband antennas in our laboratory. The antennas were separated by a wave absorber to decouple the antennas (measured S_{21} of –50 dB). The distance from the antennas to the surface of water was 1 m. It was calculated from the largest dimension of the antenna at the center frequency [25]. Note that this distance is shorter than the distance calculated from the largest dimension of the sample. Hence, the planar wavefront is not ensured. This may affect the accuracy of the upper layer thickness determination. A vector network analyzer was open, short and load calibrated and used for reflection measurement. The method to obtain a linear polarized wave from the circular polarized wave can be explained as follows:

$\vec{E}_{receive}$ is the received wave at the receiving antenna.

$$\vec{E}_{receive} = \Gamma E_i \hat{a}_w \cdot \hat{a}_{ant} = \Gamma[E_x \hat{a}_x + jE_y \hat{a}_y] \cdot [\hat{a}_x \mp j\hat{a}_y] \quad (16)$$

where Γ is the complex reflection coefficient of the object under test, E_i is incident electric wave, $\hat{a}_w$ is polarization vector of the incident wave, and $\hat{a}_{ant}$ is polarization vector of receiving antenna. When the transmitted right-hand circular polarized wave (RHCP) is received by a right-hand circular polarized antenna (RHCP), the received wave is

$$\Gamma[(E_x + E_y)]. \quad (17)$$

The wave received by the left-hand circular polarized antenna (LHCP) can be expressed by (16).

$$\Gamma[(E_x - E_y)]. \quad (18)$$

Therefore, the linear polarized wave can be found from (17) ± (18) and then divided by 2.

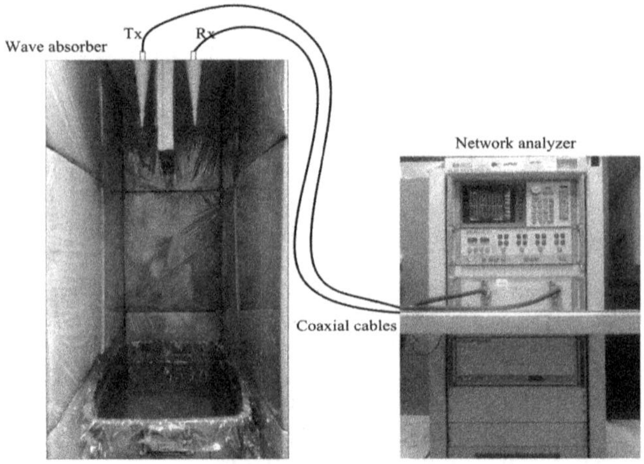

Fig. 7. Experimental setup.

Note that receiving the transmitted RHCP wave by a RHCP antenna corresponds to measuring S_{11} whereas receiving by LHCP antenna corresponds to measuring S_{21}.

This assumption is realizable since the RHCP and LHCP antennas are connected to ports 1 and 2 of the network analyzer, respectively. In addition, the reflected wave from the material under test is in the main beam direction of the antennas which polarization is almost purely circular polarization.

4.2 System Calibration

The system was calibrated by placing a conducting plate (made of copper with a thickness of 1.44 mm that does not contribute significantly to phase error) on the surface of water (see Fig. 8(a)). The total reflection coefficient (Γ_{total_PEC}) is the summation of mutual coupling (S_{21}) and reflection from the conducting plate which is assumed to be perfect conductor ($\Gamma_{in\_copper} = -1$). S_{21} can be determined from

$$S_{21} = \Gamma_{total\_PEC} - \Gamma_{in\_copper}. \tag{19}$$

After removing the conducting plate, the reflection coefficient of the material under test was measured as shown in Fig. 8(b). The total reflection coefficient is the summation of the mutual coupling of the antennas (S_{21}) and the reflection coefficient of water and saline solution interface. Substituting S_{21} in (19) one can find the reflection coefficient between saline solution and water as shown in (20).

$$\Gamma_{in\_water,5\%saline} = \\ = \Gamma_{total\_water,5\%saline} - \left(\Gamma_{total\_PEC} - \Gamma_{in\_copper}\right) \tag{20}$$

The calibrated-measured reflection coefficient can be found from the measured reflection coefficients from the material under test and the reflection of the conducting plate.

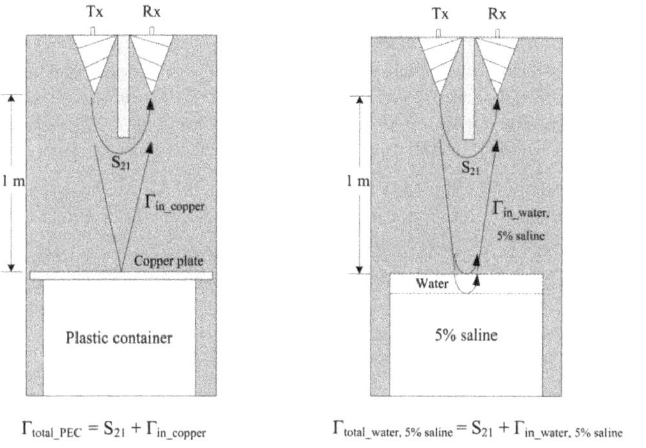

Fig. 8. Calibration process: (a) Conducting plate measurement. (b) Water and 5% saline solution measurement.

4.3 Experimental Results

The calibrated-measured results were substituted on the left side of (4b) and the dielectric constant and loss factor were determined from (12) and (13).

Fig. 9 shows of the measured dielectric properties. Figs. 9(a) and (b) are respectively for dielectric constant and loss factor when the water is 5 mm thick. Clearly, the dielectric properties approach those of water at frequencies higher than 10.7 GHz as the depth of penetration of the wave is shorter than the thickness of water.

The comparisons of the measured results show the error of thickness is in the order of 5.6%. The discrepancy can be attributed from residue transmission at the high frequency and the effect of a plastic sheet separating the two media. The error from the determined dielectric properties of the upper layer results in error in velocity of wave in the upper layer and hence the depth of the upper layer. Nevertheless, a good agreement of the determined dielectric constant can be accomplished with slight error. For the loss factor, the fluctuations around the actual values are attributed from the limited signal-to-noise ratio in the experiment. It is recommended to utilize sufficiently high signal-to-noise ratio in practical use.

Comparing the measured results in Fig. 9 to the calculation results in Fig. 4, the good agreement is obtained. The dielectric properties of the lower layer approach those of 5% saline. While those in Fig. 4 are obtained at fre-

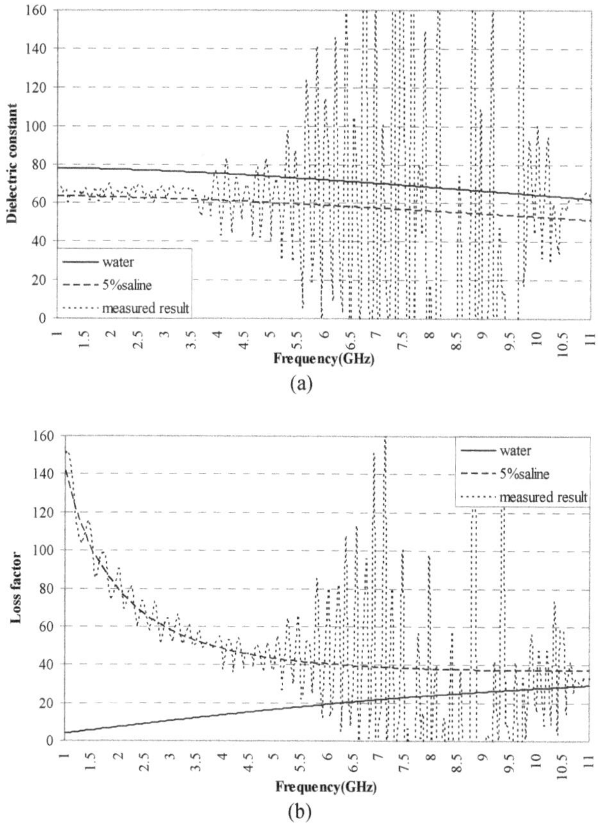

Fig. 9. Determined dielectric properties for $d = 5$ mm: (a) Dielectric constant. (b) Loss factor.

quency lower than 4.5 GHz, the results in Fig. 9 are obtained at frequency lower than 3.5 GHz. The difference results from the calculation results in Fig. 4 neglects the multiple reflections in the upper layer. In addition, the size of the measured material is not exactly the same as the one in calculation. Hence, the effect of edge diffraction is different.

5. Discussion

From the results in the previous sections, it is worth mentioning about the limitation of the system and the required bandwidth for the specific measurement.

The main objective of this work is to determine the unknown dielectric properties of the lower material at the lower frequency band. The upper material is the known material which is determined from the measured reflection coefficient at the higher frequency band. Then, from these dielectric properties at higher frequency band, the dielectric properties at lower frequency band can be found from the wideband database, which is already measured and cataloged as the priori data. With the dielectric properties at the lower frequency band, they lead to the determination of the dielectric properties of the lower material.

It is also essential to discuss about the required bandwidth of the measurement system. The bandwidth is directly related to the dielectric properties and thickness of the upper material. It is obtained from the inverse Fourier transform of the frequency domain measurement to time domain. Then, the thickness is obtained from (15). The thickness d contributes to the accuracy of the determined dielectric properties of the lower material, as seen in (7).

For the high loss material in the upper layer, the depth of penetration is shallow. For instance, when the upper layer is water the minimum d is 5 mm with error of 3.4%. For d less than 5 mm error is in excess of 10%. On the other hand, for the upper layer of 5% saline, the minimum d is 3 mm. The d thinner than 3 mm, error is excessively high. Hence, dielectric properties and thickness of the upper layer play a key role in the required bandwidth.

Furthermore, for the thick upper layer the required high frequency that wave does not penetrate to the lower layer decreases. For instance, for the same condition that the upper layer is water and the lower layer is 5% saline, the thicker upper layer requires lower frequency band. The system design for the measurement bandwidth can be accomplished by calculation using the proposed technique.

6. Conclusion

This paper has presented a dielectric properties determination technique for a stratified medium. The issue of interest in this work is a two-layer planar dielectric structure in which both the dielectric properties for the lower layer and the thickness of the upper layer are unknown. The reflection coefficients of both layers were derived and the dielectric constant and loss factor of the lower medium were extracted. The technique started with wideband measurement to obtain the frequency domain response. The steady response on high frequency exhibited the effect of only the upper layer since wave could not penetrate to the lower layer. Therefore, a single layer was considered at this frequency band and dielectric properties could be accurately determined. Then, the full band dielectric properties were utilized for determining the depth of penetration. The full band dielectric properties and the time domain response can be obtained from the inverse Fourier transform. Hence, the thickness of the upper layer was determined. Using the derived dielectric properties extraction expressions, the dielectric properties of the lower layer can be determined. The measured results in 1-11 GHz band using 5 mm-thick water and 5% saline solution validated the proposed technique at the frequency below 3.5 GHz. The good agreement is accomplished and the proposed technique can be applied for justifying material in industrial process.

Acknowledgements

This work was supported by Thailand Research Fund under the Royal Golden Jubilee Ph.D. Program, Grant no. PHD/0323/2551 and King Mongkut's Institute of Technology Ladkrabang Research Fund (Grant no. KREF115061). The authors faithfully appreciate Mr. Vichit Lohprapan for kindly proofreading the manuscript.

References

[1] GHODGAONKAR, D. K., VARADAN, V. V., VARADAN, V. K. A free-space method for measurement of dielectric constants and loss tangents at microwave frequencies. *IEEE Transactions on Instrumentation and Measurement,* 1989, vol. 37, no. 3, p. 789–793. DOI: 10.1109/19.32194

[2] ZOUGHI, R., BAKHTIARI, S. Microwave nondestructive detection and evaluation of disbounding and delamination in layered-dielectric-slabs. *IEEE Transaction on Instrumentation and Measurement,* 1990, vol. 39, no. 6, p. 1059–1063. DOI: 10.1109/19.65826

[3] VENKATESH, M. S., RAGHAVAN, G. S. V. An overview of dielectric properties measuring techniques. *Canadian Biosystems Engineering,* 2005, vol. 47, p. 7.15–7.30.

[4] SINGH, D., YAMAGUCHI, Y., YAMADA, H., SINGH, K. P. Response of microwave on bare soil moisture and surface roughness by X-band scatterometer. *IEICE Transactions on Communications,* 2000, vol. E83-B, no. 9, p. 2038–2043.

[5] KHARKOVSKY, S., AKAY, M. F., HASAR, U. C., ATIS, C. D. Measurement and monitoring of microwave reflection and transmission properties of cement-based specimens. *IEEE Transactions on Instrumentation and Measurement,* 2002, vol. 51, no. 6, p. 1210–1218. DOI: 10.1109/TIM.2002.808081

[6] THAKUR, K. P., HOLMES, W. S. Noncontact measurement of moisture in layered dielectrics from microwave reflection spectroscopy using an inverse technique. *IEEE Transactions on Microwave Theory and Techniques,* 2004, vol. 52, no. 1, p. 76–82. DOI: 10.1109/TMTT.2003.821243

[7] HASAR, U. C., WESTGAT, C. R., ERTUGRUL, M. Permittivity determination of liquid materials using waveguide measurements for industrial applications. *IET Microwave, Antennas and Propagation,* 2010, vol. 4, no. 1, p. 141–152. DOI:10.1049/iet-map.2008.0197

[8] MEHTA, P., CHAND, K., NARAYANSWAMY, D., BEETNER, D. G., ZOUGHI, R., STOECKER, W. V. Microwave reflectrometry as a novel diagnostic tool for detection of skin cancers. *IEEE Transactions on Instrumentation and Measurement,* 2006, vol. 55, no. 4, p. 1309–1316. DOI: 10.1109/TIM.2006.876566

[9] SEKIGUCHI, H., SHIRAI, H. Electromagnetic scattering analysis for crack depth estimation. *IEICE Transactions on Electronics,* 2003, vol. E86-C, no. 9, p. 2224–2229.

[10] GHASR, M. T., KHARKOVSKY S., ZOUGHI, R., AUSTIN, R. Comparison of near-field millimeter-wave probes for detecting corrosion precursor pitting under paint. *IEEE Transactions on Instrumentation and Measurement,* 2005, vol. 54, no. 4, p. 1497–1504. DOI: 10.1109/TIM.2005.851086

[11] GHASR, M. T., CARROL, B., KHARKOVSKY, S., AUSTIN, R., ZOUGHI, R. Millimeter-wave differential probe for nondestructive detection of corrosion precursor pitting. *IEEE Transactions on Instrumentation and Measurement,* 2006, vol. 55, no. 5, p. 1620–1627. DOI: 10.1109/TIM.2006.880273

[12] HUGHES, D., ZOUGHI, R. A novel method for determination of dielectric properties of materials using a combined embedded modulated scattering and near-field microwave techniques. Part I-Forward model. *IEEE Transactions on Instrumentation and Measurement,* 2005, vol. 54, no. 6, p. 2389–2397. DOI: 10.1109/TIM.2005.858132

[13] HUGHES, D., ZOUGHI, R. A novel method for determination of dielectric properties of materials using a combined embedded modulated scattering and near-field microwave techniques. Part II-Dielectric property recalculation. *IEEE Transactions on Instrumentation and Measurement,* 2005, vol. 54, no. 6, p. 2398-2401. DOI: 10.1109/TIM.2005.858133

[14] VENKATESH, M. S., RAGHAVAN, G. S. V. An overview of microwave processing and dielectric properties of agri-food materials. *Journal of Biosystems Engineering,* 2004, vol. 88, no. 1, p. 1–18. DOI: 10.1016/j.biosystemseng.2004.01.007

[15] NELSON, S. O. Agricultural applications of dielectric measurements. *IEEE Transactions on Dielectrics and Electrical Insulation,* 2006, vol. 13, no. 4, p. 688–702. DOI: 10.1109/TDEI.2006.1667726

[16] ABUDUL KHALID, M. F., RAMLI, A. S., BABA, N. H., SAAD, H. A novel preliminary study on microwave characterization of siamese mangoes ripeness at K-band. In *Proceedings of International RF and Microwave Conference.* Kuala Lumpur (Malaysia), 2008, p. 143–147.

[17] HASAR, U. C. A fast and accurate amplitude-only transmission-reflection method for complex permittivity determination of lossy materials. *IEEE Transactions on Microwave Theory and Techniques,* 2008, vol. 56, no. 9, p. 2129–2135. DOI: 10.1109/TMTT.2008.2002229

[18] ODA, M., MASE, A., UCHINO, K. Non-destructive measurement of sugar content in apples using millimeter wave reflectometry and artificial neural networks for calibration. In *Proceedings of the 25th Asia-Pacific Microwave Conference.* Melbourne (Australia), 2011, p. 1386–1389.

[19] GHASR, M. T., SIMMS, D., ZOUGHI, R. Multimodal solution for a waveguide radiating into multilayered structures-dielectric property and thickness evaluation. *IEEE Transactions on Instrumentation and Measurement,* 2009, vol. 58, no. 5, p. 1505–1513. DOI: 10.1109/TIM.2008.2009133

[20] SEAL, M. D., HYDE, M. W., HAVRILLA, M. J. Nondestructive complex permittivity and permeability extraction using a two-layer dual-waveguide probe measurement geometry. *Progress In Electromagnetics Research,* 2012, vol. 123, p. 123–142. DOI:10.2528/PIER11111108

[21] CHARVAT, G. L., KEMPEL, L. C., ROTHWELL, E. J., COLEMAN, C. M., MOKOLE, E. L. A through-dielectric radar imaging system. *IEEE Transactions on Antennas and Propagation,* 2010, vol. 58, no. 8, p. 2594–2603. DOI: 10.1109/TAP.2010.2050424

[22] OSA, K., SUMANTYO, J. T. S., NISHIO, F. An application of microwave measurement for complex dielectric constants to detecting snow and ice on road surface. *IEICE Transactions on Communications,* 2011, vol. E94-B, no. 11, p. 2987–2990. DOI: 10.1587/transcom.E94.B.2987

[23] GENNARELLI, G., RICCIO, G. A uniform asymptotic solution for the diffraction by a right-angle dielectric wedge. *IEEE Transactions on Antennas and Propagation,* 2011, vol. 59, no. 3, p. 898–903. DOI: 10.1109/TAP.2010.2103031

[24] STOGRYN, A. Equations for calculating the dielectric constant of saline water. *IEEE Transactions on Microwave Theory and Techniques,* 1971, vol. MTT-19, no. 8, p. 733–736. DOI: 10.1109/TMTT.1971.1127617

[25] KRAUS, J. D., MARHEFKA, R. J. *Antennas for All Applications.* 3rd ed. New York: McGraw-Hill, 2002.

About the Authors ...

Paiboon YOIYOD was born in Phetchaburi in 1981, Thailand. He received the Bachelors degree in Electrical Engineering with first class honor from Rangsit University and Masters degree in Telecommunications Engineering from King Mongkut's Institute of Technology Ladkrabang in 2005 and 2010, respectively. He is currently pursuing the Ph.D. degrees in Electrical Engineering. He joined in production engineer, Pioneer Manufacturing (Thailand) Co.,Ltd. in 2005 as an engineer. In 2006, he joined the Rangsit University as an engineer. His current research interests include dielectric measurements, sensors, and measurement systems.

Monai KRAIRIKSH was born in Bangkok, Thailand. He received the B.Eng., M.Eng. and D.Eng. degrees in Electrical Engineering from King Mongkut's Institute of Technology Ladkrabang (KMITL), Thailand in 1981, 1984, and 1994, respectively. He was a visiting research scholar at Tokai University in 1988 and at Yokosuka Radio Communications Research Center, Communications Research Laboratory (CRL) in 2004. He joined the KMITL and is currently a Professor at the Department of Telecommunication Engineering. He has served as the Director of the Research Center for Communications and Information Technology during 1997-2002. His main research interests are in antennas for mobile communications and microwave in agricultural applications. Dr. Krairiksh was the chairman of the IEEE MTT/AP/Ed joint chapter in 2005 and 2006. He served as the General Chairman of the 2007 Asia-Pacific Microwave Conference, and the advisory committee of the 2009 International Symposium on Antennas and Propagation. He was the President of the Electrical Engineering/Electronics, Computer, Telecommunications and Information Technology Association (ECTI) in 2010 and 2011 and was an editor-in-chief of the ECTI Transactions on Electrical Engineering, Electronics, and Communications. He was recognized as a Senior Research Scholar of the Thailand Research Fund in 2005 and 2008 and a Distinguished Research Scholar of the National Research Council of Thailand. He has been a distinguished lecturer of IEEE Antennas and Propagation Society during 2012-2014.

4

Bandwidth Efficient Root Nyquist Pulses for Optical Intensity Channels

Sabeena FATIMA, S. Sheikh MUHAMMAD, A. D. RAZA

Dept. of Electrical Engineering, National University of Computer and Emerging Sciences, Lahore, Pakistan

sabeena.fatima@nu.edu.pk, sm.sajid@nu.edu.pk, ad.raza@nu.edu.pk

Abstract. *Indoor diffuse optical intensity channels are bandwidth constrained due to the multiple reflected paths between the transmitter and the receiver which cause considerable inter-symbol interference (ISI). The transmitted signal amplitude is inherently non-negative, being a light intensity signal. All optical intensity root Nyquist pulses are time-limited to a single symbol interval which eliminates the possibility of finding bandlimited root Nyquist pulses. However, potential exists to design bandwidth efficient pulses. This paper investigates the modified hermite polynomial functions and prolate spheroidal wave functions as candidate waveforms for designing spectrally efficient optical pulses. These functions yield orthogonal pulses which have constant pulse duration irrespective of the order of the function, making them ideal for designing an ISI free pulse. Simulation results comparing the two pulses and challenges pertaining to their design and implementation are discussed.*

Keywords

Nyquist Pulse, optical intensity signaling, PSWF (Prolate Spheroidal Wave Functions), MHPF (Modified Hermite Polynomial Function), ISI (Inter Symbol Interference)

1. Introduction

The capacity of Radio Frequency (RF) systems is limited due to scarcity and cost of licensing. Therefore, to meet the need of growing data rate requirements in wireless communications optical bands are being excessively exploited. Optical signals have numerous advantages over RF signals, the most attractive ones being that they are unlicensed and have many THz of bandwidth naturally associated to them.

Wireless optical intensity channel transmit information by modulating the instantaneous optical intensity. A Light Emitting Diode (LED) or a Laser Diode (LD) converts the input electrical signal to an optical signal. A photodiode detector at the receiver end converts the incident optical intensity signal back to an electrical signal. The optical intensity channel commonly employs Intensity Modulation and Direct Detection (IM/DD) [2]. Multipath distortion arising due to the multiple reflected paths between the transmitter and the receiver in indoor environment cause severe Inter Symbol Interference (ISI), necessitating the use of pulse shaping. Diffuse links [3] are attractive as they allow user mobility and eliminate the need of alignment but at the expense of bandwidth. The diffuse indoor optical links are severely bandwidth limited.

For indoor optical applications, signaling design varies significantly from the conventional electrical channels as optical channels impose the additional requirement of signal being non-negative. Secondly optical links require that for the case of indoor optical channels the transmitted signal be Class I eye safe under all conditions. As a result the peak transmitted power of the light signal remains constrained. Received signal is corrupted by free space losses, shot noise and ambient light noise. For the case of indoor optical channels distortion due to shot noise is most significant. Shot noise is modeled as uncorrelated additive white Gaussian noise [3].

There exists a wealth of literature on design of pulses that maximize in-band energy [4, 5, 6] to mitigate the effect of ISI; however these pulses have been designed for the electrical channel. Halpern [7] proposed a design for finite duration Nyquist pulses for mean power constrained channels but not subject to amplitude constraints. Considerable information on optical intensity channels and their numerous advantages have been investigated [3, 8, 9]. However, the design of optical intensity pulse to minimize ISI still requires a much deeper study [9]. Steve Hranilovic [1] proved that all optical intensity root Nyquist pulses must be time limited to a single symbol interval. The time-limitedness of the root Nyquist pulse eliminates the possibility of finding a bandlimited pulse. Steve [1] proposed Prolate Spheroidal Wave pulse to be used since it is time limited and also has better spectral concentration among all time limited signals. For signals limited to a single symbol interval PSWF is the optimum signal choice [7, 11, 12]. From studies on pulse shaping techniques for time limited systems [13, 14, 15] Prolate Spheroidal Wave Functions (PSWF) and Modified Hermite Polynomial Function (MHPF) have emerged as potential pulse shaping functions as they provide the optimal spectral concentration. Individual discussion on these two contender pulses is widely available but no comparison between the two has been drawn. Design of a root Nyquist pulse which maximizes the in-band fractional energy has been proposed in [4, 7], in context of electrical domain. A linear combination of PSWF pulses proposed in [4] seeks to maximize the in-band fractional energy and to satisfy the Nyquist criteria. The problem was reduced to solving the linear objective function. This design can be extended to the optical domain by introducing an additional constraint of optical signal remaining non-negative.

In this work we propose root Nyquist pulse designs. The non-negativity constraint limits the root Nyquist pulses to be time-limited [1]. Although this eliminates the possibility of finding bandwidth limited root Nyquist pulses, potential still exists to find bandwidth efficient pulses for these channels. Pulses can be shaped such that most of the energy is concentrated in the band of interest [16]. Section 2 presents the discussion on root Nyquist pulse and shows that they are strictly time limited to a single symbol interval. Sections 3 and 4 focus on the discussion of PSWF and MHPF pulses. In Section 5 a comparison is drawn between the two pulses. Conclusions are presented in Section 6.

2. Optical Intensity Root Nyquist Pulses

For the case where receiver filter is matched to the transmitted pulse, to ensure zero ISI the pulse shape at the output must satisfy the Nyquist criteria [15] given by:

$$\int x(\tau)x(\tau-kT)d\tau = \delta_{k0} \quad (1)$$

where $\delta_{kl} = \{ \begin{array}{ll} 1 & k=l \\ 0 & otherwise. \end{array}$

An optical intensity root Nyquist pulse satisfies the non-negativity constraint in addition to the Nyquist criteria. All practical optical intensity root Nyquist pulses are time-limited to single symbol duration [1]. The root Nyquist pulse is required to be of small duration so that the number of interfering neighbors is small. This eliminates the possibility of obtaining band-limited root Nyquist pulses.

The search therefore reduces to bandwidth efficient pulses. Most of the energy needs to be concentrated in a finite band. The main objective is to maximize the in-band fractional energy of the time-limited root Nyquist pulse. In-band fractional energy is defined as the ratio of the energy of the pulse in the given bandwidth to the total energy of the time limited pulse. Its expression is given as [9]:

$$\frac{\int_{-W}^{W} |X(f|^2 df)}{\int_{-\infty}^{\infty} |X(f|^2 df)} = K \quad (2)$$

where K∈(0,1).

The (1-ε) fractional energy bandwidth of a transmitted symbol $x(t)$ with Fourier Transform $X(f)$ is defined as [9]:

$$W_\varepsilon(x) = W\varepsilon[0,\infty) : \int_{-W}^{W} |X(f|^2 df) \geq (1-\varepsilon)\int_{-\infty}^{\infty} |X(f|^2 df)$$

If ε is chosen such that out of band energy is small, then $x(t)$ can be thought of as efficiently band-limited to $W\varepsilon(x)$.

The prolate spheroidal wave functions (PSWF) and modified hermite polynomial functions (MHPF) give the highest spectral concentration of all time limited signals [14, 15] and are discussed in detail in the next sections.

3. Prolate Spheroidal Wave Functions

Prolate spheroidal wave functions (PSWF) have the highest spectral concentration of all time-limited signals. They are strictly time-limited to [0, T] and have maximum energy in the band of interest, i.e. [-Ω, Ω] of all unit energy functions. These functions are solution to the integral equation [15]:

$$\int_{-T/2}^{T/2} \frac{\sin\Omega(t-s)}{\pi(t-s)} \psi_i(s)ds = \lambda_i \psi_i(t). \quad (3)$$

Here ψ(t) represents the PSWF, λ is the amount of energy contained in $[-T/2, T/2]$, Ω is the bandwidth and T is the pulse duration.

PSWF are an improvement on other pulses as they satisfy the double orthogonality property. This double orthogonality property guarantees unique demodulation at the receiver.

$$\int_{-\infty}^{\infty} \psi_m(t).\psi_n(t) = \delta_{mn}, \quad (4)$$

$$\int_{-T/2}^{T/2} \psi_m(t).\psi_n(t) = \lambda_m \delta_{mn} \quad (5)$$

where $\psi_n(t)$ is the PSWF of order n.

ψ_n(t) is written in terms of prolate angular function of first kind [15] as:

$$\psi_n(t) = \psi_n(\Omega, T, t) = \frac{(2\lambda_n(c)/T)^{1/2} S_{0n}^1(c, 2t/T)}{\{\int_{-1}^{1}[S_{0n}^1(c,x)]^2 dx\}^{1/2}} \quad (6)$$

where $\{\int_{-1}^{1}[S_{0n}^1(c,x)]^2 dx\}^{1/2} = \frac{2}{2n+1}$ where S_{0n}^1 is the prolate angular function of first kind and λ_n is the fraction of the energy of ψ_n(t) that lies in the interval [-1, 1].

The prolate angular function of the first kind is given by [15]:

$$S_{0n}^1(c,t) = \begin{cases} \sum_{k=0,2,\ldots}^{\infty} d_k(c)P_k(c,t) & n \quad \text{even} \\ \sum_{k=1,3,\ldots}^{\infty} d_k(c)P_k(c,t) & n \quad \text{odd} \end{cases} \quad (7)$$

where c is the time bandwidth product given by $c = \Omega T/2$, $P_k(c,t)$ is the Legendre polynomials [13][18] and d_k(c) satisfy the recursion relation[14]:

$$\alpha_k d_{k+2}^n(c) + (\beta_k - \chi_n(c))d_k^n(c) + \gamma_k d_{k-2}^n(c) = 0 \quad (8)$$

where the coefficients α, β, γ are given by:

$$\alpha_k = \frac{(k+1)(k+2)c^2}{(2k+3)(2k+5)}, \quad (9)$$

$$\beta_k = \frac{(2k^2+2k-1)c^2}{(2k-1)(2k+3)} + k(k+1), \quad (10)$$

$$\gamma_k = \frac{(k)(k-1)c^2}{(2k-1)(2k-3)}. \quad (11)$$

In order to compute $d_k^n(c)$ the following equation [15] is solved:

$$Od^n = \chi_n d^n. \quad (12)$$

Here d^n are the eigenvector and χ_n are the eigenvalues.

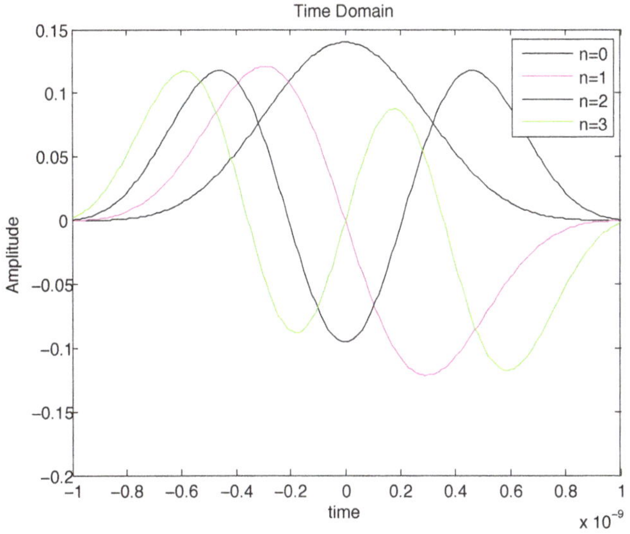

Fig. 1. Time Domain Representation of Prolate Spheroidal Wave Functions of order 0, 1, 2, 3. Note that the pulse duration is same for all orders and the number of zero crossings is equal to the pulse order.

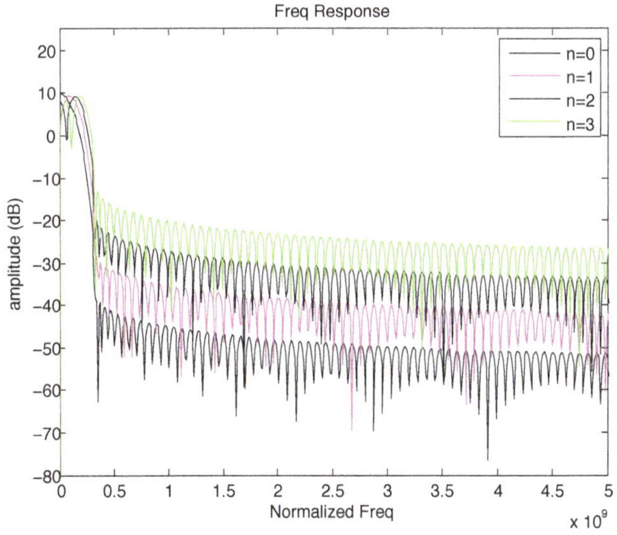

Fig. 2. Frequency Domain Representation of Prolate Spheroidal Wave Functions of order 0, 1, 2, 3.

Figures 1 and 2 represent the time and frequency characteristics for the PSWFs respectively. From the figures it can be deduced that the pulse duration and bandwidth is exactly the same for all orders n. Another interesting property being that number of zero crossings equals n. These pulses are double-orthogonal. Pulse duration and bandwidth can be bartered by utilizing the time-bandwidth product.

For wireless optical intensity channel only PSWF order 0 pulse can be utilized as it fulfills the amplitude non-negativity constraint. However, making use of the interesting property of PSWF pulses that the number of zero crossings on time axis equals the pulse order n and the pulse duration is same for all pulse orders, one approach could be to make a linear combination pulse of these orders and analyze its properties.

3.1 Time Bandwidth Product

To get a better understanding of time-bandwidth product c, PSWF order 0 pulse was plotted for different values of c. As time was kept confined to duration of 2 nanoseconds, the variation of c had a direct impact on bandwidth. The c value $\Omega T/2$ effect the width of main lobe of PSWF pulse, as the c value increases; the main lobe becomes compressed causing an expansion in frequency domain (as seen in the Figs. 3 and 4).

3.2 Linear Combination Pulse

Since we have limitations on channel bandwidth we need to concentrate most of the energy in our band of interest. We wish to maximize the in-band fractional energy which is given by:

$$\frac{\int_{-W}^{W} |X(f|^2 df)}{\int_{-\infty}^{\infty} |X(f|^2 df)} = K. \quad (13)$$

Nigam [4] proposed a linear combination of PSWF pulses which maximize the in-band fractional energy. The linear combination pulse is given as:

$$g(t) = \sum_{0}^{2N+2} a_j \psi_j(t). \quad (14)$$

Here $\psi_j(t)$ is the PSWF of j-th order; N is the number of PSWFs used and a_j are the coefficients which need to be determined.

Maximizing the in-band fractional energy reduces the problem [4] of determining the coefficients a_j to

$$\text{Maximize: } \sum_{0}^{2N-2} a_j \psi_j(0) \lambda_j \quad (15)$$

Subj to Nyquist Criteria:

$$\sum_{0}^{2N-2} a_j \psi_j(kT) = \delta_k. \quad (16)$$

This ensures that the overall pulse is zero at multiples of T except $k = 0$.

However, this pulse was designed for transmission in the electrical domain. Extending these results to the optical domain introduces an additional constraint that the transmitted pulse remains non-negative. Taking into account all

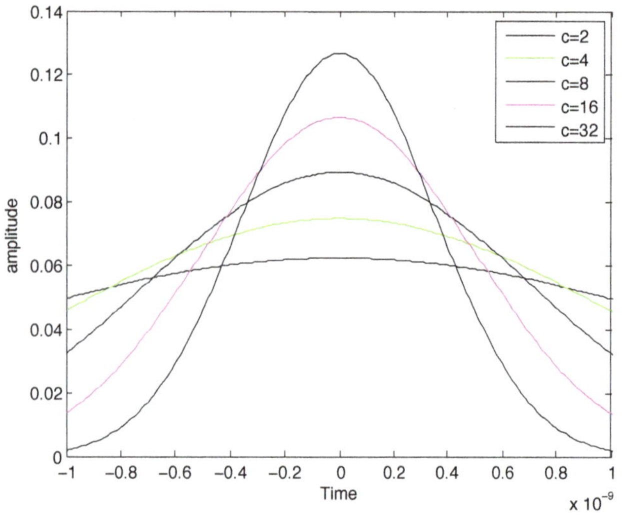

Fig. 3. Effect of varying time-bandwidth product c (time representation).

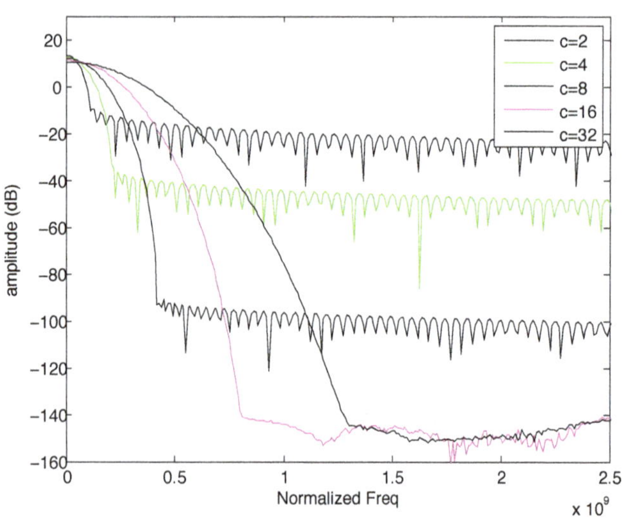

Fig. 4. Effect of varying time-bandwidth product (frequency representation). As the time-bandwidth product value increases the bandwidth occupied by the pulse increases.

these constraints the problem reduces to evaluating the linear objective function subject to linear constraints which can readily be solved by applying the simplex method. The results show that the coefficients for higher order PSWF pulse approach zero.

Firstly, we constrain the PSWF pulse to have unit value at time zero, then solving (13) and (14) for the coefficients a_j we get the values shown in Tab. 1. As can be observed the PSWF order 0 pulse makes the major contribution to the linear combination pulse, in fact up to $c = 8$ PSWF 0 pulse is the only contributor, beyond $c = 8$ other even-ordered PSWF pulse shapes come into play. This proves that PSWF order 0 pulse is the optimum pulse which fulfills the non-negativity criteria.

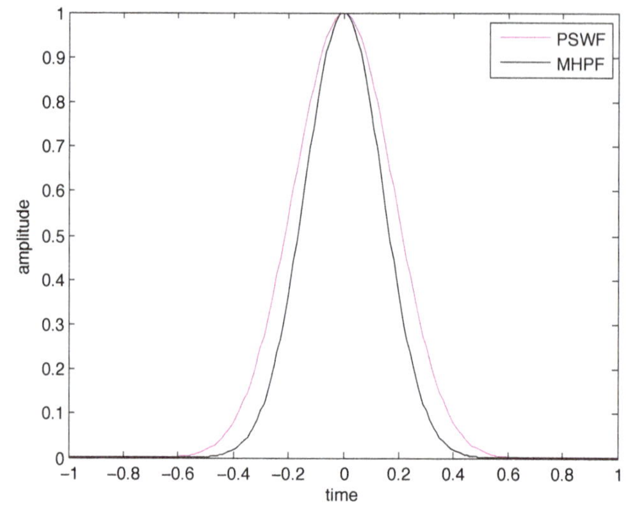

Fig. 5. Comparison between the optimal linear combination pulse and PSWF order 0 pulse proves that the optimal pulse follows the shape of the PSWF order 0.

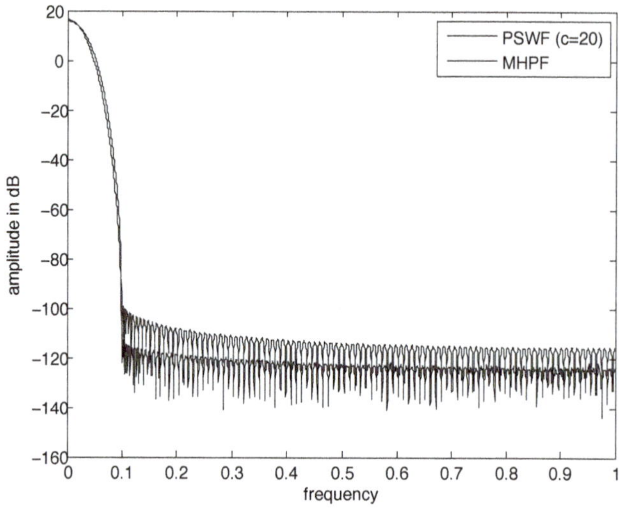

Fig. 6. No appreciable change in bandwidth between the two is observed proving that the major input is from the order 0 pulse in the optimal pulse design.

c	a_0	a_1	a_2	a_3	a_4	a_5
2	1	0	0	0	0	0
4	1	0	0	0	0	-0
8	1	0	0	0	0	-0
10	0.9396	0	0.0860	0	0	0
16	0.9442	0	0.0795	0	0	0
20	0.9479	0	0.0668	0	0.0085	-0

Tab. 1. Coefficient values computed for the linear combination pulse for different time-bandwidth product using Simplex Methods. Note in order to ensure uniformity the PSWF pulses were scaled to have unit value at time 0.

Figure 5 compares the optimal pulse with the PSWF order zero pulse, the linear combination pulse has a quicker decay rate and from the plot in Fig. 6 it can be observed that there is no significant change in band occupied by the two.

4. Modified Hermite Polynomial Function

The Hermite polynomials are defined by [15][19]:

$$h_{en}(t) = (-1)^n e^{\frac{t^2}{2}} \frac{d^n}{dt^n}(e^{\frac{-t^2}{2}}) \quad (17)$$

where $n = 0, 1, 2, .. -\infty < t < \infty$. Hermite polynomials are not orthogonal [15]. However, they can be modified to become orthogonal as follows:

$$h_n(t) = e^{\frac{-t^2}{4}} h_{en}(t) = (-1)^n e^{\frac{-t^2}{4}} \frac{d^n}{dt^n}(e^{\frac{-t^2}{2}}). \quad (18)$$

The general formula for defining Modified Hermite Polynomial Functions (MHPF) is given by [15]:

$$h_n(t) = k_n e^{\frac{-t^2}{4}} n! \sum_{i=0}^{n/2} (-\frac{1}{2})^i \frac{t^{n-2i}}{(n-2i)!i!}. \quad (19)$$

The constant k_n where $n = 0, 1, 2, \ldots$ is indicative of the energy of the pulse given by [15]:

$$k_n = \sqrt{\frac{E_n}{n!\sqrt{2\pi}}}. \quad (20)$$

Figure 7 and Figure 8 represent the time and frequency response for the MHPFs. It can be observed that the pulse duration and bandwidth is exactly the same for different orders n with the number of zero crossings in time domain being equal to n. These pulses are orthogonal to one another.

5. Comparison

The MHPF and PSWF are very different functions so drawing a comparison between the two is fairly complex. We confined the PSWF and MHPF pulse to similar time durations and observed the pulse shapes obtained.

PSWF pulse's duration and bandwidth can be bartered using a time-bandwidth product c where $c = \Omega T/2$. The MHPF pulses have no such constant c so only one pulse shape is obtained for a given pulse duration. However, the MHPF pulse time duration and bandwidth are inversely related as $\Omega T/\pi$ [20].

The curves obtained when the MHPF and PSWF generated pulse shapes have identical time domain characteristics were plotted side by side to draw a better comparison. For instance, in the current scenario when the time bandwidth product $c = 16$ with pulse duration $T = [-1, 1]$, the MHPF and PSWF pulse have identical time domain characteristics (also shown in Fig. 9). While comparing the frequency domain plots obtained it was observed that the band occupied by the MHPF pulse is 1.3 times the band occupied by PSWF pulse which is roughly equal to $\pi/2$ (Fig. 10).

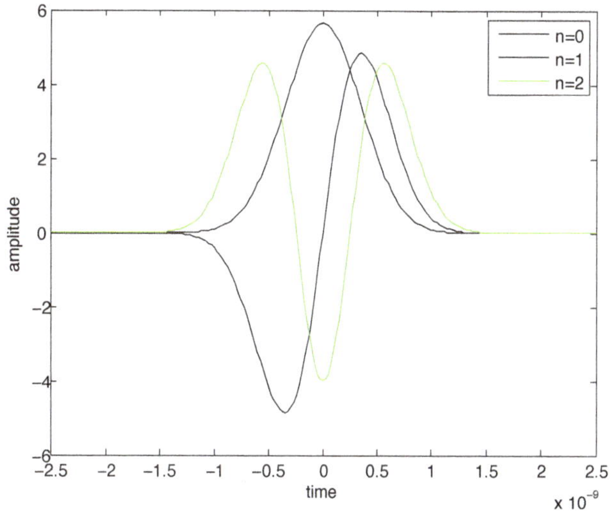

Fig. 7. Time domain representation of MHPF pulse for order 0, 1, 2. Different order pulses have identical pulse duration.

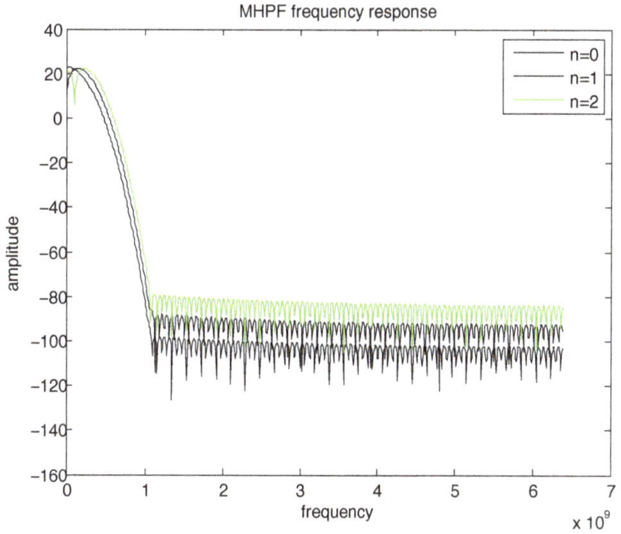

Fig. 8. Frequency representation for MHPF order 0, 1, 2 pulse. Note the constant identical bandwidth for different orders of the pulse.

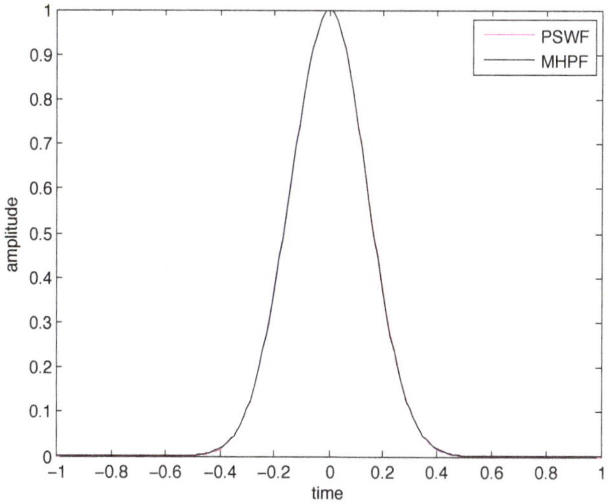

Fig. 9. PSWF (c=16) and MHPF pulse. The two pulse shapes appear to be exactly similar.

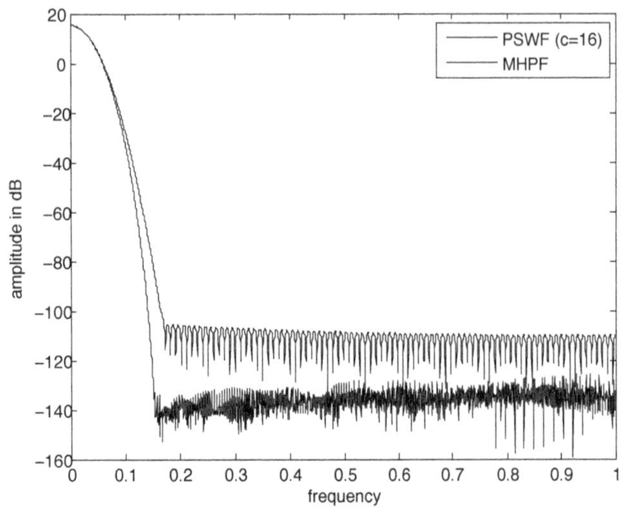

Fig. 10. Comparison between MHPF pulse and PSWF pulse ($c = 16$). Both occupy similar bandwidth as is expected from the similar time response.

Table 2 compares the band occupied by the PSWF and MHPF generated pulse while having similar parameters. c-PSWF denotes the value of time-bandwidth product to obtain a PSWF pulse shape similar to MHPF for the same time-duration. It can be observed that the band occupied by MHPF pulse is $1.2-1.3$ times the one of PSWF pulse, therefore it can be safely said that the PSWF is most efficient in terms of bandwidth occupancy.

Time	c-PSWF	BW-PSWF	BW-MHPF	Factor diff.
$[-1,1]$	16	0.154	0.196	$\simeq 1.3$
$[-\frac{3}{4},\frac{3}{4}]$	10	0.1295	0.138	$\simeq 1.1$
$[-\frac{1}{2},\frac{1}{2}]$	4	0.081	0.104	$\simeq 1.3$

Tab. 2. Comparison of band occupancy of PSWF and MHPF generated pulse shapes under similar time duraction.

To validate the benefits of the pulse design technique utilized, optical data transmission in indoor environments has been simulated. The unequalized OOK system has been adapted for validation. The input bits, assumed to be independent identically distributed (i.i.d) and uniform on [0,1] are passed through a transmitter filter whose impulse response allows pulse shaping. The block diagram of the simulative setup is shown in Fig. 11 which uses the ceiling bounce model to cater for indoor light propagation [21]. The input signal $x(t)$ is passed through a multipath channel impulse response $h(t)$ and noise $n(t)$ is added. On the receiver end matched filter detection is employed as depicted in Fig. 11. To characterize the system performance, the eye diagram analysis (Fig. 12 and 13) has been utilized. The effect of multipath propagation and intersymbol interference is catered for better with our self designed pulse as the opening of the eye (Fig. 13) on an arbitrary amplitude scale is wider compared to the scenario where a rectangular pulse is used (Fig. 12).

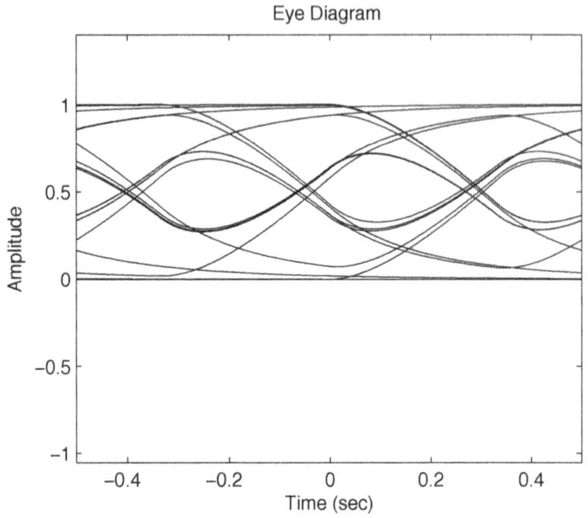

Fig. 12. Eye diagram using rectangular pulse for the Ceiling Bounce model.

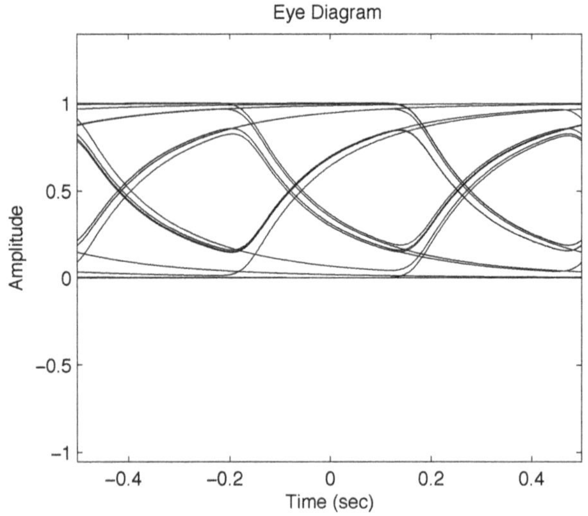

Fig. 13. Eye diagram using self designed pulse for the Ceiling Bounce model.

6. Conclusion

This paper investigates ISI free pulses satisfying the non-negative amplitude constraint of optical intensity channels for high speed optical wireless communications [22]. The optical intensity channels constrain the pulse to be non-negative; also there is a limitation on channel bandwidth. This non-negativity constraint limits the root Nyquist pulse to be time limited [1]. Modified Hermite polynomial and Prolate Spheroidal wave functions have been proposed as potential pulse shaping functions. Both are time-limited and have most of the energy concentrated in a finite band. PSWF pulses' time and bandwidth can be bartered using a constant c (the time-bandwidth product i.e. $\Omega T/2$). Varying values for c it is possible to arrive at a point where the MHPF and PSWF pulse have identical curves for any given time-duration. Comparing the bandwidth occupancy

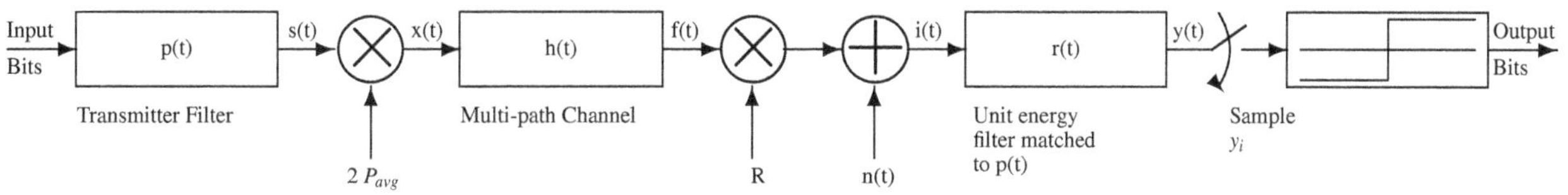

Fig. 11. Block diagram for the simulation setup.

of MHPF and PSWF under this scenario it was observed that the MHPF pulse occupies a band roughly 1.2 times the band occupied by the PSWF pulse. For a major segment (range of *c* values), the MHPF pulse occupies more than twice the band when compared to the PSWF pulse. Also from optimizing the linear objective function PSWF order 0 pulse emerges as most efficient in terms of bandwidth concentration. MHPF pulses hold the advantage of being easier to generate as they have a closed form expression, contrary to the PSWF pulse which lack a closed form solution. Since both PSWF and MHPF pulses satisfy the optical channel constraints, they are well suited for ease of implementation and fulfill system bandwidth requirements.

References

[1] HRANILOVIC, S. Minimum bandwidth optical intensity nyquist pulses. *IEEE Transactions on Communications*, 2007, vol. 55, no. 3, p. 574–583. DOI: 10.1109/TCOMM.2006.888878

[2] HRANILOVIC, S., KSCHISCHANG, F. R. Optical intensity-modulated direct detection channels: signal space and lattice codes. *IEEE Transactions on Information Theory*, 2003, vol. 49, no. 6, p. 1385–1399. DOI: 10.1109/TIT.2003.811928

[3] KAHN, J. M., BARRY, J. R. Wireless infrared communications. *Proceedings of the IEEE*, 1997, vol. 85, no. 2, p. 263–298. DOI: 10.1109/5.554222

[4] NIGAM, G., SINGH, R., CHATURVEDI, A. K. Finite duration root nyquist pulses with maximum in-band fractional energy. *IEEE Communications Letters*, 2010, vol. 14, no. 9, p. 797–799. DOI: 10.1109/LCOMM.2010.09.100314

[5] BEAULIEU, N.C., TAN, C.C., DAMEN, M.O. A better than Nyquist pulse. *IEEE Communications Letters*, 2001, vol. 5, no. 9, p. 367–368. DOI: 10.1109/4234.951379

[6] SOOD, R., XIAO, H. Root Nyquist pulses with an energy criterion. In *IEEE International Conference on Communications*. Glasgow (UK), 2007, p. 2711–2716. DOI: 10.1109/ICC.2007.450

[7] HALPERN, P. Optimum finitie duration nyquist signals. *IEEE Transactions on Communications*, 1979, vol. 27, no. 6, p. 884–888. DOI: 10.1109/TCOM.1979.1094486

[8] RAMIREZ-INIGUIEZ, R., GREEN, R. J. Indoor optical wireless communications. In *IEEE Colloquium on Optical Wireless Communications*. London (UK), 1999, p. 14/1–14/7. DOI: 10.1049/ic:19990705

[9] TAVAN, M., AGRELL, E., KAROUT, J. Bandlimited intensity modulation. *IEEE Transactions on Communications*, 2012, vol. 60, no. 11, p. 3429–3439. DOI: 10.1109/TCOMM.2012.091712.110496

[10] HRANILOVIC, S. *Wireless Optical Communication Systems*. Boston (USA): Springer Science + Business Media, 2005. DOI: 10.1007/b99592

[11] SELPIAN, D., POLLAK, H. O. Prolate spheroidal wave functions, fourier analysis and uncertainty — I. *Bell Systems Technical Journal*, 1961, p. 43–63.

[12] SELPIAN, D., POLLAK, H. O. Prolate spheroidal wave functions, fourier analysis and uncertainty — II. *Bell Systems Technical Journal*, 1962, p. 65–84.

[13] ALLEN, B., GHORASHI, S. A., GHAVAMI, M. A review of pulse design for impulse radio. In *Proceedings of IEEE Ultra Wideband Communications Technologies and System Design*. 2004, p. 93–97.

[14] GHAVAMI, M., MICHAEL, L., KOHNO, R. Hermite function based orthogonal pulses for ultra wideband communication. *Wireless Personal Multimedia Communications Symposium*, 2001.

[15] DILMAGHANI, R. S., GHAVAMI, M., ALLEN, B., AGHVAMI, H. Novel UWB pulse shaping using prolate spheroidal wave functions. In *Proceedings of IEEE Personal, Indoor and Mobile Radio Communications*. Beijing (China), 2003, vol. 1, p. 602–606. DOI: 10.1109/PIMRC.2003.1264343

[16] TAVAN, M., AGRELL, E. , KAROUT, J. Strictly bandlimited ISI-free transmission over intensity-modulated channels. In *IEEE Globecom*. Houston (TX, USA), 2011, p. 1–6. DOI: 10.1109/GLOCOM.2011.6133734

[17] PROAKIS, J. G. *Digital Communications*. 4th ed. New York: McGraw-Hill, 2001.

[18] TSAI, C. Y., JENG, S. K. Design of Legendre polynomial based orthogonal pulse generator for ultra wideband communications. In *IEEE Antennas and Propagation Society International Symposium*. 2005, vol. 2B, p. 680–683. DOI: 10.1109/APS.2005.1552105

[19] GHAVAMI, M., MICHAEL, L.B., KOHNO, R. *Ultra Wideband Signals and Systems in Communication Engineering*. 2nd ed. Wiley, 2007.

[20] OUERTANI, M., BESBES, H., BOUALLEGUE, A. Modified Hermite functions for designing new optimal UWB pulse-shapers. In *European Signal Processing Conference (EUSIPCO)*. Antalya (Turkey), 2005.

[21] GHASEEMLOOY, Z., POPOOLA, W., RAJBHANDARI, S. *Optical Wireless Communications: System and Channel Modelling with Matlab*. London: CRC Press, 2013.

[22] TAVAN, M., AGRELL, E., KAROUT, J. Indoor high-speed optical wireless communications: Recent developments. In *Proceedings of IEEE International Conference on Transparent Optical Networks (ICTON)*. Graz (Austria), 2014.

About the Authors ...

Sabeena FATIMA received her BS in Telecommunications Engineering and MS in Electrical Engineering from the National University of Computer and Emerging Science, Lahore, Pakistan, in 2009 and 2013 respectively. She joined National University of Computer and Emerging Science in the year 2009 where she currently works as lecturer in Dept of Electrical Engineering. Her research interests include optical communications, signal estimation and detection techniques, MIMO systems and adaptive signal processing.

S. Sheikh MUHAMMAD completed his Bachelors in Electrical Engineering with Honors in 2001 and received Masters in Electrical Engineering in 2003 the from University of Engineering and Technology, Lahore, Pakistan whereby he also remained in the faculty for around 3 years. He completed his PhD in Electrical Engineering in 2007 with Excellence at the Graz University of Technology conducting research on coded modulation techniques for free space optical systems. He has Guest Edited 2 special issues in Optical Wireless Communications in 2009 and 2012 and has organized regular IEEE Colloquiums on Optical Wireless in 2008 (Austria), 2010 (UK) and 2012 (Poland). He has published more than 50 peer reviewed papers in journals and conferences of repute and has chaired number of international conference sessions. He is currently working as an Associate Professor of Electrical Engineering at the National University of Computer and Emerging Sciences (FAST-NU) in Lahore and his current research revolves around application of network information theory to optical wireless and optical wave propagation through random media.

A. D. RAZA received his Bachelors in Electrical Engineering with Honors in 1973 and Masters in Electrical Engineering in 1984 from the University of Engineering and Technology, Lahore, Pakistan. He remained with Pakistan's state owned Telecom operator from 1975 to 2008 and rose to the level of Executive Vice President. He joined the National University of Computer and Emerging Sciences in 2008 and is pursuing his doctoral studies besides his teaching responsibilities as an Assistant Professor in the Department of Electrical Engineering. His current research interest revolves around application of network information theory to optical wireless networks.

5

A Wilkinson Power Divider with Harmonic Suppression and Size Reduction using High-low Impedance Resonator Cells

Mohsen HAYATI, Ashkan ABDIPOUR, Arash ABDIPOUR

Electrical Engineering Dept., Faculty of Engineering, Razi University, Tagh-E-Bostan, Kermanshah-67149, Iran

mohsen_hayati@yahoo.com, Ashkan_abdipour@yahoo.com, Arash.abdipour@yahoo.com

Abstract. *A miniaturized Wilkinson power divider using high-low impedance resonator cells is designed and fabricated. The proposed power divider occupies 23.7 % of the conventional structure circuit area at the operating frequency of 0.9 GHz and it is also able to suppress harmonics. According to the measured results at 0.9 GHz, the insertion-losses of output ports are 3.087 dB, the return-losses at all ports are more than 30 dB, and the isolation between output ports is better than 35 dB. Also, 2nd to 10th spurious frequencies are suppressed. According to the measured S_{11}, when it is less than -15 dB (from 0.65 GHz to 1.1 GHz) the fractional bandwidth of the proposed structure is 50 %. Good agreement between simulation and measured results is achieved.*

Keywords

Harmonic suppression, high-low impedance resonator, miniaturized Wilkinson power divider

1. Introduction

The power dividers are extremely important devices in microwave and millimeter-wave systems such as mixers, frequency multipliers, and the feeding network for an antenna array. There are several approaches for size reduction and harmonic suppression in the process of designing power dividers. The reported Wilkinson power divider with conventional quarter-wavelength transmission-line (TLIN) in [1], occupies a large area (especially at low operating frequencies) and it is not capable of suppressing spurious frequencies. In order to overcome these disadvantages several methods are used to reduce overall circuit size and suppress unwanted harmonics in the Wilkinson power dividers [2–7]. For instance, for size reduction and harmonic suppression, a kind of Wilkinson power divider based on standard printed-circuit-board (PCB) etching processes is reported in [2]. This structure is designed considering slow-wave loading and reduced the occupied area to 36.5% of the conventional structure at its operating frequency. In [3] and [4], microstrip electromagnetic band-gap (EBG) structure has been applied to the conventional power divider design, which led to the miniaturization and harmonic rejection in the conventional Wilkinson power divider.

Utilizing defected ground structure (DGS) can reject unwanted harmonics and decrease the occupied area in power dividers as it has been reported in [5] and [6]. Since, (DGS) and (EBG) need etching process, so their fabrication process is complex and these methods are not useable on a metal surface. The use of π-equivalent shunt-stub-based artificial transmission lines can effectively decrease the circuit size of conventional power divider [7]. This structure occupies 14.7% of the conventional power divider, but it is not capable for harmonic rejection at its operating frequency of 0.9 GHz.

In this paper, a Wilkinson power divider with harmonic suppression and size reduction is proposed. In order to reduce the circuit size of the conventional Wilkinson power divider, transmission lines with high-low impedance resonators are used instead of conventional quarter-wave-length TLIN sections. This technique not only reduces the occupied area to 23.7% of conventional one at operating frequency of 0.9 GHz, but also suppresses the second up to tenth harmonics.

2. Power Divider Design

2.1 The Procedure of Designing and the Effect of High-low Impedance Resonator on Size Reduction and Harmonic Suppression

Utilizing traditional TLIN in the structure of power dividers results in a large occupied area. Furthermore, this kind of transmission line is not able to suppress spurious frequencies. Using transmission line with loaded capacitance instead of quarter-wavelength transmission-line not only reduces the circuit size, but also can suppress unwanted harmonics. In the first step, a conventional Wilkinson power divider with an operating frequency at 0.9 GHz is designed as it is shown in Fig. 1a. In order to make capacitor loading on each $\lambda/4$ TLIN, four resonators with primary dimensions of W1 = W2 = 0.1 mm, L1 = L2 = 0.1 mm, W3 = 0.1 mm and L3 = 0.1 mm are added inside the free area of the conventional structure. These values are selected to control the effects of changing dimensions on frequency response and determine the operating frequency. The locations of the added resonators are determined with a, b, c and d in Fig. 1b. By increasing the values of W1, W2, L1 and L2 as low impedance TLINs a large loaded capacitance can be obtained. In order to reduce the occupied area of the power divider, the length of the main TLIN can decrease simultaneously, with increasing the dimensions of low impedance TLINs. Note that changing the values of variables does not have to shift the desired operating frequency, i.e. 0.9 GHz. Furthermore, adding these resonators makes a lowpass filter on the each main transmission line of the designed circuit. It appears in the insertion loss (S21), because of high order harmonics suppression in the frequency response.

The proposed power divider at 0.9 GHz and its equivalent circuit using lumped components are shown in Fig. 2a and Fig. 2b, respectively. In Fig. 2b L_b, L_c and L_d are equivalent inductors caused by the main transmission line. High-low impedance resonators are modeled by L_a, $C1$ and $C2$, where L_a determines high impedance transmission lines of these resonators. $C1$ and $C2$ show low impedance open-circuit transmission lines of resonators 1 and 2, respectively. The gaps g_1, g_2 and g_3 between low impedance open-circuit lines cause coupling effects, which are modeled by C_{g1}, C_{g2} and C_{g3}. Furthermore, C_{p1} and C_{p2} present the capacitance between the microstrip structure and the ground. *Lout* accounts for inductor of output transmission lines. Notice that the coupling capacitances between the main transmission line and open-circuit transmission lines are not included in the LC circuit as they are trivial.

Open-stub loads in the structure of high-low impedance resonators of the proposed power divider, modeled by $C1$ and $C2$ lead to a large shunt capacitance. Therefore, the circuit size of the proposed Wilkinson power divider could reduce because the propagation constant, i.e. β enhances (βproposed/βconventional is about 1.923). The relationship for β is given by:

$$\beta = \omega\sqrt{LC}\,, \tag{1a}$$

$$\beta = \frac{2\pi}{\lambda_g} \tag{1b}$$

where in (1a) L is the total inductance in per length unit of the main transmission line and high impedance line, and C depicts the total capacitance in per length unit of the main transmission line. In (1b) λ_g determines guided wavelength. Since C (the capacitance of proposed power divider) is increased in comparison with the transmission line of a conventional power divider, the propagation constant is enhanced considerably. As a result, the occupied area of circuit will be decreased [1].

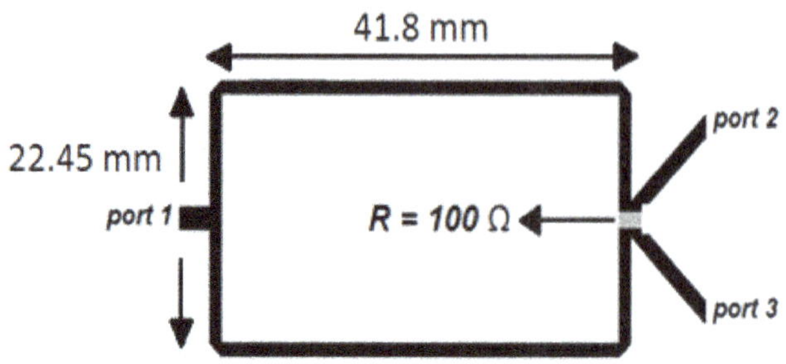

Fig. 1a. Topology of conventional Wilkinson power divider at operating frequency equal to 0.9 GHz.

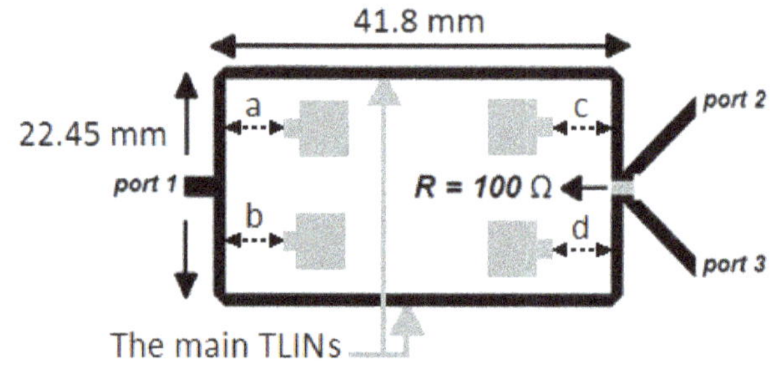

Fig. 1b. Topology of conventional Wilkinson power divider at operating frequency of 0.9 GHz with the locations of added high-low impedance resonators.

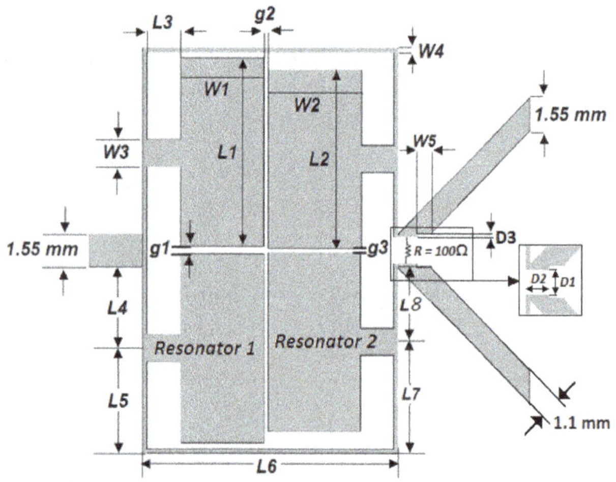

Fig. 2a. Topology of the proposed Wilkinson power divider.

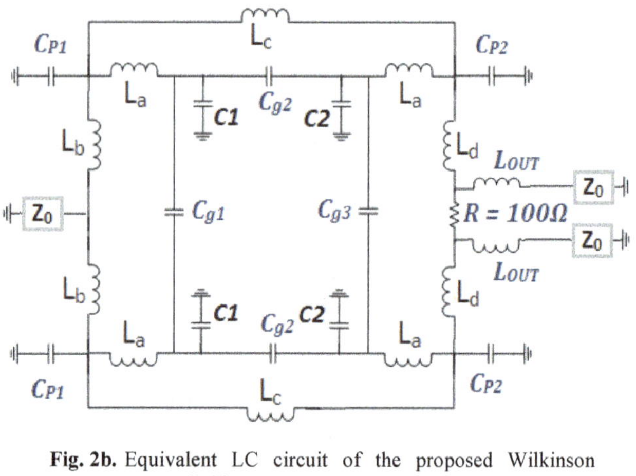

Fig. 2b. Equivalent LC circuit of the proposed Wilkinson power divider.

Moreover, based on insertion loss (S21) of the proposed structure shown in Fig. 6, optimized transmission line with high-low impedance resonators in higher frequencies has features of a lowpass filter. Spurious resonant frequencies of the resonator have been shifted by the high-low impedance resonators from the integer multiples of the basic resonant frequency [1]. So, replacing conventional quarter-wavelength transmission-line with a transmission line loaded by the high-low impedance resonators leads to harmonic suppression.

2.2 The Structure of the Proposed Power Divider

Comparison between the topology of conventional power divider in Fig. 1a and the proposed design, illustrated in Fig. 2a shows that in the proposed power divider, four microstrip high-low impedance resonator cells are used within the free area of the conventional Wilkinson power divider. The low impedance patches with rectangular shapes are microstrip open-stubs, so each of the high-low impedance resonator cells refers to a loaded capacitor. As a result, the capacitor loading not only reduces the circuit size, but also can suppress spurious frequencies.

The designed resonators, i.e. resonators 1 and 2 in both sides (left and right) have the same structure, but their rectangular patches have different dimensions. The circuit size of the proposed power divider and the conventional structure are 18.55 mm × 12 mm and 41.8 mm × 22.45 mm, respectively. It shows that the occupied area is reduced to 23.7% of the conventional power divider at operating frequency of 0.9 GHz. The type of the used 100 Ω isolation resistor is 0603, which is placed between two output ports. The dimensions of the proposed power divider shown in Fig. 2a are: W1 = 3.9, W2 = 4.3, W3 = 1.2, W4 = 0.2, W5 = 0.66, D1 = 1.15, D2 = 1.1, D3 = 0.2, L1 = 8.6, L2 = 8.2, L3 = 1.6, L4 = 3.7, L5 = 4.8, L6 = 12, L7 = 5.1, L8 = 3.4, g1 = 0.3, g2 = 0.2 and g3 = 0.15 (all in millimeter). The calculated values of inductors and capacitors of the shown LC circuit in Fig. 2b are [8]: La = 0.438 nH, Lb = 1.838 nH ,Lc = 12.38 nH, Ld = 1.68 nH, C1 = 1.36 pF, C2 = 1.43 pF, Cg1 = 25 fF, Cg2 = 128 fF, Cg3 = 42 fF,Cp1 = 0.471 pF, Cp2 = 0.466 pF. Note that the values of Cg1, Cg2 and Cg3 are achieved by tuning. Comparison between LC simulation and EM simulation results of the shown circuits in Figs. 2a and b on a substrate with permittivity of 2.2, thickness of 0.508 mm and loss tangent of 0.0009 are depicted in Figs. 3-6.

3. Simulated and Measured Results

The measured and simulated results of the proposed power divider are accomplished using Agilent's ADS Electromagnetic simulator (EM Simulator) software and HP 8720B vector network analyzer, respectively. The operating frequency of the proposed structure is located at 0.9 GHz. The designed microstrip Wilkinson power

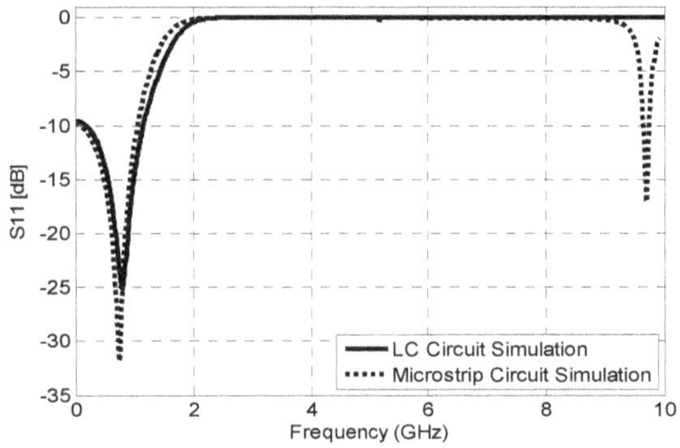

Fig. 3. Comparison between LC simulation and EM simulation results of input return-loss.

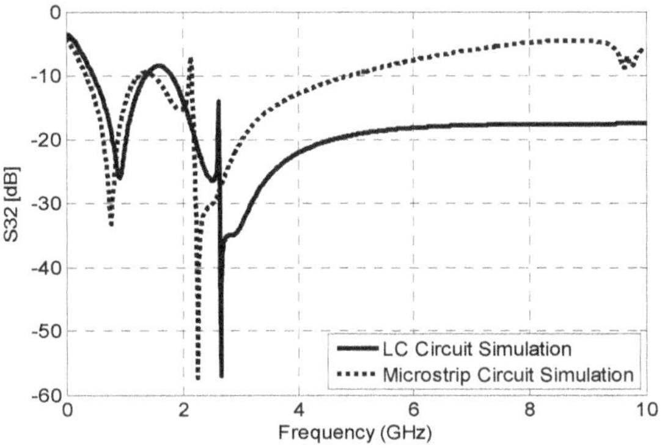

Fig. 4. Comparison between LC simulation and EM simulation results of isolation.

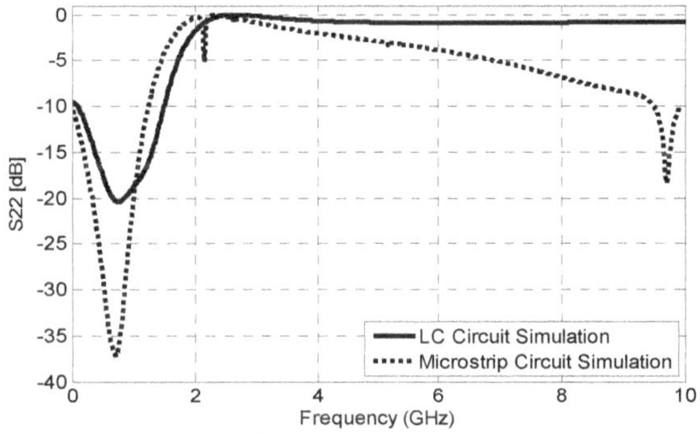

Fig. 5. Comparison between LC simulation and EM simulation results of output return-loss.

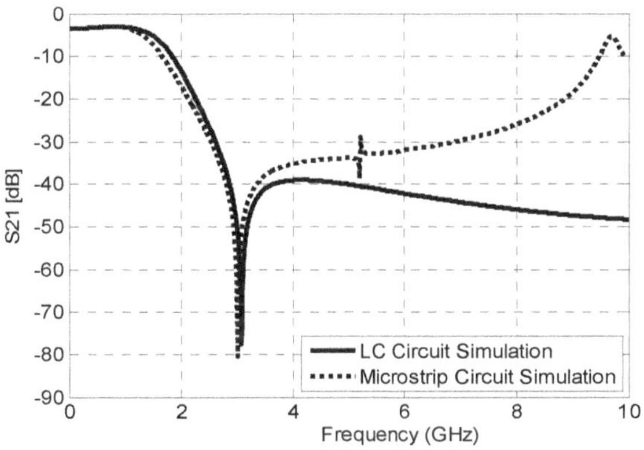

Fig. 6. Comparison between LC simulation and EM simulation results of insertion-loss.

divider is fabricated on RT/Duroid 5880 substrate with the thickness of 0.508 mm, the permittivity of 2.2 and the loss tangent of 0.0009. The results of measurement and simulation of S-parameters are illustrated in Figs. 7–11. As it is shown in Fig. 7, the measured return loss (S11) is at least –15 dB from 0.65 GHz to 1.1 GHz. In Fig. 8, the measurement shows over the frequency range 0.66–1.12 GHz the isolation (S23) is better than –15 dB. According to Fig. 9, the output return loss (S22) less than –15 dB from 0.28 GHz to 1.25 GHz is achieved. It can be observed from the measured insertion loss (S21) in Fig. 10, both even and odd spurious harmonics from 1.8 GHz to 9 GHz, i.e. second to tenth harmonics have been suppressed, where the 3rd to 10th harmonics are suppressed with a level less than –20 dB and the second harmonic is suppressed with a level better than -11 dB. It is to be noted that the suppression of higher order harmonic frequencies is related to S21 and S31. Exactly at operating frequency equal to 0.9 GHz, the measured S11, S32, and S22 are –33 dB, –38.88 dB and –48 dB, respectively. Furthermore, the measured insertion loss shows that S21 at 0.9 GHz is –3.087 dB. The characteristic impedance of all three ports are 50 Ω. Tab. 1 shows the comparison between the proposed power divider and the other published works. Based on the results of measurement shown in Fig. 11, an appropriate phase performance between two output ports around operating frequency is achieved. The measured phase difference of ports 2 and 3 as output ports is ± 0.15°. It shows that the proposed Wilkinson power divider is symmetric, so |S21| = |S31| (thus, harmonic suppression is related to both S21 and S31) and |S22| = |S33|. The photograph of the proposed Wilkinson power divider is shown in Fig. 12.

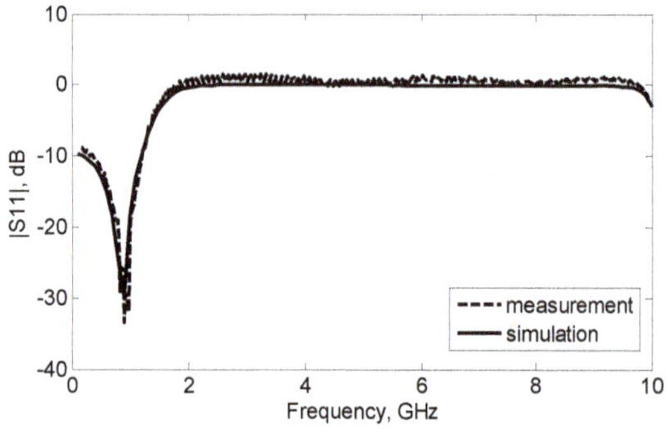

Fig. 7. Simulated and measured input return-loss.

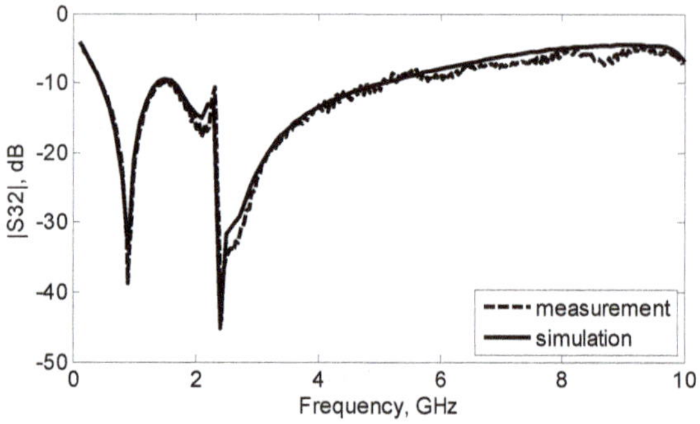

Fig. 8. Simulated and measured isolation.

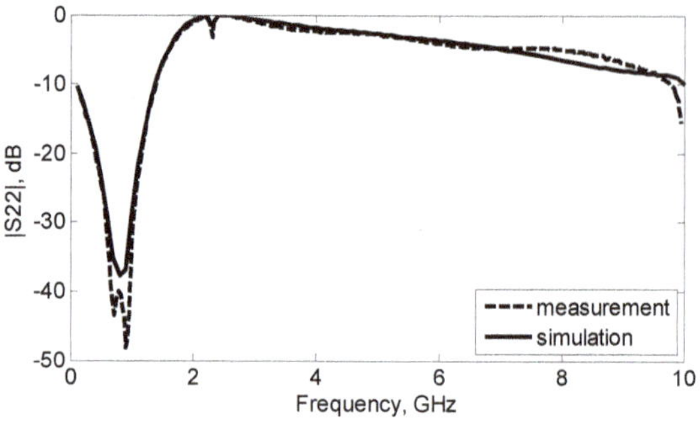

Fig. 9. Simulated and measured output return-loss.

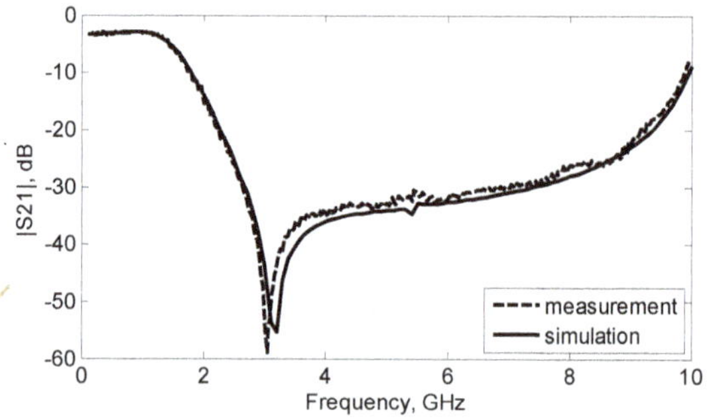

Fig. 10. Simulated and measured insertion-loss.

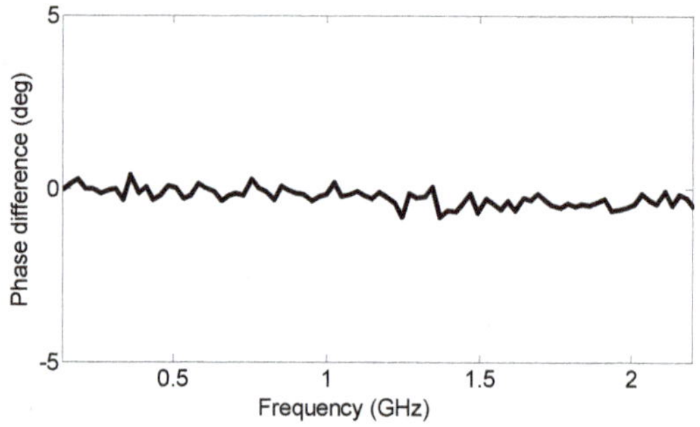

Fig. 11. Measured phase difference between S21 and S31 of the proposed power divider.

Fig. 12. Photograph of the proposed structure.

4. Conclusion

In this paper, a Wilkinson power divider using high-low impedance resonator cells for harmonic suppression and size reduction is proposed. The key features of the proposed structure are:

Ref.	Area reduction	Harmonic suppression (dB)								
		2nd	3rd	4th	5th	6th	7th	8th	9th	10th
[2]	63%	13	29	32	34	-	-	-	-	-
[3]	70%	8	32	10	12	-	-	-	-	-
[4]	39%	26	25	-	-	-		-	-	-
[5]	10%	18	15	-	-	-		-	-	-
[6]	66%	13	35	-	-	-		-	-	-
[7]	85.3%	-	-	-	-	-		-	-	-
This work	76.3%	11.3	31.5	35.5	33.2	32.4	30.1	25.9	25.9	22.4

Tab. 1. Comparison between the performance of the proposed power divider and previous works.

1- Small occupied area, i.e. 18.55 mm × 12 mm at the frequency of 0.9 GHz;

2- At operating frequency low insertion-losses of output ports (3.087 dB) and more than 30 dB return-losses at all ports are obtained. Moreover, better than 35 dB isolation and ±0.15° phase difference between output ports are achieved;

3- In the proposed structure spurious frequencies from 1.8 GHz up to 9 GHz, i.e. second to tenth harmonics are suppressed.

Therefore, the designed circuit with its operating frequency at 0.9 GHz can be used, where a power divider with small size and capable of suppress harmonics is required.

References

[1] POZAR, D. M. *Microwave Engineering*. 3rd ed. New York: Wiley, 2005, ch. 7, p. 333–337.

[2] WANG, J., NI, J., GUO, Y. X., FANG, D. Miniaturized microstrip Wilkinson power divider with harmonic suppression. *IEEE Microwave and Wireless Components Letters,* 2009, vol. 19, no. 7, p. 440–442. DOI: 10.1109/LMWC.2009.2022124

[3] LIN, C. M., SU, H. H., CHIU, J. C., WANG, Y. H. Wilkinson power divider using microstrip EBG cells for the suppression of harmonics. *IEEE Microwave and Wireless Components Letters,* 2007, vol. 17, no. 10, p. 700–702. DOI: 10.1109/LMWC.2007.905595

[4] ZHANG, F., LI, C. F. Power divider with microstrip electromagnetic band gap element for miniaturization and harmonic rejection. *Electronics Letters,* 2008, vol. 44, no. 6, p. 422–423. DOI: 10.1049/el:20083693

[5] WOO, D. J., LEE, T. K. Suppression of harmonics in Wilkinson power divider using dual-band rejection by asymmetric DGS. *IEEE Transactions on Microwave Theory and Techniques,* 2005, vol. 53, no. 6, p. 2139–2144. DOI: 10.1109/TMTT.2005.848772

[6] YANG, J., GU, C. F., WU, W. Design of novel compact coupled microstrip power divider with harmonic suppression. *IEEE Microwave and Wireless Components Letters,* 2008, vol. 18, no. 9, p. 572–574. DOI: 10.1109/LMWC.2008.2002444

[7] TSENG, C.-H., WU, C.-H. Compact planar Wilkinson power divider using pi-equivalent shunt-stub-based artificial transmission lines. *Electronics Letters,* 2010, vol. 46, no. 19, p. 1327–1328. DOI: 10.1049/el.2010.2194

[8] HONG, J.-S., LANCASTER, M. J. *Microstrip Filters for RF/Microwave Applications.* John Wiley & Sons, Inc., 2001.

About the Authors ...

Mohsen HAYATI received the BE in Electronics and Communication Engineering from Nagarjuna University, India, in 1985, and the ME and PhD in Electronics Engineering from Delhi University, Delhi, India, in 1987 and 1992, respectively. He joined the Electrical Engineering Dept., Kermanshah Branch, Islamic Azad University, Kermanshah, as a part time assistant professor in 2004. At present, he is a professor with the Electrical Engineering Dept., Kermanshah Branch, Islamic Azad University. He has published more than 155 papers in international and domestic journals and conferences. His current research interests include microwave and millimeter wave devices and circuits, application of computational intelligence, artificial neural networks, fuzzy systems, neuro-fuzzy systems, electronic circuit synthesis, modeling and simulations.

Ashkan ABDIPOUR received the B.S in Electronics Engineering from Islamic Azad University, Kermanshah Branch, Kermanshah, Iran, in 2009 and M.S degree from the Razi University, Kermanshah, Iran, in 2013. His research interests include microwave and millimeter wave devices and circuits.

Arash ABDIPOUR received the B.S in Electronics Engineering from Islamic Azad University, Kermanshah Branch, Kermanshah, Iran, in 2009 and M.S degree in Electronic Engineering from the University of Science and Research, Kermanshah Branch, Kermanshah, Iran in 2013. His research interests include microwave and millimeter wave devices and circuits.

6

A Simplified Scheme of Estimation and Cancellation of Companding Noise for Companded Multicarrier Transmission Systems

Siming PENG[1], *Zhigang YUAN*[1], *Jun YOU*[2], *Yuehong SHEN*[1], *Wei JIAN*[1]

[1]Dept. of Wireless Communications, PLA University of Science and Technology, 210014 Nanjing, China
[2]Dept. of Command Information System, PLA University of Science and Technology, 210014 Nanjing, China

lgdxpsm@gmail.com, yzhigang_cn@163.com, chunfeng22259@126.com

Abstract. *Nonlinear companding transform is an efficient method to reduce the high peak-to-average power ratio (PAPR) of multicarrier transmission systems. However, the introduced companding noise greatly degrades the bit-error-rate (BER) performance of the companded multicarrier systems. In this paper, a simplified but effective scheme of estimation and cancellation of companding noise for the companded multicarrier transmission system is proposed. By expressing the companded signals as the summation of original signals added with a companding noise component, and subtracting this estimated companding noise from the received signals, the BER performance of the overall system can be significantly improved. Simulation results well confirm the great advantages of the proposed scheme over other conventional decompanding or no decompanding schemes under various situations.*

Keywords

Multicarrier transmission systems, peak-to-average power ratio (PAPR), nonlinear companding transform (NCT), companding noise cancellation

1. Introduction

Multicarrier transmission is a promising technique in future communication systems. However, one of the major drawbacks of multicarrier systems is the inherent high peak-to-average power ratio (PAPR) of the transmitted signals. It's known that the efficiency of the high power amplifier (HPA) is directly related to the PAPR of the input multicarrier signals especially in the orthogonal frequency division multiplexing (OFDM) systems, which is applied in many important wireless communication standards such as the Third Generation Partnership Project (3GPP) Long-Term Evolution Advanced (LTE-A) standard [1], [2]. The PAPR problem still prevents OFDM from being adopted in the uplink of wireless communication standards [1].

Up to now, many works have been conducted to deal with this high PAPR problem [3]. Such as the iterative clipping and filtering [4], [5], [6], coding [7], Partial Transmission Sequence (PTS) [8], [9], [10], Selective Mapping (SLM) [11], Tone Reservation (TR) [12], companding transform (CT) [13]–[20] and so on. Among them, the clipping and companding may be the simplest two methods, since they can be employed directly to the multicarrier systems without any restrictions on the number of subcarriers and frame format and so on. However, due to the clipping often introduces serious in-band distortions as well as out-of-band radiation, consequently, the bit-error-rate (BER) performance of the system is greatly degraded. On the contrary, the companding transform can not only achieve more effective PAPR reduction but also better BER performance than the clipping method. Hence, it attracts more and more researchers' attention in recent years.

Since companding transform is an extra predistortion process employed on the original signals, hence, the introduced companding noise may also degrade the system performance to some extend. Conventional decompanding operation can approximately remove the companding noise at the receiver [13], [14], [15], [16], but the channel noise will be amplified by the decompanding function simultaneously, and consequently, the system performance will not be so optimistic especially under low signal-to-noise ratio (SNR) region. In order to avoid amplifying the channel noise caused by the decompanding operations, in [17], the authors proposed to abandon the decompanding operation at the receiver, and although a great BER performance improvement can be achieved, however, there is still a relative large gap of BER performance away from the performance bound. In [19], the authors proposed an iterative receiver to estimate and cancel the companding noise. By referring to the Bussgang theorem, the companding signals are regarded as the summation of a useful attenuated input replica and an uncorrected nonlinear distortion noise, and then removed from the receiver. Although the BER performance can be greatly improved in this scheme, however, due to the inherent complex expression of the companding function, the accurate attenuation coefficient of the input signals will difficultly to be determined, and consequently, greatly restrains the effectiveness for general companding transforms.

In this paper, a simplified scheme of estimation and cancellation of nonlinear companding noise is proposed. By expressing the companded signals as the summation of the companding noise added to the original signals, and then, subtracting the estimated companding noise from the receiver, the above difficulties of analysis and calculation of the attenuation coefficient of the original signals can be well resolved. Moreover, a significant BER performance improvement than conventional operations, such as the decompanding or no decompanding, at the receiver can be achieved simultaneously. It also shows that the presented scheme is robust in various practical situations.

The rest of this paper is organized as follows. In Sec. 2, a typical multicarrier system model is described and the PAPR problem is formulated briefly. The theoretical analysis of the proposed scheme is presented in Sec. 3. In the next section, the overall BER performance of a typical exponential companding (EC) equipped with the proposed companding noise cancellation scheme is evaluated and followed by the conclusion summarized in Sec. 5.

2. System Model

Fig. 1 shows the block diagram of a typical multicarrier transmission system with companding transform. Let us denote the data symbols $X_k, k = 0,1,\ldots,N-1$, as a vector $\mathbf{X} = [X_0, X_1, \ldots, X_{N-1}]^T$ with N subcarriers, where $(\cdot)^T$ is the matrix transpose operation. The complex baseband representation of a multicarrier signal is given by

$$x(t) = \frac{1}{\sqrt{N}} \sum_{k=0}^{N-1} X_k e^{j2\pi k\Delta f t}, 0 \leqslant t \leqslant NT_s \quad (1)$$

where $j = \sqrt{-1}$, Δf is the subcarrier interval, and NT_s is the useful data block period. In general, the subcarriers are chosen to be orthogonal (i.e. $\Delta f = 1/NT_s$).

Generally, the PAPR of multicarrier signals $x(t)$ is defined as the ratio between the maximum instantaneous power and its average power, i.e.

$$PAPR = \frac{\max\limits_{0 \leqslant t \leqslant NT} \left[|x(t)|^2\right]}{1/NT \int\limits_0^{NT} |x(t)|^2 dt}. \quad (2)$$

To better approximate the PAPR of continuous-time OFDM signals, the OFDM signals samples are obtained by L times oversampling. L-times oversampled time-domain samples can be achieved by performing a LN-point IFFT of the data block with $(L-1)N$ zero-padding, i.e.

$$\mathbf{X}_p = \left[X_0, \ldots, X_{\frac{N}{2}-1}, \underbrace{0, \ldots, 0}_{(L-1)N}, X_{\frac{N}{2}}, \ldots, X_{N-1} \right]^T. \quad (3)$$

Therefore, the oversampled IFFT output can be expressed as

$$x(n) = \frac{1}{\sqrt{N}} \sum_{k=0}^{N-1} X_k e^{j\frac{2\pi nk}{LN}}, 0 \leq n \leq LN-1. \quad (4)$$

The corresponding PAPR computed from the L-times oversampled time domain OFDM signal sample is defined as

$$PAPR = \frac{\max\limits_{0 \leq n \leq LN-1} \left[|x(n)|^2\right]}{\mathbb{E}\left[|x(n)|^2\right]} \quad (5)$$

where $\mathbb{E}[\cdot]$ denotes the expectation operator.

Assume that the input information symbols are statistically independent and identically distributed. Based on the central limit theory, $x(n)$ can be approximated as a complex Gaussian process with zeros mean and variance σ^2 when the number of sub-carriers N is large enough (e.g. $N \geqslant 64$). Thus, the amplitude $|x(n)|$ follows a Rayleigh distribution with the probability distribution function (PDF) as

$$f_{|x_n|} = \frac{2x}{\sigma^2} \exp(-\frac{x^2}{\sigma^2}), x \geqslant 0. \quad (6)$$

From (6), we can see that the peak power of $|x_n|$ can take a value much large than its average power. That is to say, the multicarrier signal has a large PAPR, which leads to a negative impact on the system performance. While, by reallocating the power or the statistics of multicarrier signal reasonable, the companding transform can well resolve this high PAPR problem.

The fundamental principle of nonlinear companding transform can be described as follows [14]. The original signal $x(n)$ is companded before converted into analog waveform and amplified by the HPA. The companded signal is denoted as

$$y_n = \zeta(x_n) \quad (7)$$

where $\zeta(\cdot)$ is the companding function which only changes the amplitude of x_n. When passing through the AWGN channel, the transmitted signals can be recovered by the corresponding decompanding function $\zeta^{-1}(\cdot)$, i.e.

$$\tilde{x}_n = \zeta^{-1}(y_n + w_n) \approx x_n + \zeta^{-1}(w_n) \quad (8)$$

where w_n is the channel noise.

It has been pointed in [17] that the BER performance of companded multicarrier systems can be improved by carefully design the companding function $\zeta(\cdot)$ [13]–[20], however, due to the amplified channel noise (by decompanding operation) or unprocessed companding noise (by no decompanding operation), this modification may be limited in practice. Hence, there is considered to be a trade-off between PAPR reduction and BER performance for companding transform. However, in the subsequent works, we will show that this problem can be well resolved by the proposed companding noise cancellation scheme.

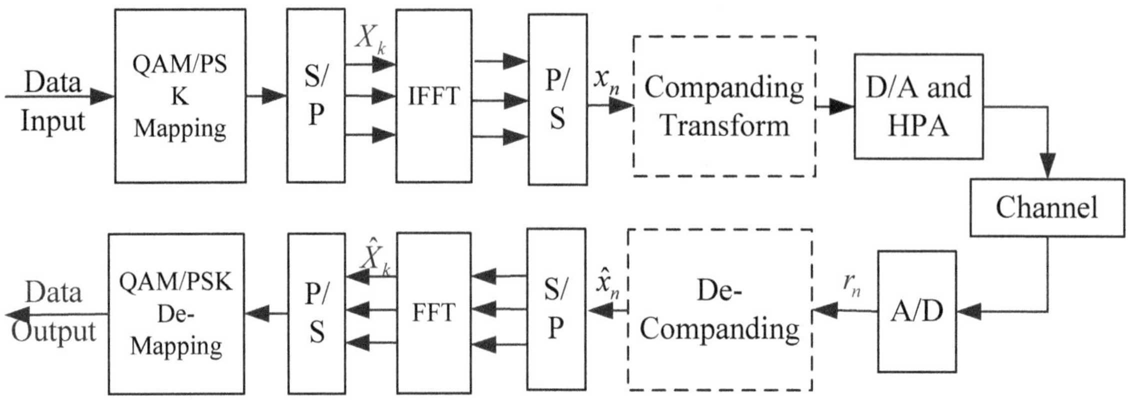

Fig. 1. Block diagram of typical multicarrier transmission system with companding transform.

3. Algorithm Formulation

In this section, we will review the basic concepts of conventional Bussgang based companding noise cancellla-tion scheme briefly, and then, the proposed scheme will be presented subsequently.

3.1 Bussgang Theorem Based Scheme

According to the Bussgang theorem [19], the companded multicarrier signal $y(n)$ can be modeled as the aggregate of an attenuated signal component and companding noise d_n, i.e.

$$y(n) = \alpha x(n) + d_n, n = 0, 1, \ldots, LN-1 \tag{9}$$

where α is the attenuate coefficient, and which is a time invariant for stationary input processes.

From (9), the attenuate coefficient α can be calculated as

$$\alpha = \frac{\mathbb{E}\{y(n)x^*(n)\}}{\mathbb{E}\{x(n)x^*(n)\}} = \frac{1}{\sigma^2}\int_0^\infty x \cdot \zeta(x) \cdot f_{|x|}(x)dx \tag{10}$$

where $f_{|x|}(x)$ is as defined in (6) and $x^*(t)$ is the complex conjugate of $x(t)$.

In the receiver, by making full use of the received signal $y(n)$ and reconstructing the companding process as that of the transmitter, the estimated companding noise component can be calculated as

$$\hat{d}_n = \hat{y}(n) - \alpha\hat{x}(n), n = 0, 1, \ldots, LN-1 \tag{11}$$

where $\hat{y}(n), \hat{x}(n)$ are the reconstructed companded signal and the detected signal at the receiver, respectively.

However, it should be noted that the companding function $\zeta(\cdot)$ in (10) often has a complex expression (for example, see the companding functions in [14], [15] and [17]), which make the coefficient α difficult to be accurately calculated for general companding transforms. Moreover, the precision of α will have a great impact on the ultimate system performance. Hence, the effectiveness of conventional estimation and cancellation of companding noise will be greatly restrained in practice.

3.2 Proposed Scheme

Different from (9), if the attenuation of original signals is regarded as caused by the companding noise, then the output of the compander can be simply expressed as the original signals added with an extra companding noise component, i.e.

$$y(n) = x(n) + \hat{d}_n, n = 0, 1, \ldots, LN-1 \tag{12}$$

where $\hat{d}_n$ is the equivalent companding noise.

Hence, the received signal can be expressed as

$$\begin{aligned} r(n) &= h(n) * y(n) + w_n \\ &= h(n) * x(n) + h(n) * \hat{d}_n + w_n, n = 0, 1, \ldots, LN-1 \end{aligned} \tag{13}$$

where '$*$' is the convolution operation and $h(n)$ is the impulse response of the transmitting channel.

Assume the detected symbol is $\tilde{X}_k$(or equivalently $\tilde{x}(n)$), and the output of the compander at the receiver is $\tilde{y}(n)$, i.e.

$$\tilde{y}(n) = \tilde{x}(n) + \tilde{d}_n, n = 0, 1, \ldots, LN-1 \tag{14}$$

where $\tilde{d}_n$ is the companding noise regenerated at the receiver.

Since $\tilde{y}(n)$, $\tilde{x}(n)$ are all achievable at the receiver, hence, the companding noise can be estimated as

$$\tilde{d}_n = \tilde{y}(n) - \tilde{x}(n), n = 0, 1, \ldots, LN-1. \tag{15}$$

According to current channel estimation $\tilde{h}(n)$, then, $\tilde{h}(n) * \tilde{d}_n$ is subtracted from the current channel observation $r(n)$ to obtain the refined channel observation $\hat{r}(n)$, i.e.

$$\begin{aligned}\hat{r}(n) &= r(n) - \tilde{h}(n) * \tilde{d}_n \quad n = 0,1,\ldots,LN-1 \\ &= h(n) * x(n) + (h(n) * \hat{d}_n - \tilde{h}(n) * \tilde{d}_n) + w_n.\end{aligned} \tag{16}$$

It can be seen from (16) that the channel estimation error between $\tilde{h}(n)$ and $h(n)$ may affect the ultimate result of $\hat{r}(n)$, however, this error may also affect the estimation precision of $\tilde{d}_n$ (due to the decision error of $\tilde{X}_k$). On the other hand, when $\tilde{h}(n)$ severe deviates from $h(n)$, then, the channel estimation error will become the dominate interference component. Hence, we will mainly consider the approximate ideal channel estimation, i.e. $\tilde{h}(n) \approx h(n)$ in the subsequent works.

When $\tilde{d}_n \to \hat{d}_n$, then we will have the asymptotic ideal[1] result, i.e.

$$\hat{r}_n \to h(n) * x(n) + w_n, n = 0,1,\ldots,LN-1. \tag{17}$$

The above processes can be summarized in Fig. 2.

Comparing (9) and (12), we can have the conclusions that

- The proposed scheme is more simple than that of the conventional estimation and cancellation of companding noise scheme which bases on the Bussgang theorem [19]. Moreover, since the calculation of the attenuated coefficient α can be avoided in the presented scheme, hence, it can be easily adopted to general companding schemes.
- The novel scheme is more robust than that of the conventional scheme. This is due to the estimated companding noise in the presented scheme will not be affected by the precision of α as that in the conventional scheme.

4. Simulation Results

To evaluate the overall system performance, computer simulations are performed based on an OFDM system with $N = 256$ subcarriers and the input bit stream is modulated by Quaternary Phase Shift Keying (QPSK) and 16 Quadrature Amplitude Modulation (16QAM). The AWGN channel as well as frequency-selective multi-path fading channel [15] is considered in the simulations. Moreover, the oversampling factor $L = 4$ and a cycle prefix with the length of $1/4$ symbol is employed to mitigate the inter-symbol interference. Furthermore, for the convenience of comparison, a typical nonlinear companding function, i.e. the EC scheme proposed in [13] with companding function[2]

$$\zeta(x) = \mathrm{sgn}(x)\sqrt[d]{\nu[1-\exp(-\frac{|x|^2}{\sigma^2})]} \tag{18}$$

where

$$\nu = \left(\frac{\mathbb{E}\left[|x|^2\right]}{\mathbb{E}\left[\sqrt[d]{\left[1-\exp(-\frac{|x|^2}{\sigma^2})\right]^2}\right]}\right)^{\frac{d}{2}} \tag{19}$$

and d is the companding degree (the PAPR performance of EC according to [13] with different companding degree d is summarized in Tab. 1), is equipped with several conventional operations at the receiver, including conventional decompanding (DC) [13], [14], [15], [16], [20], no decompanding (NDC) [17], [18] as well as the proposed companding noise cancellation (CNC) operations are all considered in the simulations.

	EC		Original signals
	$d=1$	$d=2$	
PAPR(dB), CCDF=10^{-3}	4.98	3.21	11.38

Tab. 1. PAPR performance of EC with different companding degrees.

4.1 BER in AWGN Channel

Figure 3 shows the BER performance of OFDM system with QPSK modulation employing EC and equipped with different operations at the receiver over AWGN channel. In this figure, the curve of "Performance bound" is the signals transmitted without living through any nonlinear distortions. Given that the BER at $P_e = 10^{-5}$, we can find that the companded signals are suffering from a serious nonlinear distortions especially when with conventional decompanding operations at the receiver, and there is a 3.23 dB ε_b/N_0 gap between EC-DC with $d = 1$ and the performance bound and more than 6 dB ε_b/N_0 gap for EC-DC with $d = 2$. While, although there is a significant BER improvement when with no decompanding operation at the receiver, however, there is still a gap of 0.78 dB ε_b/N_0 at EC-NDC with $d = 1$ and 3.03 dB ε_b/N_0 at EC-NDC with $d = 2$ away from the performance bound. However, when with the proposed companding noise cancellation scheme, the BER gap between EC-CNC and the performance bound can be effectively restored to no more than 0.29 dB regardless of the companding degree d.

Figure 4 shows the BER performance of EC with 16QAM modulation and different operations at the receiver over AWGN channel. It's obvious that the overall BER performance has suffering a more serious degradation than that of with QPSK modulation. Furthermore, when with EC-DC and $d = 1$, there is a 5.58 dB ε_b/N_0 gap away from the performance bound. While, we can also find that when with high ε_b/N_0, the conventional decompanding operation will trend to outperform that with no decompanding operation at

[1] As we will show later, this is due to the inherent estimated errors of the detected signals $\tilde{x}(n)$, and this error is unavoidable but can be improved by some other techniques, such as the channel coding.

[2] It's obvious that (10) is not appropriate for this companding function, since the coefficient ν in (18) can only be calculated numerically, and this phenomenon is widely exits in some other companding functions (see [14]–[20]).

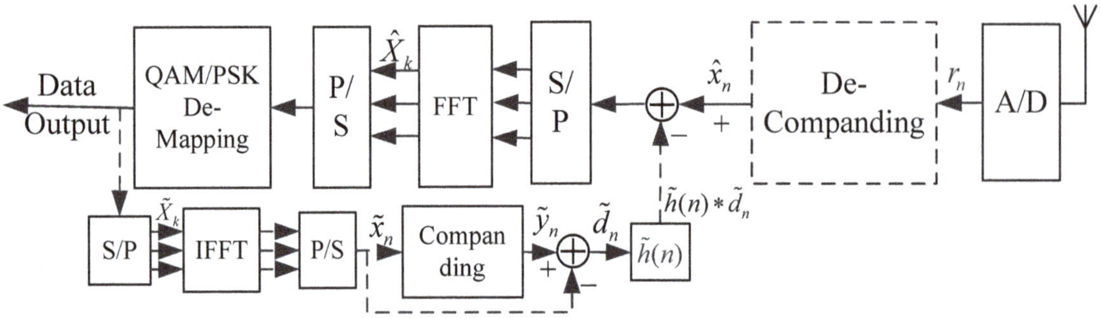

Fig. 2. Block diagram of the proposed scheme for companding transform. Note that when the SNR is low, it's prefer to with no decompanding at the receiver, and then $\hat{x}(n) = r(n)$.

the receiver. This is due to that although the decompanding operation may amplifier the channel noise, while, when ε_b/N_0 is high enough (e.g. $\varepsilon_b/N_0 > 18$ dB), the companding noise will become the dominant interference component and must be removed from the received signals. When with the proposed companding noise cancellation scheme, there is a 1.09 dB ε_b/N_0 gap between EC-CNC with $d = 2$ and the performance bound, while the relative gap is only 0.16 dB for EC-CNC with $d = 1$, which further demonstrates the great advantages of the proposed scheme over other conventional operations.

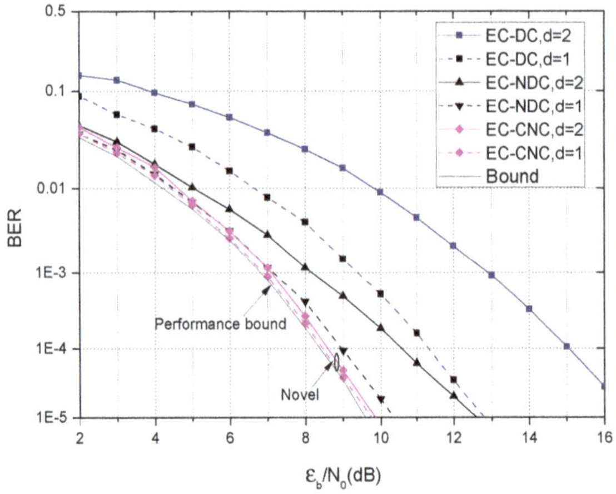

Fig. 3. BER performance of QPSK modulation and EC equipped with different operations over AWGN channel.

4.2 BER with HPA over AWGN Channel

For most wireless communication systems, the HPA is widely used to provide adequate transmit power. In this paper, the solid state power amplifier (SSPA) model described in [21] is considered in the subsequent works. Fig. 5 shows the BER performance of different operations at the receiver with SSPA and $IBO = 0$ dB over AWGN channel. It is seen that after passing through the SSPA, the BER performance has been degraded to some extent. In this figure, the "Original signals" is the signals without companding transform and directly transmitted to the SSPA, consequently, result in 3.33 dB ε_b/N_0 degradation compared to the performance bound as shown in Fig. 3. Moreover, we can also find that the EC-CNC with $d = 2$ outperforms the original signals 2.55 dB and 0.16 dB better than EC-CNC with $d = 1$.

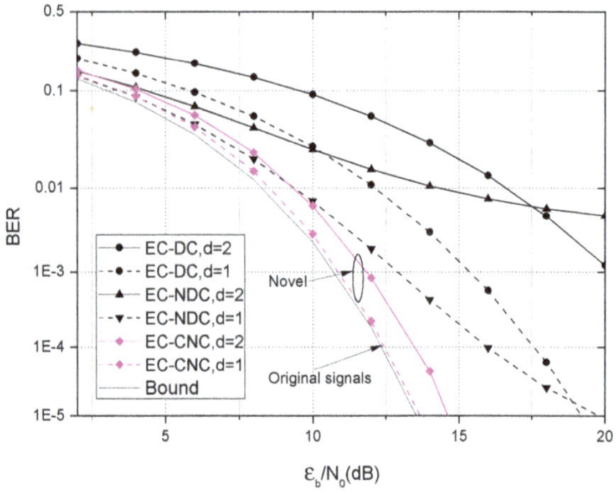

Fig. 4. BER performance 16QAM modulation of EC equipped with different operations at the receiver over AWGN channel.

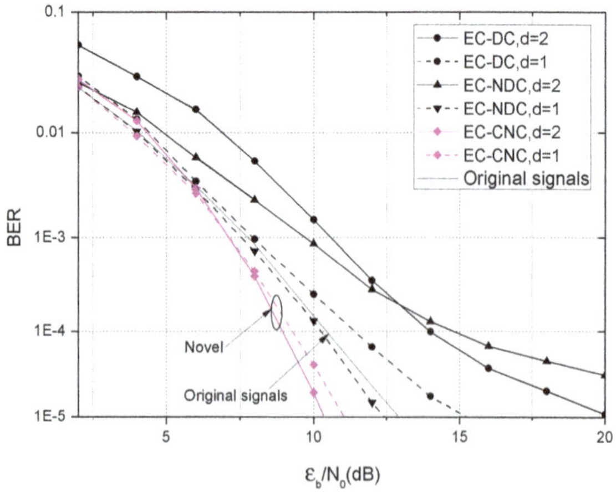

Fig. 5. BER performance of EC with HPA over AWGN channel with different operations at the receiver.

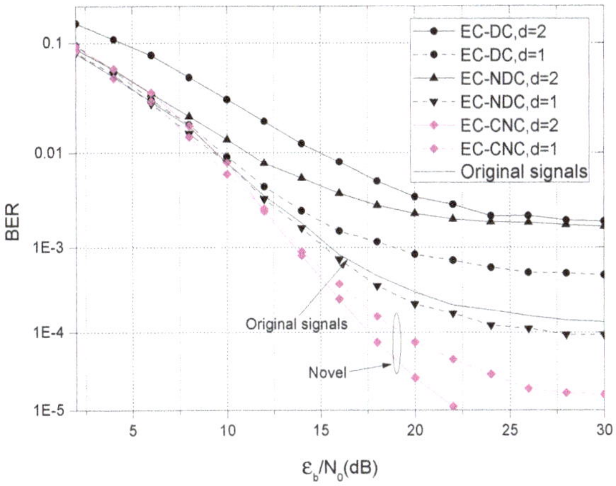

Fig. 6. BER performance of the receiver with HPA and equipped with various operations over fading channel.

4.3 BER with HPA over Fading Channel

Figure 6 shows the BER performance of EC with SSPA and equipped with different operations at the receiver over fading channel. In this figure, the IEEE 802.16 fading channel model described in [15] is considered in the simulations. From it we can see that different with other conventional operations at the receiver, the EC equipped with CNC greatly outperforms the original signals. Furthermore, comparing Fig. 6 and Fig. 5, we can also find that the advantages of the proposed scheme are more significant in practical wireless communication systems than that of in the AWGN channel.

4.4 BER with Iterative Filtering

Since the companding operation is a type of nonlinear process that may lead to out-of-band radiation, hence, an iterative companding and filtering technique may need to remove the out-of-band radiation and peak regrowth. However, it's known that the filtering may lead to the degradation of the BER performance, hence, it's necessary to estimate the practical BER performance of companding transform when considering the iterative filtering technique.

In this subsection, we mainly employ filtering in the baseband signals in the frequency domain [4], and the filter is based on the rectangular window, and can be defined as

$$\mathbf{Y}_f = \mathbf{H}_f \mathbf{Y}_c \tag{20}$$

where $\mathbf{Y}_c$ is the companded signal in frequency domain, $\mathbf{Y}_f$ is the filtered signals, and $\mathbf{H}_f$ is the frequency response of the filter defined by

$$H_f(k) = \begin{cases} 1, 0 \leqslant k \leqslant N/2-1, L-N/2 \leqslant k \leqslant L-1 \\ 0, \text{otherwise.} \end{cases} \tag{21}$$

Figure 7 shows the BER performance of EC over AWGN channel with iterative filtering at the transmitter and equipped different operations at the receiver with 2 iterations. Note that the receiver also performs the same iterative companding and filtering process as that at the transmitter. From it we can see that although the SNR gap between EC-CNC and the performance bound is slightly increased to 0.2 dB at $d = 1$ and 0.98 dB at $d = 2$ compared with Fig. 3, however, it's obvious that the proposed scheme still greatly outperforms other conventional schemes, which well verifies the robustness of the proposed scheme under various situations.

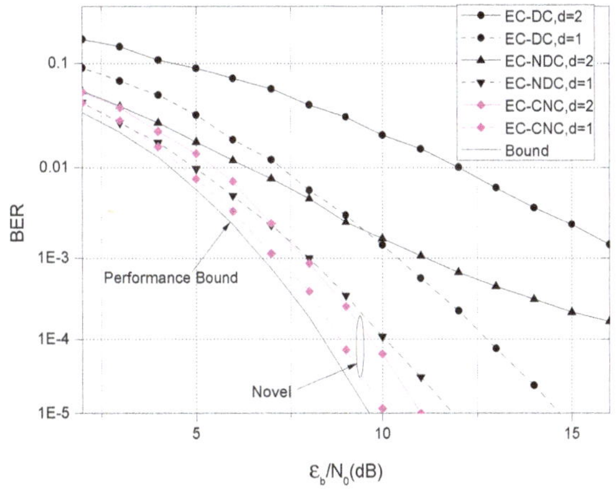

Fig. 7. BER performance of EC with iterative filtering and equipped with various operations at the receiver over AWGN channel.

5. Conclusion

Nonlinear companding noise in companded multicarrier systems greatly degrades the BER performance of the overall system. In this paper, a simplified but effective scheme of estimation and cancellation of companding noise is proposed to enhance the BER performance of the companded multicarrier systems. By expressing the companding process as the original signals added with a companding noise component and removing the estimated companding noise from the receiver, a great BER performance improvement can be achieved compared to conventional operations. Although with a slightly increased complexity (due to additional IFFT/FFT and companding operations at the receiver), we believe that the proposed scheme may well resolve the embarrassment, i.e. the trade-off problem between PAPR reduction and BER performance, for the conventional companding schemes.

Acknowledgements

This work was supported by the National Natural Science Foundation of China (grant nos 60772083 and 61201242); The Natural Science Foundation of Jiangsu Province (grant no. BK2012057) and the PLA University Preresearch Foundation (KYTYZLXY1208).

References

[1] WUNDER, G., FISHCHER, R. F. H., BOCHE, H., ET AL. The PAPR problem in OFDM transmission: new direction for long-lasting problem. *IEEE Signal Processing Magazine*, 2013, vol. 51, no. 11, p. 130–144. DOI: 10.1109/MSP.2012.2218138

[2] POLAK, L., KRATOCHVIL, T. Exploring of the DVB-T/T2 performance in advanced mobile TV fading channels. In *Proc. 36th International Conference on Telecommunictaions and Signal Processing (TSP2013)*, Rome (Italy), 2013, p. 768–772. DOI: 10.1109/TSP.2013.6614042

[3] HAN, S. H., LEE, J. H. An overview: peak-to-average power ratio reduction techniques for multicarrier transmission. *IEEE Wireless Communications*, 2004, vol. 12, no. 2, p. 56–65. DOI: 10.1109/MWC.2005.1421929

[4] ZHU, X., PAN, W., LI, H., ET AL. Simplified approach to optimized iterative clipping and filtering for PAPR reduction of OFDM signals. *IEEE Transactions on Communications*, 2013, vol. 61, no. 5, p. 1891–1901. DOI: 10.1109/TCOMM.2013.021913.110867

[5] CHEN, H., HAIMOVICH, A., M. Iterative estimation and cancellation of clipping noise for OFDM signals. *IEEE Communications Letters*, 2003, vol. 7, no. 7, p. 305–307. DOI: 10.1109/LCOMM.2003.814720

[6] XIA, L., LI, Z., YOUXI, T., ET AL. Analysis of the performance of iterative estimation and cancellation of clipping non-linear distortion in OFDM. In *Proceeding of Future Generation of Communications and Networking*, 2007, p. 1–5. DOI: 10.1109/FGCN.2007.68

[7] SLMANE, S., B. Reduction the peak-to-average power ratio of OFDM signals through precoding. *IEEE Transactions on Vehicular Technology*, 2007, vol. 56, no. 2, p. 686–695. DOI: 10.1109/TVT.2007.891409

[8] QI, X., LI, Y, HUANG., H. A low complexity PTS scheme based on tree for PAPR reduction. *IEEE Communications Letters*, 2012, vol. 16, no. 9, p. 1486–1488. DOI: 10.1109/LCOMM.2012.072012.121228

[9] VARAHRAM, P., ALI, M., B. A low complexity partial transmit sequence for peak to average power ratio reduction in OFDM systems. *Radioengineering*, 2011, vol. 20, no. 3, p.677–682.

[10] MARSALEK, R. On the reduced complexity interleaving method for OFDM PAPR reduction. *Radioengineering*, 2006, vol. 15, no. 3, p. 49–53.

[11] JIANG, T., NI, C., GUAN, L. A novel phase offset SLM scheme for PAPR reduction in Alamouti MIMO-OFDM systems without side information. *IEEE Signal Processing Letters*, 2013, vol 20, No. 4, p. 383–386. DOI: 10.1109/LSP.2013.2245119

[12] WANG, L., TELLAMBURE, C. Analysis of clipping noise and tone reservation algorithms for peak reduction in OFDM systems. *IEEE Transactions on Vehicular Technology*, 2008, vol. 57, no. 3, p. 1675–1694. DOI: 10.1109/TVT.2007.907282

[13] JIANG, T., YANG, Y., SONG, Y., H. Exponential companding technique for PAPR reduction in OFDM systems. *IEEE Transactions on Broadcasting*, 2005, vol. 51, no. 2, p. 244–248. DOI: 10.1109/TBC.2005.847626

[14] WANG, Y., GE, J., WANG, L., ET AL. Nonlinear companding transform for reduction of peak-to-average power ratio in OFDM systems. *IEEE Transactions on Broadcasting*, 2013, vol. 59, no. 2, p. 369–375. DOI: 10.1109/TBC.2012.2219252

[15] JENG, S., S., CHEN, J., M. Effective PAPR reduction in OFDM systems based on a companding technique with trapezium distribution. *IEEE Transactions on Broadcasting*, 2010, vol. 56, no. 2, p. 258–262. DOI: 10.1109/TBC.2011.2112237

[16] PENG, S., SHEN, S., YUAN, Z., ET AL. A novel nonlinear companding transform for PAPR reduction in lattice-OFDM systems. *Frequenz*, 2014, vol. 69, no. 2, p. 461–469. DOI: 10.1515/freq-2013-0169

[17] HOU, J., GE, J., ZHAI, D.,ET AL. Peak-to-average power ratio reduction of OFDM signals with nonlinear companding scheme. *IEEE Transactions on Broadcasting*, 2010, vol. 56, no. 2, p. 258–262. DOI: 10.1109/TBC.2010.2046970

[18] PENG, S., SHEN, Y., YUAN, Z., ET AL. PAPR reduction of LOFDM signals with an efficient nonlinear companding transform. In *Proceeding of International Conference on Wireless Communications and Signal Processing*, WCSP'13, Hangzhou, 2013, p. 1–6. DOI: 10.1109/WCSP.2013.6677217

[19] JIANG, T., YAO, W., SONG, Y., ET AL. Two novel nonlinear companding schemes with iterative receiver to reduce PAPR in multicarrier modulation systems. *IEEE Transactions on Broadcasting*, 2006, vol. 52, no. 2, p. 268–273. DOI: 10.1109/TBC.2006.872992

[20] PENG, S., SHEN, Y., YUAN, Z. PAPR reduction of multi-carrier systems with simple nonlinear companding transform. *Electronic Letters*, 2014, vol. 50, no. 6, p. 473–475. DOI: 10.1049/el.2013.4216

[21] COSTA, E., MIDRIO, M., PUPOLIN, S. Impact of amplifier nonlinearities on OFDM transmission systems performance. *IEEE Communications Letters*, 1999, vol. 3, no. 2, p. 37–39. DOI: 10.1109/4234.749355

About the Authors...

SIMING PENG was born in Hubei, China Republic in 1990. He received his bachelor's degree from Wuhan University of Science and Technology in 2012. He is currently a M.Sc. candidate at the Department of Wireless Communications, College of Communications Engineering, PLA University of Science and Technology. His research interests include multicarrier communications and signal processing.

ZHIGANG YUAN was born in Hebei, China Republic in 1980. He received his Ph.D. degree in Communication and Information System from PLA University of Science and Technology in 2008. He is currently a lecture at the same university. His interests are signal processing and lattice multi-carrier communication theory.

JUN YOU was born in Jiangsu Province. He received the M.S. and Ph.D. degree from Institute of Communications Engineering, PLA University of Science and Technology, Nanjing, China, in 2001 and 2005, respectively. He is currently an associate professor in the College of Command Information System, PLAUST. His interest is fast signal processing.

YUEHONG SHEN was born in Hubei, China Republic in 1959. He received his Ph.D. degree in Communication Engineering from Nanjing University of Science and Technique in 1999. And now, he is a professor and doctor advisor of Wireless Department at the Institute of Communication Engineering, PLA University of Science and Technology,

China. His interests include blind and source separation in communication systems.

WEI JIAN was born in Henan, China Republic in 1976. He received his Ph.D. degree in Communication and Information Engineering from PLA University of Science and Technology in 2006. His interests cover communication theory and high speed wireless communication systems.

Electromagnetic Scattering and Statistic Analysis of Clutter from Oil Contaminated Sea Surface

Cong-hui QI, Zhi-qin ZHAO

School of Electronic Engineering, University of Electronic Science and Technology of China, Xiyuan Ave 2006, Chengdu, Sichuan, China

qiconghui0826@163.com, zqzhao@uestc.edu.cn

Abstract. *In order to investigate the electromagnetic (EM) scattering characteristics of the three dimensional sea surface contaminated by oil, a rigorous numerical method multilevel fast multipole algorithm (MLFMA) is developed to preciously calculate the electromagnetic backscatter from the two-layered oil contaminated sea surface. Illumination window and resistive window are combined together to depress the edge current induced by artificial truncation of the sea surface. By using this combination, the numerical method can get a high efficiency at a less computation cost. The differences between backscatters from clean sea and oil contaminated sea are investigated with respect to various incident angles and sea states. Also, the distribution of the sea clutter is examined for the oil-spilled cases in this paper.*

Keywords

Oil contaminated sea, sea clutter, MLFMA, resistive loading

1. Introduction

Detection of oil spills is a long-term objective in sea remote sensing [1], [2]. The oil contaminated sea water is a threat to the marine environment. The vertical polarization of high frequency surface wave radar (HFSWR) has the feature of small energy attenuation, all weather operating and over-horizon. Most importantly, targets can be detected beyond the range of visibility. Thus HFSWR can be used to detect oil spills in far distance. When the sea water is contaminated by oil, the sea dynamics as well as the surface tension are changed [3]. Accordingly, the EM scattering characteristics from the oil contaminated sea surface are changed, such as radar echo and sea clutter distributions. The difference between backscatters from clean sea and oil contaminated sea is a useful feature and desirable in Synthetic Aperture Radar (SAR) imagery simulation of ocean scene which is often used to detect oil pollution on the ocean surface. Therefore, how to quantify the scattering differences is an open problem for the radar engineering.

Many researches have been made to examine the EM backscatter from the oil contaminated sea surface, but most of them are limited to two dimensional (2D) sea surface or focus on approximate method [3], [4]. However, the hypothetical 2D model lames their applications to three dimensional (3D) cases in practical engineering. Meanwhile, the oil contaminated sea surface is a two-layered dielectric problem. Therefore, the simulation in computational electromagnetics (CEM) becomes much complicate than the homogenous dielectric medium case, and needs further researches. It is our motivation.

In this paper, the electric field integral equation (EFIE) on the interfaces is derived using the wave equation combining the boundary condition. In order to exactly simulate the EM backscatter from the sea surface, the rigorous numerical multilevel fast multipole algorithm (MLFMA) [5], [6] is applied to obtain reliable results. In order to accelerate the EM simulation process when using MLFMA at low-grazing-angle (LGA) cases, a combination model of Thorsos illumination window [7] and resistive loading window [8] is applied to avoid the edge effect induced by artificial truncation of the sea surface in range and azimuth directions. These treatments can greatly improve the computation efficiency thus the simulation of the sea clutter is possible using relatively less computation resources. It is noted that this proposed method is more universal than these existing analytical methods, and unlimited by the 3D sea states or larger incident angles.

This rigorous numerical method and its treatments make the Monte Carlo simulation at LGA incidence possible. Then, the electromagnetic backscatter from the oil contaminated sea surfaces is computed with the variation of incident angles and sea states. As a comparison, the backscatter reduction and the statistic characteristics of the two kinds of sea clutter are analyzed compared with the clean sea surface. Some interesting conclusions are summarized in our experiments.

2. Sea Model

The sea surface can be generated using a spectral method which considers it as a superposition of harmonics.

The amplitudes of the harmonics are proportion to a certain wind-dependent surface-roughness spectrum $W(K, \varphi)$. Then the sea surface can be obtained by inverse fast Fourier transform (IFFT). In this paper, the 2D Pierson-Moskowitz (PM) spectrum is adopted to generate the 3D sea surface [9]. The PM spectrum is defined by

$$W(K_w,\varphi)=\frac{\alpha}{2K_w^4}\exp\left\{-\frac{\beta g^2}{K_w^2U^4}\right\}\cos^4(\frac{\varphi-\varphi_w}{2}) \quad (1)$$

where K_w is the spatial wavenumber, U is the wind speed at the height of 19.5 m above the sea surface, two constants $\alpha = 8.1 \times 10^{-3}$ and $\beta = 0.74$. The angle φ is measured in the horizontal x-y plane with respect to the x-axis and φ_w is the wind direction. In our simulation, the wind direction is towards positive x-axis.

When the sea surface is contaminated by oil, the sea surface tension as well as the sea movement is changed. In the research of Lombardini et al. [10], it is demonstrated that the height of oil contaminated sea surface is damped comparing with the clean sea surface. This damping effect can be expressed by an attenuation coefficient, called Marngoni viscous damping coefficient [11]. In this paper, the oil layer thickness of the contaminated sea is 1 mm.

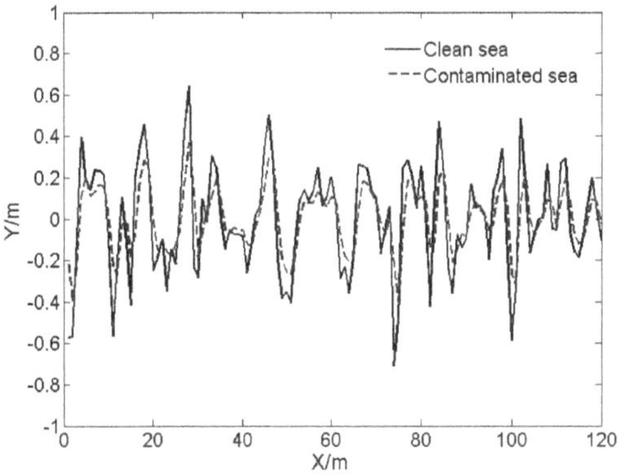

Fig. 1. Comparison of clean sea surface and contaminated sea surface.

Figure 1 shows the comparison of the geometrical profile between the clean sea surface and the contaminated sea surface. The simulation is performed at the wind speed of $U = 7$ m/s which is considered as the average wind speed over the ocean. It is observed that the surface heights of the contaminated sea are indeed damped comparatively to the clean sea surface. Moreover, the small capillarity-scale variations are strongly damped, which implies a strong damping of the surface slopes.

3. EM Modeling

Clean sea water is homogeneous dielectric medium with a relative permittivity ε_1 and permeability μ_1. However, when the sea water is contaminated by oil whose relative permittivity is ε_2 and permeability is μ_2 (generally,

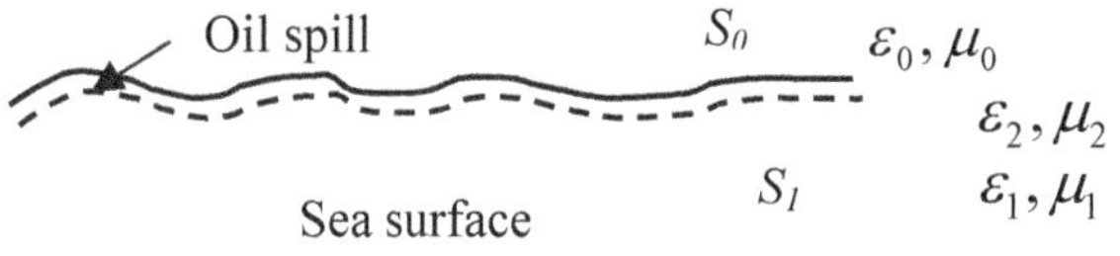

Fig. 2. Geometry of two-layered medium.

$\mu_1 = \mu_2 = 1$), the contaminated sea surface becomes a two-layered dielectric material with its thickness d, as shown in Fig. 2.

Generally, the HFSW can work at LGA for some extreme cases. Therefore, the multipath effect is complicated for EM simulation. In this section, the rigorous numerical method MLFMA and its treatments are proposed to get the backscatters from the two-layered medium problem. The computation efficiency can be greatly improved, thus makes the Monte Carlo simulation of the sea surface possible.

3.1 MLFMA

Using the wave equation and the boundary conditions, EFIE can be derived as functions of the electric equivalent electric current $\boldsymbol{J}$ and equivalent magnetic current $\boldsymbol{M}$ which exist on the interface S_0 between the oil and air as well as the interface S_1 between oil and water. The two sets of surface currents on the boundaries S_0 and S_1 ($\boldsymbol{J}_0$, $\boldsymbol{M}_0$, $\boldsymbol{J}_1$, $\boldsymbol{M}_1$) can be determined using the following numerical simulation procedure.

$$-ik_0Z_0\int_{S_0}\left\{\boldsymbol{J}_0G_0+\frac{1}{k_0^2}\nabla'\cdot\boldsymbol{J}_0\nabla G_0\right\}ds'+\oint_{S_0}\boldsymbol{M}_0\times\nabla G_0ds'-\frac{1}{2}\boldsymbol{M}_0$$
$$=-E^i,\bar{r}\in S_0 \quad (2.a)$$

$$ik_1Z_1\int_{S_0}\left\{\boldsymbol{J}_0G_1+\frac{1}{k_1^2}\nabla'\cdot\boldsymbol{J}_0\nabla G_1\right\}ds'-\oint_{S_0}\boldsymbol{M}_0\times\nabla G_1ds'$$
$$-ik_1Z_1\int_{S_1}\left\{\boldsymbol{J}_1G_1+\frac{1}{k_1^2}\nabla'\cdot\boldsymbol{J}_1\nabla G_1\right\}ds'+\oint_{S_1}\boldsymbol{M}_1\times\nabla G_1ds'-\frac{1}{2}\boldsymbol{M}_0=0,$$
$$\bar{r}\in S_0 \quad (2.b)$$

$$ik_1Z_1\int_{S_0}\left\{\boldsymbol{J}_0G_1+\frac{1}{k_1^2}\nabla'\cdot\boldsymbol{J}_0\nabla G_1\right\}ds'-\oint_{S_0}\boldsymbol{M}_0\times\nabla G_1ds'$$
$$-ik_1Z_1\int_{S_1}\left\{\boldsymbol{J}_1G_1+\frac{1}{k_1^2}\nabla'\cdot\boldsymbol{J}_1\nabla G_1\right\}ds'+\oint_{S_1}\boldsymbol{M}_1\times\nabla G_1ds'-\frac{1}{2}\boldsymbol{M}_1=0,$$
$$\bar{r}\in S_1 \quad (2.c)$$

$$-ik_2Z_2\int_{S_0}\left\{\boldsymbol{J}_1G_2+\frac{1}{k_2^2}\nabla'\cdot\boldsymbol{J}_1\nabla G_2\right\}ds'+\oint_{S1}\boldsymbol{M}_1\times\nabla G_2ds'-\frac{1}{2}\boldsymbol{M}_1$$
$$=0,\bar{r}\in S_1 \quad (2.d)$$

where $\oint_{S_0} ds'$ denotes principle integration, Z_n is the characteristic impedance, k_n is the wave number and

$G_n = g_n(r,r') = 4\pi e^{ik_n|r-r'|}/|r-r'|$ is the Green function. The subscript n = 1, 2, 3 denotes the three cases in free space, sea water and oil medium, respectively.

Here the equivalent impedance boundary condition is used to deal with the boundary condition scattering problem [12]. It means the electric current $\boldsymbol{J}$ and equivalent magnetic current $\boldsymbol{M}$ satisfy $n' \times \boldsymbol{M}_0 = Z_1 \boldsymbol{J}_0$ and $n' \times \boldsymbol{M}_1 = Z_2 \boldsymbol{J}_1$, where n' is the outer normal vector at each field point. Therefore, the unknown number in (2) will be halved. The unknown equivalent electric current can be obtained by the Galerkin's method in method of moment (MoM), as following.

$$\begin{bmatrix} [Z_{11}^{nm}][Z_{12}^{nm}] & 0 & 0 \\ [Z_{21}^{nm}][Z_{22}^{nm}] & [Z_{23}^{nm}][Z_{24}^{nm}] \\ [Z_{31}^{nm}][Z_{32}^{nm}] & [Z_{33}^{nm}][Z_{34}^{nm}] \\ 0 \quad 0 & [Z_{43}^{nm}][Z_{44}^{nm}] \end{bmatrix} \bullet \begin{bmatrix} [I_1^n] \\ [I_2^n] \\ [I_3^n] \\ [I_4^n] \end{bmatrix} = \begin{bmatrix} [V_1^m] \\ [0] \\ [0] \\ [0] \end{bmatrix}. \quad (3)$$

To investigate the statistic characteristics of the backscatter of the sea surface, Monte Carlo simulation is necessary. However, it is rather computationally expensive and time-consuming. In order to accelerate the computing process, MLFMA is used to solve the above equation. In the process of MLFMA, the interactions between the elements are classified as near-region and the far-region. The near-region matrix elements are calculated directly using MoM, while the far-region elements are acquired by using MLFMA.

$$\sum_{l} Z^{nm} I^n = \sum_{l \in NR} Z^{nm} I^n + \sum_{l \in FR} Z^{nm} I^n \quad (4)$$

where NR represents the near-region and FR represents the far-region, l means the basis function, I^n are the expansion coefficients needed to solve using MoM, and Z^{nm} are the elements of the impedance matrix. More details can be referenced in [5].

3.2 Treatments

Numerical methods can guarantee sufficient precision when solving EM computing problems even at LGA. However, it will bring edge diffraction which is induced by the artificial truncation to the sea surface. The electric current $\boldsymbol{J}$ will change suddenly at the edge of the truncated surface. This phenomenon is called edge effect. In the research of EM scattering in those numerical methods, how to avoid the edge effect is an important issue. The most popular way to suppress the edge effect is using the illumination tapered window (Thorsos window) [7] which is got by modulating a tapered function to incident electric field, as following.

$$p(x,z) = e^{-j\frac{2(x-z\tan\theta_i)^2/g^2-1}{(kg\cos\theta_i)^2}k\hat{k}\cdot r'} e^{-\frac{(x-z\tan\theta_i)^2}{g^2}} \quad (5)$$

where g is a constant that controls the width of the illumination beam, θ_i is the incident angle.

However, according to the studies by Jin [13], the length L of rough surface has to satisfy the condition: $L \geq 24\lambda/(\cos\theta_i)^{1.5}$, where λ is the EM wavelength. Accordingly, at LGA the scale of the computing sea surface is very large. For example, if $\theta_i = 85°$, then $L \geq 932\lambda$. And it leads to a large amount of unknown number on the sea surface. Therefore, instead of the illumination window, the resistive loading window is applied to suppress the edge effect under the plane wave illumination. It can be accomplished by simply adding $\boldsymbol{RJ}$ to the EFIE formula, taking (2a) as an example, as following

$$\overline{\overline{L}}(\boldsymbol{J}) + R\boldsymbol{J} = -E^i, \vec{r} \in S_0 \quad (6)$$

where R is the resistance, and $\overline{\overline{L}}(\cdot)$ is the matrix-vector multiply operator and denotes the procedure in left side of (2a). The resistance is loaded within the resistive loading area, which was discussed in detail in reference [14]. Once the resistive loading method is used, the length L of sea surface only needs to be as large as 10 times of the correlation length of the rough surface. It is much smaller than the condition of Thorsos window case.

However, it is found that the resistive loading will bring the convergence problem because that the impedance matrix Z^{nm} is a non-diagonally dominant matrix for this two-layered dielectric medium scattering problem in our simulations. Since the resistance R will attenuate the value of self-impedance element, the condition number of the final discretized impedance matrix will become worse. Therefore, more computational CPU time will be taken in the impedance matrix iteration processes.

A method combining the two different windows is proposed to accelerate its convergence. Resistive loading window is applied in range direction while a Gaussian illumination window is used in azimuth direction. In azimuth direction, the incident plane wave is modulated by a Gaussian function $p(y) = e^{-y^2/g^2}$ instead of the Thorsos window. Meanwhile, an important work in range direction is to extend the original sea surface with smoothly curved sections (for example, we used radius of 10λ) that join to planar sections which are angled 30° down from the horizon. So that all the points on the extension surfaces are shadowed by the original rough surface. The resistive loading of the edges has the advantage of not requiring the surface to be modified at LGA.

As a comparison, a 600 m (L)×600 m (W) PM sea surface realization at U = 3 m/s was generated and simulated under those system parameters: radar frequency 15 MHz, incident angle 60°, vertical polarization. As shown in Fig. 4, the convergence iteration is 228 steps for the traditional resistive window [8] in range and azimuth directions whereas only 50 steps are needed for the proposed combination of the two different windows.

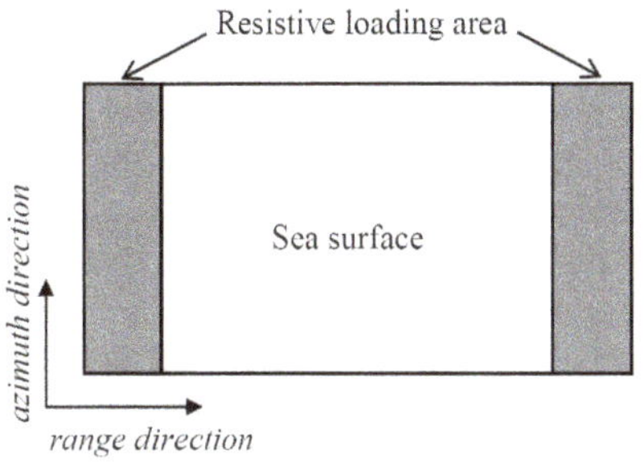

Fig. 3. The scheme of two windows.

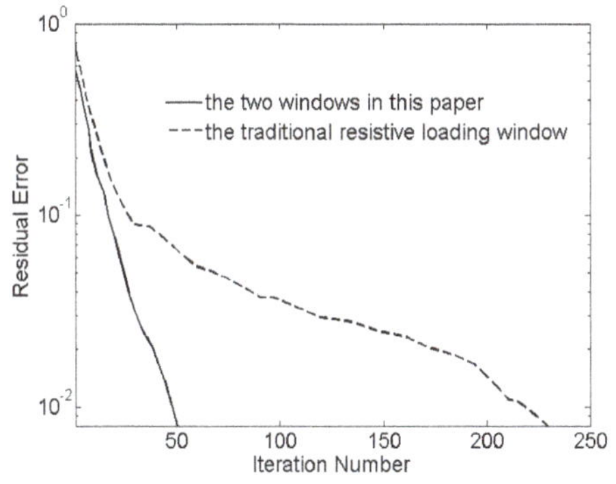

Fig. 4. Convergence comparison.

Therefore, the proposed method to suppress the edge effect is more effective and can greatly reduce the CPU time, which is a big improvement in Monte Carlo simulation.

4. Numerical Simulations

4.1 Comparison of Backscatters

By using this proposed method, some experiments are presented in this section. The HFSWR operates at the frequency of 15 MHz, vertical polarization. Monte Carlo simulation is applied to get average radar backscatter. A total number of 1000 sea surfaces are generated for each sea state. The simulation sea area is 600 m long with a width of 600 m. The simulation is performed when wind speed is 3 m/s and 9 m/s, corresponding to Douglas sea state 2 and 4.

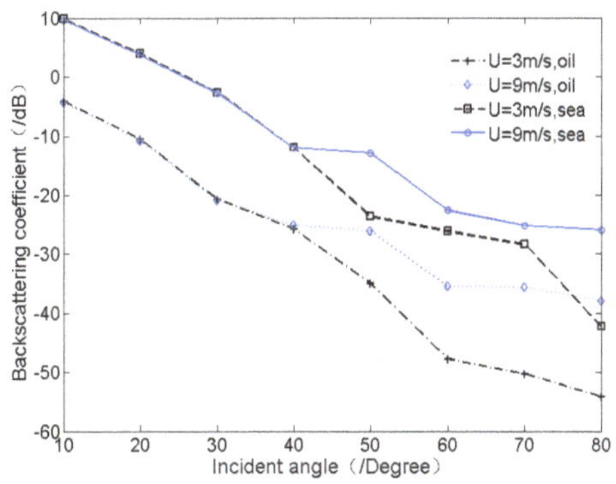

Fig. 5. Backscatter coefficients with various wind speeds.

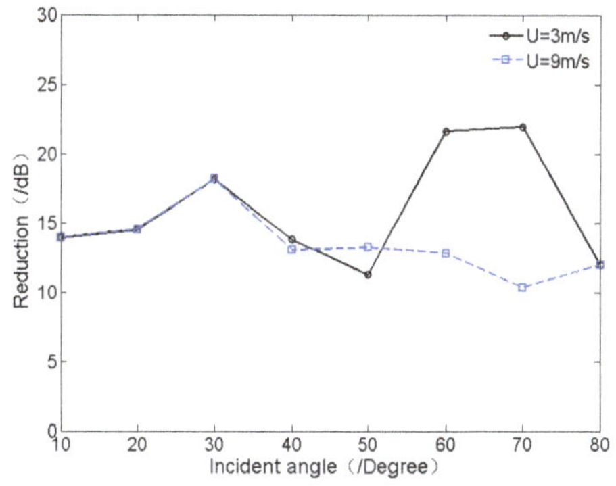

Fig. 6. Backscatter coefficient reduction.

As shown in Fig. 5, the radar backscatters of the contaminated sea surface is much lower than that of the clean sea surface. The reduction of the backscatter coefficient corresponding to different sea states is shown in Fig. 6. A reduction of more than 10 dB at moderate incident angles is found through our simulation, and even 15 dB at LGA angle. In SAR systems, the oil polluted area is seen as "dark" sea. Capillary waves on the sea surface reflect electromagnetic energy. However, when the oil layer on the sea surface, the capillary waves are dampened. As a consequence, the backscatter is much weaker than clean sea surface.

4.2 Sea Clutter Distribution

By using the radar echo data acquired in the above experiments, the distribution of the sea clutter is also examined. For radar system, statistical models (Rayleigh, Lognormal, Weibull and K-distributions) are often used to describe the sea clutter. In this paper, the probability density functions (PDF) of the backscatter of the sea clutter for clean sea and oil-contaminated sea are studied. The parameters of Rayleigh, Lognormal and Weibull distribution are estimated by the maximum likelihood (ML) method. It is a standard approach to parameter estimation and provides optimum estimates in the sense that these estimates are the most probable parameter values. However, ML is computationally expensive to estimate the parameters of K distribution. A new method based on higher order and fractional moments is proposed in these references [15], [16]. It can significantly reduce the computational requirement. Details about the formulas of these four distributions and the estimation methods are summarized in reference [16]. Take this sea state of $U = 9$ m/s as an example, the sea clutter distribution is shown for different incident angles in Fig. 7.

Kolmogorov-Smirnov test (K-S test) is used to find the best fitting distribution of the sea clutter among the four distributions. K-S test has the advantage of making no assumption about the distribution of the sea data. It is nonparametric and distribution free. The K-S test statistic quantifies a distance between the empirical distribution function of the sample and the cumulative distribution function of the reference distribution, or between the

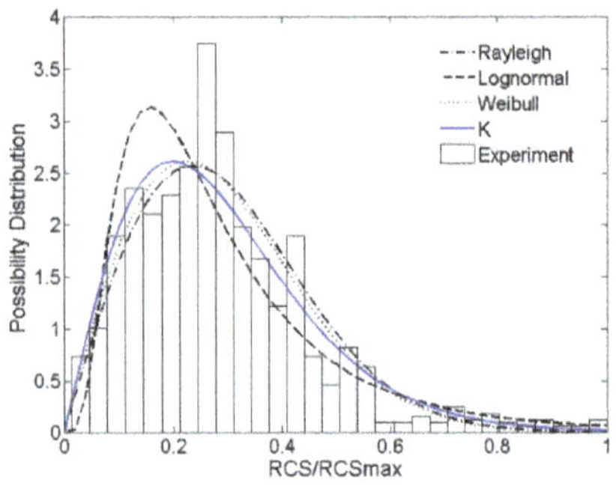

(a) Clean sea surface at incident angle of 30°

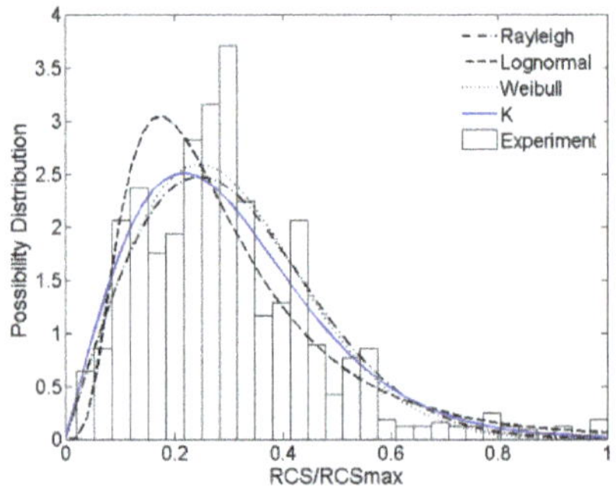

(b) Oil contaminated sea at incident angle of 30°

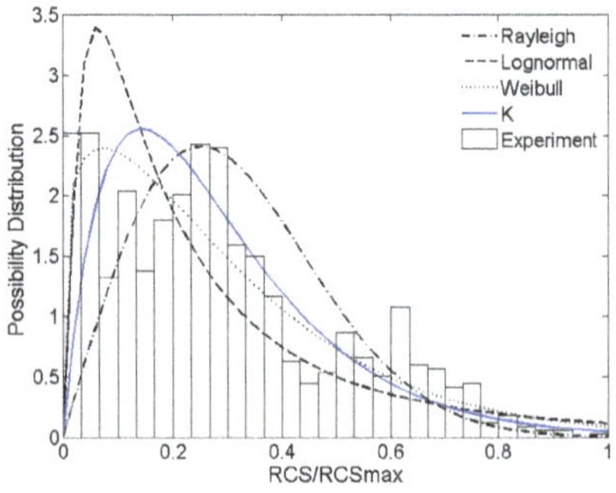

(c) Clean sea at incident angle of 80°

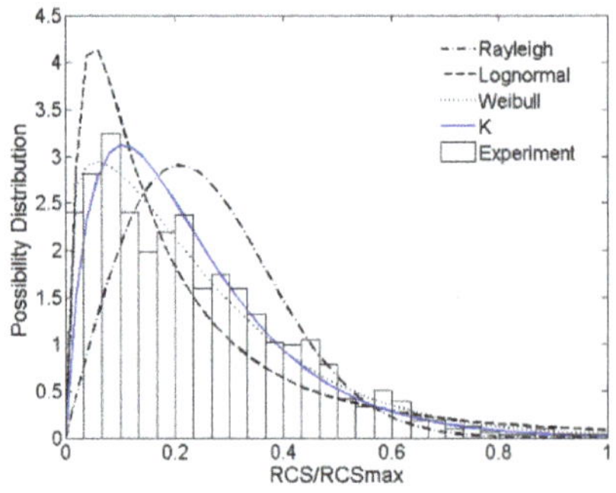

(d) Oil contaminated sea at incident angle of 80°

Fig. 7. The backscatter distribution of sea clutter.

empirical distribution functions of two samples. The K-S test statistic is defined as *De* which is shown

$$De = \max_{1 \le i \le N} \left| F(x_i) - F_t(x_i) \right| = \max_{1 \le i \le N} \left| n / N - F_t(x_i) \right| . \quad (7)$$

$F_t(x_i)$ is the cumulative distribution of the distribution being tested and $F(x_i)$ is the cumulative distribution of the simulated sea clutter. *De* means the maximum deviation between the cumulative distribution of the tested distribution and the statistic distribution of the simulation result. The best fit distribution is with the smallest *De*. Tab. 1 is the K-S test of the sea clutter distribution corresponding to Fig. 7. And the sea clutter distributions are also examined under different sea states in Tab. 2 and Tab. 3.

De	Rayleigh	Lognormal	Weibull	*K*
Sea /30°	**0.0419**	0.1140	0.0451	0.0597
Sea /80°	0.1281	0.1446	**0.0614**	0.0914
Oil /30°	0.0493	0.1156	**0.0444**	0.0609
Oil/ 80°	0.1530	0.1106	**0.0397**	0.0566

Tab. 1. K-S test.

Incident angle	10°	20°	30°	40°	50°	60°	70°	80°
State 2 (3 m/s)	L	K	K	L	W	W	L	W
State 4 (9 m/s)	K	R	W	W	W	W	W	W

Tab. 2. The distribution of sea clutter for contaminated sea.

Incident angle	10°	20°	30°	40°	50°	60°	70°	80°
State 2 (3 m/s)	K	K	K	L	W	L	L	W
State 4 (9 m/s)	K	K	R	W	W	W	W	W

Tab. 3. The distribution of sea clutter for clean sea.

As can be seen from Tab. 2 and Tab. 3, the distribution of clutter from oil contaminated sea surface is different from that of clean sea surface. For most cases, the sea clutter satisfies K distribution and Weibull distribution. When the wind speed above the sea surface is moderate (3 m/s), K distribution describes the sea clutter quite well at small incident angle. As the incident angle getting large, it is Weibull distribution for the contaminated sea whereas it is Lognormal distribution for the clean sea. When the wind speed gets higher (9 m/s), Weibull distribution will be dominant at LGA. The distribution of clutter from oil contaminated sea is first mentioned in this paper. Rigorous numerical EM simulation method can ensure sufficient accuracy. Moreover, with the treatments this method can obtain a high efficiency. Consequently, Monte Carlo simulation can be performed on the electric large problem. The real radar data from HFWSR system is usually very expensive. Instead of the real data from the radar, the simulation results by using rigorous numerical computational electromagnetic method as shown in this paper is workable as well. Moreover, it is more cost-effective and available.

5. Conclusions

In this paper, the sea clutter distribution is examined when there is oil pollution on the sea surface. In HFSWR system, the observation area is very large; accordingly the accurate EM simulation of the large sea surface is rather difficult. The integral equations both on the clean sea surface and the oil contaminated sea surface are presented. Numerical method with treatments on it can ensure sufficient accuracy to get reliable results of the backscatter from sea surface. Meanwhile, the computational resources are saved a lot, and then Monte Carlo simulation can be performed. It is found that the backscatter from oil contaminated sea surface is much weaker than that from clean sea surface. That is because the oil damped the capillary waves. Accordingly, the reflected electromagnetic energy is weaker than before. This effect can be seen in SAR image that the oil contaminated area is dark. The distribution of the sea clutter which is an important statistic characteristic is investigated by using the simulated results. Differences of the sea clutter distribution are found between the oil contaminated sea surface and the clean sea surface.

Acknowledgements

This work was supported by the National Natural Science Foundation of China under grant 61171044; the Fundamental Research Funds for the Central Universities under grant ZYGX2012YB006; and Research Fund for the Doctoral Program of Higher Education of China under grant 20120185110024.

References

[1] SOLBERG, A. H. S. Remote sensing of ocean oil-spill pollution. *Proceedings of the IEEE*, 2012, vol. 100, no. 10, p. 2931–2945. DOI: 10.1109/JPROC.2012.2196250

[2] FINGAS, M., BROWN, C. Review of oil spill remote sensing. *Marine Pollution Bulletin*, 2014. DOI: 10.1016/S1353-2561(98)00023-1

[3] PINEL, N., DECHAMPS, N., BOURLIER, C. Modelling of the bistatic electromagnetic scattering from sea surfaces covered in oil for microwave applications. *IEEE Transactions on Geoscience and Remote Sensing*, 2008, vol. 46, no. 2, p. 385–392. DOI: 10.1109/TGRS.2007.902412

[4] GHANMI, H., KHENCHAF, A., COMBLET, F. Numerical modeling of electromagnetic scattering from sea surface covered by oil. *Journal of Electromagnetic Analysis and Applications*, 2014, vol. 6, p. 15–24. DOI: 10.4236/jemaa.2014.61003

[5] SONG, J., LU, C.-C., CHEW, W. C. Multilevel fast multipole algorithm for electromagnetic scattering by large complex objects. *IEEE Transactions on Antennas and Propagation*, 1997, vol. 45, no. 10, p. 1488–1493. DOI: 10.1109/8.633855

[6] YANG, W., ZHAO, Z., QI, C., NIE, Z. Electromagnetic modeling of breaking waves at low grazing angles with adaptive higher order hierarchical Legendre basis functions. *IEEE Transactions on Geoscience and Remote Sensing*, 2011, vol. 49, no. 1, p. 346–352. DOI: 10.1109/TGRS.2010.2052817

[7] THORSOS, E. I. The validity of the Kirchhoff approximation for rough surface scattering using a Gaussian roughness spectrum. *Journal of the Acoustical Society of America*, 1988, vol. 83, p. 78 to 92. DOI: DOI: 10.1121/1.396188

[8] ZHAO, Z., WEST, J. C. Resistive suppression of edge effects in MLFMA scattering from finite conductivity surfaces. *IEEE Transactions on Antennas and Propagation*, 2005, vol. 53, p. 1848–1852. DOI: 10.1109/TAP.2005.846810

[9] LI, X., XU, X. Scattering and Doppler spectral analysis for two-dimensional linear and nonlinear sea surfaces. *IEEE Transactions on Geoscience and Remote Sensing*, 2011, vol. 49, p. 603–611. DOI: 10.1109/TGRS.2010.2060204

[10] LOMBARDINI, P., FISCELLA, B., TRIVERO, P., CAPPA, C., GARRETT, W. Modulation of the spectra of short gravity waves by sea surface films: slick detection and characterization with a microwave probe. *Journal of Atmospheric and Oceanic Technology*, 1989, vol. 6, p. 882–890. DOI: 10.1175/1520-0426(1989)006<0882:MOTSOS>2.0.CO;2

[11] GADE, M., ALPERS, W., HUHERFUSS, H., WISMANN, V. R., LANGE, P. A, On the reduction of the radar backscatter by oceanic surface films: Scatterometer measurements and their theoretical interpretation. *Remote Sensing of Environment*, 1998, vol. 66, p. 52–70. DOI:10.1016/S0034-4257(98)00034-0

[12] GLISSON A. W. Electromagnetic scattering by arbitrarily shaped surface with impedance boundary conditions. *Radio Science*, 1992, vol. 27, no. 6, p. 935–943. DOI: 10.1029/92RS01782

[13] LIU, P., JIN, Y. Q. Numerical simulation of bistatic scattering from a target at low altitude above rough sea surface under an EM-wave incidence at low grazing angle by using the finite element method. *IEEE Transactions on Antennas and Propagation*, 2004, vol. 52, no. 5, p. 1205–1210. DOI: 10.1109/TAP.2004.827497

[14] ZHAO, Z., WEST, J. C. Resistive treatment of edges in MLFMA LGA scattering from finite conductivity 2D surfaces. In *IEEE Antennas and Propagation Society International Symposium*, 2002, p. 264–267. DOI: 10.1109/APS.2002.1016974

[15] ISKANDER, D. R., ZOUBIR, A. M. Estimation of the parameters of the K-distribution using higher order and fractional moments [radar clutter]. *IEEE Transactions on Aerospace and Electronic Systems*, 1999, vol. 35, p. 1453–1457. DOI: 10.1109/7.805463

[16] FARINA, A., GINI, F., GRECO, M., VERRAZZANI, L. High resolution sea clutter data: statistical analysis of recorded live data. *IEE Proceedings-in Radar, Sonar and Navigation*, 1997, vol. 144, no. 3, p. 121–130. DOI: 10.1049/ip-rsn:19971107

About the Authors ...

Cong-hui QI was born in Hebei, China. She received her M.Sc. from University of Electronic Science and Technology of China in 2009. Her research interests include computational electromagnetic, backscatter from rough surface and SAR image.

Zhi-qin ZHAO was born in Hunan, China. He received B.S. and M.S. degrees in Electronic Engineering from the University of Electronic Science and Technology of China, Sichuan, and the Ph.D. degree in Electrical Engineering from Oklahoma State University, Stillwater, in 1990, 1993, and 2002, respectively. His research interests include radar signal processing and computational electromagnetics.

A Wideband Direct Data Domain Genetic Algorithm Beamforming

Hassan ELKAMCHOUCHI, Mohamed HASSAN

Dept. of Electrical Engineering, University of Alexandria, Alhuria Street, Alexandria, Egypt

helkamchouchi@hotmail.com, engmohamedmokhtar1@yahoo.com

Abstract. *In this paper, a wideband direct data-domain genetic algorithm beamforming is presented. Received wideband signals are decomposed to a set of narrow sub-bands using fast Fourier transform. Each sub-band is transformed to a reference frequency using the steering vector transformation. So, narrowband approaches could be used for any of these sub-bands. Hence, the direct data-domain genetic algorithm beamforming can be used to form a single 'hybrid' beam pattern with sufficiently deep nulls in order to separate and reconstruct frequency components of the signal of interest efficiently. The proposed approach avoids most of drawbacks of already-existing statistical and gradient-based approaches since formation of a covariance matrix is not needed, and a genetic algorithm is used to solve the beamforming problem.*

Keywords

Wideband array, direct data-domain, genetic algorithm, signal of interest, matrix pencil method

1. Introduction

An adaptive array is able to electronically steer its main lobe to any desired direction and place deep pattern nulls to directions of interference sources. That way, the antenna could adaptively minimize the interference power while maintaining the array gain in the direction of the target signal [1–3].

Statistical methods of adaptive antennas are computationally intensive processes and require stationary data to construct a covariance matrix. Direct data-domain (D^3) methods could overcome drawbacks of statistical techniques by processing data on snapshot-by-snapshot basis without constructing a covariance matrix. Hence, D^3 methods could handle non-stationary environments and coherent interferers [4].

Genetic algorithms (GAs) might be more efficient than gradient-based methods for nulling a linear antenna array since the gradient-based methods have following disadvantages:

- The methods are highly sensitive to starting points when the number of variables, and hence the size of the solution space, increases.
- The methods frequently converge to local suboptimum solutions.
- The methods require a continuous and differentiable objective function.
- The methods require piecewise linear cost approximation (for linear programming).
- The methods have problems with convergence and algorithm complexity (for non-linear programming).

Future wireless systems have to utilize wideband smart antennas to meet high speed data transmission while avoiding undesired interference [15]. Beamforming techniques used in narrowband systems are inappropriate for wideband systems due to the limited ability of tracking a desired user or forming nulls in directions of interfering sources over a large frequency band [5], [6]. Some earlier work has been done to solve such problems. Existing concepts of wideband beamforming exhibit disadvantages which could be summarized as follows:

- Different array patterns are used for different frequencies [12–14]. Obviously, such an approach is quite cumbersome.
- Gradient-based wideband beamforming [1], [7], [16] has the already described disadvantages.
- Statistical methods depend on the formation of a covariance matrix [17]. Therefore, stationary data are expected to estimate the covariance matrix. In case of non-stationary data, resulting errors in the covariance matrix reduce the ability to handle coherent interferers [4].
- In [7] it is mentioned that "although the main beam is directed to the signal of interest (SOI) direction and the jammers are nulled correctly but the nulls are not deep enough"; same drawback appears in [1]. Such disadvantage affects the accuracy of SOIs reconstruction. In addition, in [1] and [7], nulls' depths and DOAs estimations' accuracy are found to be frequency dependent. Therefore, selecting one beam pattern corresponding to single sub-band does not assure sufficient nulls' depths to cancel interferers in all other sub-bands.

The proposed wideband direct data-domain genetic algorithm (WD^3GA) beamforming relies on decomposing the received wideband signals into a set of narrow sub-bands by using fast Fourier transform (FFT) and transform all sub-bands to a reference frequency by using the steering vector transformation [5]. Any of these sub-bands could be used by narrowband techniques for direction of arrival (DOA) estimation and beamforming. The investigated narrowband direct data-domain adaptive nulls genetic algorithm (D^3ANGA) beamformer [9] is used to cancel jammers' frequency components in each sub-band. Finally, inverse fast Fourier transform could be used to retrieve the SOI in time domain.

WD^3GA beamforming has the following *unique* set of characteristics:

- WD^3GA beamforming uses only a single hybrid array beam pattern; thus the method complexity is reduced.
- GA is used to solve the beamforming problem. Hence, the gradient-based methods' drawbacks are avoided.
- Both FFT and covariance-matrix-based techniques require recorded samples of the received signals [4]. However, WD^3GA beamforming doesn't make any assumption about the statistics of the environment. Therefore, data non-stationarity has a little effect on the method performance.
- The problem of frequency dependent estimations of DOAs and nulls' depths [1], [7] could be solved by taking into consideration that in D^3ANGA beamforming [9], the determination of the nulls' depths and the DOAs estimation are done before and independent on the beamforming algorithm. Hence, for wideband beamforming, the nulls' depths could be selected prior to the beamforming algorithm to be proportional to strongest interferers' frequency components and the most accurate estimated DOAs could be selected priory. Hence, one 'hybrid' reconstruction array beam pattern combines both sufficient deep nulls and accurate estimated DOAs could be formed.

2. Steering Vector Transformation

For a wideband antenna array of N elements and d spacing between adjacent elements, consider $q+1$ uniformly spaced directions covering a pre-specified angular azimuth region $\mathbf{\Phi}_\mathrm{q}$ [5], [7] where,

$$\mathbf{\Phi}_\mathrm{q} = \left[\phi_0, \phi_1, \ldots, \phi_q\right] \quad (1)$$

In this paper, all coming signals are considered to be in the azimuth plane ($\theta = 90°$). The steering vector transformation is based on transforming the array steering matrix at the kth frequency f_k to another array steering matrix at a pre-specified reference frequency f_o using a transformation matrix $\boldsymbol{T}_\mathrm{q}(k)$ for the angular region $\mathbf{\Phi}_\mathrm{q}$ such that:

$$\boldsymbol{A}\left(\mathbf{\Phi}_\mathrm{q}, f_o\right) = \boldsymbol{T}_\mathrm{q}(k)\boldsymbol{A}\left(\mathbf{\Phi}_\mathrm{q}, f_k\right) \quad (3)$$

where $\boldsymbol{A}\left(\mathbf{\Phi}_\mathrm{q}, f_o\right)$ and $\boldsymbol{A}\left(\mathbf{\Phi}_\mathrm{q}, f_k\right)$ are the array steering matrices for the angular region $\mathbf{\Phi}_\mathrm{q}$ at the kth frequency f_k and at the reference frequency f_o respectively. The array steering matrix could be computed as follows:

$$\boldsymbol{A}\left(\mathbf{\Phi}_\mathrm{q}, f\right) = \left[\boldsymbol{a}(\phi_0, f), \boldsymbol{a}(\phi_1, f), \ldots, \boldsymbol{a}(\phi_q, f)\right] \quad (4)$$

where $\boldsymbol{a}(\phi, f)$ is the steering vector defined by:

$$\boldsymbol{a}(\phi, f) = \left[1,\ e^{\frac{2\pi f d}{c}\cos(\phi)}, \ldots,\ e^{\frac{2\pi f (N-1) d}{c}\cos(\phi)}\right]^T \quad (5)$$

where T denotes the transpose of the vector. Equation (3) could be solved for $\boldsymbol{T}_\mathrm{q}(k)$ using the least squares method which yields the solution:

$$\boldsymbol{T}_\mathrm{q}(k) = \boldsymbol{A}\left(\Phi_\mathrm{q}, f_o\right)\boldsymbol{A}\left(\Phi_\mathrm{q}, f_k\right)^H\left(\boldsymbol{A}\left(\Phi_\mathrm{q}, f_k\right)\boldsymbol{A}\left(\Phi_\mathrm{q}, f_k\right)^H\right)^{-1} \quad (6)$$

where the H superscript represents the conjugate transpose of a complex matrix. The processed input voltage vector at the kth frequency which has been transformed to the reference frequency f_o, $\boldsymbol{x}(f_o)$, could be written as:

$$\boldsymbol{x}(f_o) = \boldsymbol{T}_\mathrm{q}(k)\boldsymbol{x}(f_k) \quad (7)$$

where $\boldsymbol{x}(f_k)$ is the input voltage vector at the kth frequency. Using this transformation, a single narrowband beamformer tuned at the reference frequency f_o could be used.

3. Matrix Pencil Method

The MP method is a narrowband D^3 method to estimate the DOA of various signals impinging on an antenna array; the signals' complex amplitudes could be estimated as well [4], [8].

For a uniformly linear array composed of $N+1$ element, the voltage induced in the array n^{th} element, x_n, could be written as:

$$x_n = \sum_{k=1}^{P} s_k e^{\left(\frac{j2\pi n d \cos(\phi_k)}{\lambda}\right)} + \xi_n = \sum_{k=1}^{P} s_k a_k^n + \xi_n \quad (8)$$

where ξ_n is the noise at the nth array element, P is the number of incident signals, S_k is the complex amplitude of the kth incident signal, λ is the wave length, d is the distance between two adjacent elements, ϕ_k is the DOA of the kth signal, and a_k are the poles to be estimated.

The poles a_k could be estimated by constructing and processing a Hankel matrix as illustrated in [4], [8], then the DOAs of various signals could be obtained as follows:

$$\phi_k = \cos^{-1}\left[\frac{\lambda \ln\left(a_{es\,k}\right)}{j2\pi d}\right] \quad (9)$$

where $a_{es\,k}$ is the kth estimated pole. The complex amplitudes vector of the P signals, $\boldsymbol{AMP}$, could be obtained by:

$$AMP = \left(P_0^H P_0\right)^{-1} P_0^H x \tag{10}$$

where P_0 is the matrix containing the pole of each incident signal at each antenna element and x is a vector containing the induced voltages at the array elements.

4. D^{3}ANGA Beamforming

D^{3}ANGA is a narrowband beamformer [9]. Consider a linear antenna array with uniformly spaced N elements. Hence, N complex weights are used for beamforming. Given the DOAs and the strengths of all coming signals, a genetic algorithm (GA) is used to find the optimal values of these weights in order to fulfill the algorithm objectives [10], [11] which are as follows:

- Minimizing beam the pattern average value to minimize the pattern side lobes level.
- Maximizing the pattern value in the direction of the SOI (P_S) to radiate maximum possible power in this direction.
- Placing deep nulls in directions of the interferers. In addition, nulls' depths are selected to be proportional to interference incident signals' intensities.

Hence, the fitness function could be written as:

$$Fit = w\sum_{i=1}^{i=J}\left\|\frac{P_i}{P_s}\right| - N_i\right| + \left|\frac{P_{av}}{P_s}\right| \tag{11}$$

where J is the number of the interferer (jammer) signals, P_i is the array beam pattern complex value in the direction of the ith jammer, P_S is the array beam pattern complex value in the direction of the SOI, N_i is the ith normalized pattern null value corresponding to the ith jammer, w is the weighting factor used to balance GA optimization between the two terms of the fitness function. | | denotes the absolute (magnitude) of the complex quantities and P_{av} is the pattern average value which could be calculated by:

$$P_{av} = \frac{\sum_{\theta}\sum_{\phi}\left|P(\theta,\phi)\right|}{N_P} \tag{12}$$

where N_P is the number of points at which the pattern values are calculated. The pattern value at any direction, $P(\theta,\phi)$, could be computed by

$$P(\theta,\phi) = W^{\mathrm{T}} A(\theta,\phi) \tag{13}$$

where W is the complex weights' vector to be estimated by the GA and $A(\theta,\phi)$ is the steering vector in the direction of (θ,ϕ). Assuming all coming signals are in the azimuth plane ($\theta = 90°$), $A(\theta,\phi)$ could be expressed as:

$$A = \left[1,\ e^{j2\pi\frac{d}{\lambda}cos\phi}, e^{j2\pi\frac{2d}{\lambda}cos\phi}, \ldots, e^{j2\pi\frac{(N-1)d}{\lambda}cos\phi}\right]^{\mathrm{T}} \tag{14}$$

where d is the space between array adjacent elements and λ is the wavelength corresponding to the operating frequency. The ith normalized pattern null value, N_i, corresponding to the ith jammer could be computed by:

$$N_i = (|S_i| * C)^{-1} \tag{15}$$

where S_i is the ith jammer intensity and C is the cancelling factor (CF) used to make the nulls' depths sufficient to cancel interference signals efficiently.

In this paper, all the pattern values and all the obtained complex weights are normalized with respect to P_S. The GA-estimated complex weights could be used to separate and reconstruct the SOI using:

$$RSOI = W_n^T x \tag{16}$$

where $RSOI$ is the reconstructed SOI, W_n is the normalized weights vector, and x is the received signals' vector.

5. Genetic Algorithm Components

GA is a powerful optimization technique based on the concept of natural selection and natural genetics [10]. GA repeatedly modifies a population of individual solutions. At each step the GA selects individuals at random from the current population to be parents and uses them to produce the children of the next population. Over successive generations, the population "evolves" toward an optimal solution which is considered to be the solution which gives the minimum of the fitness function [10], [11]. In this paper, GAs are implemented based on the built-in genetic algorithm of R2013a MATLAB software package. The basic GA components, used in this paper, are reviewed briefly as follows [10]:

- Genetic representation of solution: Real number encoding is used to represent individual solutions or chromosomes.
- Population initialization: uniform random initialization is used and the population size is selected to be ten times the number of the antenna array elements taking into consideration that two chromosomes are used to represent each array element complex weight one for the real part and the other for the imaginary part [11].
- Evaluation of the fitness function: The GA should find the global minimum of the fitness function.
- Fitness scaling: The ranking method in which the scaling of raw scores is based on the rank of each individual instead of its score is used [11].
- Selection methods: Stochastic uniform selection method is used. Stochastic uniform selection method lays out a line in which each parent corresponds to a section of the line of length proportional to its scaled value. The algorithm moves along the line in steps of equal size. At each step, the algorithm allocates a parent from the section it lands on. The first

step is a uniform random number less than the step size. Also, Elitism forces GA to retain some number of the best individuals at each generation [10], [11]. In this paper, the most fit two chromosomes survive directly to the next generation as elite chromosomes.

- Genetic operators: genetic operators are used to produce new individuals. Crossover and mutation are the most frequently used genetic operators and are described as follows:

1- The crossover operator is the exchange of genes between parent's chromosomes to produce offspring. The scattered crossover method is used. In this method, crossover is done by creating a random binary vector and selecting the genes where the vector's elements are ones from the first parent, and the genes where the vector's elements are zeros from the second parent, and combines the genes to form the child [10], [11]. The fraction of each population, other than elite children, that are made up of crossover children is set to 0.8. The remaining chromosomes are mutation children.

2- Mutation is done by the addition of a random number which is chosen from a Gaussian distribution to each entry of the parent vector.

- Termination condition: a maximum number of 500 generations is used to terminate GA.

6. The Proposed WD^{3}GA

WD^{3}GA beamformer is implemented by a narrowband decomposition structure whereby each signal received at each array element is transformed into its frequency domain components using FFT. Each narrow frequency sub-band is transformed to a reference frequency using the steering vector transformation. The transformed sub-bands could be processed by narrowband techniques. Hence, they are sent to the MP stage, as shown in Fig. 1. In the MP stage, the DOAs as well as the complex magnitudes of all incident signals' frequency components are estimated.

One array beam pattern is used for reconstructing all SOI frequency components from all sub-bands. Two parameters determine the accuracy of reconstruction of the SOI frequency components: 1- the DOA estimation accuracy of all coming signals; 2- the pre-specified pattern nulls' depths. The estimated DOAs should be as accurate as possible, so it is recommended to select the DOAs estimated using the nearest sub-band to the reference frequency f_o since it is observed that the nearest sub-band to the reference frequency has the minimum error in the steering vector transformation [5]. The error in the steering vector transformation could be calculated as follows:

$$\boldsymbol{Error} = \boldsymbol{A}\left(\boldsymbol{\Phi}_q, f_o\right) - \boldsymbol{T}_q(k)\boldsymbol{A}\left(\boldsymbol{\Phi}_q, f_k\right). \quad (17)$$

When the reference frequency is selected to be one of the considered signals' frequencies, the steering vector transformation is not required for the corresponding sub-band. Hence, the corresponding steering vector transformation error is zero. For the array beam pattern nulls' depths, it is recommended to select the deepest nulls corresponding to the strongest estimated interferers' frequency components by the MP method in order to cancel all the interferers' frequency components in all sub-bands efficiently. So, one hybrid reconstruction beam pattern combines both accurate estimated DOAs and deep nulls and could be used for the estimation of all SOI frequency components then inverse FFT could be used to retrieve the SOI in time domain.

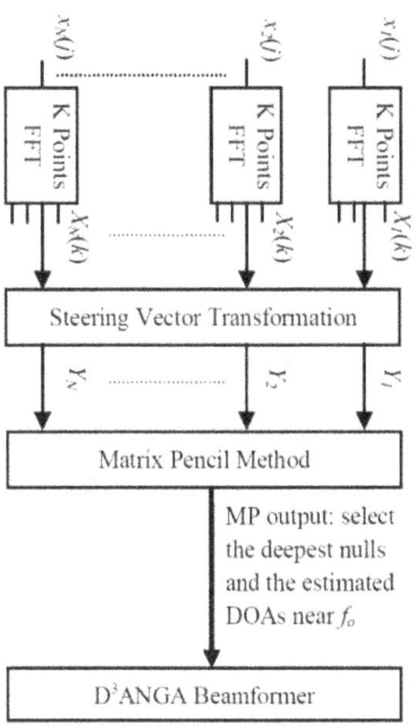

Fig. 1. The proposed WD^{3}GA beamforming structure.

7. Numerical Examples

Consider a uniform linear array of 6 elements. The received wideband signals are decomposed into their narrowband components at 6 frequencies within the design frequency band [3 GHz – 4 GHz], these frequencies are: 3, 3.2, 3.4, 3.6, 3.8, and 4 GHz. The inter-element spacing used equals half wavelength corresponding to the maximum frequency within the design frequency band in order to avoid spatial aliasing [5], [6]. Consider one SOI and one jammer, with arrival angles of $\phi = 120°$ and $\phi = 100°$ respectively. All SOI frequency components' amplitudes are equal to 1 V/m. The signal to noise ratio is 30 dB and the reference frequency is chosen to be 4 GHz. This configuration and values are used in the following two examples; the difference is in the jammer strength.

7.1 Example 1: Constant Jammers Frequency Components Magnitudes

In the 1st example, all jammer frequency components' magnitudes are 40 dB over the corresponding SOI frequency components. Fig. 2 and Fig. 3 show the estimated

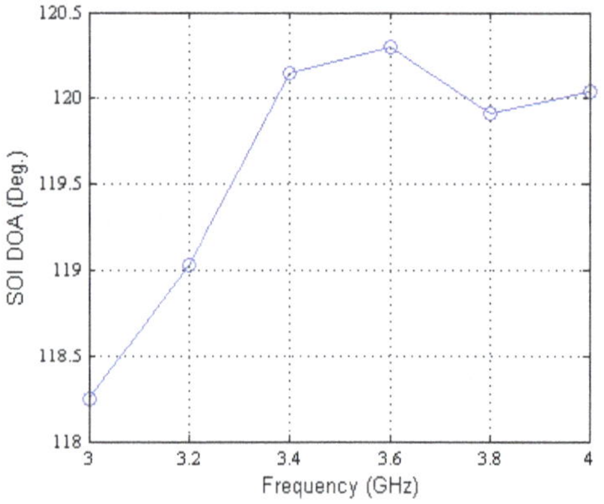

Fig. 2. SOI estimated DOA.

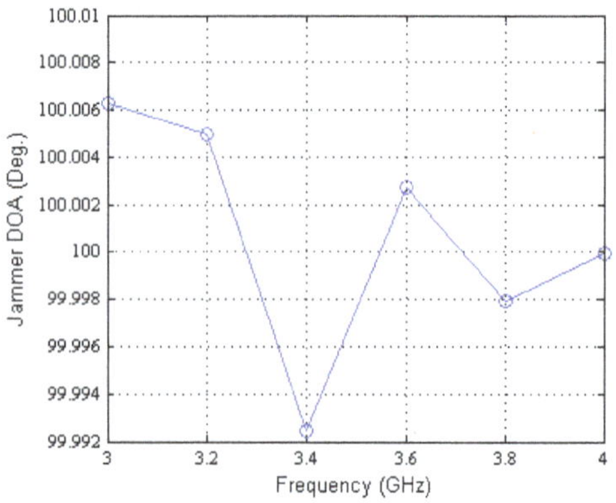

Fig. 3. Jammer estimated DOA.

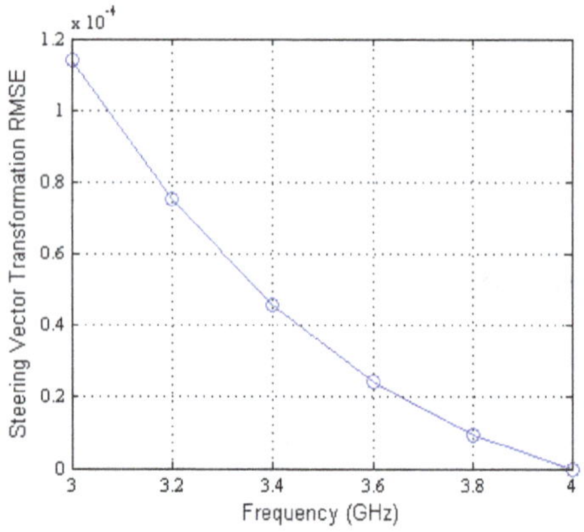

Fig. 4. Steering vector transformation RMSE.

DOA for both the SOI and the jammer signal, respectively, corresponding to each frequency sub-band. Obviously, the estimated DOAs become more accurate as the frequency is approaching the reference value, 4 GHz. This could be explained by noticing the decreasing steering vector transformation RMSE values shown in Fig. 4.

Fig. 5 shows six generated beam patterns corresponding to the six sub-bands. It is clear that the array beam patterns are maintained at 0 dB in the direction of the SOI and the nulls are placed correctly in the direction of the jammer. Fig. 6 shows the detailed 'zoomed' nulls' patterns. Since nulls' patterns are very sensitive to the errors in the estimated jammer DOA, the nulls' pattern values in the jammer direction become deeper as the frequency becomes closer to the reference frequency.

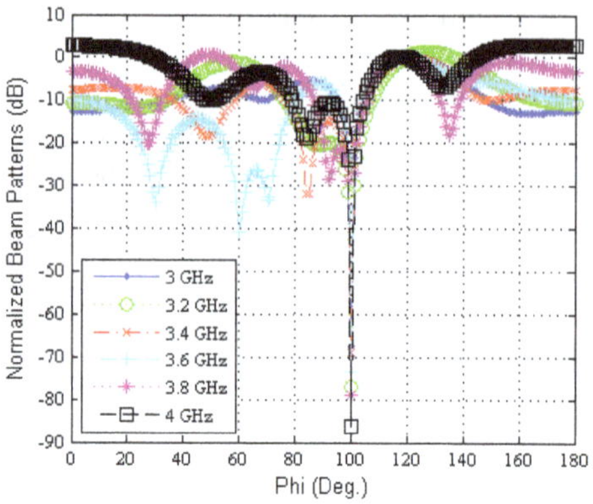

Fig. 5. Normalized array beam patterns.

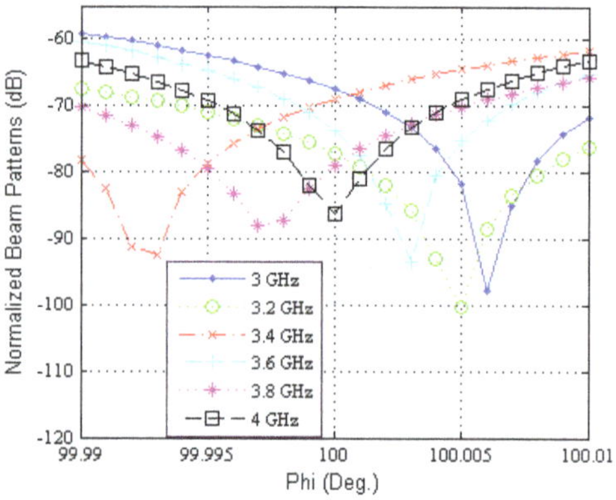

Fig. 6. Normalized detailed nulls' patterns.

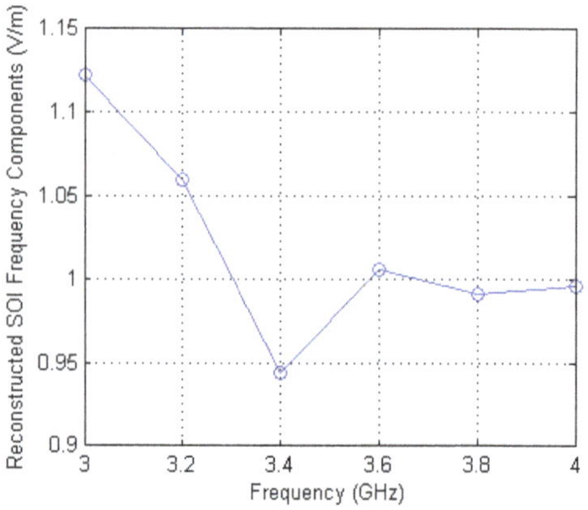

Fig. 7. Reconstructed SOI frequency components (V/m).

Fig. 7 shows the estimated magnitudes of the SOI frequency components using each corresponding beam pattern. Note that the estimation error is large at the lower frequencies and decreases as the reference frequency is approached. To improve the SOI frequency components' reconstruction accuracy, a hybrid reconstruction beam pattern is used. The hybrid pattern combines the most accurate estimated DOAs, i.e. DOAs estimated at the reference frequency, and the deepest nulls corresponding to the strongest estimated interferer frequency component since all the jammer frequency components magnitudes are equal, the already formed beam pattern corresponding to the highest frequency, f_o, is used as a single reconstruction beam pattern by which all the SOI frequency components' magnitudes are accurately estimated as shown in Fig. 8.

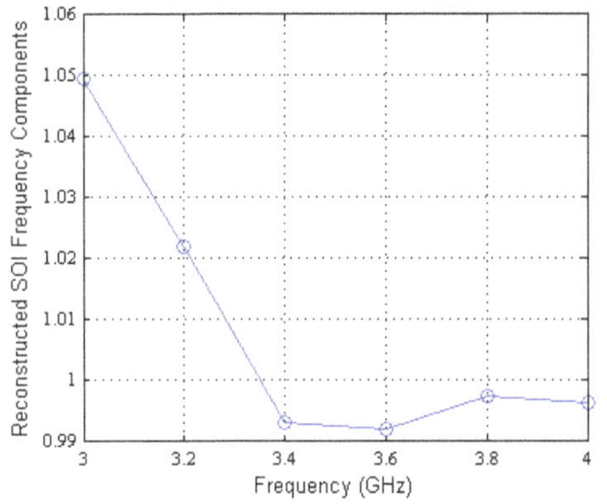

Fig. 8. Reconstructed SOI frequency components (V/m).

7.2 Example 2: Varying Jammers' Frequency Components Magnitudes

In the 2nd example, all jammer frequency components' magnitudes are 60 dB over the corresponding SOI frequency components except the last two jammer frequency components, i.e. at 3.8 GHz and 4 GHz, which are only 40 dB over the corresponding SOI frequency components' strengths. In this case, the beam pattern corresponding to the highest frequency sub-band has a null in the jammer direction but its depth is not sufficient to cancel the high jammer frequency components' strengths at the other sub-bands. Fig. 9 and Fig. 10 show the beam pattern corresponding to the highest frequency sub-band and the corresponding estimated SOI frequency components respectively. In Fig. 10, it is obvious that the estimations at the last two frequencies are acceptable but the estimations at the other frequencies, whereby the jammer frequency components' strengths are high, are not acceptable. These bad estimations could be improved by using a hybrid reconstruction beam pattern which combines the accurate DOA estimations at the reference frequency and the null depth corresponding to the strongest estimated jammer frequency component. The hybrid reconstruction pattern is shown in Fig. 11 and the corresponding estimated SOI frequency components are shown in Fig. 12. The improvement in the SOI frequency components estimations is clear as the null shown in Fig. 11 is deep enough to eliminate the strong jammer frequency components efficiently.

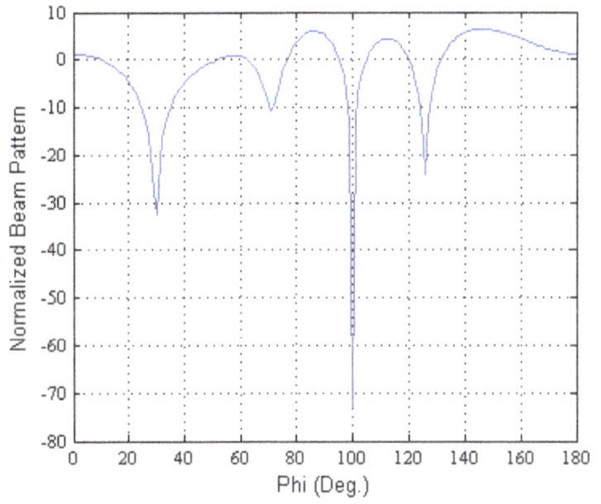

Fig. 9. Normalized beam pattern.

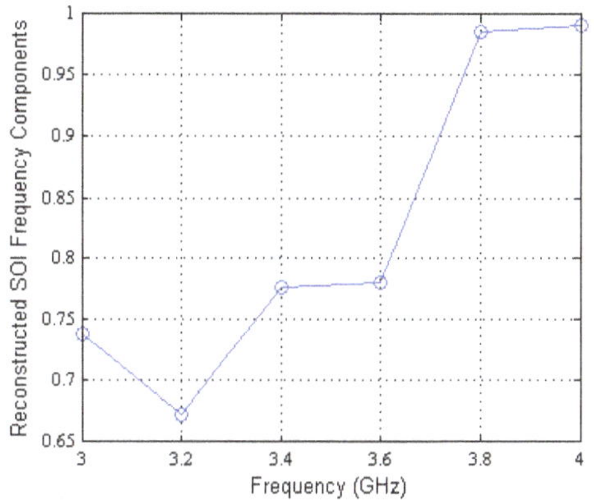

Fig. 10. Reconstructed SOI frequency components (V/m).

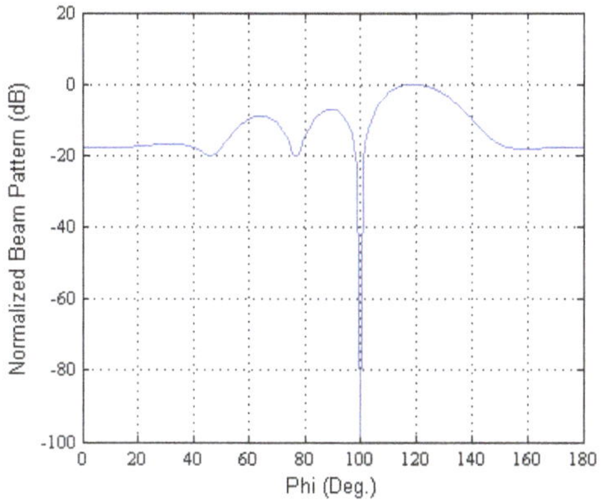

Fig. 11. Hybrid reconstruction beam pattern.

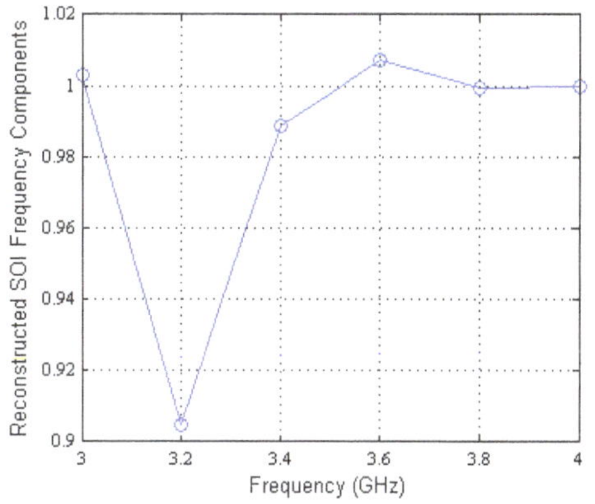

Fig. 12. Reconstructed SOI frequency components (V/m).

8. Conclusion

This paper presents a new wideband direct data domain genetic algorithm (WD^{3}GA) beamforming. The proposed beamformer is based on decomposing the received wideband signal into its narrow sub-bands using FFT. The beamformer is using a single hybrid reconstruction beam pattern which combines both sufficient deep nulls and accurate DOA estimations.

References

[1] ELKAMCHOUCHI, H., MOHAMED, D., ALI, W. D^{3}LS STAP approach on wideband signals using uniformly spaced real

elements. *International Journal of Computer Applications,* 2011, vol. 22, no. 4, p. 42–47. DOI: 10.5120/2568-3530

[2] SVENDSEN, A., GUPTA, I. The effect of mutual coupling on the nulling performance of adaptive antennas. *IEEE Antennas and Propagation Magazine,* June 2012, vol. 54, no. 3, p. 17–38. DOI: 10.1109/MAP.2012.6293947

[3] ADVE, R., SARKAR, T. Compensation for the effects of mutual coupling on direct data domain adaptive algorithms. *IEEE Transactions on Antennas and Propagation*, January 2000, vol. 48, no. 1, p. 86–94. DOI: 10.1109/8.827389

[4] SARKAR, T., WICKS, M., SALAZAR-PALMA, M., BONNEAU, R. *Smart Antennas*. Wiley-IEEE Press, 2003.

[5] SHABAN, M., KISHK, S. Steering vector transformation technique for the design of wideband beamformer. In *The Proceeding of 27th National Radio Science Conference*. Menouf (Egypt), 2010.

[6] LIU, W., WEISS, S. *Wideband Beamforming Concepts and Techniques.* Wiley (Wiley Series on Wireless Communication and Mobile Computing), 2010. 302 p. ISBN: 978-0-470-71392-1

[7] ELLATIF, W. *Smart Antennas: Space Time Adaptive Processing Based on Direct Data Domain Least Squares Using Real Elements. Doctoral Dissertation.* Electrical Engineering, Alexandria University, 2011.

[8] SARKAR, T., PEREIRA, O. Using the matrix pencil method to estimate the parameters of a sum of complex exponentials. *IEEE Antennas and Propagation Magazine*, May 1995, vol. 37, no. 1, p. 48–55. DOI: 10.1109/74.370583

[9] ELKAMCHOUCHI, H., HASSAN, M. Space time adaptive processing using real array elements based on direct data domain adaptive nulls genetic algorithm beam forming. In *Proceedings of the International Conference on Electronics and Communication System.* Coimbatore (India), 2014, vol. 2, p. 183–187. DOI: 10.1109/ECS.2014.6892563

[10] HASSAN, M. A *Proposed Fault Identification Scheme in Systems Using Soft Computing Methodologies. M.Sc. Dissertation.* Electrical Engineering, Alexandria University, 2008.

[11] *Genetic Algorithm and Direct Search Tool Box User's Guide*. The MathWorks, Inc., 2006.

[12] LIU, W., WEISS, S., HANZO, L. A generalized side lobe canceller employing two-dimensional frequency invariant filters. *IEEE Transactions on Antennas and Propagation,* 2005, vol. 53, no. 7, p. 2339–2343. DOI: 10.1109/TAP.2005.850759

[13] LIU, W., WEISS, S. A new class of broad arrays with frequency invariant beam patterns. In *Proceeding of International Conference on Acoustics, Speech, and Signal Processing ICASSP 2004.* Montreal (Canada), 2004, vol. 2, p. 185–188. DOI: 10.1109/ICASSP.2004.1326225

[14] MOGHADDAM, P., AMINDAVAR, H. Direction of arrival estimation: a new approach. In *Proceeding of Signal Processing Conference.* Nordic (Sweden), 2000.

[15] MONTHIPPA, U., BIALKOWSKI, M. E. A wideband smart antenna employing spatial signal processing. *Journal of Telecommunication and Information Technology*, 2007, no. 1, p. 13–17.

[16] REN, Y., HE, H., ZHANG, Y., ZHANG, K. An amplitude-only direct data domain least square algorithm of wideband signals based on the uniform circular array. In *Proceeding of the 5th International Conference on Wireless Communications, Networking and Mobile Computing (WiCom 2009).* Beijing (China), 2009, p. 1–4. DOI: 10.1109/WICOM.2009.5302753

[17] MANI, V.V.; BOSE, R. Genetic algorithm based smart antenna design for UWB beamforming. In *IEEE International Conference on Ultra-Wideband ICUWB 2007.* Singapore, 2007, p. 442–446. DOI: 10.1109/ICUWB.2007.4380985

About the Authors ...

Hassan ELKAMCHOUCHI was born in Egypt, 1943. He received his Ph.D. degree in Communication Engineering in 1972 from Alexandria University. He is now a professor for Alexandria University, Egypt. His research interests include adaptive antennas, optimization algorithms, radars, electromagnetic theory, cryptography and digital communication.

Mohamed HASSAN was born in Egypt. He received his M.Sc. from Alexandria University in 2008. His research interests include adaptive antennas, optimization algorithms, radars, electromagnetic theory and modeling.

9

A 1.2 V and 69 mW 60 GHz Multi-channel Tunable CMOS Receiver Design

Ahmet ONCU

Dept. of Electrical and Electronics Engineering, Bogazici University, Istanbul, 34342, Turkey

ahmet.oncu@boun.edu.tr

Abstract. *A multi-channel receiver operating between 56 GHz and 70 GHz for coverage of different 60 GHz bands worldwide is implemented with a 90 nm Complementary Metal-Oxide Semiconductor (CMOS) process. The receiver containing an LNA, a frequency down-conversion mixer and a variable gain amplifier incorporating a band-pass filter is designed and implemented. This integrated receiver is tested at four channels of center frequencies 58.3 GHz, 60.5 GHz, 62.6 GHz and 64.8 GHz, employing a frequency plan of an 8 GHz-intermediate frequency (IF). The achieved conversion gain by coarse gain control is between 4.8 dB and 54.9 dB. The millimeter-wave receiver circuit is biased with a 1.2 V supply voltage. The measured power consumption is 69 mW.*

Keywords

60 GHz, CMOS, integrated circuit, receiver design, low-power

1. Introduction

Research on the Complementary Metal-Oxide Semiconductor (CMOS) millimeter-wave transceiver operating at 60 GHz has been driven by the availability of the unlicensed bands in many countries, including Europe (57 ÷ 66 GHz), the United States (57 ÷ 64 GHz) and Japan (59 ÷ 66 GHz) as shown in Fig. 1. The available 9 GHz wide millimeter-wave band is also divided into approximately 2 GHz wide four sub-channels. This has resulted in many enabling design blocks and various modeling techniques reported in the literature [1–5]. Potential applications include low-power short-distance consumer applications such as wireless High-Definition Multimedia Interface (HDMI) for high-definition television (HDTV) video streaming and high data-rate wireless personal area networks. The distance limit is due to the 60 GHz electromagnetic waves being attenuated more as a result of oxygen absorption than at other frequencies. However, taking advantage of the wider multi-gigahertz bandwidth, short-distance applications can allow high data rates while allowing more frequency reuse in a limited area with minimal interference. A consequence is that the system offers higher security since signals cannot travel far beyond the intended recipients. The motivation of this work is to develop further such a system using the CMOS process technology. Traditional developments of the 60 GHz system have been largely confined to process technologies such as GaAs or SiGe. With price-sensitive services, lower production cost and possible integrations consumer electronics driving developments, CMOS technology offers the advantages of an accessible foundry with the digital baseband. The present state-of-the art CMOS process nodes at sub-100nm demonstrate device f_t that exceeds 400 GHz [6], thereby providing reasonable gain and other design margins for 60 GHz systems.

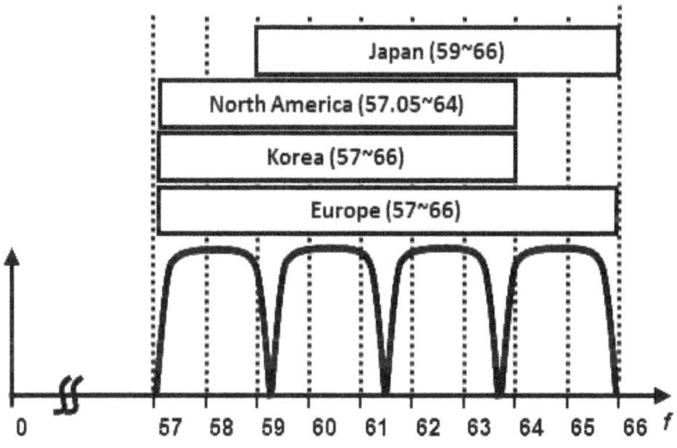

Fig. 1. Worldwide 60 GHz regulations with sub-channels.

A critical building block in the 60 GHz system is the millimeter-wave receiver, which includes the low-noise amplifier (LNA), down conversion mixer, band-pass filter and variable gain amplifier. In this work, a wideband CMOS receiver circuit that covers all 60 GHz bands and selects the sub-channels is proposed. In the following sections the design of the proposed receiver circuit and the performance of the implemented receiver are presented.

2. Design of Wideband Receiver

A block diagram of the proposed multi-channel tunable receiver based on the enhanced wide-band LNA is shown in Fig. 2. The receiver consists of an LNA, down-conversion mixer, an intermediate frequency (IF) variable gain amplifier (VGA) integrating a band-pass filter (BPF)

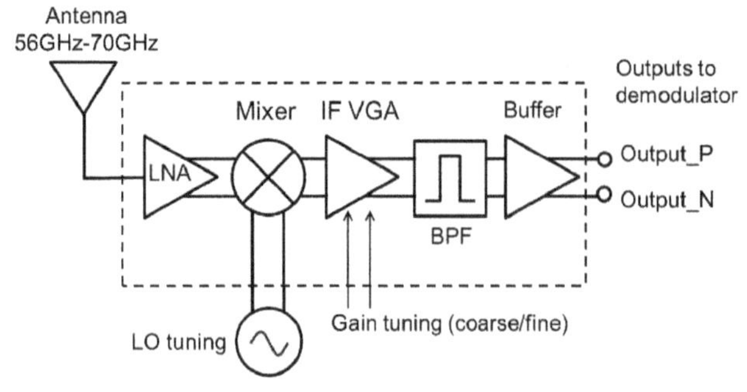

Fig. 2. Proposed receiver architecture.

and an output buffer. To test the performance of the proposed receiver chain, LO signal is externally applied, but in a practical application local oscillator will be implemented on the chip. A wideband LNA with active balun in the first stage to cover the all 60 GHz bands as reported in [9] is designed for this multi-channel receiver. Its schematic is shown in Fig. 3.

The LNA contains an internal balun. It has a single ended input and two outputs with 180° phase difference. The designed wideband LNA has 20 dB gain and 14 GHz bandwidth from 56 GHz to 70 GHz. It covers all 60 GHz bands sub-channels. The noise figure of the LNA is 6.8 dB at 60 GHz.

The main purpose of this receiver is to receive all four sub-channels of the license free 60 GHz bands from 57 GHz to 66 GHz. The LNA can amplify the signals from 56 GHz to 70 GHz frequency range. Therefore it is suitable for our application. The intermediate frequency is chosen as 8 GHz to be able to select the sub-channels and not to have an image rejection filter to have less complex receiver architecture. Since the frequency gap between the desired signal and the image will be 16 GHz, the image signal can be filtered out by the receiver front-end. For lower frequency bands, the image frequency will be far from the receiver's front-end coverage frequency band. When the desired frequency goes higher, image frequency goes closer to the coverage band.

When the receiver receives the 66 GHz signal, the highest frequency of the 60 GHz band sub-channel, the image frequency will be at 49 GHz in the worst case. It is outside of the LNA bandwidth. Therefore, the image frequency will be filtered by LNA. In practical wireless communication, an antenna will be connected to receiver input and it will also contribute to filter out the image signal.

Figure 4 shows the circuit of the implemented mixer. A double-balanced design is realized by the four transistors M10–M13 with each differential set of RF and LO inputs applied 180° out of phase. A modification made to this circuit from the standard Gilbert-cell configuration is the input of the *RF* signal directly into the source of the MOSFETs, instead of first passing through the gates of common source amplifiers. This current mode interface between the LNA and the mixer alleviates the low headroom limitation of the advanced CMOS process to improve the linearity and high-speed operation with a limited supply voltage

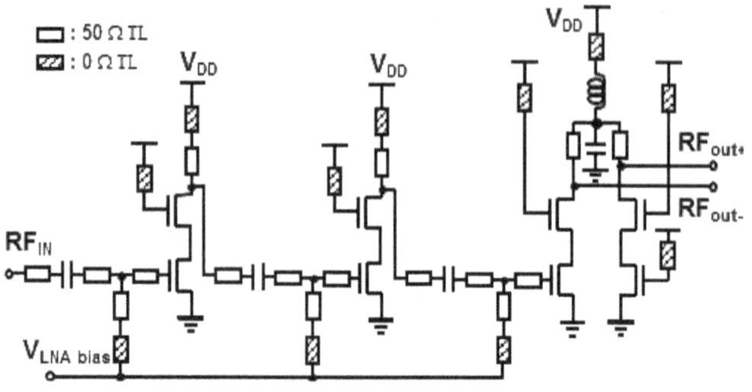

Fig. 3. Schematic of the LNA.

magnitude. A tail current control is then implemented by M9 to provide the appropriate DC current for the differential pairs. To avoid the noise contribution from M9 and its current source, a large bypass capacitance Cbypass is added [16]. Without the capacitor the noise from the M9 and the noise coming from the bias V_{b1} will inject to the mixer tail current in common mode. Due to the non-idealities in pair transistors, M10-M13, the noise will appear at the mixer output that will reduce receiver the sensitivity. A center tapped inductor shown by L5 and L6 are added to resonate out the parasitic capacitance at the source terminals of M10–M13. Thus, a high input impedance at the RF port from 56 GHz to 70 GHz can be achieved. These inductors can also reduce the signal loss caused by the parasitic capacitances. Inductors L1–L4 are chosen instead of resistors for gain and linearity purposes since an increase in resistance results in a decrease in the voltage headroom at the output node. Thus, there is a trade-off between conversion gain and linearity. In practice, L1–L4 and L5-L6 are implemented as single center-tap inductors in order to save chip area. The inductances L1–L4 and the parasitic capacitances of M10–M13 comprises two resonating tanks. The bandwidth of the tanks is equal to the bandwidth of the mixer. The bandwidth of the tank is narrower than the desired one. Therefore, in this design two resistors R1 and R2 are added to ensure 2 GHz bandwidth at 8 GHz IF center frequency.

In comparison with a conventional mixer, the proposed down-conversion mixer aims to achieve lower noise

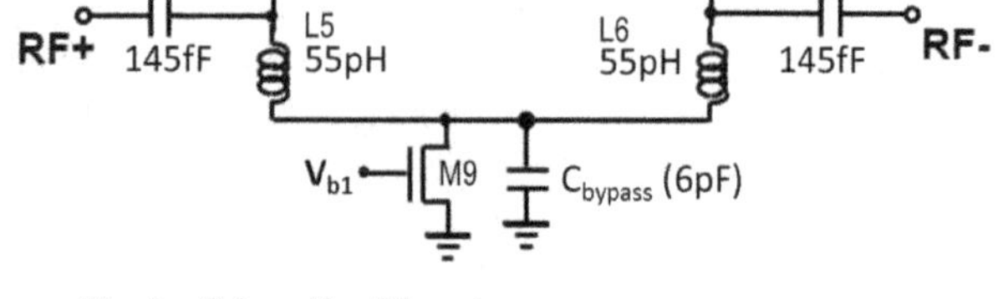

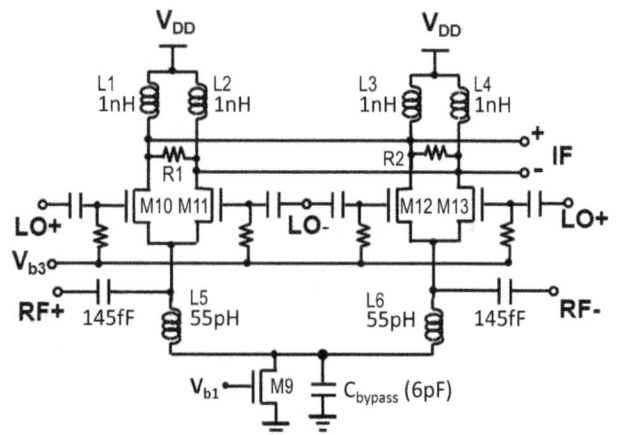

Fig. 4. Schematic of the mixer.

because of the absence of the noise generated from the transconductance stage in a conventional mixer. In addition, a DC current through this transconductance stage is not required. Thus the DC current can be reduced to improve the noise performance further. These considerations were necessary to reduce noise for stringent wideband and high frequency operations.

Figure 5 shows the circuit of the VGA and the BPF implemented. The VGA topology is a two-stage differential cascode. Due to the transconductance characteristic of the MOSFET, coarse tuning of the gain can be made in several discrete steps by selecting values of V_{b3} thereby directly setting M16, M17, M20 and M21 bias points. If desired, fine tuning can be made through V_{b2} at the gate of M14 and M15 to control the DC current through the differential cascodes.

The band-pass filter is implemented by two RLC circuits, each in parallel with the output of an amplifier stage. Fixed inductor sizes are used with capacitances to center the BPF at IF. The capacitances are tunable by varactors designed with a tuning range of 10%. Using this method, the 3-dB bandwidth can be controlled and the flatness of the BPF response can be adjusted with the proper selection of the value of R3 and R4 since the fractional bandwidth F_b of each parallel RLC can be characterized by (1). Note that the quality factor of the resonant tank is the inverse of F_b.

$$F_b = \frac{1}{R}\sqrt{\frac{L}{C}}. \quad (1)$$

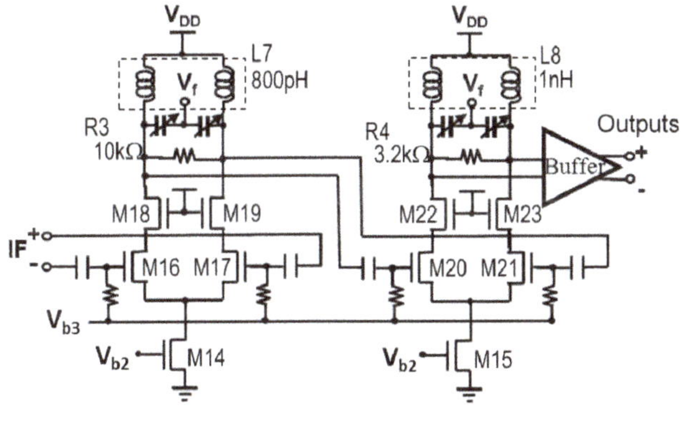

Fig. 5. Schematic of VGA, BPF and buffer block.

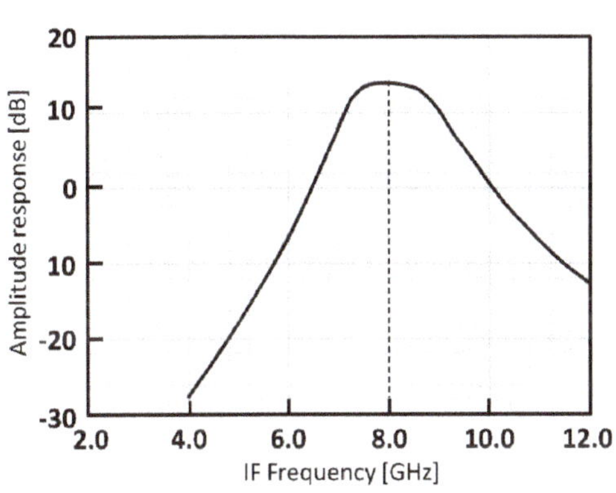

Fig. 6. Simulated results of IF filtering by BPF.

Equation (1) assumes a high-Q inductor. With a realistic CMOS inductor model [17], simulations were performed. Figure 6 shows the simulated results of the IF filtering by the BPF. The BPF is centered at 8 GHz with a bandwidth of 1.8 GHz. The simulated channel rejection at this bandwidth is 14.9 dB. Tuning is available with a range of 1.2 GHz. For measurement considerations, the unity gain open-drain buffers are used to drive the 50 Ω measurement system. It can drive the 50 Ω instruments and obtain the desired output signal. The buffer consumes 12.3 mW.

3. Experimental Result

Figure 7 shows the micrograph of the complete wideband receiver chip design. The LNA, mixer, VGA and band-pass filter are indicated on the micrograph. The receiver was fabricated using the 90 nm 1P8M CMOS process.

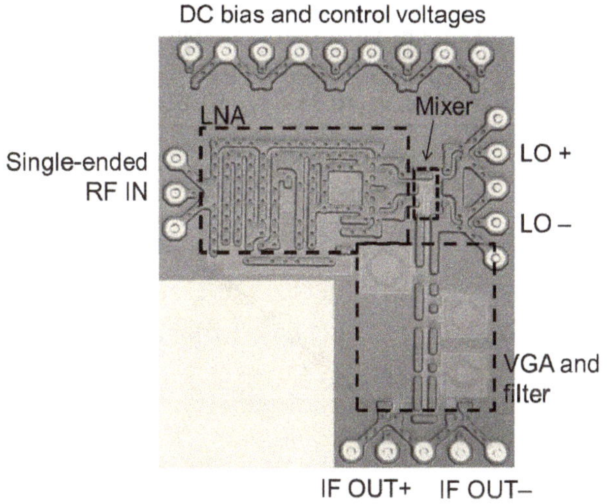

Fig. 7. Micrograph of wideband receiver chip.

The fabricated chip was tested by using a probe station. To test the performance of the fabricated receiver circuit the LO signal was applied from an external single ended millimeter-wave signal source by using an on chip GSG probe. But in practice the differential ended LO signal can be applied on the chip. It would reduce even order nonlinearities. In this measurement, the LO signal was injected at P_{LO} = -1 dBm at four frequencies corresponding to four different channels to evaluate the multi-channel characteristics of the receiver. The IF output was maintained at 8 GHz. Setting the gain control voltages to V_{b2} = 0.65 V and V_{b3} = 0.35 V, the IF output power response as a function of the RF input frequency is shown in Fig. 8. In this figure, a gain of 23 dB is observed with an input-referred 1 dB-compression point at -29 dBm. To measure the nonlinear characteristic of the receiver, two tone millimeter wave signal was applied from two external sources trough the RF input of the receiver by using a GSG probe. The measured fundamental and third order responses are shown in Fig. 9. The input referred third order interception point was measured to be a -20.5 dBm. The measured noise figure of the whole receiver is 8.1 dB. These performances can be acceptable for indoor short-range communications operating at low-power levels.

In such a scenario, adjacent channel interferences at 60 GHz are limited due to attenuation and the absence of large transmitter signals on the same chip, even though the receiver operates wideband. Depending on the linearity requirements of the modulation scheme, the trade-off between current drain for lower power consumption and linearity may be further adjusted. Figure 10(a) shows the results of the coarse tuning of the VGA gain.

As shown in Fig. 10, the coarse gain tuning of the VGA block is set by V_{b3} between 0.2 V and 0.5 V, corresponding to a minimum gain of 4.8 dB and a maximum of 54.9 dB. The coarse tuning demonstrates the limit of the gain obtainable by further fine tunings using V_{b2}. Results shown in Fig. 10(a) was obtained when V_{b2} is set to 0.65 V. It is noted that the gain flatness deteriorates at higher gain settings with a span of several gigahertz, resulting in changing 3-dB bandwidths. This is likely a result of the VGA input impedance changing due to bias changes, causing the mixer-VGA wideband inter-stage matching network. For a system with narrowband channels, it is possible to adjust the band-pass filter bandwidth to achieve better gain flatness over different gain settings. Alternatively, in our envisioned pulse communication system that typically employs On-Off-Keying (OOK) or Amplitude-Shift Keying (ASK), such gain control characteristics may be sufficient.

Measurement of the receiver channel selection by LO tuning is shown in Fig. 10(b). The LO frequency is selected at four frequencies of 50.3 GHz, 52.5 GHz, 54.6 GHz and 56.8 GHz. With the IF at 8 GHz, the corresponding RF-input frequencies for channels 1 to 4 are 58.3 GHz, 60.5 GHz, 62.6 GHz and 64.8 GHz, respectively. The LO power is set to -1 dBm. It is noted that the frequency response at the third channel (CH3) is about 2 dB lower than the other channels. It is believed that this performance is caused by the limited gain flatness of the LNA. Its gain can be compensated by VGA gain tuning. This proposed receiver reports a considerable gain-bandwidth performance. The proposed wide-band single receiver can select all four 60 GHz band sub-channels and its gain can be controlled to improve the communication performance. In Tab. 1, recently reported wireless 60-GHz

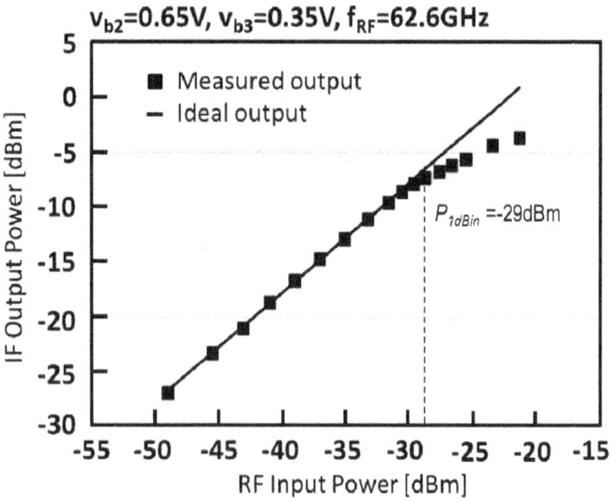

Fig. 8. Conversion gain characteristics of receiver.

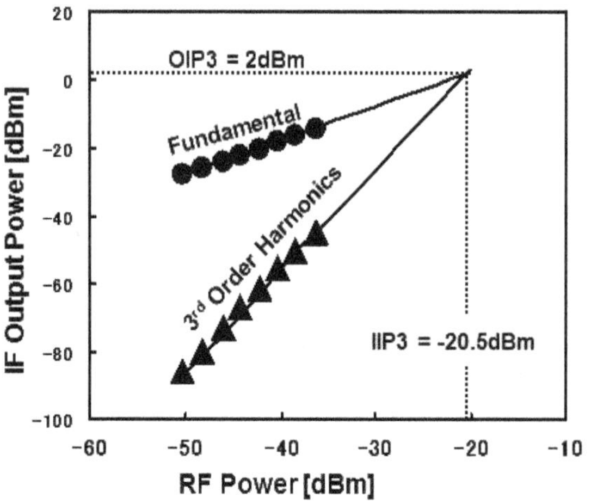

Fig. 9. Measured IIP3 of the receiver.

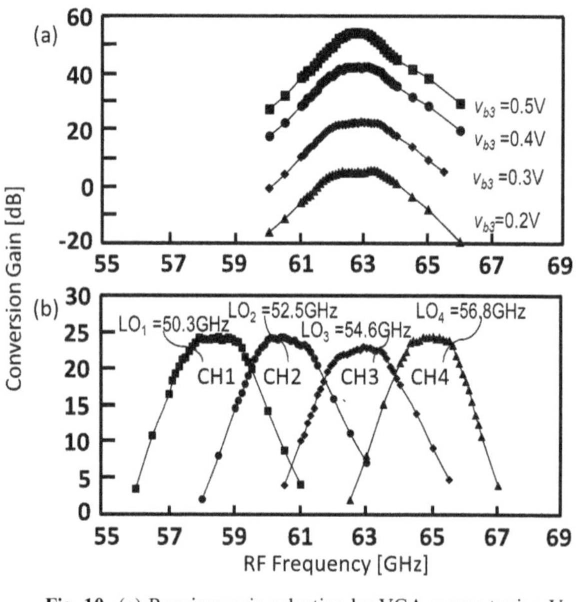

Fig. 10. (a) Receiver gain selection by VGA coarse tuning V_{b3} (V_{b2} set at 0.65 V). (b) Receiver channel selection by LO tuning (PLO = -1 dBm).

	Tech.	Freq. [GHz]	Gain [dB]	Power [mW]
This work	90nm CMOS	56.0-70.0	4.8-54.9	69.0
[18]	65nm CMOS	57.5-66.5	10.0-19.0	74.0
[19]	65nm CMOS	54.0-61.0	4.7-17.3	81.5
[20]	65nm CMOS	55.0-68.0	14.0-35.5	75.0
[21]	40nm CMOS	58.0	20.0	41.0
[22]	90nm CMOS	57-61.0	19.8-22.0	36.0
[23]	65nm CMOS	55.0-65.0	14.7	151.0
[24]	45nm CMOS	56.0-67.0	26.0	11.0-21.0
[25]	130nm SiGe	57.0-65.0	14.0	18.0

Tab. 1. Comparison of recent works.

band millimeter-wave receivers are compared. This proposed receiver using the enhanced LNA has the largest gain-bandwidth among them.

4. Conclusion

A multi-channel receiver operating between 56 GHz and 70 GHz for coverage of all four 60 GHz band sub-channels worldwide has been implemented with a 90 nm CMOS process. The receiver comprises of the millimeter-wave LNA, frequency mixer, VGA and bandpass filter. This enables the receiver to operate at wideband with reference measurements showing a 23 dB gain through LO injections of -1 dBm at each of the four channels tested, corresponding to receiver RF of 58.3 GHz, 60.5 GHz, 62.6 GHz and 64.8 GHz. Tuning of the VGA allows a range of gain to be selected with measured results between 4.8 dB and 54.9 dB. It requires a 1.2 V supply voltage and consumes 69 mW power. The proposed millimeter-wave multi-channel CMOS receiver can be used for low-voltage, low-power consuming and high-speed wireless communication applications.

Acknowledgements

The chip in this work was realized in 90nm CMOS process through Silicon Library Inc. Author thanks Prof. Minoru Fujishima for accessing millimeter-wave measurement facilities, Dr. Lai Chee Hong and Mr. Pong for their valuable discussion on low power-LNA and layout designs.

References

[1] DOAN, C. H., EMAMI, S., NIKNEJAD, A. M., BRODERSEN, R. W. Design of CMOS for 60GHz applications. In *IEEE International Solid-State Circuits Conference ISSCC Digest of Technical Papers*. 2004, p. 440–441. DOI: 10.1109/ISSCC.2004.1332783

[2] REYNOLDS, S. K., FLOYD, B. A., PFEIFFER, U. R., BEUKEMA, T., GRZYB, J., HAYMES, C., GAUCHER, B., SOYUER, M. A silicon 60-GHz receiver and transmitter chipset for broadband communications. *IEEE Journal on Solid-State Circuits*, 2006, vol. 41, no. 12, p. 2820–2831. DOI: 10.1109/JSSC.2006.884820

[3] LAI, I. C. H., KAMBAYASHI, Y., FUJISHIMA, M. 60-GHz CMOS down-conversion mixer with slow-wave matching transmission lines. In *Proceedings of the IEEE Asian Solid-State Circuits Conference ASSCC 2006*. Hangzhou (China), 2006, p. 195–198. DOI: 10.1109/ASSCC.2006.357884

[4] RAZAVI, B. A millimeter-wave CMOS heterodyne receiver with on-chip LO and divider. In *IEEE International Solid-State Circuits Conference ISSCC Digest of Technical Papers*, 2007, p. 188–189. DOI: 10.1109/ISSCC.2007.373357

[5] EMAMI, S., WISER, R. F., ALI, E., FORBES, M. G., GORDON, M. Q., GUAN, X., LO, S., MCELWEE, P. T., PARKER, J., TANI, J. R., GILBERT, J. M., DOAN, C. H. A 60GHz CMOS phased-array transceiver pair for multi-Gb/s wireless communications. In *IEEE ISSCC Digest of Technical Papers*, 2004, p. 164–166.

[6] YASUTAKE, N., OHUCHI, K., FUJIWARA, M., ADACHI, K., HOKAZONO, A., KOJIMA, K., AOKI, N., SUTO, H., WATANABE, T., MOROOKA, T., MIZUNO, H., MAGOSHI, S., SHIMIZU, T., MORI, S., OGUMA, H., SASAKI, T., OHMURA, M., MIYANO, K., YAMADA, H., TOMITA, H., MATSUSHITA, D., MURAOKA, K., INABA, S., TAKAYANAGI, M., ISHIMARU, K., ISHIUCHI, H. A hp22 nm node low operating power (LOP) technology with sub-10 nm gate length planar bulk CMOS devices. In *VLSI Technology Symposium Digest of Technical Papers*, 2004, p. 84–85. DOI: 10.1109/VLSIT.2004.1345407

[7] VARONEN, M., KARKKAINEN, M., KANTANEN, M., HALONEN, K. Millimeter-wave integrated circuits in 65-nm CMOS. *IEEE Journal of Solid-State Circuits*, 2008, vol. 43, no. 9, p. 1991–2002. DOI: 10.1109/JSSC.2008.2001902

[8] COSTANTINI, A., LAWRENCE, B., MAHON, S., HARVEY, J., MCCULLOCH, G., BESSEMOULIN, A. Broadband active and passive balun circuits: Functional blocks for modern millimeter-wave radio architectures. In *Proceedings of the 1st European Microwave Integrated Circuits Conference*. Manchester (UK), 2006, p. 421–424. DOI: 10.1109/EMICC.2006.282672

[9] NATSUKARI, Y., FUJISHIMA, M. 36 mW 63 GHz CMOS differential low-noise amplifier with 14 GHz bandwidth. In *CORD Conference Proceedings*. June 2009, p. 252–253.

[10] RAZAVI, B. A 60GHz direct-conversion CMOS receiver. In *IEEE International Solid-State Circuits Conference ISSCC Digest of Technical Papers*. San Francisco (USA), 2005, p. 400–401. DOI: 10.1109/ISSCC.2005.1494038

[11] MARCU, C., CHOWDHURY, D., THAKKAR, C., KONG, L. K., TABESH, M., PARK, J. D., WANG, Y., AFSHAR, B., GUPTA, A., ARBABIAN, A., GAMBINI, S., ZAMANI, R., NIKNEJAD, A. M., ALON, E. A 90nm CMOS low-power 60GHz transceiver with integrated baseband circuitry. In *IEEE International Solid-State Circuits Conference ISSCC Digest of Technical Papers*. 2009, p. 314–315.

[12] PINEL, S., SARKAR, S., SEN, P., PERUMANA, B., YEH, D., DAWN, D., LASKAR, J. A 90nm CMOS 60GHz radio. In *IEEE International Solid-State Circuits Conference ISSCC Digest of Technical Papers*. San Francisco (USA), 2008, p. 130 131. DOI: 10.1109/ISSCC.2008.4523091

[13] LEE, J., HUANG, Y., CHEN, Y., LU, H., CHANG, C. A low-power fully integrated 60GHz transceiver system with OOK modulation and on-board antenna assembly. In *IEEE International Solid-State Circuits Conference ISSCC Digest of Technical Papers*. San Francisco (USA), 2009, p. 316 –318. DOI: 10.1109/ISSCC.2009.4977435

[14] WEYERS, C., MAYR, P., KUNZE, J. W., LANGMANN, U. A 22.3dB voltage gain 6.1dB NF 60GHz LNA in 65nm CMOS with differential output. In *International Solid-State Circuits Conference ISSCC Digest of Technical Papers*. San Francisco (USA), 2008, p. 192 -606. DOI: 10.1109/ISSCC.2008.4523122

[15] CHAO-SHIUN WANG, J. W. H. A 0.13μm CMOS fully differential receiver with on-chip baluns for 60GHz broadband wireless communications. In *Proc. of Custom Integrated Circuits Conference CICC 2008*. San Jose (USA), 2008, p. 479–482. DOI: 10.1109/CICC.2008.4672125

[16] CHEN, P.-H., CHEN, M.-C., KO, C.-L., WU, C.-Y. An integrated CMOS front-end receiver with a frequency tripler for V-band applications. *IEICE Transactions on Electronics*, 2010, vol. E93–C, no. 6, p. 877–883. DOI: 10.1587/transele.E93.C.877

[17] BLASCHKE, V., VICTORY, J. A scalable model methodology for octagonal differential and single-ended inductors. In *Proceedings*

of Custom Integrated Circuits Conference CICC 2006. San Jose (USA), 2006, p. 717 –720. DOI: 10.1109/CICC.2006.320897

[18] SILIGARIS, A., RICHARD, O., MARTINEAU, B., MOUNET, C., CHAIX, F., FERRAGUT, R., DEHOS,C., LANTERI, J., DUSSOPT, L., YAMAMOTO, S. D., PILARD, R., BUSSON, P., CATHELIN, A., BELOT, D., VINCENT, P. A 65nm CMOS fully integrated transceiver module for 60GHz wireless HD applications. In *IEEE International Solid-State Circuits Conference ISSCC Digest of Technical Papers 2011.* San Francisco (USA), 2011, p. 162–164. DOI: 10.1109/ISSCC.2011.5746264

[19] OKADA, K., MATSUSHITA, K., BUNSEN, K., MURAKAMI, R., MUSA, A., SATO, T., ASADA, H., TAKAYAMA, N., LI, N., ITO, S., CHAIVIPAS, W., MINAMI, R., MATSUZAWA, A. A 60GHz 16QAM/8PSK/QPSK/BPSK direct-conversion transceiver for IEEE 802.15.3c. In *IEEE International Solid-State Circuits Conference ISSCC 2011 Digest of Technical Papers.* San Francisco (USA), 2011, p. 160–162. DOI: 10.1109/ISSCC.2011.5746263

[20] VECCHI, F., BOZZOLA, S., POZZONI, M., GUERMANDI, D., TEMPORITI, E., REPOSSI, M., DECANIS, U., MAZZANTI, A., SVELTO, F. A wideband mm-wave CMOS receiver for Gb/s communications employing interstage coupled resonators. In *IEEE International Solid-State Circuits Conference ISSCC 2010 Digest of Technical Papers.* San Francisco (USA), 2010, p. 220–221. DOI: 10.1109/ISSCC.2010.5433953

[21] KAWASAKI, K., AKIYAMA, Y., KOMORI, K., UNO, M., TAKEUCHI, H., ITAGAKI, T., HINO, Y., KAWASAKI, Y., ITO, K., HAJIMIRI, A. A millimeter-wave intra-connect solution. In *IEEE International Solid-State Circuits Conference ISSCC 2010 Digest of Technical Papers.* San Francisco (USA), 2010, p. 414 to 415. DOI: 10.1109/ISSCC.2010.5433831

[22] PARSA A., RAZAVI, B. A new transceiver architecture for the 60-GHz band. *IEEE Journal of Solid-State Circuits,* 2009, vol. 44, no. 3, p. 751–762. DOI: 10.1109/JSSC.2008.2012368

[23] TOMKINS, A., AROCA, R. A., YAMAMOTO, T., NICOLSON, S. T., DOI, Y., VOINIGESCU, S. P. A zero-IF 60 GHz 65 nm CMOS transceiver with direct BPSK modulation demonstrating up to 6 Gb/s data rates over a 2 m wireless link. *IEEE Journal of Solid-State Circuits,* 2009, vol. 44, no. 8, p. 2085–2099. DOI: 10.1109/JSSC.2009.2022918

[24] BORREMANS, J., RACZKOWSKI, K., WAMBACQ, P. A digitally controlled compact 57-to-66GHz front-end in 45nm digital CMOS. In *IEEE International Solid-State Circuits Conference ISSCC2009 Digest of Technical Papers.* San Francisco (USA), 2009, p. 492–493. DOI: 10.1109/ISSCC.2009.4977523

[25] NATARAJAN, A., TSAI, M.-D., FLOYD, B. 60GHz RF-path phase-shifting two-element phased-array front-end in silicon. In *Dig. Symposium on VLSI Circuit.* Kyoto (Japan), 2009, p. 250 to 251.

About the Author ...

Ahmet ONCU was born in Istanbul, Turkey in 1979. He received the B.S. degree in Physics and the B.S. degree in Electrical and Electronics Engineering from Middle East Technical University (METU), Ankara, Turkey, in 2001 and 2002, respectively. He received the M.S. degree in Microwave Engineering from the Technical University of Munich, Germany, in 2004. He received the PhD degree in Frontier Sciences from the University of Tokyo, Japan, in 2008. Currently he is an assistant professor at the Department of Electrical and Electronics Engineering, Bogazici University, Istanbul, Turkey. Dr. Oncu received FP7 Marie Curie Reintegration Grant for (project no: 268232) (UWB-IR) Study on Low-power Multi-Gbps Ultra-Wideband Impulse Radio at License-free UWB and 60GHz bands in 2010. His research interests are in designs of high-speed low-power CMOS analog and RF integrated circuits.

Enhanced Model of Nonlinear Spiral High Voltage Divider

Václav PAŇKO [1,3], Stanislav BANÁŠ [1,3], Richard BURTON [4], Karel PTÁČEK [2,3], Jan DIVÍN [1,3], Josef DOBEŠ [1]

[1]Dept. of Radio Engineering, Czech Technical University in Prague, Technická 2, 166 27 Praha 6, Czech Republic
[2]Dept. of Microelectronics, Brno University of Technology, Technicka 3058/10, 61600 Brno, Czech Republic
[3]ON Semiconductor, SCG Czech Design Center, 1. maje 2594, 75661 Roznov p. R., Czech Republic
[4]ON Semiconductor, 5005 East McDowell Road, Phoenix, AZ 85008, USA

vaclav.panko@onsemi.com, stanislav.banas@onsemi.com, richard.burton@onsemi.com, karel.ptacek@onsemi.com, jan.divin@onsemi.com, dobes@fel.cvut.cz

Abstract. *This paper deals with the enhanced accurate DC and RF model of nonlinear spiral polysilicon voltage divider. The high resistance polysilicon divider is a sensing part of the high voltage start-up MOSFET transistor that can operate up to 700 V. This paper presents the structure of a proposed model, implemented voltage, frequency and temperature dependency, and scalability. A special attention is paid to the ability of the created model to cover the mismatch and influence of a variation of process parameters on the device characteristics. Finally, the comparison of measured data vs. simulation is presented in order to confirm the model validity and a typical application is demonstrated.*

Keywords

High voltage start-up MOSFET, pinch-off, high voltage spiral divider, statistical modeling

1. Introduction

Nowadays, the power consumption is one of the most important integrated circuit parameters. High voltage power start-up MOSFET transistor described in this paper is used to minimize the power consumption [1, 2]. It is designed to provide initial current directly from the high voltage source. This MOSFET transistor charges up the regulator voltage on an external capacitor to about 14 V. The main goal is to minimize power consumption of the circuit that is directly connected to the rectified DC high voltage source. This high voltage can be up to 400 V for a 230 V AC supply and 700 V for switcher applications using power factor correction.

The HV start-up MOSFET is fabricated in an analog 1 μm CMOS technology. The simplified structure of this MOSFET is depicted in Fig. 2. The source and drain are formed from a low-doped Nwell and are contacted by N+ diffusion. The drain drift area contains a floating P doped resurf diffusion (ptop) fabricated before field oxide. The MOSFET channel is created from Pwell not isolated from the P-substrate and it is covered by polysilicon gate. This drain-gate-source structure is rotary symmetrical around vertical axis in the center of the drain. It means that the drain is created in the shape of a circle and the gate and the source in the shape of an annulus.

The drain is located in the center of the device and contains rounded bonding pad. A drain bonding wire is connected directly to this bonding pad and this is only one possible way how the drain can be connected. The oxide breakdown is much lower (about 100 V) than maximum allowed drain voltage. The drain can be biased up to 700 V and this makes integrated direct sensing of the high drain voltage impossible. Hence, the high resistance polysilicon spiral voltage divider is used for sensing of high drain voltage. The spiral is connected to the drain and continues spirally toward the gate. How the polysilicon spiral divider is connected to other device components is depicted in the schematic symbol of HV MOSFET in the Fig. 1 (terminals d, tap1, tap2).

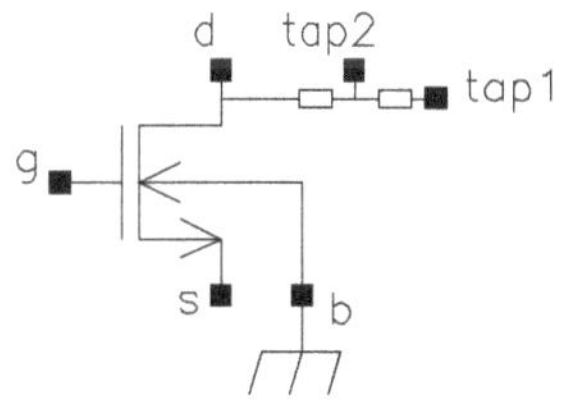

Fig. 1. Schematic symbol of HV start-up MOSFET.

The spiral divider is designed to have the electric field distribution as much similar as possible as the drain drift area under it. This ensures the voltage between divider and silicon does not exceed oxide breakdown voltage. The polysilicon spiral divider has a big impact on a distribution of electric field in low doped drain drift area. And on the contrary, the strong electric field in low doped drain drift area causes a lot of parasitic effects that have a big influence on DC and RF device characteristics. These attributes make the modeling of this start-up MOSFET complicated, especially the divider ratio voltage and frequency dependency. The divider is usually modeled by the simple RC network, but there exist the operation areas where such simple model is not sufficient.

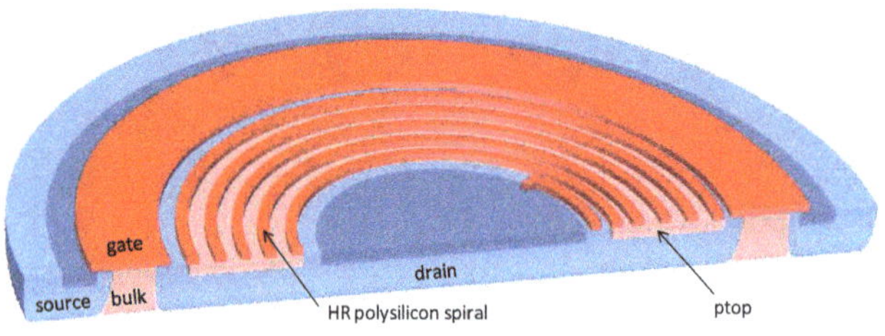

Fig. 2. The simplified 3D structure of HV start-up MOSFET transistor.

2. Spiral Divider Modeling

For the purpose of the equivalent lumped element circuit creation the polysilicon spiral is divided into several separate spiral elements. This division is shown in Fig. 3(a) where each spiral element has a different color. For better lucidity only the first four turns are depicted in this figure. The equivalent 3D circuit in Fig. 3(b) is obtained if these spiral elements are uncoiled to parallel plains. The 3D equivalent circuit in Fig. 3(b) can be redrawn for better lucidity to the 2D equivalent circuit, which is depicted in Fig. 4.

2.1 Spiral Element Length

The spiral divider of the HV MOSFET transistor is a special case of the Archimedes spiral [3]. The radius r of the spiral is increased in one turn by a radius increment Δr. The basic equation defined in polar coordinates for the radius is

$$r = r_0 + \varphi \frac{\Delta r}{2\pi} \tag{1}$$

where r_0 is an initial radius of the spiral and φ is an actual angle circumscribed by the spiral.

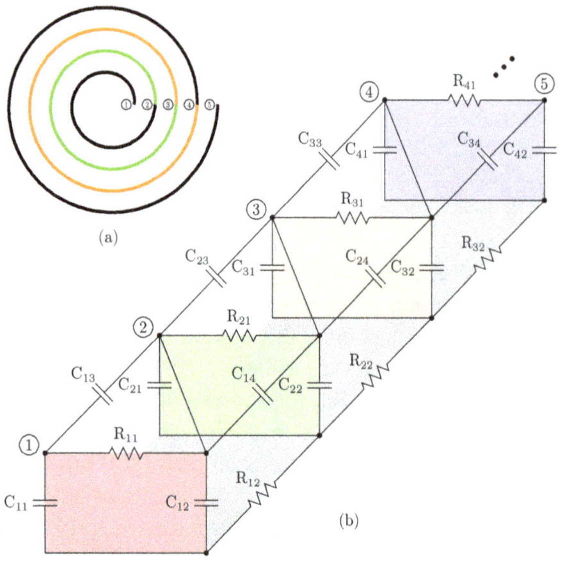

Fig. 3. Equivalent lumped 3D circuit of first four spiral poly subsegments and ptop: (a) colored spiral subsegment, (b) equivalent circuit. Colors from (a) match (b).

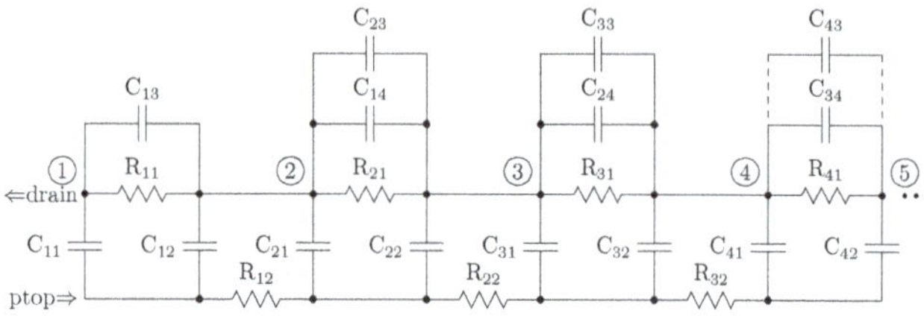

Fig. 4. Equivalent lumped 2D circuit of first four spiral poly subsegments and ptop.

The curve length can be calculated in the following way. If $f(\phi)$ is the function of the curve in polar coordinates (ϕ is an angle) then the length L of the curve is defined as

$$L = \int_0^{\varphi} \sqrt{[f(\phi)]^2 + \left(\frac{\mathrm{d}f(\phi)}{\mathrm{d}\phi}\right)^2}\, \mathrm{d}\phi. \tag{2}$$

For the spiral defined in polar coordinates by (1) the spiral length L is obtained by substituting the equation (1) into (2), and by solving this integral the spiral length is

$$\begin{aligned} L = {} & \frac{\Delta r}{4\pi} \ln \left(\frac{r_0 + \varphi\frac{\Delta r}{2\pi} + \sqrt{\left(r_0 + \varphi\frac{\Delta r}{2\pi}\right)^2 + \frac{\Delta r^2}{4\pi^2}}}{r_0 + \sqrt{r_0^2 + \frac{\Delta r^2}{4\pi^2}}} \right) + \\ & + \left(\frac{r_0\pi}{\Delta r} + \frac{\varphi}{2}\right) \sqrt{\left(r_0 + \varphi\frac{\Delta r}{2\pi}\right)^2 + \frac{\Delta r^2}{4\pi^2}} - \\ & - \frac{r_0\pi}{\Delta r}\sqrt{r_0^2 + \frac{\Delta r^2}{4\pi^2}}. \end{aligned} \tag{3}$$

When $\Delta r << r_0$ then equation (3) can be simplified to

$$L = r_0\varphi + \frac{\Delta r}{4\pi}\varphi^2. \tag{4}$$

2.2 Divider Ratio Modeling

Model of a similar device has been published in [4] but without ratio scalability and statistical modeling. These two important model abilities have been developed and implemented into the new model that is introduced in this paper. The divider ratio is dependent on the drain and source voltage V_D and V_S. The voltage dependency caused by depletion effects in the ptop and nwell layers is modeled by Verilog-A code using nonlinear functions. The increasing of the drain voltage causes the depletion of the ptop and nwell and when the ptop is fully depleted under the spiral polysilicon divider then it causes a change of the ratio voltage dependency slope as is depicted in Fig. 5. The geometrical ratio is based on (4) and can be expressed as

$$\begin{aligned} ratio_{geom} &= \frac{L_1 + L_2}{L_2} = \\ &= \frac{4\pi r_0(\varphi_{tap1} - \varphi_D) + \Delta r(\varphi_{tap1}^2 - \varphi_D^2)}{4\pi r_0(\varphi_{tap1} - \varphi_{tap2}) + \Delta r(\varphi_{tap1}^2 - \varphi_{tap2}^2)} \end{aligned} \tag{5}$$

where φ_D, φ_{tap1} and φ_{tap2} are drain, tap1 and tap2 angles on the spiral, L_1 and L_2 are long and sensing part of the spiral.

The normalized ratio is modeled as

$$ratio_{norm} = \frac{ratio_{el}}{ratio_{geom}} = \frac{V_D/V_{tap2}}{(L_1+L_2)/L_2} = \tag{6}$$

$$= \begin{cases} 1+(\beta_{D1}+\beta_{D2})V_D+\beta_S V_S & \text{for } V_D \leq V_P \\ 1+\beta_{D2}V_D+\beta_S V_S & \text{for } V_D > V_P \end{cases}$$

where $ratio_{el}$ is electrical ratio, β_{D1}, β_{D2} and β_S are drain and source voltage dependency model parameters, V_D, V_S and V_{tap2} are voltages on pins drain, source and tap2, and V_P is ptop pinch-off voltage. The voltage dependency coefficient β_{D2} is temperature dependent and can be expressed as

$$\beta_{D2} = \beta_{D2Tnom}\left[1+\alpha_D\left(T-T_{nom}\right)\right] \tag{7}$$

where T is temperature, model parameter β_{D2Tnom} represents value of β_{D2} at nominal temperature T_{nom} and α_D is temperature coefficient.

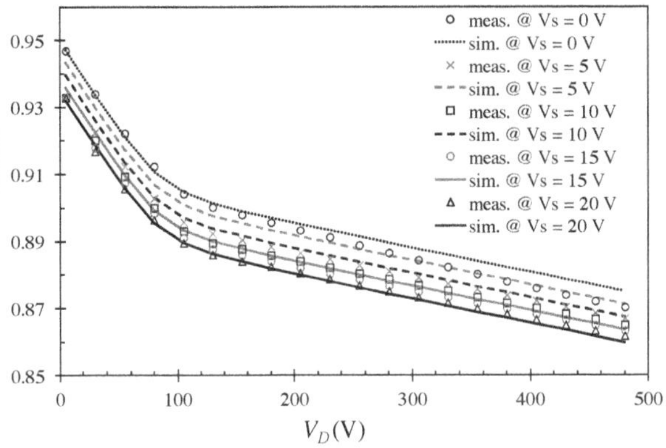

Fig. 5. Drain and source voltage dependency of normalized ratio.

The model is scalable by an editable model argument $ratio_{geom}$ that is refined to

$$ratio'_{geom} = \left(ratio_{mult}\, ratio_{geom} - ratio_{\Delta}\right)\delta_{ratio} \tag{8}$$

where $ratio_{mult}$ and $ratio_{\Delta}$ are model fitting parameters for ratio scalability and δ_{ratio} is relative statistical mismatch model parameter.

2.3 Divider Dynamical Modeling

The AC response is modeled by a distributed RC network. The magnitude and phase of the normalized ratio are depicted in Figs. 6 and 7. The AC measurement setup is described in Appendix. The full macromodel circuit of HV start-up MOSFET is shown in Fig. 12.

The high resistance polysilicon spiral segments are modeled by the Verilog-A code. Each resistor segment is modeled as

$$R_{seg} = \frac{R_{tot}}{N_{seg}}\left(1-\frac{1}{ratio_{norm}\, ratio'_{geom}}\right) \tag{9}$$

where N_{seg} is number of divider segments (excluding sense segment), R_{tot} is total spiral divider resistance and is calculated as

$$R_{tot} = \frac{R_{SH}L_{tot}\left(1+\alpha_{p1}\Delta_T+\alpha_{p2}\Delta_T^2\right)}{W+\delta_W} \tag{10}$$

where R_{SH} is polysilicon spiral sheet resistance, L_{tot} is a total spiral divider length, $\Delta_T = T - T_{nom}$, W is spiral segment width, δ_W is absolute statistical process model parameter, α_{p1} and α_{p2} are temperature coefficients dependent on R_{SH}

$$\begin{aligned} \alpha_{p1} &= \alpha_{rsh1}R_{SH}+\alpha_{rp1} \\ \alpha_{p2} &= \alpha_{rsh2}R_{SH}+\alpha_{rp2} \end{aligned} \tag{11}$$

where α_{rsh1}, α_{rsh2}, α_{rp1}, and α_{rp2} are polysilicon temperature coefficients. The resistance of the sense segment is

$$R_{sense} = \frac{R_{tot}}{ratio_{norm}\, ratio'_{geom}}. \tag{12}$$

The capacitances are modeled by the Verilog-A code and are voltage dependent similarly as resistances. The voltage dependency is caused by depleting effects of very low doped drift drain area due to the high electric field. Each capacitor segment is modeled [4] as

$$C_{seg} = \frac{C_{tot}}{N_{seg}}\left(\frac{1}{\pi}\arctan\left(\frac{V_P-V_D}{2}\right)+c_P\right) \tag{13}$$

where c_P is the pinch-off capacitance coefficient model parameter and C_{tot} is total spiral divider capacitance and is calculated as

$$C_{tot} = L_{tot}(W+\delta_W)C_{pa}+2L_{tot}C_{fr}+C_c \tag{14}$$

where C_{pa} is polysilicon (field oxide) capacitance per unit area, C_{fr} is fringe capacitance per length and C_c is capacitance model fitting parameter.

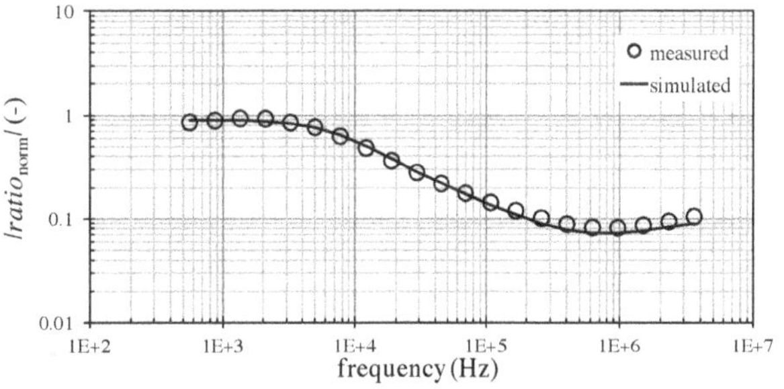

Fig. 6. Magnitude of normalized divider ratio.

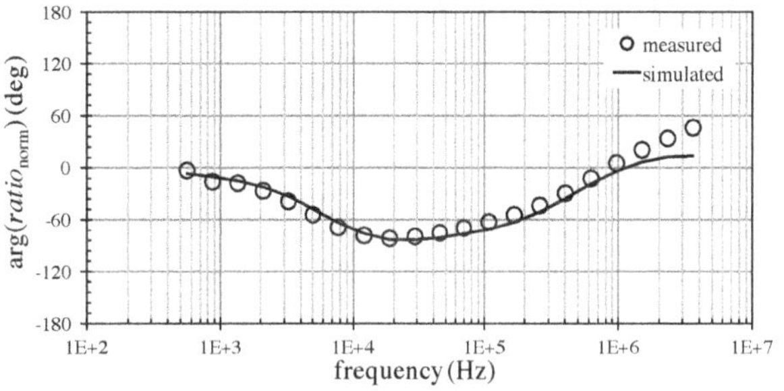

Fig. 7. Phase of normalized divider ratio.

2.4 Divider Statistical Modeling

This HV start-up MOSFET was placed on process control monitoring test chip where this device is measured on each fabricated wafer in a standard production. Data from this test chip are used for statistical process control and also for statistical modeling.

The mismatch modeling [5], [6] is implemented only to the divider ratio. The parameter δ_{ratio} is relative statistical mismatch model parameter and is defined as

$$\delta_{ratio} = 1 + \sigma_{ratio} \cdot VAR_{MATCH\_RATIO} \quad (15)$$

where σ_{ratio} is relative standard deviation of the divider ratio and VAR_{MATCH_RATIO} is random variable of mean 0 and standard deviation 1 that represents the normalized Gaussian distribution for modeling the stochastic variations. The histogram of measured and simulated voltage V_{tap2} and box plot are depicted in Figs. 8 and 9 (one lot typically contains from 20 to 30 wafers and one wafer typically contains 5 test chips). The number of measured devices was 3825. The standard deviation σ_{ratio} is equal to the standard deviation of measured electrical parameter V_{tap2} at $V_D = 100\,\text{V}$, $V_S = V_G = V_{tap1} = 0\,\text{V}$.

The influence of process parameters variation on the device parameters is implemented through master variables by using mapping equations:

$$R_{SH} = R_{SH\_nominal} + \sigma_{RSH} \cdot VAR_{RSH}, \quad (16)$$

$$\delta_W = \delta_{W\_nominal} + \sigma_{DW} \cdot VAR_{DW}, \quad (17)$$

$$C_{pa} = C_{pa\_nominal} + \sigma_{CPA} \cdot VAR_{CPA}, \quad (18)$$

$$C_{fr} = C_{fr\_nominal} + \sigma_{CFR} \cdot VAR_{CFR}, \quad (19)$$

where $R_{SH\_nominal}$, $\delta_{W\_nominal}$, $C_{pa\_nominal}$ and $C_{fr\_nominal}$ are nominal values, σ_{RSH}, σ_{DW}, σ_{CPA} and σ_{CFR} are the standard deviations, and VAR_{RSH}, VAR_{DW}, VAR_{CPA} and VAR_{CFR} are master random variables of mean 0 and standard deviation 1 that represents the normalized Gaussian distribution for modeling the stochastic variations.

As an example, the histogram of measured and simulated electrical process parameter DW is depicted in Fig. 10 and the box plot in Fig. 11. The number of measured devices was 29257. The standard deviations of device parameters σ_{RSH}, σ_{DW}, σ_{CPA} and σ_{CFR} can be calculated from the standard deviations of these measured electrical process parameters by using forward and backward propagation of variances [7].

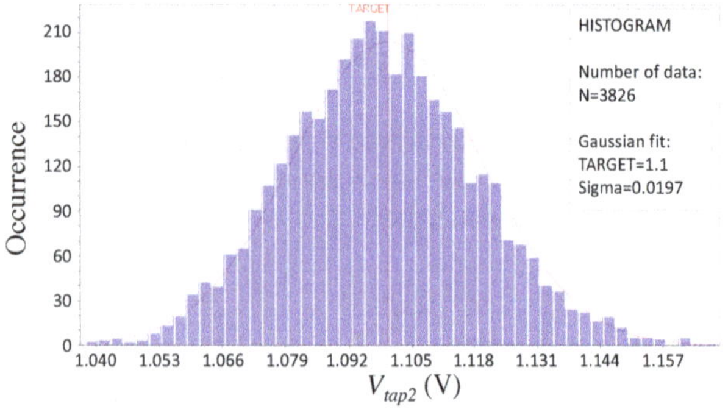

Fig. 8. The histogram of measured and simulated electrical parameter V_{tap2}. The red curve represents modeled Gaussian distribution.

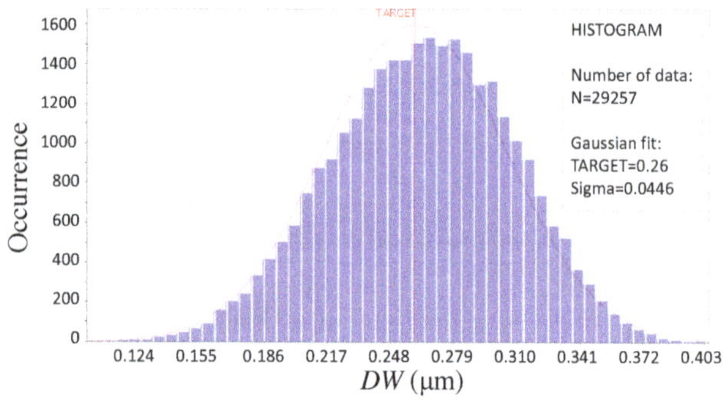

Fig. 10. The histogram of measured and simulated electrical process parameter DW. The red curve represents modeled Gaussian distribution.

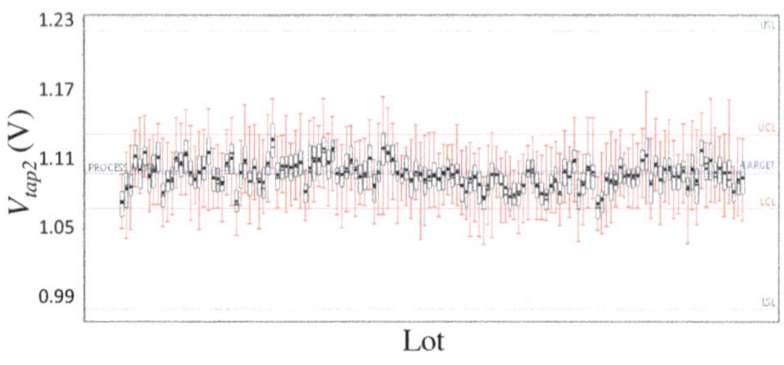

Fig. 9. The boxplot of measured electrical parameter V_{tap2}. The green lines define upper and lower specification limit (USL and LSL) while the red lines define upper and lower control limit (UCL and LCL).

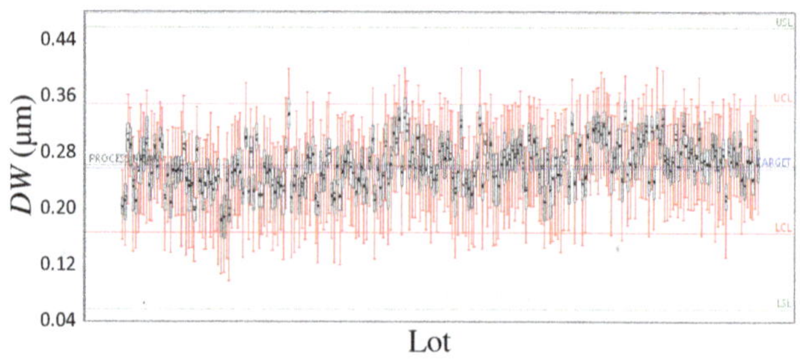

Fig. 11. The boxplot of measured electrical process parameter DW. The green lines define upper and lower specification limit (USL and LSL) while the red lines define upper and lower control limit (UCL and LCL).

3. HV Start-Up MOSFET Application

The AC/DC convertor has been selected as an example of typical application of HV start-up MOSFET with polysilicon spiral divider. The simplified AC/DC convertor circuit is depicted in Fig. 13. The HV start-up MOSFET subblock is modeled by the circuit in Fig. 12 and by the equations introduced in this paper.

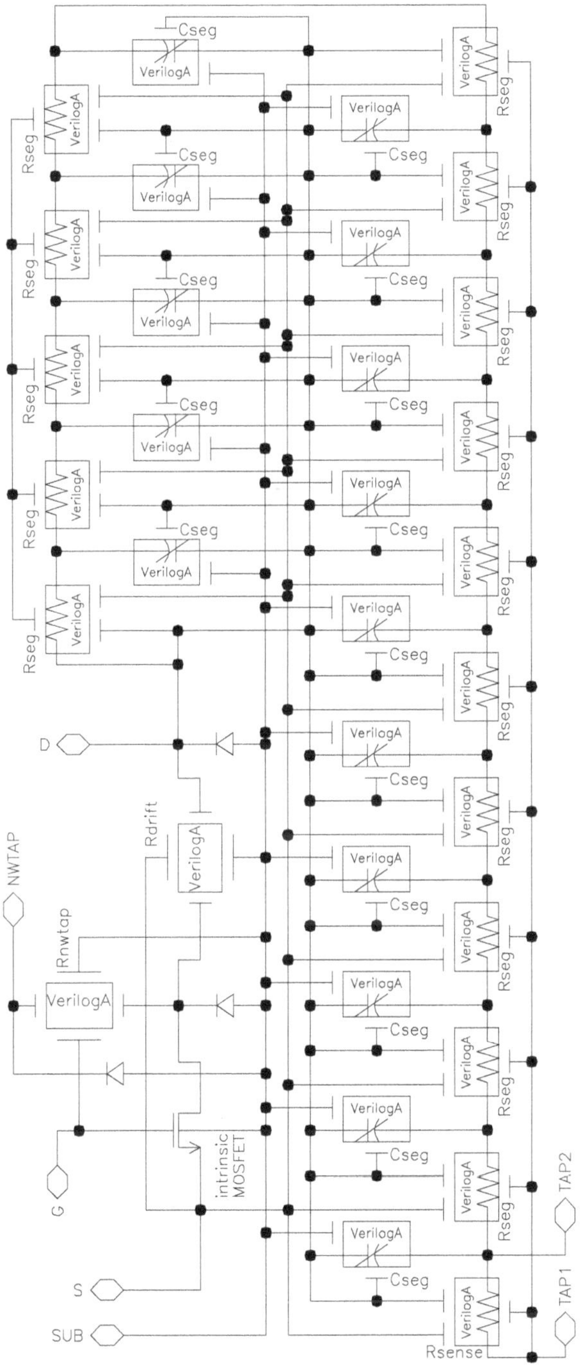

Fig. 12. The full macromodel circuit of HV start-up MOSFET with polysilicon spiral divider.

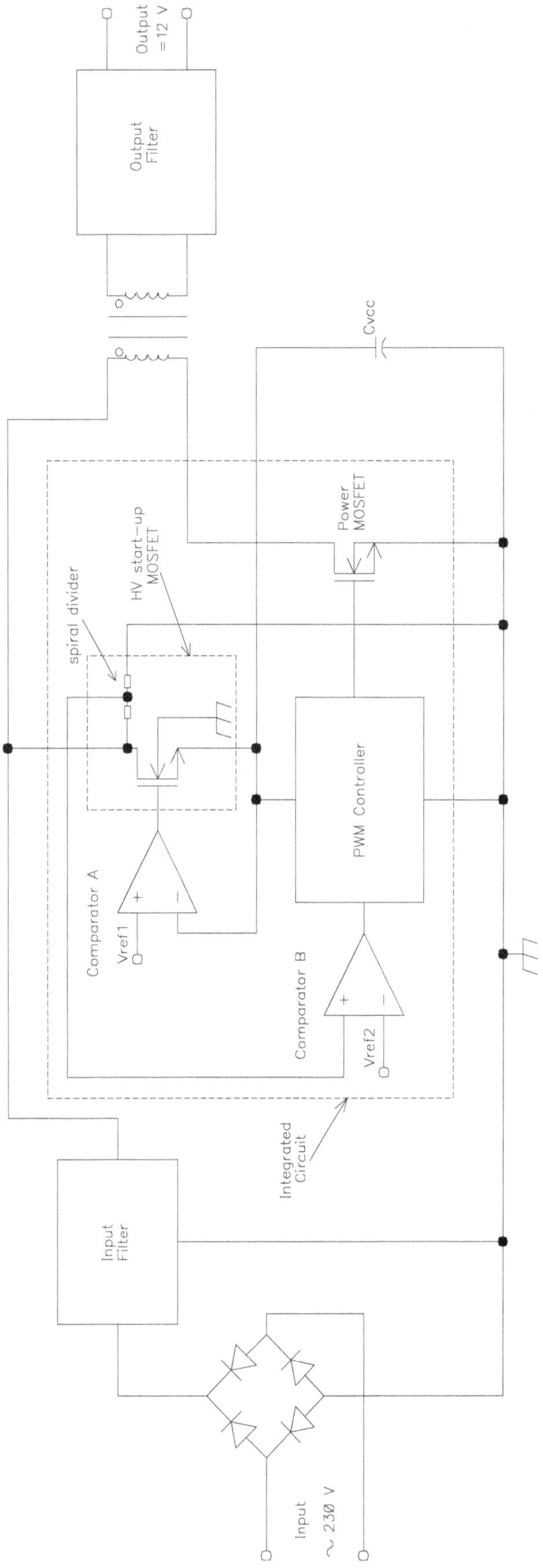

Fig. 13. The simplified circuit of AC/DC convertor. The HV start-up MOSFET subblock is modeled by the circuit in Fig. 12.

The voltage reference V_{ref1} determines the voltage (in Comparator A) to which the external capacitor C_{vcc} is charged up and basically determines the voltage at which the HV start-up MOSFET is switched off. This solution decreases the power consumption in comparison with older solution where the external capacitor C_{vcc} was permanently charged through an external resistor. The solution with HV spiral divider enables to switch off the charging in case the external capacitor C_{vcc} is charged up enough. The voltage reference V_{ref2} determines when the whole AC/DC convertor is switched on based on direct sensing of high voltage by the spiral divider described in this paper. The high input voltage is divided by the spiral divider and compared in Comparator B with the voltage reference V_{ref2}. It means that AC/DC convertor is turned off if the input high voltage is lower than defined value (depends on product specification, e.g. 112 V). See [12] or [13] for more applications.

The circuits that use HV start-up MOSFET with spiral divider allow designing applications with many features such as:

- The dynamic self-supply,
- No need of auxiliary winding [13],
- Low standby-power,
- High voltage sensing,
- Brown-out protection [12],
- Line overvoltage protection [12].

4. Conclusions

The enhanced accurate DC and RF model of nonlinear spiral polysilicon voltage divider in high voltage start-up MOSFET transistor has been created and is presented in this paper. The modeling results are compared with measured data and the maximal relative model error of divider ratio is less than 1.1 %. The intrinsic MOSFET is modeled by the standard BSIM3v3 model described in [8, 9, 10, 11].

A special attention is paid to the ability of the created model to cover the influence of a variation of process parameters on the device characteristics. The statistical process variation model is created based on measurement about 30000 devices and mismatch model is based on measurement about 3000 devices. It has to be pointed out that Monte Carlo simulations represent the most powerful tool to verify the robustness of a designed circuit over natural process and mismatch variations.

The big advantage of this model are smooth derivatives of simulated characteristics. The simulation speed is acceptable and any convergency issue was not observed during the verification realized on several real designs.

Appendix

The AC measurement setup is depicted in Fig. 14. The drain is biased to DC 50, 100, 300, and 400 V and through C_{bias} to AC source. Pin tap1 of the spiral is connected to the ground and pin tap2 is measured output. Gate, source and bulk are connected to the ground.

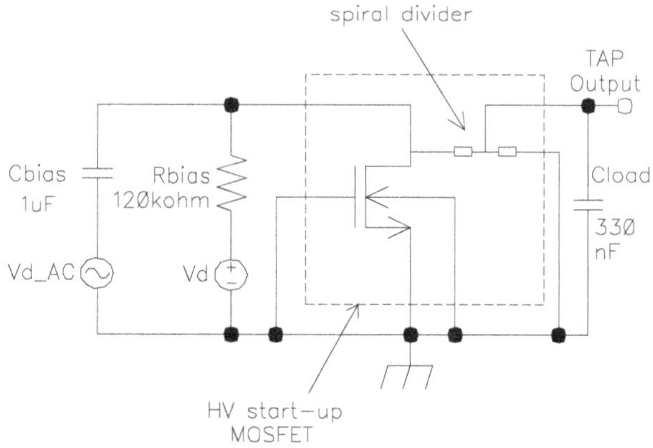

Fig. 14. The AC measurement setup.

Acknowledgements

This paper has been supported by ON Semiconductor company and also by the Grant Agency of the Czech Technical University in Prague, grant No. SGS14/082/OHK3/1T/13 and grant No. SGS13/206/OHK3/3T/13.

References

[1] HALL, J., QUDDUS, M. T., BURTON, R., OIKAWA, K., CHANG, G. *High Voltage Sensor Device and Method Therefor.* US patent 955943. 2011.

[2] HALL, J., QUDDUS, M. T. *Method of Sensing a High Voltage.* US Patent 8349625. 2013.

[3] WEISSTEIN, E. W. Archimedes Spiral. *MathWorld – A Wolfram Web Resource.* [Online] Cited 2014-12-02. Available at: http://mathworld.wolfram.com/ArchimedesSpiral.html

[4] PANKO, V., BANAS, S., PTACEK, K., BURTON, R., DOBES, J. An accurate DC and RF modeling of nonlinear spiral polysilicon voltage divider in high voltage MOSFET transistor. In *Proceedings of the IEEE 11th International Conference on Solid-State and Integrated Circuit Technology (ICSICT).* Xi'an (China), 2012. DOI: 10.1109/ICSICT.2012.6467635

[5] PAPATHANASIOU, K. A designer's approach to device mismatch: Theory, modeling, simulation techniques, applications and examples. *Analog Integrated Circuits and Signal Processing,* 2006, vol. 48, no. 2, p. 95–106. DOI: 10.1007/s10470-006-5367-2

[6] DRENNAN, P. G., McANDREW, C. C. Understanding MOSFET mismatch for analog design. *IEEE Journal of Solid-State Circuits,* 2003, vol. 38, no. 3, p. 450–456. DOI: 10.1109/JSSC.2002.808305

[7] McANDREW, C.C. Statistical modeling using backward propagation of variance (BPV). *Compact Modeling*. Springer Netherlands, 2010, p. 491–520. DOI: 10.1007/978-90-481-8614-3_16

[8] LIU, W. *MOSFET Models for SPICE Simulation: Including BSIM3v3 and BSIM4*. Wiley – IEEE Press, 2001.

[9] YTTERDAL, T., CHENG, Y., FJELDLY, T. A. *Device Modeling for Analog and RF CMOS Circuit Design*. Wiley, 2003.

[10] LIOU, J., ORTIZ-CONDE, A., GARCÍA-SÁNCHEZ, F. *Analysis and Design of MOSFETs: Modeling, Simulation, and Parameter Extraction*. Kluwer Academic Publishers, 1998. DOI: 10.1007/978-1-4615-5415-8

[11] MASSOBRIO, G., ANTOGNETTI, P. *Semiconductor Device Modeling with SPICE*, 2nd edition. McGraw-Hill, 1993.

[12] ON Semiconductor. *Design of a 65 W Adapter Utilizing the NCP1237 PWM Controller*. Application note AND8461/D. October 2014.

[13] ON Semiconductor. *Designing Converters with the NCP101X Family*. Application note AND8134/D. October 2003.

About the Authors ...

Václav PAŇKO was born in Czech Republic in 1979. He received his bachelor's degree in Electronics and Communications in 2006 and master's degree in Radioelectronics in 2008, both at the Czech Technical University in Prague, where he currently works toward the Ph.D. degree in Radioelectronics. In 2000, he joined ON Semiconductor, where he is presently a Senior Modeling and Characterization Engineer. His main interests include device modeling and parameter extraction, behavioral models development, parasitic effects modeling, and modeling and test chip design tools development.

Stanislav BANÁŠ was born in 1970. He received his M.Sc. degree in Electrical Engineering from the Technical University in Brno in 1994. Last year of his study he spent in scholarship in CNRS institute in Grenoble, where he was interested in optoelectronic properties of hydrogenated amorphous silicon. From 1996 he works as a modeling and characterization engineer in Motorola Czech Design Center in Roznov, later transferred to ON Semiconductor SCG Czech Design Center in Roznov. From 2012 he studies for Ph.D. in Technical University in Prague. His research interests include the modeling of high-voltage semiconductor components.

Richard BURTON received the Ph.D. degree in Electrical Engineering at Carnegie Mellon University in Pittsburg in 1994 simulating, designing, and fabricating integrated optoelectronics. From 1994 to 1996 he developed MESFET and HEMT technologies for cell phone power amplifiers. From 1996 to 2006 he developed manufacturable and reliable InGaP HBT technologies for OC-192 fiber optic communications PHY interfaces and highly rugged PAs for multi-band cellular communications. In 2006 he joined ON Semiconductor focusing on high reliable ultra-high voltage IC technology development for AC-DC off-line applications. His research interests include advanced technology development with focus on manufacturability and reliability.

Jan DIVÍN was born in Valašské Meziříčí, Czech Republic, in 1986. He works as characterization engineer of the ON Semiconductor company. He is a post gradual student of Czech Technical University in Prague at the Department of Radio Engineering. He has a M.Sc. in Electronics and Communication from the Brno University of Technology. His Ph.D. study is devoted to characterization of new model types of radio-frequency semiconductor devices.

Karel PTÁČEK received his M.Sc. degree in Electrical Engineering from the Brno University of Technology, Czech Republic, in 2003. Currently, he works as a design engineer at ON Semiconductor, Czech Republic. His research interests include designing of monolithic high voltage devices as well as their ESD protections. He works on his Ph.D. thesis with the topic of communication between galvanically isolated high voltage and low voltage parts of an integrated circuit.

Josef DOBEŠ received the Ph.D. degree in microelectronics at the Czech Technical University in Prague in 1986. From 1986 to 1992, he was a researcher of the TESLA Research Institute, where he performed analyses on algorithms for CMOS Technology Simulators. Currently, he works at the Department of Radio Electronics of the Czech Technical University in Prague. His research interests include the physical modeling of radio electronic circuit elements, especially RF and microwave transistors and transmission lines, creating or improving special algorithms for the circuit analysis and optimization, such as time- and frequency-domain sensitivity, poles-zeros or steady-state analyses, and creating a comprehensive CAD tool for the analysis and optimization of RF and microwave circuits.

11

Low-Profile Fully-Printed Multifrequency Monopoles Loaded with Complementary Metamaterial Transmission Line

Xue LI [1,2], *Jinwen TIAN* [1]

[1] Inst. for Pattern Recognition and Artificial Intelligence, Huazhong University of Science and Technology, Wuhan 430074, China
[2] Defense Forces Academy, Zhengzhou 450052, China

ghoul_gargoyle@163.com, jwtian@mail.hust.edu.cn

Abstract. *The design of a new class of multifrequency monopoles by loading a set of resonant-type complementary metamaterial transmission lines (CMTL) is firstly presented. Two types of CMTL elements are comprehensively explored: the former is the epsilon negative (ENG) one by loading complementary split ring resonators (CSRRs) with different configurations on the signal strip, whereas the latter is the double negative (DNG) one by incorporating the CSRRs and capacitive gaps. In both cases, the CMTLs are considered with different number of unit cells. By cautiously controlling the geometrical parameters of element structure, five antenna prototypes coving different communication standards (GSM, UMTS, DMB and WiMAX) are designed, fabricated and measured. Numerical and experimental results illustrate that the zeroth-order resonance frequencies of the ENG and DNG monopoles are in desirable consistency. Moreover, of all operating frequencies the antennas exhibit fairly good impedance matching performances better than -10 dB and quasi-omnidirectional radiation patterns.*

Keywords

Metamaterial transmission line, planar monopole, multi-frequency antenna, complementary, zeroth-order resonator (ZOR)

1. Introduction

Over recent years, there has been a renewed interest in using artificial metamaterial (MTM) transmission line (TL) in the design of microwave devices and components [1]-[20] due to its abnormal electromagnetic (EM) properties that are hardly realized in nature. As to the planar microstrip monopoles [9]-[20], these antennas in general can be classified into three categories according to the loading manner and working mechanism of the MTM TL elements. The monopoles of the first category are based on the double negative (DNG) MTM TLs [9]-[16] which are also termed as composite right/left handed (CRLH) TLs. These antennas can be engineered with broadband or multi-frequency operation because the operating modes can be arbitrarily controlled by the well-conducted dispersion curve of CRLH elements. The second category is the monopole or dipole antennas built on the meta-surface made of periodically arranged left handed (LH) particles [17]. Through this type of loading, the antenna performances are improved in terms of both enhanced radiation behavior and broadened impedance matching bandwidth. The third category is the monopole by introducing LH MTMs [18] or split ring resonators (SRRs) [19] along it or even by embedding complementary SRRs (CSRRs) in the monopoles [20]. In this regard, the miniaturization and band-notch characteristic in an ultrawide operation band were achieved in virtue of the subwavelength resonance of the loaded elements.

Although compactness and multifunction are realized in aforementioned monopoles made of DNG MTM TLs, the mostly reported structures are confined to the nonresonant-type TL elements by using chip components which are restricted to low frequency operation and are not efficient radiators [9]-[12]. As to the rarely reported distributed TL monopoles [13], [14], the shunt inductors are commonly realized by grounded vias which would degrade the antenna gain due to the metallic losses. Moreover, further miniaturization is still a pressing task since the compactness is of great importance to portable and handheld antennas. These issues make an improved and alternative strategy that can be easily characterized and experimentally implemented a pressing task. The goal of this paper is thus to explore a novel avenue in the implementation of fully printed monopoles with simultaneous compact and multifunctional feature. The fundamentals and working mechanism of the monopoles using resonant-type complementary MTM TL (CMTL) will be firstly introduced. Then the multifrequency monopoles made of epsilon negative (ENG) and DNG CMTLs by etching the CSRRs on the signal strip is proposed, characterized and eventually fabricated.

2. Resonant-type Microstrip-fed ENG and DNG Monopoles

2.1 Fundamentals and Theoretical Background

The crucial obstacle of pushing the resonant-type CMTL element such as CSRRs in the monopole application is the structure incompatibility because the CSRRs commonly require a ground plane while the host monopole is unable to afford. In this paper, the resonant-type structures utilized for the monopoles are inspired from [16], [21], where the CSRRs are etched on the signal strip for filter and divider applications. Fig. 1 plots the sketch of the conventional and proposed conceptual microstrip-fed monopoles, respectively and the corresponding equivalent circuit model. As a first step towards the design of a DNG monopole, four essential parameters are defined: the phase φ_{CMTL} induced by the CMTL element, the phase φ_{Mono} caused by the host monopole, resonant modes (indices) n and the number of utilized CMTL elements N. Similar to the nonresonant-type CRLH monopole, the overall possible eigenfrequencies of the DNG monopoles should satisfy the following resonant condition [13], [16]

$$\varphi_{\text{total}} = N \times \varphi_{\text{CMTL}} + \varphi_{\text{Mono}} = n\pi/2, \quad n = 0, \pm 1, \pm 3, \ldots \qquad (1)$$

where φ_{total} is the phase shift across the entire antenna. Observation from (1) indicates that the relation of the φ_{Mono} and φ_{CMTL} plays an important role in determination of the resonant modes. Note that the stop band of the ENG TL element in the guided wave design can also be employed for a radiated wave design in the case of an open circuit (infinitely large series impedance) or a resonance (often referred to as zeroth-order resonance, ZOR) of the shunt branch in the equivalent circuit model. In both ENG and DNG cases, the current has no place to go except through the shunt branch of the circuit.

2.2 Monopoles by Loading Single CMTL Element

Figure 2 shows the flow chart of the formation of the proposed DNG CMTL element evolved from [21]. As can be seen, the element is composed of CSRRs etched on the square patch in the upper metallic strip and a square ring with or without two splits in the top and bottom region. The square ring connected to the feedline is formed by introducing an additional gap (closed slot) outside the CSRRs. The CSRRs still response to the axial electric-field component and thus afford the resonant negative permittivity. The resonant effect of the CSRRs is accounted by the parallel tank formed by L_p and C_p in the circuit model. The capacitive gaps at both sides are modeled by the capacitance C_g, whereas the inductive effect of the host TL is modeled by the inductance L_s. Both C_g and L_s contribute to the negative permeability. The DNG CMTL element can be ENG CMTL element by closing the splits of the square ring (defined as a hybrid ENG resonator) or by removing the gap (CSRRs left only). Since the length of the gap has been extended significantly relative to [21], a larger C_g and in turn enhanced LH characteristics will be engineered. In this particular design, since the ground plane of the CMTL element is not placed beneath the CSRRs but is coplanar with the host monopole, the fringing capacitances of the gaps C_f are weakened and thus are neglected for convenience. Therefore, the proposed CMTL element can be easily integrated with the planar monopoles of which the ground plane remains unaltered. The C_f has nothing to do with the resonant modes but results in slightly reduced resonant frequencies, which will be corroborated by extensive calculations in the upcoming section.

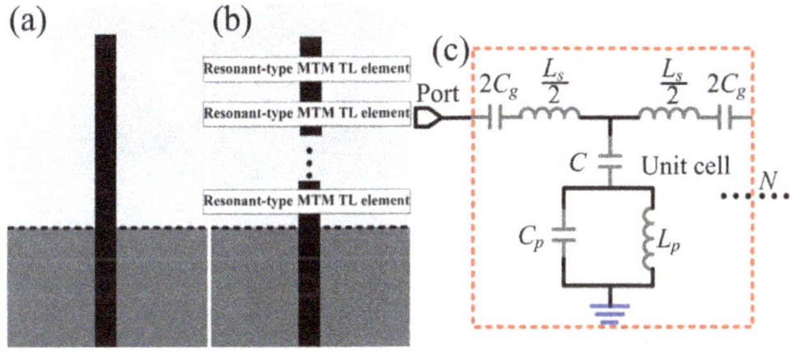

Fig. 1. Sketch of microstrip-fed monopoles: (a) Conventional monopole. (b) Conceptual monopole loaded with resonant-type CMTL elements and (c) equivalent circuit model.

Fig. 3 portrays the layout of the resulting monopoles based on CSRRs-loaded CMTL elements with different configurations. The CMTL element is loaded at the end of

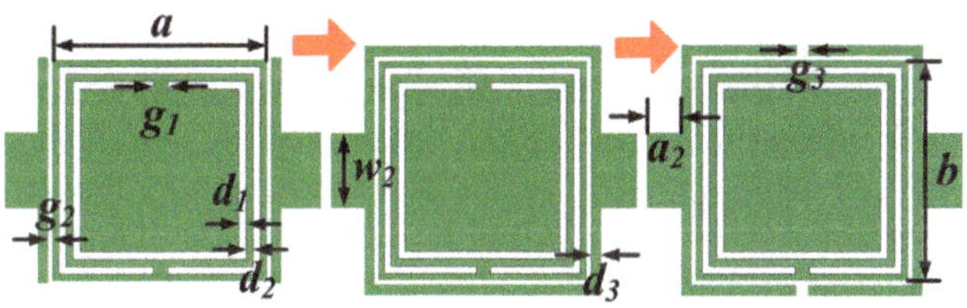

Fig. 2. Flow chart of the formation of the proposed ENG (the middle one) and DNG (the last one) CMTL element evolved from previous CRLH element (the first one) [21].

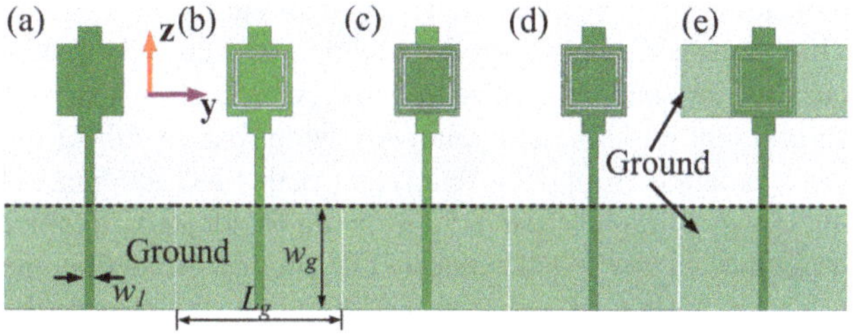

Fig. 3. Layouts of the proposed single-cell microstrip-fed monopoles. (a) Conventional antenna (Case 1); (b) ENG antenna (Case 2); (c) hybrid ENG antenna (Case 3); (d) DNG antenna (Case 4); (e) hybrid ENG antenna with ground underneath the CMTL element (Case 5). The geometrical parameters of these antennas (in millimeter: mm) are $a = 11.5$, $b = 10.8$ $g_1 = 0.72$, $g_2 = 0.2$ $g_3 = 0.4$, $d_1 = d_2 = 0.36$, $d_3 = 0.2$, $w_1 = 1.6$, $w_2 = 4$, $L_g = 30$, $w_g = 18$, and $a_2 = 1.5$.

the host monopole. A total of six cases are considered for comprehensive analysis. They are orderly the conventional monopole without CMTL loading, the ENG monopoles loaded by the CSRRs only or by aforementioned hybrid resonator, the DNG monopole, the hybrid ENG monopole with a ground plane, and the monopole with the gap only (not shown here for brevity of contents). In the fifth case, the ground beneath the hybrid CMTL element is not connected to that of the host monopole.

The design flow starts from the conventional microstrip-fed monopole where the ground plane $L_g \times w_g$ is finely designed to obtain a desirable impedance matching over a wide bandwidth and a normal monopolar radiation with high efficiency. The length of the host monopole is designed to operate at GSM band (centered at 1.8 GHz). All designs are conducted in the commercial full-wave finite-element-method (FEM) EM field simulator Ansoft HFSS (Version of 13.2) and are all layouts are built on the commonly available 1mm-thick F4B substrate with dielectric constant of $\varepsilon_r = 2.2$ and loss tangent of $\tan\delta = 0.001$. Fig. 4 depicts the simulated reflection coefficients S_{11} of the proposed single-cell MTM-inspired monopoles. To illustrate the effects of the gap, the results of the monopole with the gap only are also provided.

Following Fig. 4, we conclude that the fundamental reflection dip (around GSM 1.8 GHz) corresponds to the operating frequency f_0 of the host monopole, whereas the second reflection dip (covering the Satellite Digital Mobile Broadcasting (DMB) band, 2605÷2655 MHz) corresponds to the ZOR frequency f_{M1} of the CSRRs while the third reflection dip (around 4.65 GHz, f_{S2}) corresponds to the resonance of the gap in hybrid DNG antenna case. As previously discussed, C_f is negligible when the ground is removed, which finds strong support from the almost constant trend of the reflection response except for the slightly reduced operating frequencies in case 5 due to the enhanced C_f. A further comparison between case 3 and the case with the gap only also indicates that the CSRRs interact with the gap, leading to the slightly reduced f_{S2} in case 3. Moreover, the uniform E-field distributions (not shown for brevity of contents) are clearly observed in CSRRs at f_{M1} while are not shown at residual frequencies, indicating a ZOR mode and strong radiation of CSRRs. This feature distinguishes the monopoles from any previous ones by loading chip components which are always not an efficient radiator. The relatively wider bandwidth of the DNG antenna around f_{M1} relative to both ENG antennas is due to that the excited $n = +1$ mode is located in close proximity to the $n = 0$ mode (ZOR mode). Note that the $n = +1$ mode never occurs in the ENG case. Moreover, the $f_0 = 2.14$ GHz (UMTS band) in the DNG antenna case has been shifted upwards. This is because the CMTL element provides less phase shift than conventional monopole. This means that the DNG monopole affords slightly shorter actual length. Most importantly, the ZOR frequency is observed almost the same in all ENG and DNG cases. This is because the ZOR frequency is only dependent on the L_p and C_p relative to the dimensions of the CSRRs.

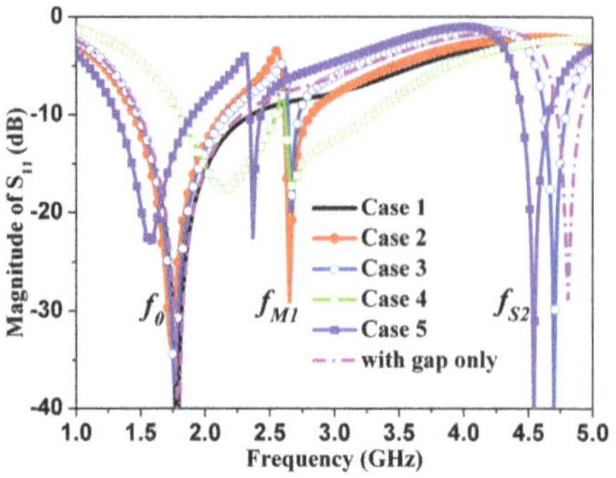

Fig. 4. Simulated reflection coefficients of the proposed single-cell MTM-inspired antennas.

To examine the far-field radiation characteristics, Fig. 5 depicts the 3-D patterns of the MTM-inspired antennas at different frequencies for an intuitionistic view. Following the figure, the monopolar or quasi-monopolar patterns are clearly observed in all cases. The little broken uniformity of the patterns in xoy plane at f_{S2} is due to the radiation of the gap which also radiates and thus facilitates the spatial power to be re-synthesized. The detailed performances of these antennas can be referred to Tab. 1. The relatively low efficiency of the ENG antennas at f_{M1} with respect to that at f_0 and f_{S2} is due to that f_{M1} is located very close to the low efficiency dip (not shown for brevity) at which no radiation occurs. However, it is improved in the DNG case due to double negative permittivity and permeability around f_{M1}, which enables strong radiation. The low antenna gain in all cases is due to the omnidirectional patterns and small ground dimensions of the host monopole. Further simulation results indicate that the gain is enhanced by an average of 0.7 dB when L_g increases per 10 mm in some specific range.

For verification, we have fabricated three antenna prototypes (case 2, case 3 and case 4, see orderly in Fig. 6) whose footprints occupy the same area of $1 \times 30 \times 52.6$ mm^3. Fig. 6 compares the simulated and measured (through a N5230C vector network analyzer) reflection coefficients while Fig. 7 gives the measured radiation patterns in two

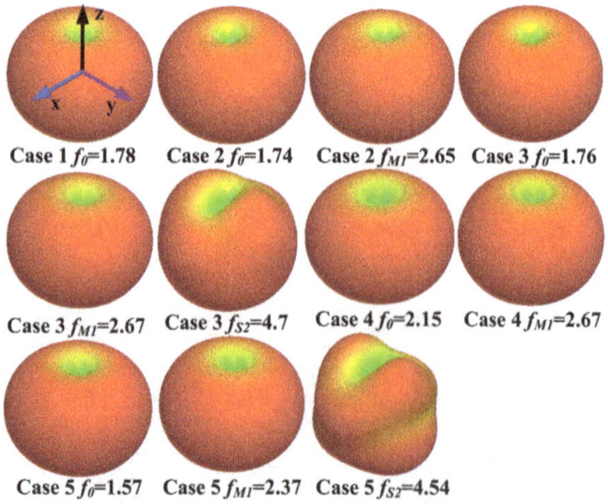

Fig. 5. Simulated 3-D radiation patterns of the single-cell MTM-inspired monopoles at different operating frequencies.

Antennas	Antenna gain (dB)			BW (GHz)			Radiation efficiency (%)		
	f_0	f_{M1}	f_{S2}	f_0	f_{M1}	f_{S2}	f_0	f_{M1}	f_{S2}
Case 1	-1.44	-	-	0.9	-	-	97.5	-	-
Case 2	-1.64	0.59	-	0.7	0.21	-	98.3	82.6	-
Case 3	-1.44	0.38	0.8	0.7	0.09	0.17	97.3	75.8	89.1
Case 4	0.23	0.64	-	0.75	0.67	-	95.8	88.1	-

Tab. 1. Comparison of performances of the single-cell MTM-inspired monopoles. Note: the bandwidth (BW) is defined by the -10 dB reflection coefficients. The efficiency is defined as the radiated power divided by the accepted power at the center frequencies shown in Fig. 5.

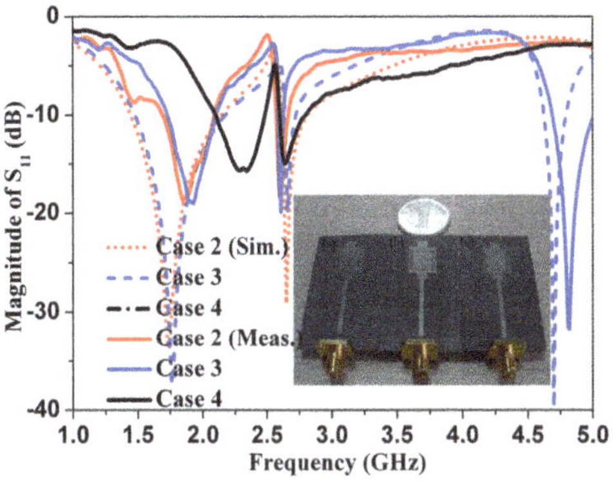

Fig. 6. Simulated and measured reflection coefficients of the fabricated antennas.

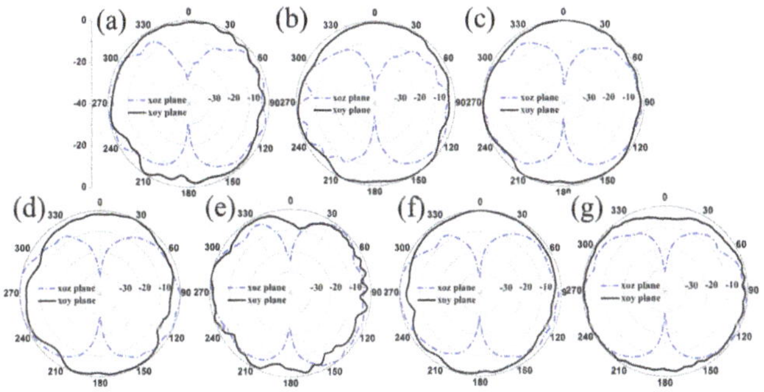

Fig. 7. Measured co-polarized patterns of the (a)-(b) ENG, (c)-(e) hybrid ENG and (f)-(g) DNG antenna at different frequencies. (a) f_0 = 1.85 GHz, (b) f_{M1} = 2.6 GHz, (c) f_0 = 1.92 GHz, (d) f_{M1} = 2.61 GHz, (e) f_{S2} = 4.81 GHz, (f) f_0 = 2.28 GHz, (g) f_{M1} = 2.64 GHz.

principle planes through a far-field measurement system in an anechoic chamber. As can be seen, a good agreement of results between simulation and measurement is achieved.

The slight frequency upwards in the measurement is attributable to the nonideal substrate that is utilized and also to the tolerances that are inherent in fabrication process. Across all operating frequencies, the measured S_{11} is better than -10 dB. From Fig. 7, the quasi-monopolar radiations are clearly illustrated at all selected frequencies from the nearly null radiation at the broadside direction in *xoz* plane while quasi-omnidirectional patterns in *xoy* plane. Thus far, the multifrequency operation of the proposed antennas by loading single CMTL element has been unambiguously demonstrated.

2.3 Monopoles by Loading Dual CMTL Elements

In this section, we will explore the effects of the number of the CMTL elements on the antenna performances. For simplicity, dual CMTL elements are considered without loss of generality. Fig. 8 plots the proposed monopoles by loading dual-cell CMTL elements with the same geometrical parameters and footprints as those shown in Fig. 3. The reflection coefficients are compared in Fig. 9. As can be seen, the impedance matching and bandwidth are almost without deterioration when additional element is loaded for ENG antennas. By contrary, the matching of the DNG antenna deteriorates significantly especially for the not well excited resonance of the host monopole. Nevertheless, the ZOR frequency still exhibits in all ENG and DNG

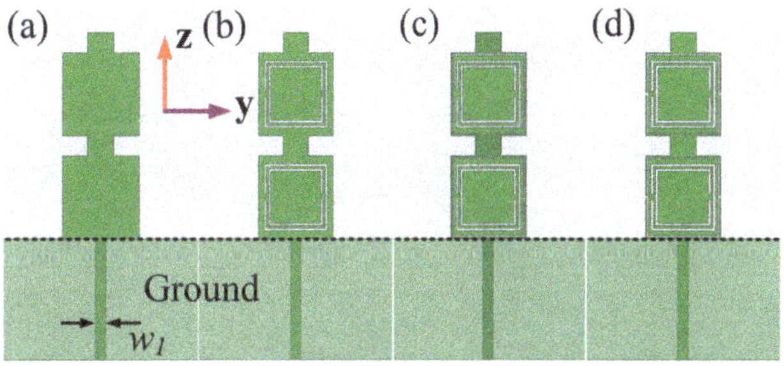

Fig. 8. Layouts of the proposed monopoles by loading dual-cell CMTL elements. (a) Conventional antenna; (b) ENG antenna; (c) Hybrid ENG antenna; (d) DNG antenna.

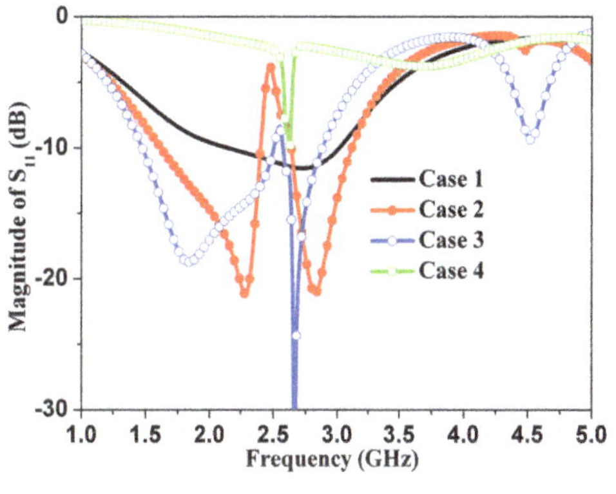

Fig. 9. Simulated reflection coefficients of the proposed dual-cell MTM-inspired monopoles.

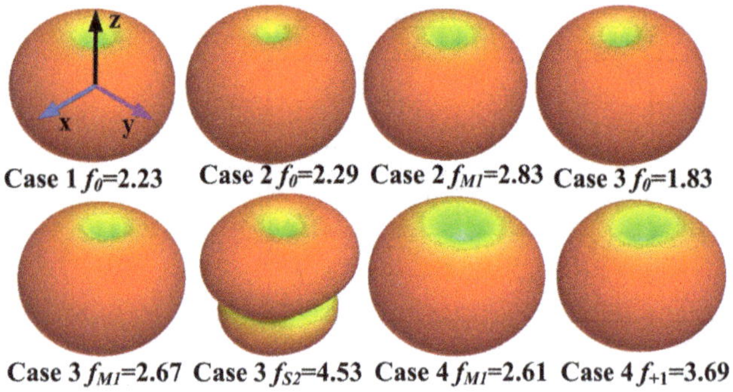

Fig. 10. Simulated 3-D radiation patterns of the dual-cell MTM-inspired monopoles at different operating frequencies.

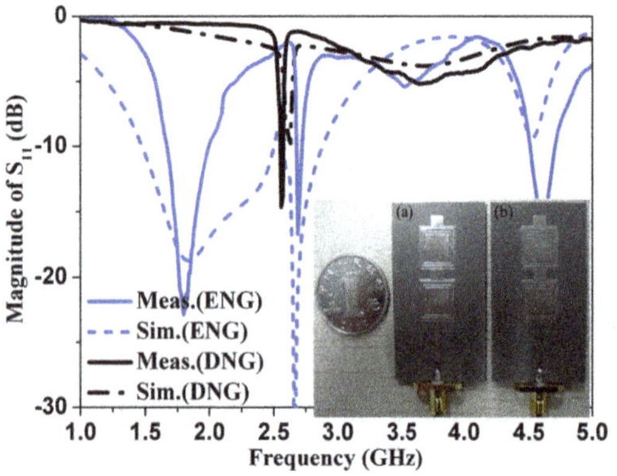

Fig. 11. Simulated and measured reflection coefficients of the dual-cell monopoles.

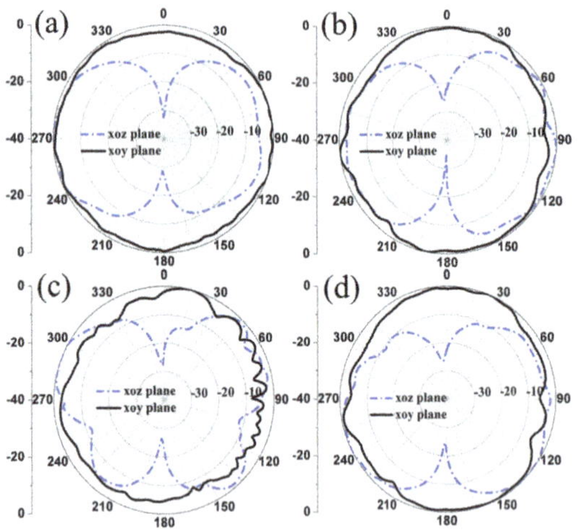

Fig. 12. Measured co-polarized patterns of the (a)-(b) hybrid ENG and (c)-(d) DNG antennas at different frequencies. (a) f_0 = 1.8 GHz, (b) f_{M1} = 2.69 GHz, (c) f_{S2} = 4.61 GHz, (d) f_0 = 2.56 GHz.

antennas around 2.6 GHz, further demonstrating that ZOR mode is independent while the input impedance is seriously dependent on the number of MTM elements [22]. In DNG case, the $n = +1$ mode around 3.5 GHz (WiMAX band) is weakly excited. Fig. 10 gives the 3-D far-field radiation patterns. Almost the same phenomena are expected as those shown in Fig. 5 except for the antenna gain which has been considerably improved when additional CMTL element is loaded. The reason is because that the CSRRs contribute to the major radiation of the monopole as previously discussed.

For verification, the hybrid ENG and DNG antennas are fabricated. Fig. 11 plots the comparison of reflection coefficients between simulations and measurements while Fig. 12 depicts the measured radiation patterns in two principle planes. A desirable agreement of results is also observed. The reason for slight deviations especially for the narrower bandwidth in the measurement case is due to that the CMTL dimensions of prototypes deviate from those in the simulation model. Nevertheless, the measured S_{11} is better than -10 dB over all operating frequencies. From Fig. 12, the measured typical monopolar patterns further confirm the effectiveness of dual-cell elements in the monopole design.

3. Conclusion

The use of resonant-type ENG and DNG CMTLs in the design of low-profile multifrequency monopoles is presented. Results reveal that at all operating frequencies the monopoles exhibit quasi-monopolar radiation patterns and fairy good impedance matching performances with return loss better than 10 dB. Moreover, the ZOR frequency is independent on the number of unit cells and is observed as the same for both ENG and DNG antennas. These antennas feature compact, low profile, completely uniplanar, and are without any metallic vias and lumped loadings, rendering easy avenue toward monopoles with multifunction and high integration by using simple photolithography. Other variations of CSRRs and geometrical parameters can be explored for arbitrary manipulation of frequency ratio and further improvement of performances.

References

[1] ELEFTHERIADES, G. V., BALMAIN, K. G. *Negative Refraction Metamaterials: Fundamental Principles and Applications.* Hoboken, NJ: Wiley, 2005.

[2] XU, H.-X., WANG, G.-M., LIANG, J.-G., PENG, Q. Novel CRLH TL based on fractal geometry and series power divider application. *Acta Physica Sinica,* 2012, vol. 61, no. 7, p. 074101.

[3] XU, H.-X., WANG, G.-M., LIANG, J.-G. Novel designed CSRRs and its application in tunable tri-band bandpass filter based on fractal geometry. *Radioengineering*, 2011, vol. 20, no. 1, p. 312 to 316.

[4] CALOZ, C., ITOH, T. *Electromagnetic Metamaterials: Transmission Line Theory and Microwave Applications: The Engineering Approach.* Hoboken, NJ: Wiley, 2006.

[5] XU, H.-X., WANG, G.-M., ZHANG, C.-X., WANG, X. Characterization of composite right/left handed transmission line. *Electronics Letters*, 2011, vol. 47, no. 18, p. 1030–1032. DOI: 10.1049/el.2010.3707

[6] MARQUES, R., MARTIN, F., SOROLLA, M. *Metamaterials with Negative Parameters: Theory, Design, and Microwave Applications.* Hoboken, NJ: Wiley, 2008.

[7] XU, H.-X., WANG, G.-M., ZHANG, C.-X., LIANG, J.-G. Novel design of compact microstrip diplexer based on fractal-shaped composite right/left handed transmission line. *Journal of Infrared and Millimeter Waves*, 2011, vol. 30, no. 5, p. 390–396. DOI: 10.3724/sp.j.1010.2011.00390

[8] XU, H.-X., WANG, G.-M., CHEN, X., LI, T.-P. Broadband balun using fully artificial fractal-shaped composite right/left handed transmission line. *IEEE Microwave and Wireless Components Letters*, 2012, vol. 22, no. 1, p. 16–18. DOI: 10.1109/LMWC.2011.2173929

[9] KIM, D., KIM, M. Narrow-beamwidth T-shaped monopole antenna fabricated from metamaterial wires. *Electronics Letters*, 2008, vol. 44, no. 3, p. 180–182. DOI: 10.1049/el:20082854

[10] ANTONIADES, M. A., ELEFTHERIADES, G. V. A folded-monopole model for electrically small NRI-TL metamaterial antennas. *IEEE Antennas and Wireless Propagation Letters*, 2008, vol. 7, p. 425–428. DOI: 10.1109/LAWP.2008.2008773

[11] JI, J. K., KIM, G. H., SEONG, W. M. Bandwidth enhancement of metamaterial antennas based on composite right/left-handed transmission line. *IEEE Antennas and Wireless Propagation Letters*, 2010, vol. 9, p. 36–39. DOI: 10.1109/LAWP.2010.2041628

[12] ZHU, J., ANTONIADES, M. A., ELEFTHERIADES, G. V. A compact tri-band monopole antenna with single-cell metamaterial loading. *IEEE Transactions on Antennas and Propagation*, 2010, vol. 58, no. 4, p. 1031–1038. DOI: 10.1109/TAP.2010.2041317

[13] IBRAHIM, A., SAFWAT, A. M. E., EL-HENNAWY, H. Triple-band microstrip-fed monopole antenna loaded with CRLH unit cell. *IEEE Antennas and Wireless Propagation Letters*, 2011, vol. 10, p. 1547–1550. DOI: 10.1109/LAWP.2011.2181813

[14] ANTONIADES, M. A., ELEFTHERIADES, G. V., ROGERS, E. S. A broadband dual-mode monopole antenna using NRI-TL metamaterial loading, *IEEE Antennas and Wireless Propagation Letters*, 2009, vol. 8, p. 258–261. DOI:10.1109/LAWP.2009.2014402

[15] KOKKINOS, T., FERESIDIS, A. P. Low-profile folded monopoles with embedded planar metamaterial phase-shifting lines. *IEEE Transactions on Antennas and Propagation*, 2009, vol. 57, p. 2997–3008. DOI: 10.1109/TAP.2009.2028605

[16] XU, H.-X., WANG, G.-M., LV, Y.-Y., QI, M.-Q., GAO, X., GE, S. Multifrequency monopole antennas by loading metamaterial transmission lines with dual-shunt branch circuit. *Progress in Electromagnetics Research*, 2013, vol. 137, p. 703–725.

[17] ELSHEAK, D. N., ISKANDER, M. F., ELSADE, H. A., ABDALLAH, E. A., ELHENAWY, H. Enhancement of ultra-wideband microstrip monopole antenna by using unequal arms V-shaped slot printed on metamaterial surface. *Microwave and Optical Technology Letters*, 2010, vol. 52, no. 10, p. 2203–2209. DOI: 10.1002/mop.25447

[18] PALANDOKEN, M., GREDE, A., HENKE, H. Broadband microstrip antenna with left-handed metamaterials. *IEEE Transactions on Antennas and Propagation*, 2009, vol. 57, no. 2, p. 331–338. DOI: 10.1109/TAP.2008.2011230

[19] BARBUTO, M., BILOTTI, F., TOSCANO, A. Design of a multifunctional SRR-loaded printed monopole antenna. *International Journal of RF and Microwave Computer-Aided Engineering*, 2012, vol. 22, no. 4, p. 552–557. DOI: 10.1002/mmce.20645

[20] LIU, J., GONG, S., XU, Y., ZHANG, X., FENG, C., QI, N. Compact printed ultra-wideband monopole antenna with dual band-notched characteristics. *Electronics Letters*, 2008, vol. 44, no. 12, p. 710–711. DOI: 10.1049/el:20080931

[21] GIL, M., BONACHE, J., MARTIN, F. Synthesis and applications of new left handed microstrip lines with complementary split-ring resonators etched on the signal strip. *IET Microwaves Antennas and Propagation*, 2008, vol. 2, no. 4, p. 324–330. DOI:10.1049/iet-map:20070225

[22] XU, H.-X., WANG, G.-M., GONG, J.-Q. Compact dual-band zeroth-order resonance antenna. *Chinese Physics Letters*, 2012, vol. 29, no.1, p. 014101-1–014101-4. DOI: 10.1088/0256-307X/29/1/014101

About the Authors ...

Xue LI was born in Henan province of China. He received his MS degree from the Ordnance Engineering University, Shijiazhuang, China, in 2003, and MS degree from the PLA Information Engineering University, Zhengzhou, China, in 2006. He is now pursing his PhD degree. His research interests include computer vision and pattern recognition.

Jinwen TIAN was born in the Hebei province of China. He received his BS degree from the Daqing Petroleum Institute, Daqing, China, in 1983, and MS and PhD degrees from Huazhong University of Science and Technology, Wuhan, China, in 1994 and 1998, respectively. His current interest includes remote sensing image information processing, streaming media technology and its applications, the wavelet transform theory and its applications, image data compression, target detection and recognition, augmented reality and computer software simulation, etc.

12

Design of Passive Analog Electronic Circuits using Hybrid Modified UMDA Algorithm

Josef SLEZAK, Tomas GOTTHANS

Dept. of Radio Electronics, Brno University of Technology, Technická 12, 616 00 Brno, Czech Republic

xsleza08@stud.feec.vutbr.cz, gotthans@feec.vutbr.cz

Abstract. *Hybrid evolutionary passive analog circuits synthesis method based on modified Univariate Marginal Distribution Algorithm (UMDA) and a local search algorithm is proposed in the paper. The modification of the UMDA algorithm which allows to specify the maximum number of the nodes and the maximum number of the components of the synthesized circuit is proposed. The proposed hybrid approach efficiently reduces the number of the objective function evaluations. The modified UMDA algorithm is used for synthesis of the topology and the local search algorithm is used for determination of the parameters of the components of the designed circuit. As an example the proposed method is applied to a problem of synthesis of the fractional capacitor circuit.*

Keywords

Evolutionary algorithm, optimization, estimation of distribution algorithm, EDA, fractional capacitor

1. Introduction

In many areas of the engineering applications Estimation of Distribution Algorithms (EDA) have proved excellent capabilities of dealing with various kinds of problems of different nature. However, in the area of the evolutionary electronics utilization of the EDA algorithms has not been sufficient. Several papers focused on the evolutionary design of the whole analog circuit (the topology and the parameters of the components) have been published. In [2] Zinchenko published method of synthesis of the passive analog circuits using Univariate Marginal Distribution Algorithm (UMDA) algorithm however the method has two drawbacks. Parallel connection of the components in the branches of the circuits is not possible and with increasing number of the nodes of the designed circuit the number of the components is increasing more than necessary. Another paper focused on the evolutionary synthesis of the passive analog circuits using EDA algorithms was published by Torres [3] however the method employs UMDA algorithm only for determination of the number of the resistors, capacitors and inductors in the designed circuit. Minimal Switching Graph Problem solved using hybrid EDA algorithm method was presented in [4]. The method employs the UMDA to sample start search points and a hill-climbing algorithm to find local optimum of the search space.

The proposed EDA algorithm solves the drawbacks of the method [2] mentioned above. The maximum number of the nodes and the maximum number of the components of the synthesized circuit can be set independently. Also encoding of the parallel components is possible. Furthermore the proposed method employs efficient hybrid approach which significantly reduces the number of the objective function evaluations. For the same optimization problem the number of the evaluations of the objective function of the proposed method is almost four times lower than for simulated annealing algorithm [12].

2. Hybrid Modified UMDA Algorithm

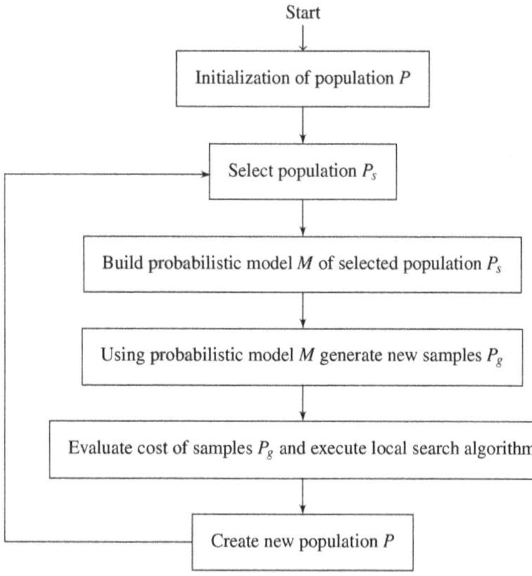

Fig. 1. Principal flowchart of the proposed method [1].

The flowchart of the proposed method is presented in Fig. 1. Population P is formed of binary vectors of length 135 bits which are initialized randomly with seeding of 10 bits ($n_c = 10$). Parameters storage e_{ps} is formed of a vector of length 135 consisting of real numbers in the range $\langle 0,1 \rangle$. Parameters storage e_{ps} is dynamically optimized during the whole synthesis process and it is adapted to the selected topologies in the selection phase of the algorithm. The vector includes a component value for every single admittance of the used expanded fully connected admittance network (see Sec. 6). During initialization phase e_{ps} is set randomly with uniform distribution.

In the next step selected population P_s is formed of the good individuals of the previous population P.

Probabilistic model M of selected population P_s is built. For more information on building of the probabilistic model in UMDA algorithm please refer to [5].

Probabilistic model M built in the previous step is used for generation of new samples of solutions P_g. The new samples are generated using Stochastic Universal Sampling method (SUS) and are repaired using the repairing method described in Sec. 3.

The generated samples are evaluated using the topological information stored in the individuals of P_g and the parameters of the components stored in e_{ps}. If the condition of execution of the local search algorithm is fulfilled, the local search algorithm tries to optimize the parameters of the current solution. If the accuracy of the current solution is improved, then storage of the parameters e_{ps} is updated according to the results of the local search algorithm. Detailed description of the cost evaluation phase and the local search algorithm is presented in Sec. 4.

In the next step new population P is formed of the best individuals of P_g and selected population P_s. The described process is repeated until one of the termination criteria of the algorithm is met.

3. The Unitation Constraints

Generally desired specifications of an analog circuit are easier to reach using an analog circuit of higher complexity. Due to the fact the evolutionary analog circuits synthesis methods tend to evolve analog circuits with complexity as large as possible. Without restriction of the number of the components of the evolved circuit its complexity becomes higher than necessary.

Therefore the number of the components of the evolved circuit should be restricted to a user define value. Since in the proposed encoding method (section 6) the number of the components of the encoded circuit is determined by the number of the "ones" of the binary characteristic vector c the restriction of the number of the components leads to a problem with unitation constraints [6].

Definition 1. Let's define vector $x = (x_1, x_2, \ldots, x_n) \in \Omega$. Then the unitation value of x is defined as

$$u(x) := \sum_{i=1}^{n} x_i. \tag{1}$$

Value of unitation function $u(x)$ depends only on the number of the "ones" in an input binary vector x. The unitation values of two vectors with the same numbers of "ones" are equal.

A problem with unitation constraints is defined as solution e in which unitation value $u(e)$ (the number of the "ones" in solution e) is restricted to a defined number [6].

As was described above the analog circuit synthesis problem has to be viewed as a problem with unitation constraints [6]. Modification of Factorized Distribution Algorithm (FDA) [7] which enables solving problems with unitation constraints was described in [6]. Modification of the UMDA algorithm which is able to handle the problems with unitation constraints is proposed in the text bellow. Pseudocode of the original UMDA algorithm [5] is presented in Fig. 2.

step0: Set $k = 1$. Generate $n_i \gg 0$ points randomly.
step1: Select $n_s \leq n_i$ points. Compute the marginal frequencies $r_{k;i}(x_i)$ of the selected set.
step2: Generate n_i new points according to the distribution $q_{k+1}(x) = \prod_{i=1}^{n} r_{k;i}(x_i)$. Set $k = k+1$.
step3: If not terminated, go to **step1**.

Fig. 2. Pseudo-code of the original UMDA algorithm [1].

In [6] to handle unitation constraints problems only generation phase (sampling) of FDA algorithm was modified. The presented approach was adopted also in the proposed modification of the UMDA algorithm. The UMDA algorithm was implemented using toolbox MATEDA 2.0 [8]. Pseudo-code of the modified UMDA algorithm is presented in Fig. 3.

step0: Set $k = 1$. Generate $n_i \gg 0$ points randomly.
step1: Select $n_s \leq n_i$ points. Compute the marginal frequencies $r_{k;i}(x_i)$ of the selected set.
step2: Generate n_i new points according to the distribution $q_{k+1}(x) = \prod_{i=1}^{n} r_{k;i}(x_i)$. Set $k = k+1$.
step3: With regard to unitation constraints repair generated points.
step4: If not terminated, go to **step1**.

Fig. 3. Pseudo-code of the modified UMDA algorithm [1].

One additional step (**step3**) was added to the original UMDA algorithm. In **step3** the generated samples are repaired to satisfy the desired unitation constraints.

The repairing function is applied to every single individual of the population of the generated samples in **step2**. Only n_c "ones" with the highest marginal frequencies $r_{k;i}$ of every generated sample are accepted. The rest of the "ones" of the samples are set to zero. If the number of the "ones" of the sample is equal or lower than n_c then the sample is accepted without any modification and no repairing is performed. This way the number of the "ones" (which corresponds to the number of the components of the encoded analog circuit) of every generated sample never exceeds n_c.

Note that the repairing function is applied only for the part of the encoding vector e which encodes the topology of the solution (characteristic vector c in Fig. 13).

4. The Local Search Algorithm

Evaluation of the cost values and using of the local search algorithm will be discussed in the section. Principal flowchart of this phase is presented in Fig. 4.

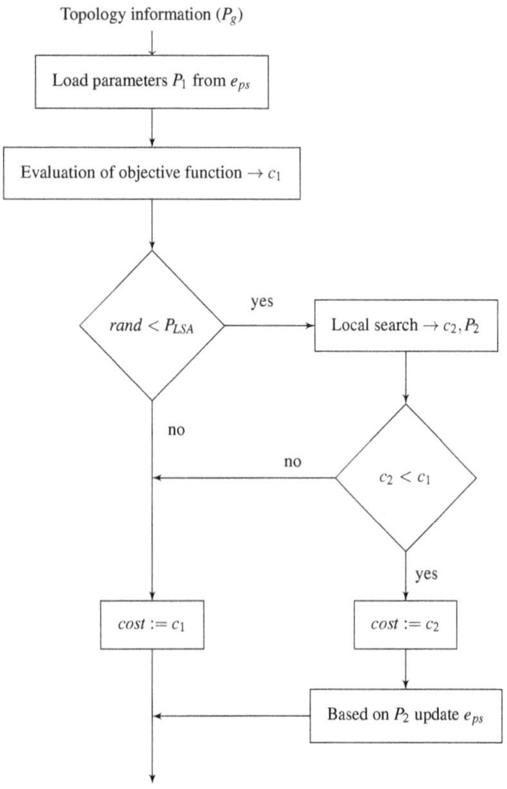

Fig. 4. Evaluation of the cost value and employing of the local search algorithm [1].

The following procedure which is described bellow is performed for every single individual $P_g(i)$ of the population of the generated samples P_g.

Based on the topology information stored in the binary vector of individual $P_g(i)$ appropriate set of parameters P_1 is loaded from parameters storage e_{ps} and cost value c_1 of individual $P_g(i)$ is evaluated.

If the condition of execution of the local search algorithm (LSA) is fulfilled ($rand < P_{LSA}$) the LSA tries to improve accuracy of individual $P_g(i)$. The probability of execution of the LSA is set to $P_{LSA} = 0.02$. The initial point of the search of the LSA is set to parameters set P_1. The number of the objective function evaluations of the LSA is set to $MaxFunEvals = 100$. The results of the LSA are the cost value of the optimized solution c_2 and set of the optimized parameters P_2.

If the LSA is successful in improving of the accuracy of individual $P_g(i)$ and its cost value was improved ($c_2 < c_1$) then value $cost$ of individual $P_g(i)$ is set to c_2 ($cost := c_2$) and the appropriate parameters of parameters storage e_{ps} are updated according to parameters set P_2.

If the condition of execution of the LSA was not fulfilled ($rand > P_{LSA}$) or the LSA was not able to improve the cost value of individual $G(i)$ ($c_2 \geq c_1$) the resulting value $cost$ of individual $P_g(i)$ is set to c_1 ($cost := c_1$).

After performing of the described process cost value $cost$ of individual $P_g(i)$ and updated parameters storage e_{ps} are obtained. During the run of the algorithm the parameters stored in e_{ps} are adapted to the topological information of the good individuals selected in the selection phase of the algorithm (Fig. 1). This way the information about the topology stored in P_g and the information about the parameters stored in e_{ps} are mutually optimized and the whole synthesis process is directed towards the promising areas of the solution space.

5. Application of the Method

A problem of synthesis of a fractional capacitor circuit was adopted from [9] and will be used for demonstration of the synthesis capabilities of the proposed method. The goal is to synthesize a circuit with input impedance (2)

$$Z_{in} = s^{-0.6}. \tag{2}$$

In Fig. 5 there is a circuit realization of function (2) as presented in [9]. For the rest of the section, the circuit will be called original approximation circuit.

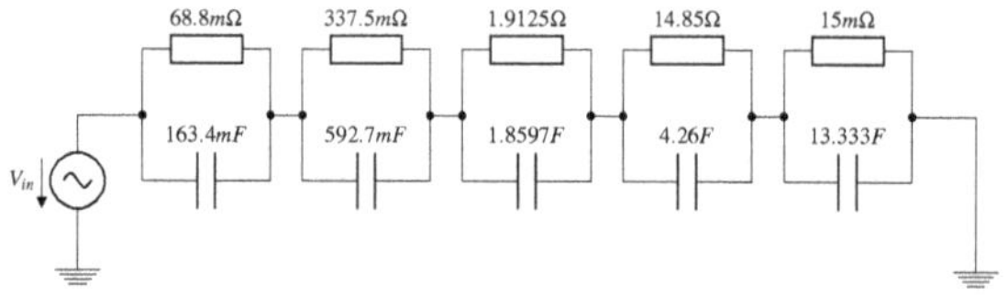

Fig. 5. Schematic of the original approximation circuit [1].

Comparison of the magnitude and the phase of Z_{in} of the original approximation circuit and (2) is presented in Fig. 6 and Fig. 7 respectively. Deviation of the magnitude and the phase of Z_{in} of the original approximation circuit and (2) is presented in Fig. 8 and Fig. 9 respectively.

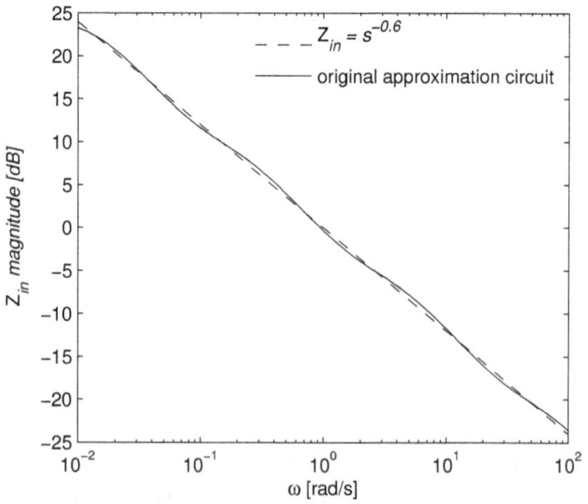

Fig. 6. Comparison of the magnitude characteristics of Z_{in} regarding (2) and the original approximation circuit [1].

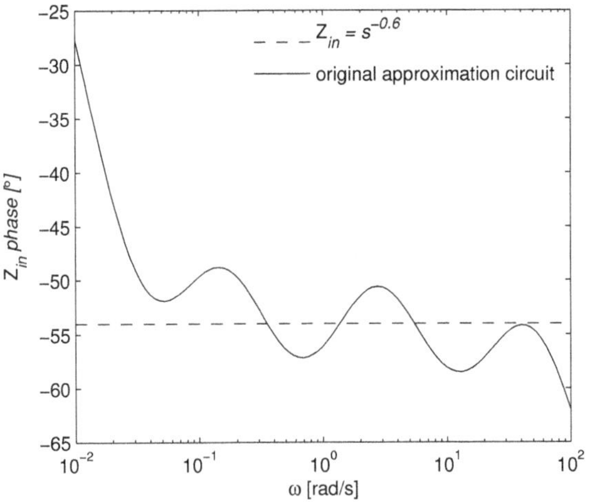

Fig. 7. Comparison of the phase characteristics of Z_{in} closer (2) and the original approximation circuit [1].

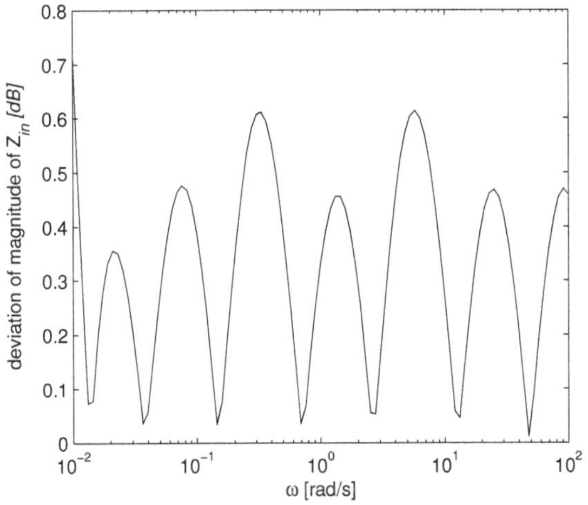

Fig. 8. Deviation of the magnitude of Z_{in} of the original approximation circuit [1].

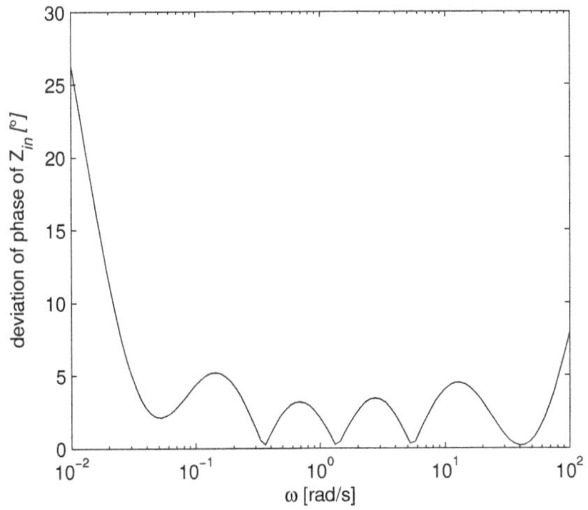

Fig. 9. Deviation of the phase of Z_{in} of the original approximation circuit [1].

The highest deviations of the magnitude and the phase of Z_{in} of the original approximation circuit are presented in Tab. 1.

Magnitude			
ω[rad/s]	0.01	100	0.33
Δ_m[dB]	0.71	0.46	0.61

Phase			
ω[rad/s]	0.01	100	0.14
Δ_p[○]	26.3	7.98	5.2

Tab. 1. The highest deviations of the magnitude and the phase of Z_{in} of the original approximation circuit.

The following sections will be focused on synthesis of the fractional capacitor circuit problem using the proposed hybrid modified UMDA method.

6. The Encoding Method

The used encoding method is based on the idea of fully connected admittance network. For chosen number of the nodes the fully connected admittance network is formed by connecting the admittances between all combinations of the nodes of the network. The number of the admittances of the fully connected admittance network with n_n nodes can be calculated according to (3)

$$n_{adm} = \binom{n_n}{2} = \frac{n_n!}{2!(n_n-2)!}. \tag{3}$$

Every single admittance of the fully connected admittance network can be replaced by resistor, capacitor, inductor or their parallel combination. Therefore the largest circuit which can be for chosen number of the nodes n_n obtained is the circuit where every single admittance of the fully connected admittance network is replaced by parallel combination of resistor, capacitor and inductor. In this paper such circuit is denoted as expanded fully connected admittance network and includes $3n_{adm}$ components. Example of the expanded fully connected admittance network is presented in Fig. 10.

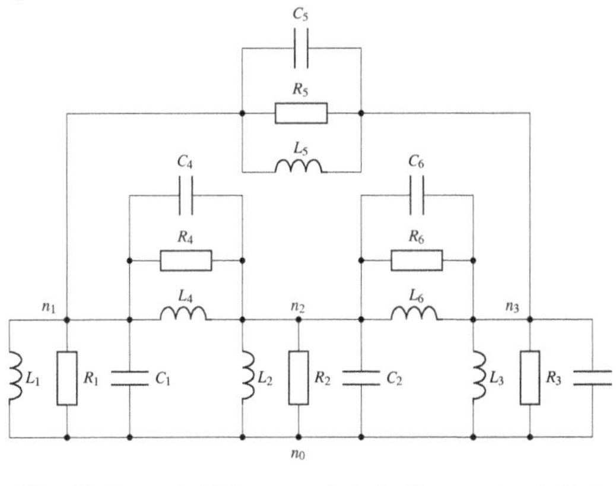

Fig. 10. Expanded fully connected admittance network N_c (n_c = 4) [1].

The expanded fully connected admittance network can be represented using complete multigraph with three multiple edges at the most [10]. Complete multigraph G_c corresponding to expanded fully connected admittance network

N_c is presented in Fig. 11. Nodes n_0 to n_3 of network N_c correspond to vertices v_0 to v_3 of complete multigraph G_c. Branches of network N_c correspond to edges of complete multigraph G_c. For example edges $e_5(1), e_5(2), e_5(3)$ (on complete multigraph G_c) correspond to components L_5, R_5, C_5 (in network N_c) respectively. Then the problem of searching of the topology of the analog RLC circuits can be defined as searching of subgraph G_s on complete multigraph G_c [10].

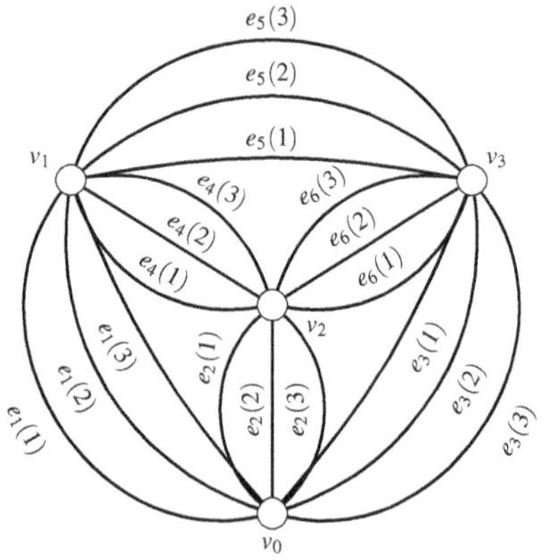

Fig. 11. Complete multigraph G_c representing expanded fully connected admittance network N_c [1].

Subgraph G_s can be encoded using binary characteristic vector c of length $3\,n_{adm}$. Every single bit of characteristic vector c represents including or not including of the corresponding edge of the complete multigraph G_c in subgraph G_s. For example complete multigraph G_c is encoded using characteristic vector c of length 18 bits where $c(i) = 1$ for $i \in \{1, 2, \ldots, 18\}$. Example of subgraph G_s, corresponding analog circuit and its characteristic vector c are presented in Fig. 12.

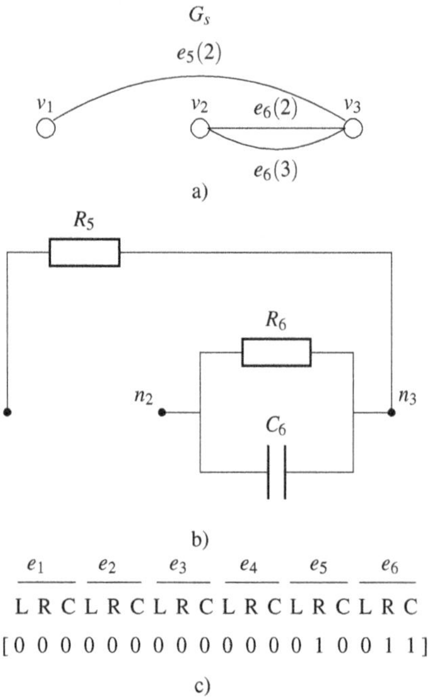

Fig. 12. a) Example of subgraph G_s b) analog circuit corresponding to G_s c) characteristic vector c of G_s [1].

Encoding vector e of every single solution is formed of two parts. The first one represents the topology and is encoded using characteristic graph c approach as described above. The number of the nodes of the expanded fully connected admittance network was experimentally chosen $n_n = 10$. To enable direct comparison of the accuracy of the circuits synthesized using the proposed method and the original approximation circuit presented in [9] the maximum number of the components of the synthesized circuit was set to the number of the components of the original approximation circuit ($n_c = 10$). According to (3) the topology is encoded using binary characteristic vector c of length 135 bits ($3\,n_{adm}$).

The second part of encoding vector e represents information about the parameters of the components and is represented by vector of real numbers p of length n_c.

Schematic of the used encoding vector e is presented in Fig. 13. The topological information is encoded using binary characteristic vector c = {b1, b2, b3, ..., b135} and the parameters (the values of the components) are encoded using vector of real numbers p = {dbl1, dbl2, db3, ..., dbl10}.

$$\overbrace{\qquad c \qquad}\ \overbrace{\qquad p \qquad}$$
e = [b1 b2 b3 ... b135] [dbl1 dbl2 dbl3 ... dbl10]

Fig. 13. Schematic diagram of encoding vector e [1].

Based on the components selected in the topological part of the information (characteristic vector c) corresponding values of the parameters are loaded from storage of the parameters e_{ps} and copied to vector p. For example let's assume that the topology is encoded using characteristic vector $c(j) = 1$ for $j \in \{1, 3, 15, 18, 28, 52, 78, 92, 107, 115\}$. Based on the information in characteristic vector c corresponding parameters of storage of the parameters e_{ps} will be loaded to vector of the parameters p as follows: $p(k) = e_{ps}(j)$ for $k \in \{1, 2, \ldots, N_c\}$ and $j \in \{1, 3, 15, 18, 28, 52, 78, 92, 107, 115\}$.

Based on the parameters in vector p the values of the components are calculated using formula (4)

$$v = \frac{2 \times 10^6}{1 + e^{(-1.4(10r - 14))}} \tag{4}$$

where r are values of the parameters loaded from vector of the parameters p. Formula (4) was formed to map the values of the parameters stored in vector p to suitable range of the values of the components. Since the parameters in vector p are set in the range <0,1> corresponding values of the components for the lowest $r = 0$ and for the highest $r = 1$ are $v_{min} = 0.0061$ and $v_{max} = 7.3685 \times 10^3$ respectively. Note that formula (4) is used for all three types of components RLC. Since the used angular frequency range is from 0.01 rad/s to 100 rad/s, the values of the components are set to non-realistic values.

7. The Objective Function

Cost value *cost* is according to (7) computed as weighted summation of the magnitude and the phase differences. Difference of magnitude Δ_m is according to (5) calculated as weighted absolute value of differences of desired magnitude function f_{md} and magnitude of current solution f_{mc} over $m = 101$ frequency points in the range 0.01 rad/s to 100 rad/s. Similarly difference of phase Δ_p is according to (6) calculated as weighted absolute value of differences of desired phase function f_{pd} and phase of current solution f_{pc}.

$$\Delta_m = \frac{1}{m}\sum_{i=1}^{m} w_{dm}(i)|f_{md}(i) - f_{mc}(i)| \quad (5)$$

$$\Delta_p = \frac{1}{m}\sum_{i=1}^{m} w_{dp}(i)|f_{pd}(i) - f_{pc}(i)| \quad (6)$$

$$cost = \Delta_m w_{cm} + \Delta_p w_{cp}. \quad (7)$$

Weights w_{cm} and w_{cp} were set to 1 and 2 respectively. Setting of weights w_{dm}, w_{dp} is presented in Tab. 2. All weight coefficients were set experimentally.

angular frequency range:	w_{dm}	w_{dp}
0.01 rad/s to 0.0398 rad/s	1.3	1.3
0.0437 rad/s to 20.8930 rad/s	1	1
22.9087 rad/s to 100 rad/s	1.3	1.3

Tab. 2. Setting of weights w_{dm} and w_{dp}.

Frequency responses of the current solution f_{mc} and f_{pc} are obtained using nodal analysis method implemented in Matlab.

8. Settings of the Proposed Algorithm

The goal of the synthesis is to design a circuit which approximates function (2). The only information supplied to the system is desired magnitude and phase characteristics (2), maximum number of the used components ($n_c = 10$), maximal number of the nodes ($n_n = 10$) and types of the used components (resistors, capacitors, inductors).

The number of the objective function evaluations (evals) required by the proposed algorithm consists of the number of the evaluations required for calculation of the cost values of all individuals of the population ($PopEvals$) and the number of the evaluations required by the used local search algorithm ($LSevals$). Population size was set $PopSize = 200$ and number of generations was set $MaxGen = 200$. Therefore $PopEvals = 40e3$. The local search method requires $MaxFunEvals = 100$ evaluations in each its run and it is executed with probability $P_{LSA} = 0.02$ (2% Lamarckian approach [11]). The condition of execution of LSA is tested during every single objective function evaluation. Therefore $LSevals = 80e3$ and the total number of the objective function evaluations required by the proposed algorithm is 120e3 ($PopEvals + LSevals$). Local search algorithm was realized using Matlab function *fmincon*.

All parameters of the synthesized problem and settings of the proposed algorithm are summarized in Tab. 3.

maximum number of the nodes (n_n)	10 (0 to 9)
maximum number of the components (n_c)	10
used types of the components	R,L,C
negative resistors	not allowed
angular frequency range	0.01 rad/s to 100 rad/s
number of the points (m)	101 (25 points/decade)
population size ($PopSize$)	200
number of generations ($MaxGen$)	200
probability of execution of LSA (P_{LSA})	0.02
w_{cm}	1
w_{cp}	2

Tab. 3. Settings of the proposed algorithm.

The algorithm UMDA is realized using Matlab toolbox MATEDA 2.0 [8]. MATEDA initialization file is presented in Fig. 14

```
PopSize = 200; n = 135; cache  = [0,0,0,0,0];
Card = 2*ones(1,n); MaxGen = 200; MaxVal = -1e-2;
stop_cond_params = {MaxGen,MaxVal};
Cliques = CreateMarkovModel(n, 0);
edaparams{1} = {'seeding_pop_method','seedingFract',25};
edaparams{2} = {'learning_method','LearnFDA',{Cliques}};
edaparams{3} = {'sampling_method','SampleFDAmodif',{PopSize}};
edaparams{4} = {'stop_cond_method','maxgen_maxval',stop_cond_params};
[AllStat,Cache]=RunEDAfractal(PopSize, n, F, Card, cache, edaparams);
```

Fig. 14. Configuring of MATEDA 2.0 toolbox for realization of the proposed algorithm [1].

The whole program flow of the UMDA algorithm is realized using the functions of MATEDA 2.0 toolbox. The only exception is the sampling phase of the algorithm which is modified to enable dealing with the problems with unitation constraints. All the parameters in Tab. 2 and Tab. 3 were set experimentally after high number of experiments. The experiments were performed on a computer with processor AMD Athlon II X2 245, 6GB RAM and operational system Centos 6.5.

9. Experiments and The Solutions

There were executed 20 instances of the proposed algorithm. Average running time of a single execution was 11 min. The cost values of the solutions are presented in Tab. 4.

id of run	1	2	3	4	5
cost value	3.805	2.442	3.805	2.431	2.428
id of run	6	7	8	9	10
cost value	2.437	2.443	3.653	2.433	2.436
id of run	11	12	13	14	15
cost value	2.432	3.662	5.197	5.663	5.353
id of run	16	17	18	19	20
cost value	6.071	3.805	5.691	6.072	2.429

Tab. 4. Results of 20 runs of the proposed algorithm.

As can be seen in Tab. 4 the best solution was achieved in run 5 with cost value 2.428. The schematic diagram is presented in Fig. 15. Schematic diagrams of another three good solutions (run 4, run 11, run 20) are presented in Fig. 16 to Fig. 18.

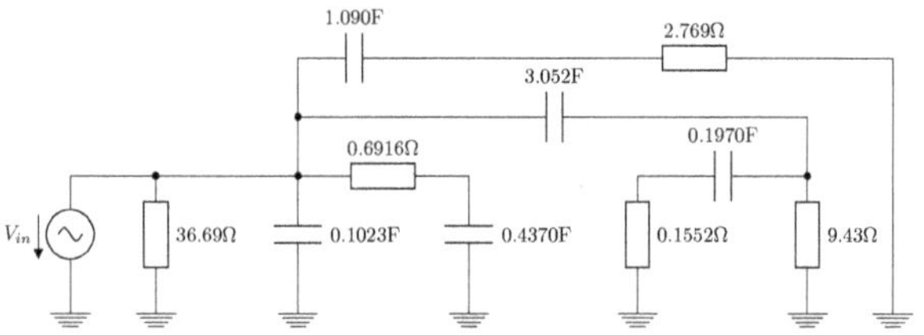

Fig. 15. Schematic of the circuit synthesized in run 5 [1].

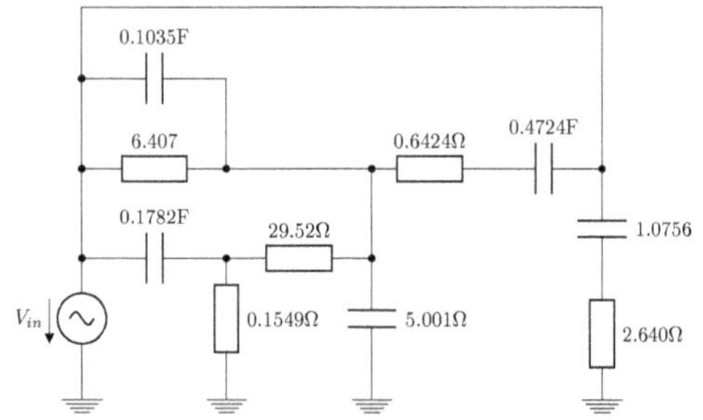

Fig. 16. Schematic of the circuit synthesized in run 4 [1].

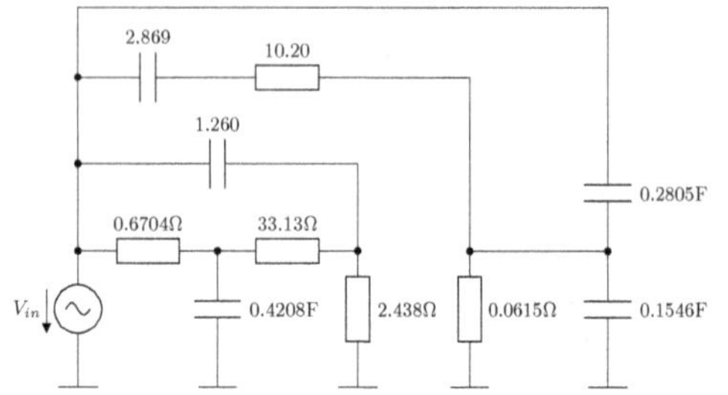

Fig. 17. Schematic of the circuit synthesized in run 11 [1].

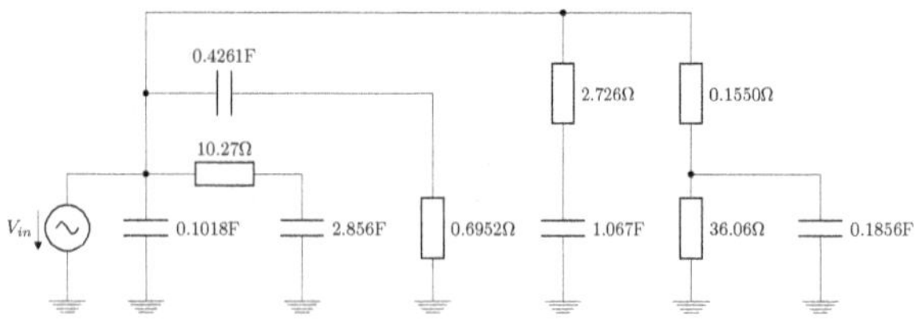

Fig. 18. Schematic of the circuit synthesized in run 20 [1].

Comparison of the magnitude and the phase characteristics of Z_{in} of the best found approximation circuit and desired function (2) are presented in Fig. 19 and Fig. 20 respectively.

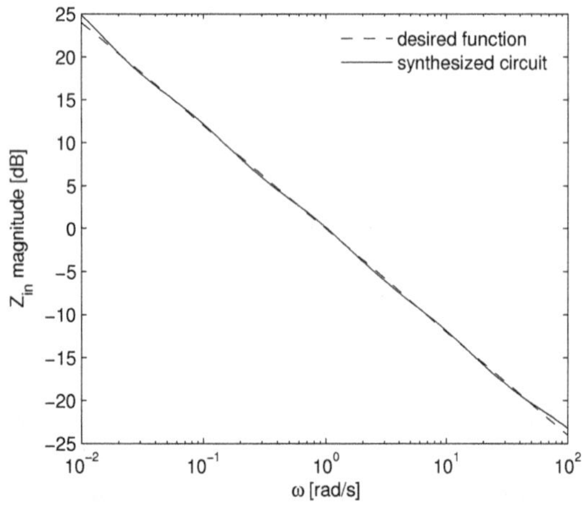

Fig. 19. Comparison of the magnitude of Z_{in} of the best found approximation circuit and (2) [1].

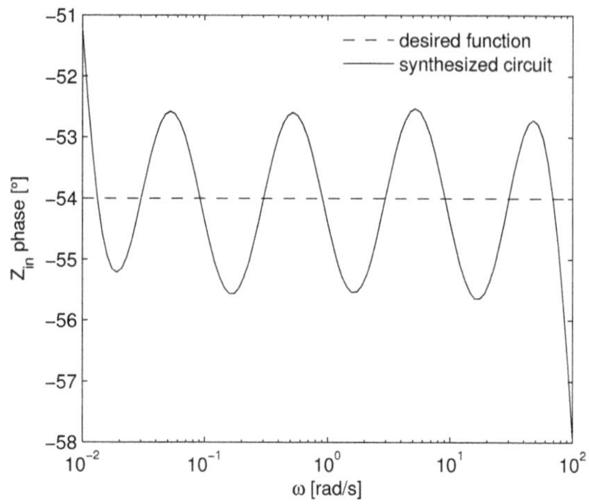

Fig. 20. Comparison of the phase of Z_{in} of the best found approximation circuit and (2) [1].

Absolute values of the deviations of the magnitude and the phase of Z_{in} of the best synthesized circuit are presented in Fig. 21. and Fig. 22 respectively.

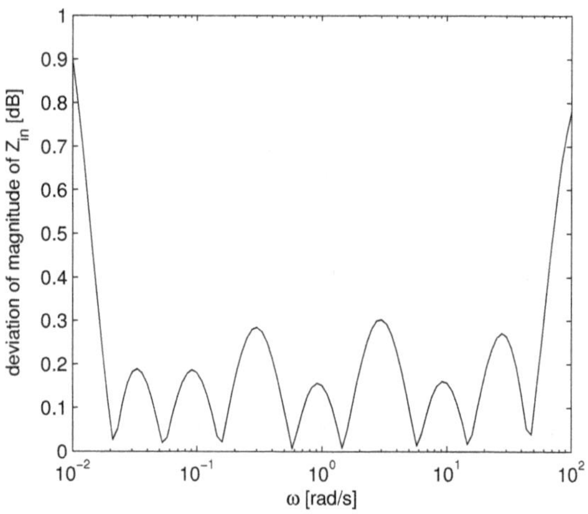

Fig. 21. Deviation of the magnitude of Z_{in} of the best found approximation circuit [1].

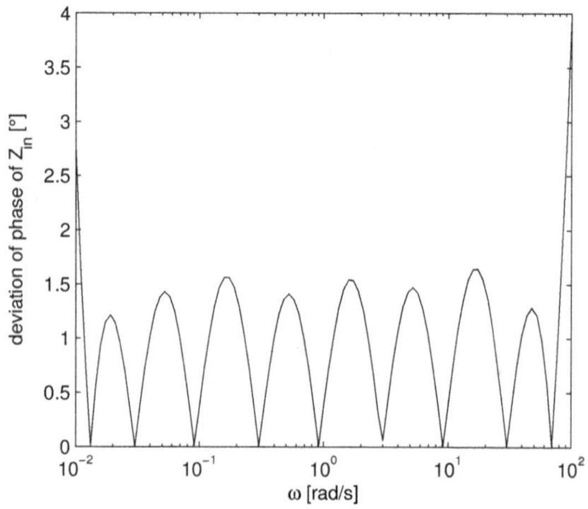

Fig. 22. Deviation of the phase of Z_{in} of the best found approximation circuit [1].

Three highest deviations of the magnitude and the phase characteristics of Z_{in} of the best synthesized circuit are summarized in Tab. 5.

Magnitude			
$\omega[rad/s]$	0.01	100	3.02
$\Delta_m[dB]$	0.9	0.78	0.30

Phase			
$\omega[rad/s]$	0.01	100	17.4
$\Delta_p[\circ]$	2.78	3.97	1.64

Tab. 5. The highest deviations of the magnitude and the phase of Z_{in} of the best synthesized approximation circuit.

As can be seen in Fig. 19 to Fig. 22 the maximum deviations of the magnitude and phase responses are located at the boundaries of the used frequency range. At these frequencies the unoptimized areas of the frequency response (0 to 10^{-2} rad/s and 10^2 to ∞ rad/s) affect behavior of the circuit in the area where the optimization was performed. The zeros and poles diagram of the synthesized circuit is presented in Fig. 23. All coefficients of the denominator of the approximation function of Z_{in} are positive therefore stability of the synthesized circuits is guaranteed.

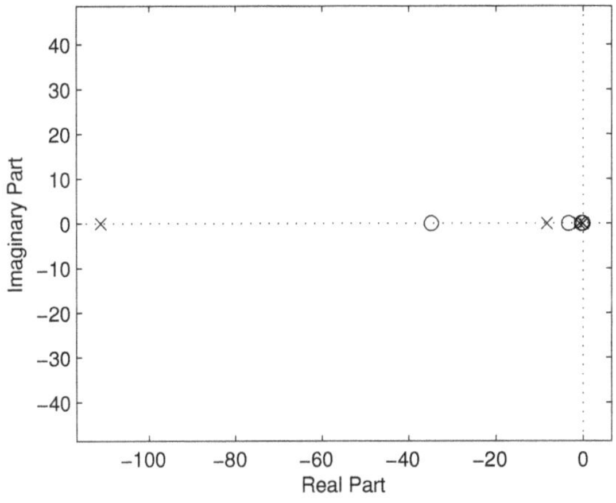

Fig. 23. The zeros and poles diagram of the best synthesized circuit [1].

Although the probability of using of all three component types (resistors, capacitors, inductors) was equal during the synthesis process, none of the circuits presented in Fig. 15 to Fig. 18 include any inductors. As the proposed algorithm was constrained to use only $n_c = 10$ components, it seems that using only capacitors and resistors allows the method to reach lower cost values than in solutions where inductors are included.

10. Comparison of The Results

In the section the best synthesized approximation circuit obtained using the proposed algorithm will be compared to the original approximation circuit designed in [9] by a classical method of the analog circuits design.

Since the proposed evolutionary synthesis method was configured to use the same circuit complexity (10 components at the most) as the original approximation circuit, accuracy of both circuits can be directly compared.

Except the deviations at the boundaries of the used frequency range (as was commented above) for the original approximation circuit the highest deviation of the magnitude is $\Delta_m = 0.61$ dB at angular frequency 0.33 rad/s. For the best solution of the proposed method the highest deviation of the magnitude is $\Delta_m = 0.27$ dB at angular frequency 0.30 rad/s. Thus in terms of deviation of the magnitude the accuracy of the synthesized circuit is more than twice better than the original approximation circuit. Comparison of the deviations of the magnitude of Z_{in} for both circuits is presented in Tab. 6.

original circuit	ω[rad/s]	0.01	100	0.33
	Δ_m[dB]	0.71	0.46	0.61
synthesized circuit	ω[rad/s]	0.01	100	0.30
	Δ_m[dB]	0.93	0.92	0.27

Tab. 6. Comparison of the deviations of the magnitude of Z_{in} of the original approximation circuit and the best synthesized circuit.

The highest phase deviation inside the used frequency range is for original circuit $\Delta_p = 5.2^\circ$ at angular frequency 0.14 rad/s and for the synthesized circuit it is $\Delta_p = 1.5^\circ$ at angular frequency 0.58 rad/s. Thus phase accuracy of the best synthesized circuit is more than three times better than the original approximation circuit. Comparison of the maximum deviations of the phase of Z_{in} is presented in Tab. 7.

original circuit	ω[rad/s]	0.01	100	0.14
	$\Delta_p[\circ]$	26.3	7.98	5.2
synthesized circuit	ω[rad/s]	0.01	100	0.58
	$\Delta_p[\circ]$	3.0	4.2	1.5

Tab. 7. Comparison of the deviations of the phase of Z_{in} of the original approximation circuit and the best synthesized circuit.

In [12] the same problem of synthesis of the fractional capacitor circuit was solved using simulated annealing method. The method was able to reach solutions of the same accuracy however the number of required evaluations of the objective function was almost four times higher. Comparison of accuracy of the best solutions and numbers of evaluations of the objective function of the proposed EDA method and simulated annealing method [12] is presented in Tab. 8.

simulated annealing	best cost	evals
method [12]	2.45	440e3
the proposed	best cost	evals
EDA method	2.43	120e3

Tab. 8. Comparison of the proposed EDA method and simulated annealing method presented in [12].

11. Conclusion

Automated analog circuit synthesis approach based on hybrid evolutionary method employing modified UMDA algorithm and a local search algorithm was presented in the paper. Used hybrid approach enables to employ specialized methods for both sub problems of different nature (synthesis of the topology and determination of the parameters).

Synthesis of the topology which is combinational optimization problem was solved using modified UMDA algorithm. Determination of the parameters which is continuous optimization problem was solved using a local search algorithm. The principle of the method is based on mutual interaction of synthesis of the topology phase (modified UMDA algorithm) and determination of the parameters phase (the local search algorithm) of the desired solution. Modification of the UMDA algorithm which allows solving problems with unitation constrains was proposed in the paper.

The proposed method was verified on the problem of synthesis of the fractional capacitor circuit introduced in [9]. Presented experiments have shown that the proposed algorithm is capable to synthesize solutions with accuracy overperforming solutions obtained using a classical method of the analog circuit design given in [9]. Accuracy of the magnitude of Z_{in} of the best obtained solution was more than twice better than the original approximation circuit. Accuracy of the phase of Z_{in} of the best obtained solution was more than three times better than the original approximation circuit.

In [12] the problem of fractional capacitor circuit realization was solved using simulated annealing method (SA). The accuracy of the circuits synthesized using SA was the same as the solutions produced using the proposed EDA method.

However SA required much higher number of the objective function evaluations. While the proposed EDA method required 120e3 objective function evaluations for synthesis of the same problem SA required 440e3 objective function evaluations.

For demonstration purposes the presented algorithm was verified using nodal analysis circuit simulator implemented in Matlab. Another improvement of the efficiency of the algorithm can achieved using Model Order Reduction Techniques [13].

Acknowledgements

The presented research was financed by the Czech Ministry of Education in frame of the National Sustainability Program, the grant LO1401 INWITE. For the research, infrastructure of the SIX Center was used.

References

[1] SLEZÁK, J. *Evolutionary Synthesis of Analog Electronic Circuits Using EDA Algorithms*. Ph.D. Thesis, Brno: Brno University of Technology, The Faculty of Electrical Engineering and Communication, 2014. 123 pages.

[2] ZINCHENKO, L., MÜHLENBEIN, H., KUREICHIK, V., MAHNING, T. Application of the univariate marginal distribution algorithm to analog circuit design. In *Proceedings of NASA/DoD Conference on Evolvable Hardware*, 2002, p. 93–101. DOI: 10.1109/EH.2002.1029871

[3] TORRES, A., PONCE, E. E., TORRES, M. D., DIAZ, E., PADILLA, F. Comparison of two evolvable systems in the automated analog circuit synthesis. In *Proceedings of Eighth Mexican International Conference on Artificial Intelligence (MICAI 2009)* Mexico, 2009, p. 3–8, ISBN 978-0-7695-3933-1. DOI: 10.1109/MICAI.2009.25

[4] TANG, M., LAU, R. Y. K. A hybrid estimation of distribution algorithm for the minimal switching graph problem. In *Proceedings of International Conference on Computational Intelligence for Modelling, Control and Automation and International Conference on Intelligent Agents, Web Technologies and Internet Commerce*, Austria 2005, p. 708–713. DOI: 10.1109/CIMCA.2005.1631347

[5] MÜHLENBEIN, H., PAAß, G. From recombination of genes to the estimation of distributions I. Binary parameters. *Lecture Notes in Computer Science 1411: Parallel Problem Solving from Nature - PPSN IV*, p. 178–187. DOI: 10.1007/3-540-61723-X_982

[6] SANTANA, R., OCHOA, A., SOTO, M. R. Factorized distribution algorithms for functions with unitation constraints. In *Proceedings of the Third Symposium on Adaptive Systems (ISAS-2001)*, 2001, p. 158–165.

[7] MÜHLENBEIN, H., MAHNIG, T., OCHOA, A. Schemata, distributions and graphical models in evolutionary optimization. *Journal of Heuristics*, 1999, p. 215–247.

[8] SANTANA, R., ECHEGOYEN, C., MENDIBURU, A., BIELZA, C., LOZANO, J. A., LARRAÑAGA, P., ARMAÑANZAS, R., SHAKYA, S. MATEDA: A suite of EDA programs in Matlab. *Technical Report EHU-KZAA-IK-2/09*, University of the Basque Country, 2009.

[9] CARLSON, G. E., HALIJAK, C. A. Approximation of fractional capacitors $(1/s)^{(1/n)}$ by a regular Newton process. *IEEE Transactions on Circuit Theory*, 1964, vol. 11, no. 2, p. 210–213, ISSN 0018-9324. DOI: 10.1109/TCT.1964.1082270

[10] BALAKRISHNAN, V. K. *Graph Theory*, McGraw-Hill, 1997, ISBN 0-07-005489-4.

[11] EL-MIHOUB, T. A., HOPGOOD, A. A., NOLLE, L., BATTERSBY, A. Hybrid genetic algorithms: A review. *Engineering Letters*, 2006, vol. 13, no. 2, p. 124-137, ISSN: 1816-093X.

[12] SLEZAK, J., GOTTHANS, T., DRINOVSKY, J. Evolutionary Synthesis of Fractional Capacitor Using Simulated Annealing Method. *Radioengineering*, 2012, vol. 21, no. 4, p. 1252–1259, ISSN 1210-2512.

[13] FELDMANN, P. Model order reduction techniques for linear systems with large numbers of terminals. In *Proceedings of Design, Automation and Test in Europe Conference and Exhibition*, 2004, vol. 2.1, p. 944-947. DOI: 10.1109/DATE.2004.1269013

About the Authors...

Josef SLEZAK was born in Zlin, Czech Republic, in 1982. He received the MSc. degree at the Brno University of Technology in 2007. In 2014 he received Ph.D. degree from the Brno University of Technology. His research interests include evolutionary synthesis of analog electronic circuits, design automation and optimization.

Tomas GOTTHANS was born in Brno, Czech Republic, in 1985. He received the MSc. degree at the Brno University of Technology (BUT) in 2010. In 2014 he received Ph.D. degree from the Université Paris-Est (UPE) and from the Brno University of Technology. He was working in the École Supérieure d'Ingénieurs en Électronique et Électrotechnique de Paris (ESIEE) in the ESYCOM Laboratory. He is presently a researcher in the Sensor, Information and Communication Systems (SIX) research laboratory, BUT. His research interests include programming, wireless communications and non-linear phenomenons.

13

Fast Implementation of Transmit Beamforming for Colocated MIMO Radar

Gollakota Venkata Krishna SHARMA [1], *Konduri RAJA RAJESWARI* [2]

[1] Dept. of ECE, GITAM Institute of Technology, GITAM University, India
[2] Dept. of ECE, College of Engineering Autonomous, Andhra University, India

gvksarma@gitam.edu, krrau@yahoo.com

Abstract. *Multiple-input Multiple-output (MIMO) radars benefit from spatial and waveform diversities to improve the performance potential. Phased array radars transmit scaled versions of a single waveform thereby limiting the transmit degrees of freedom to one. However MIMO radars transmit diverse waveforms from different transmit array elements thereby increasing the degrees of freedom to form flexible transmit beampatterns. The transmit beampattern of a colocated MIMO radar depends on the zero-lag correlation matrix of different transmit waveforms. Many solutions have been developed for designing the signal correlation matrix to achieve a desired transmit beampattern based on optimization algorithms in the literature. In this paper, a fast algorithm for designing the correlation matrix of the transmit waveforms is developed that allows the next generation radars to form flexible beampatterns in real-time. An efficient method for sidelobe control with negligible increase in mainlobe width is also presented.*

Keywords

Multiple-input multiple-output radar, transmit beamforming, fast implementation, waveform diversity, zero-lag correlation matrix

1. Introduction

MIMO radars employ multiple transmit antennas and multiple receive antennas thereby benefiting from increased degrees of freedom due to spatial diversity. These radars further have the flexibility to transmit diverse waveforms from their different transmit antennas thereby benefiting from waveform diversity. This spatial diversity and waveform diversity together can be used to improve many aspects of radar system performance [1]. MIMO configurations employing widely separated antennas [2] called "Statistical MIMO" radars improve target detection capabilities due to spatial diversity. MIMO radars can also employ closely spaced antennas [3] called "Colocated MIMO" radars to improve interference rejection capability, parameter identifiability and resolution performance due to increased virtual aperture. In this paper we deal with colocated MIMO radar only.

Conventionally MIMO radars employ orthogonal signals to obtain the phase delay for each transmitting/receiving antenna pair, thus increasing the degrees of freedom for beamforming. However MIMO radars can also employ partially correlated signals to form flexible transmit beampatterns with better performance than phased array beampatterns for a given number of transmit antenna elements. The advantages of MIMO radars (employing diverse waveforms) in forming flexible transmit beampaterns over phased array counterparts are well illustrated in [4]. While phased arrays employ only spatial diversity to form transmit beams, MIMO radars employ both spatial and waveform diversity to form transmit beams with enhanced flexibility.

The transmit beampattern of a colocated MIMO radar depends on the zero-lag correlation matrix of different transmit waveforms [5–6]. Orthogonal transmit waveforms result in uniform illumination of power in all directions which is very helpful in search applications. Fully correlated waveforms (perhaps also scaled by a complex constant) result in directive beams used in tracking applications. Partially correlated waveforms with a specific zero-lag correlation matrix can be used to form a wide range of beampatterns.

To manage the complexity, the design of diverse waveforms to achieve a desired transmit beampattern is considered in the literature in two stages. In the first stage, the zero-lag correlation matrix $\mathbf{R} = \mathbf{X}\mathbf{X}^{H}$ of the transmit waveform matrix $\mathbf{X}$ is designed. In the second stage, the design of signal waveform matrix $\mathbf{X}$ having a zero-lag correlation matrix $\mathbf{R}$ (obtained from the previous stage) and meeting some practically motivated constraints is considered. Low complexity algorithms for the latter problem exist in the literature [11–14]. Most of the existing algorithms in the literature addressing the former problem [4–10] are based on computationally complex optimization methods and hence difficult to solve in real-time. Design of signal correlation matrix for specific transmit beampattern was first addressed in [5], [6] as an optimization problem solved using gradient search method. In [4], the design of correlation matrix for designing different transmit beampatterns along with minimization of crossbeam pattern is formulated as an SDP optimization problem and solved using an efficient cyclic algorithm proposed therein. In [7], an eigenvalue decomposition method is used to achieve specific illumination pattern. Many target tracking applications require flexible transmit beampatterns generated in real-time. Previous works on signal correlation matrix design for approximation of desired transmit beampatterns employed optimization methods which are difficult to solve in real-time or highly complex to implement in hardware.

This paper considers the real-time computation of the zero-lag signal correlation matrix **R** that approximates a desired beampattern. The main contributions of this paper are (a) closed-form solution for computing the zero-lag correlation matrix and (b) FFT based algorithm for computing the zero-lag correlation matrix. The advantage of the proposed algorithm is that the proposed algorithm is easily amenable to hardware implementation since efficient architectures for FFT algorithms exist [18].

Section 2 describes MIMO radar signal model and defines a complete beampattern that includes the specification of transmit and cross beampattern. Section 3 derives the closed form solution and presents a fast implementation of the solution using 2D FFT algorithm. Section 4 presents the numerical results and Section 5 concludes the paper.

2. MIMO Radar Signal Model

Consider a monostatic MIMO radar that contains M transmitters with the antenna elements configured as uniform linear arrays. We assume a point target and also that the target and transmitters lie in the same 2-D plane as shown in Fig. 1. Let d_T represent the spacing between consecutive transmitters. Let θ be the target angle with respect to the broadside direction and λ is the carrier wavelength of the transmitted waveforms. Let $\{u_m(t)\}$, $m \in \{0,1,\ldots,M-1\}$ represent the M transmitter waveforms. All the transmit antennas transmit waveforms simultaneously in time. We further assume that the transmitter waveforms are narrowband and the baseband signal waveforms are not modified because of Doppler effect [15]. The correlation between two transmit waveforms $u_m(t)$ and $u_{m'}(t)$ at zero time-lag is defined as

$$r_{m,m'} = \int_0^{T_0} u_m(t)u_{m'}^*(t)dt \tag{1}$$

and $\mathbf{R} = [r_{m,m'}]_{M\times M}$ represents the zero-lag correlation matrix of the M transmit waveforms.

2.1 Transmit Beampattern

The baseband signal at the target location can be described by the expression

$$\sum_{m=0}^{M-1} e^{-j2\pi fm} u_m(t) \triangleq \mathbf{a}^H(f)\mathbf{u}(t) \tag{2}$$

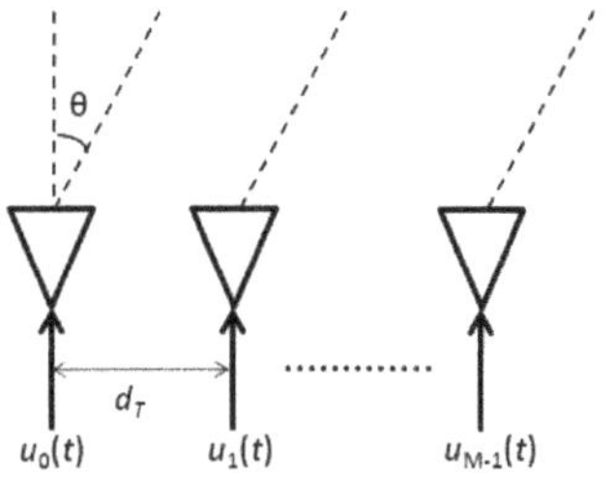

Fig. 1. Transmitter model.

where $f = d_T \sin(\theta)/\lambda$ is the spatial frequency of the target,

$$\mathbf{u}(t) = \left[u_0(t)u_1(t)u_2(t)\ldots u_{M-1}(t)\right]^T \tag{3}$$

is the vector of M transmit waveforms and $\mathbf{a}(f)$ is the array steering vector given by

$$\mathbf{a}(f) = \left[e^{j2\pi f.0} \quad e^{2\pi f.1} \quad \cdots \quad e^{j2\pi f(M-1)}\right]^T. \tag{4}$$

With typical transmitter spacing of $d_T = \lambda/2$, the spatial frequency f is in [–½, ½]. The spatial distribution of power of the transmit signals is called the *transmit beampattern* and is given by [4],

$$P(f) = \mathbf{E}[\mathbf{a}^H(f)\mathbf{u}(t)\mathbf{u}^H(t)\mathbf{a}(f)] = \mathbf{a}^H(f)\mathbf{Ra}(f)$$
$$= \sum_{m=0}^{M-1}\sum_{m'=0}^{M-1} r_{m,m'} e^{j2\pi f(m-m')}. \tag{5}$$

Consider $P(f)$ for a phased array radar case. The $M \times 1$ transmit signal vector $\mathbf{u}(t)$ is given by $\mathbf{u}(t) = \mathbf{a}(f_0)\, u(t)$ where $f_0 = d_T \sin(\theta_0)/\lambda$ with θ_0 denoting the steered direction. Then, $\mathbf{R} = \mathbf{a}(f_0)\,\mathbf{a}^H(f_0)$ assuming unit power signal $u(t)$ and

$$P(f) = \mathbf{a}^H(f)\mathbf{a}(f_0)\mathbf{a}^H(f_0)\mathbf{a}^H(f) = \left|\mathbf{a}^H(f)\mathbf{a}(f_0)\right|^2. \tag{6}$$

Note that the transmit gain attains maximum value in the direction θ_0 and is decreased at $\theta \neq \theta_0$. Now, consider $P(f)$ with orthogonal signals. Then, $\mathbf{R} = \mathbf{I}$, and

$$P(f) = \mathbf{a}^H(f)\mathbf{a}(f) = M \tag{7}$$

this implies that the beampattern is omnidirectional. Thus, the traditional beamforming results in a focused beampattern while the beampattern of MIMO with orthogonal signals is uniform in all directions. In some applications, it might be desirable to synthesize a beampattern that is between these two extremes so that wide focus areas can be formed without wasting power in the directions that are of no interest. This can be achieved by adjusting the correlation matrix **R** of the transmitted waveforms.

2.2 Crossbeampattern

The crosscorrelation between the signals backscattered to the radar by any two targets (at location parameters f and f'), called the *crossbeampattern* is given by

$$c(f,f')=\mathbf{E}\left[\mathbf{a}^H(f)\mathbf{u}(t)\mathbf{u}^H(t)\mathbf{a}(f')\right]=\mathbf{a}^H(f)\mathbf{R}\mathbf{a}(f') = \sum_{m=0}^{M-1}\sum_{m'=0}^{M-1} r_{m,m'}e^{j2\pi(fm-f'm')}, f\neq f' \quad (8)$$

In practical applications, it is desirable to minimize $c(f,f')$ for improving the quality of adaptive localization techniques [16]. In phased array beamforming, the signals backscattered to the radar from two targets (at location parameters f and f') are fully coherent, which in particular makes the adaptive localization techniques inapplicable [16].

2.3 Complete Beampattern

The complete beampattern $s(f,f')$ as a function of spatial frequencies f and f' can be written as

$$s(f,f')=\mathbf{a}^H(f)\mathbf{R}\mathbf{a}(f')=\sum_{m=0}^{M-1}\sum_{m'=0}^{M-1} r_{m,m'}e^{j2\pi(fm-f'm')} \quad (9)$$

We note that transmit beampattern $P(f)$ is the complete beampattern evaluated along the $f=f'$ line i.e., $P(f)=s(f,f)$. The crossbeampattern $c(f,f')$ is the complete beampattern $s(f,f')$ at $f\neq f'$. From (9), we see that the complete beampattern is the scaled two-dimensional Fourier transform of the zero-lag correlation matrix **R**. We also notice that with typical transmitter spacing $d_T=\lambda/2$, $s(f,f')$ is periodic w. r. t f,f' with period 1.

2.4 Problem Formulation

The objective in transmit beampattern design, is to design **R** so that the transmit power is directed in desired directions f and $c(f,f')$ is minimized. The specification of complete beampattern $s(f,f')$ (rather than just the transmit beampattern $P(f)$ captures both these goals. We state the design problem as follows: Given a desired beampattern function $s_d(f,f')$, find the matrix $\mathbf{R}=[r_{m,m'}]_{M\times M}$ that closely approximates the relation

$$s_d(f,f')=\sum_{m=0}^{M-1}\sum_{m'=0}^{M-1} r_{m,m'}e^{j2\pi(fm-f'm')}. \quad (10)$$

Any practically achievable signal correlation matrix **R** will be positive semi-definite and Hermitian (i.e. $r_{m,m'}=r^*_{m,m'}$). **R** will have the property if we choose the desired beampattern function $s_d(f,f')$ that is real valued and symmetric about the $f=f'$ line. Further the total transmit energy can be constrained to P by multiplying **R** by the factor $P/\mathrm{tr}\{\mathbf{R}\}$.

3. Closed Form Solution

We now proceed to derive a closed form solution to design the correlation matrix **R** to realize the desired complete beampattern $s_d(f,f')$ given by

$$s_d(f,f')=\sum_{m=0}^{M-1}\sum_{m'=0}^{M-1} r_{m,m'}e^{j2\pi(fm-f'm')}. \quad (11)$$

Multiplying (11) on both sides by $e^{-j2\pi fk_1}e^{j2\pi f'k_2}$ and integrating w.r.t f and f' over one period (0,1) we have

$$\int_0^1\int_0^1 s_d(f,f')e^{-j2\pi fk_1}e^{j2\pi f'k_2}df df' = \int_0^1\int_0^1\sum_{m=0}^{M-1}\sum_{m'=0}^{M-1} r_{m,m'}e^{j2\pi f(m-k_1)}e^{-j2\pi f'(m'-k_2)}df df'. \quad (12)$$

Interchanging the order of integration and summation we have

$$\sum_{m=0}^{M-1}\sum_{m'=0}^{M-1} r_{m,m'}\left(\int_0^1 e^{j2\pi f(m-k_1)}df\right)\left(\int_0^1 e^{-j2\pi f'(m'-k_2)}df'\right)=r(k_1,k_2) \quad (13)$$

In the above equation, the relation $\int_0^1 e^{-j2\pi f(n-k)}df=\delta(n-k)$ has been used. The closed form solution for the zero-lag correlation matrix $\mathbf{R}=[r_{m,m'}]_{M\times M}$ is given by

$$r_{m,m'}=\int_0^1\int_0^1 s_d(f,f')e^{-j2\pi(fm-f'm')}df df'. \quad (14)$$

A similar solution is presented in [17]. However the difficulty is in the evaluation of the integral in (14). In mission critical target tracking applications, computation of **R** from (14) is not possible in real-time. This problem is analogous to FIR filter design using Fourier series method wherein the desired unit sample response $h_d(n)$ of the FIR filter to the designed one is calculated from the desired frequency response $H_d(e^{j\omega})$ as [18].

$$h_d(n)=\frac{1}{2\pi}\int_{-\pi}^{\pi}H_d(e^{j\omega})e^{j\omega n}d\omega. \quad (15)$$

An approximation to $h_d(n)$ can be obtained by sampling $H_d(e^{j\omega})$ and using the inverse discrete Fourier transform to compute

$$\tilde{h}_d(n)=\frac{1}{K}\sum_{k=0}^{K-1}H_d\left(e^{j(2\pi/K)k}\right)e^{j(2\pi/N)kn}, \quad 0\le n\le N-1 \quad (16)$$

where N is the length of the FIR filter. If $K>>N$, $\tilde{h}_d(n)$ can be expected to be a good approximation to $h_d(n)$. In a similar way, the integral in (14) can be approximated as

$$\tilde{r}_{m,m'}=\frac{1}{K}\frac{1}{K}\sum_{k=0}^{K-1}\sum_{k'=0}^{K-1}s_d\left(\frac{k}{K},\frac{k'}{K}\right)e^{-j2\pi\left(\frac{km}{K}-\frac{k'm'}{K}\right)}, \quad 0\le m\le M-1,\ 0\le m'\le M-1. \quad (17)$$

If $K>>M$, $\tilde{r}_{m,m'}$ can be expected to be a good approximation to $r_{m,m'}$. (17) can be efficiently computed using 2-D inverse fast Fourier transform (FFT). This allows the correlation matrix to be computed in real-time in contrast to previously proposed optimization based methods. The steps of the algorithm are presented below

1. Given the desired complete beampattern function $s_d(f, f')$ (specified over a large grid of points over f and f') as $\hat{s}_d(k,k') = s_d(f,f')$ at $(f,f') = \left(\frac{k}{K}, \frac{k'}{K}\right)$ with $k,k' \in \{0,1,..,K-1\}$ where $K >> M$. Typically $K = 3M$ is adequate.
2. Find the 2-D inverse fast Fourier transform of $\hat{s}_d(k,k')$ to obtain $\hat{S}_d(m, m')$.
3. The desired correlation matrix can be obtained as $r_{m,m'} = \hat{S}_d\left(\left(-m\right)_M, m'\right)$ over $0 \le m,\ m' \le (M-1)$. Here $()_M$ indicates modulo M operation.

In Step 3 above, the reflection property of 2-D Fourier transform viz. If $x(n_1, n_2) \to X(\omega_1, \omega_2)$ then $x(n_1, -n_2) \to X(\omega_1, -\omega_2)$, is used.

4. Design Results

Two scenarios are considered for the simulation. The first scenario considers the maximum power design at known target locations and the second scenario considers beampattern matching design.

4.1 Maximum Power Design for Known Target Locations

In practice, to obtain the prior knowledge about the target locations, orthogonal waveforms [19], 20] are used for MIMO probing. This is referred to as "initial probing". After we get the target location estimates with this probing and spatial spectrum estimation techniques [21], we can optimize the transmitted beampattern to direct the transmit power in these known target locations. We define a simple parameterized expression for the desired beampattern function. Assume that there are K targets in the range of the radar, located at angles θ_k, corresponding to spatial frequencies $\{f_k\}$, $k = 0,1, \ldots, K-1$ with $f_k = d_T \sin(\theta_k)/\lambda$. We would like to concentrate the energy in the vicinity of each target. The desired beampattern $s_d(f, f')$ is expressed as

$$s_d(f,f') = \sum_{k=1}^{K} s_{d,k}(f,f') * g(f,f') \tag{18}$$

where

$$s_{d,k}(f,f') = \begin{cases} 1 & f = f' = f_k \\ 0 & \text{otherwise} \end{cases}$$

$$g(f,f') = \frac{1}{f_x f_y \sqrt{1-\rho^2}} .\exp\left\{ \frac{-1}{2(1-\rho^2)} \left[\frac{f^2}{f_x^2} - \frac{2\rho f f'}{f_x f_y} - \frac{f'^2}{f_y^2} \right] \right\} \tag{19}$$

and * represents convolution.

The function $g(f, f')$ is used to reduce the sidelobe level in the resulting beampattern by providing smooth tapering on all sides. When $f_x = f_y$ and $0 < \rho < 1$ the contour of $g(f, f')$ reduces to an ellipse rotated by angle $\pi/4$ i.e. along the line $f = f'$. This choice makes $s_d(f, f')$ symmetric along the $f = f'$ line, resulting in a zero-lag correlation matrix **R** that is positive semi-definite and Hermitian. The choice of ρ affects the resolution along the $f = f'$ line and $f \neq f'$ directions. The choice of $\rho = 0$ and $\rho = 1$ makes the contour of desired beampattern a circle and ellipse respectively at each desired direction along the $f = f'$ line. The function in (19) provides more degrees of freedom to control sidelobe levels compared to a tapering window approach used in [11]. The values f_x, f_y can be increased so that energy in $\hat{S}_d(m, m')$ can be spread uniformly along m,m', so that $r_{m,m} \approx$ constant thus satisfying the uniform elemental power constraint [4]. As an example, the desired beampattern function for three targets located at $f_0 = 0.25$, $f_1 = 0.4$ and $f_2 = 0.75$ is shown in Fig. 2. The other parameters are chosen as $f_x = f_y = 0.02$ and $\rho = 0.2$.

The 2-D inverse FFT coefficients $\hat{S}_d(m, m')$ of the desired beampattern function are shown in Fig. 3. We see that most of the energy is contained in the coefficients with $0 \le m \le 15$ and $0 \le m' \le 15$.

The beampattern obtained for different values of M is shown in Fig. 4. Note that $M = 5$ does not provide enough degrees of freedom for synthesizing three distinct beams. We also note that $M = 20$ does not offer distinct advantages over $M = 15$. The transmit beampattern for different values of M is shown in Fig. 5.

One of the practically motivated constraint is to have the transmit waveforms with uniform elemental power constraint to limit the transmit signals having wildly varying magnitudes thereby limiting signal distortion. The diagonal elements of the computed correlation matrix represent the power of the transmit signals emitted by various transmitters. With $\rho = 0.4$, the spread of the energy in the resulting correlation matrix **R**, is evenly distributed resulting in a maximum to minimum element ratio of 1.2 thus nearly satisfying the uniform elemental power constraint.

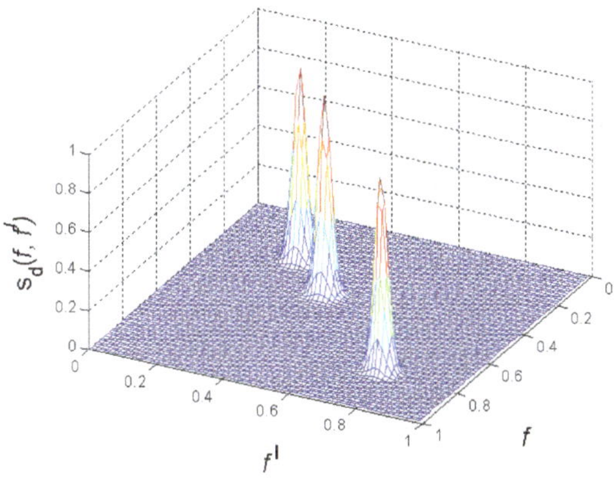

Fig. 2. Desired beampattern function $s_d(f, f')$.

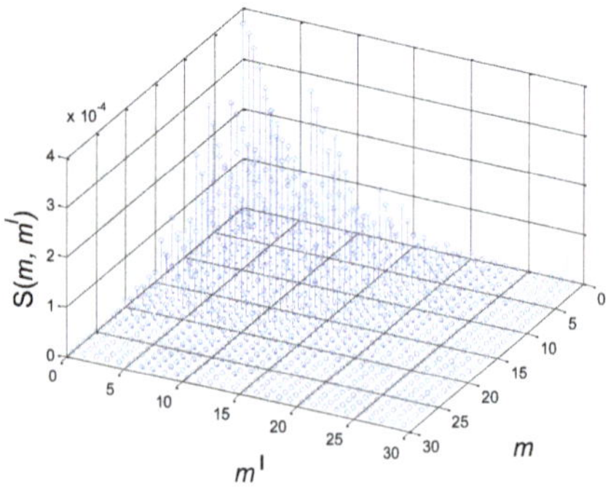

Fig. 3. 2-D inverse FFT coefficients $\hat{S}_d(m, m')$ of the desired beampattern function.

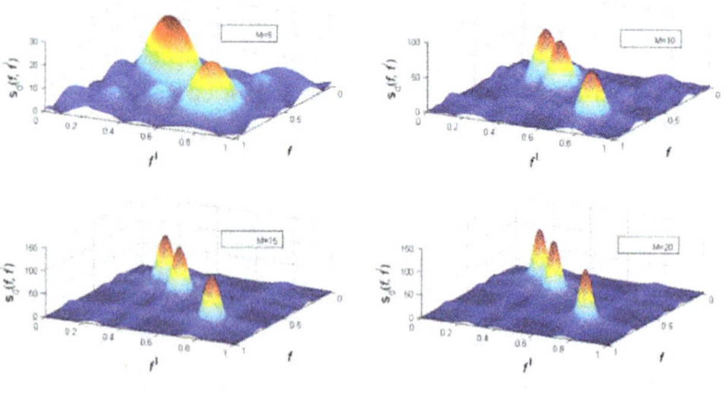

Fig. 4. Beampattern function with M=5, 10, 15 and 20.

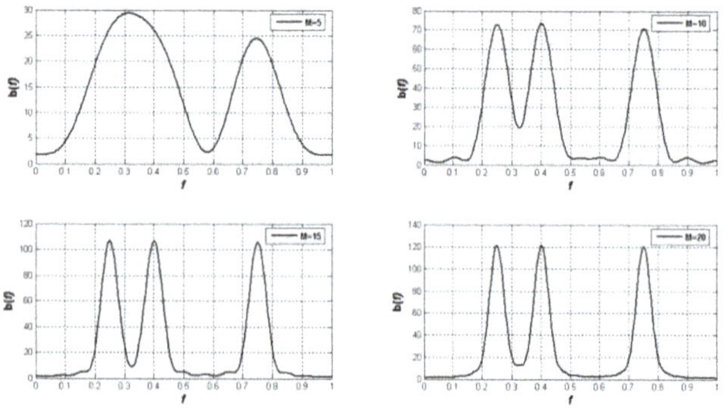

Fig. 5. Transmit beampattern with M=5, 10, 15 and 20.

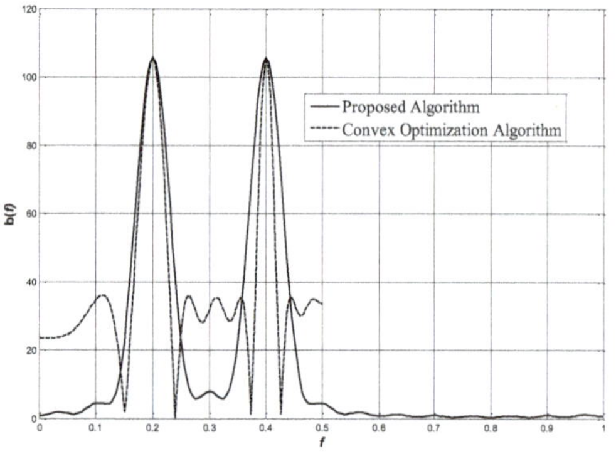

Fig. 6. Transmit beampattern evaluated using the algorithm in [4] and the proposed algorithm.

For comparison, the correlation matrices $\mathbf{R}_i$ and $\mathbf{R}_{ii}$ for achieving maximum power at spatial frequencies $f_1 = 0.2$ and $f_2 = 0.4$ are computed using (i) convex optimization method proposed in [4], and (ii) the proposed FFT based algorithm. The other parameters are chosen as $M = 15$, $f_x = f_y = 0.02$ and $\rho = 0.2$. The transmit beampattern of corresponding correlation matrices is plotted in Fig. 6. It can be noticed that the proposed algorithm achieves the low sidelobe levels with small increase in mainlobe width.

4.2 Beampattern Matching Design

Consider a desired beampattern $\phi(\theta)$ given by

$$\phi(\theta) = \begin{cases} \mu_k & \theta \in [\theta_k - \Delta_k, \theta_k + \Delta_k], k = 1,2,\ldots,K \\ 0 & \text{otherwise} \end{cases} \quad (20)$$

with $K = 3$, $\theta_1 = -30°$, $\theta_2 = 0°$, $\theta_3 = 40°$, $\Delta_k = 10°$, $\mu_1 = 0.5$, $\mu_2 = 1$, and $\mu_3 = 0.75$. Here θ_k represents the location of the kth target. Δ_k, μ_k represent the beamwidth and transmit gain in the direction of the kth target respectively. The desired beampattern is obtained by convolution of ideal beampattern with Gaussian tapering given by (20) with $f_x = f_y = 0.015$ and $\rho = 0$. The desired beampattern for the three targets is shown in Fig. 7. The beampattern approximation for different values of M is shown in Fig. 8. The transmit beampattern (in dB) as a function of angle (in degrees) for different values of M is shown in Fig. 9. From the above figures we notice that with $M = 20$, the degrees of freedom are sufficient to closely approximate the desired beampattern. We also notice that $M = 25$ does not offer improvement over $M = 20$.

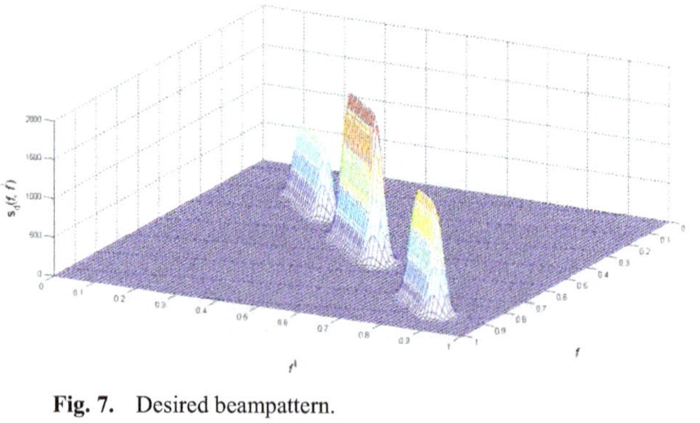

Fig. 7. Desired beampattern.

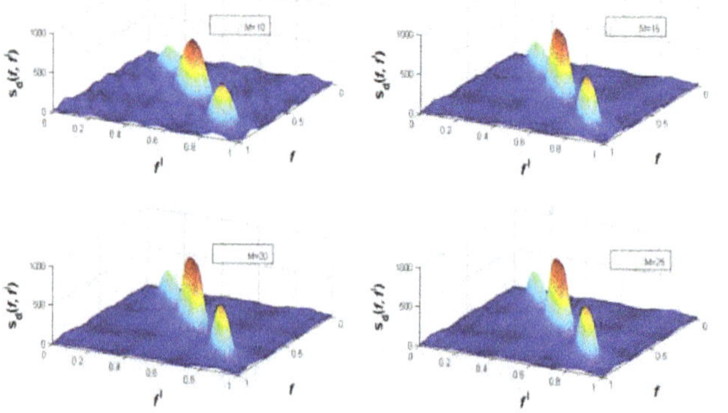

Fig. 8. Beampattern function with M = 10, 15, 20 and 25.

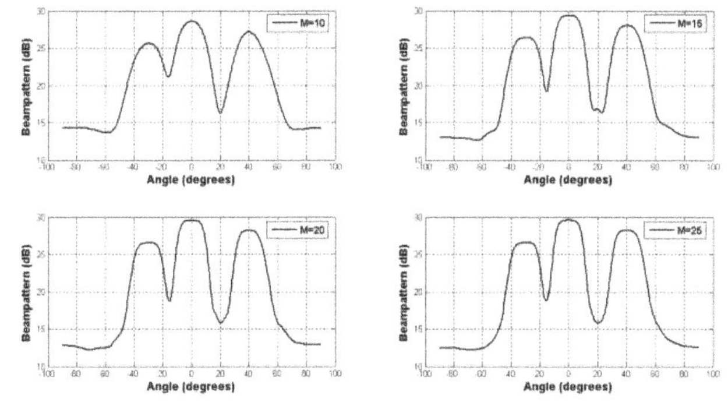

Fig. 9. Transmit beampattern with $M=10$, 15, 20 and 25.

4.3 Computational Complexity and Comparison with Previous Works

FFT is the only operation utilized in the proposed algorithm. The complexity of the N-point FFT is known to be $O(N\log N)$ [18]. Since 2-D FFT algorithm is a separable transform, computing $\hat{S}_d(m, m')$ from $\hat{s}_d(k,k')$ in Step 2 of the algorithm above requires computing M 1-D FFTs along the rows of $\hat{s}_d(k,k')$ followed by M 1-D FFTs along the columns. Hence, the overall complexity of the proposed algorithm is $O(M^2\log M)$. As a comparison, the SQP algorithms proposed in [4], [6], [8] have a complexity of $O(\log(1/\eta)M^{3.5})$ for an accuracy of η [13]. For comparison, the number of computations required for calculating the above correlation matrix using (i) convex optimization method proposed in [4] and (ii) the proposed FFT based algorithm are evaluated using MATLABv7. The number of complex computations using both the algorithms (with $M = 20$, $K = 60$ grid points over $f, f' \in \{0,1\}$) are found to be 28317904 and 826712 respectively. Roughly, in this case the FFT based algorithm speeds up the computation by a factor of 34.

In terms of the calculated transmit beampattern, it has been found that the solution given by the iterative methods [4–6], [9], [10] and the proposed algorithm in this paper achieve identical transmit beampatterns (under similar set of input conditions), with possibly a tradeoff between mainlobe width, sidelobe level and out-of-band roll off.

5. Conclusions

In this paper, a fast calculation of signal correlation matrix that approximates a desired transmit beampattern of colocated MIMO radar based on 2-D inverse FFT is proposed. Simulation results show correct operation of the proposed algorithm. While practically motivated constraints are easy to incorporate into the previously developed convex optimization based methods, they are difficult to solve in real time. However the proposed FFT based algorithm allows the correlation matrix to be computed in real-time and is easily amenable to hardware implementation [21].

Previous literature [1] considered windowing of zero-lag correlation matrix to achieve low sidelobe level at the cost of increased mainlobe width. This paper achieved sidelobe control of beampattern function by considered the desired beampattern tapered by a two-dimensional Gaussian density function. Results show low sidelobe levels with negligible increase in mainlobe width. The sidelobe levels and mainlobe width of the crossbeam pattern can be controlled by adjusting the parameters of the function $g(f, f')$. If extremely low sidelobe levels are desired then the tapering windows approach [22] can be applied on the computed correlation matrix.

References

[1] LI, J., STOICA, P. (Eds.) *MIMO Radar Signal Processing*. New York: Wiley, 2008.

[2] HAIMOVICH, A. M., BLUM, R. S., CIMINI, L. J. MIMO radar with widely separated antennas. *IEEE Signal Processing Magazine*, 2008, vol. 25, no. 1, p. 116–129. DOI: 10.1109/MSP.2008.4408448

[3] LI, J., STOICA, P. MIMO radar with colocated antennas. *IEEE Signal Processing Magazine*, Sept. 2007, vol. 24, no. 5, p. 106 to 114. DOI: 10.1109/MSP.2007.904812

[4] LI, J., STOICA, P., XIE, Y. On probing signal design for MIMO radar. *IEEE Transactions on Signal Processing*, Aug. 2007, vol. 55, no. 8, p. 4151–4161. DOI: 10.1109/TSP.2007.894398

[5] FUHRMANN, D., SAN ANTONIO, G. Transmit beamforming for MIMO radar systems using partial signal correlation. In *Proc. 38th Asilomar Conference on Signals, Systems and Computers*. Pacific Grove, (CA, USA), Nov. 2004, vol. 1, p. 295–299. DOI: 10.1109/ACSSC.2004.1399140

[6] FUHRMANN, D., SAN ANTONIO, G. Transmit beamforming for MIMO radar systems using signal cross-correlation. *IEEE Transactions on Aerospace Electronic Systems*, Jan. 2008, vol. 44, no. 1, p. 171–186. DOI: 10.1109/TAES.2008.4516997

[7] SHADI, K., BEHNIA, F. Transmit beampattern synthesis using eigenvalue decomposition in MIMO radar. In *8th International Conference on Information, Communication and Signal Processing* (ICICS 2011). Singapore, December 2011. DOI: 10.1109/ICICS.2011.6174302

[8] AITTOMAKI, T., KOIVUNEN, V. Signal covariance matrix optimization for transmit beamforming in MIMO radars. In *Conf. Record of the Forty-First Asilomar Conference on Signals, Systems and Computers (2007, ACSSC)*. Nov. 2007, p. 182–186. DOI: 10.1109/ACSSC.2007.4487191

[9] AHMED, S., THOMPSON, J. S., PETILLOT, Y. R., MULGREW, B. Unconstrained synthesis of covariance matrix for MIMO radar transmit beampattern. *IEEE Transactions on Signal Processing*, 2011, vol. 59, no. 8, p. 3837–3849. DOI: 10.1109/TSP.2011.2153200

[10] AITTOMAKI, T., KOIVUNEN, V. Low complexity method for transmit beamforming in MIMO radar. In *Proc. IEEE International Conference on Acoustics, Speech, and Signal Processing (ICASSP)*. Honolulu (HI, USA), Apr. 2007, vol. 2, p. 305–308. DOI: 10.1109/ICASSP.2007.366233

[11] HE, H., STOICA, P., LI, J. Designing unimodular sequence sets with good correlations – including an application to MIMO radar. *IEEE Transactions on Signal Processing*, Nov. 2009, vol. 57, no. 11, p. 4391–4405. DOI: 10.1109/TSP.2009.2025108

[12] LI, J., STOICA, P., ZHU, X. MIMO radar waveform synthesis. In *IEEE Radar Conference 2008 (RADAR'08)*. Rome (Italy), May 2008, 6 p. DOI: 10.1109/RADAR.2008.4721118

[13] AHMED, S., THOMPSON, J. S., PETILLOT, Y. R., MULGREW, B. Finite alphabet constant-envelope waveform design for MIMO radar. *IEEE Transactions on Signal Processing*, Nov. 2011, vol. 59, no. 11, p. 5326–5337. DOI: 10.1109/TSP.2011.2163067

[14] MATHELIER, B., KIRAN, D., S., REDDY, V. U. Synthesis of waveforms from zero-lag cross-correlation matrix with specified constraints and power levels. In *International Conference on Signal Processing and Communications SPCOM 2012*. Bangalore, 2012. DOI: 10.1109/SPCOM.2012.6289999

[15] PEEBLES, P. Z. *Radar Principles*. Wiley IEEE Press, 1998.

[16] XU, L., LI, J., STOICA, P. Radar imaging via adaptive MIMO techniques. In *Proceedings of 14th European Signal Processing Conference (EUSIPCO'06)*. Florence (Italy), Sep. 2006.

[17] SRINIVAS, A., REDDY, V. U. Transmit beamforming for colocated MIMO radar. In *International Conference on Signal Processing and Communications SPCOM*. Bangalore, July 2010, 5 p. DOI: 10.1109/SPCOM.2010.5560551

[18] OPPENHEIM, A. V., SCHAFER, R. W. *Digital Signal Processing*. Prentice Hall, 1975.

[19] SHARMA, G.V.K., RAJA RAJESWARI, K. Four phase orthogonal code design for MIMO radar systems. In *IEEE National Conference on Communications, NCC-2012*. IIT Kharagpur (India), Feb. 2012, 4 p. DOI: 10.1109/NCC.2012.6176764

[20] RAO, H. M., SHARMA, G. V. K., RAJA RAJESWARI, K. Orthogonal phase coded waveforms for MIMO radars. *International Journal of Computer Applications*, Feb. 2013, vol. 63, no. 6, p. 31–35. DOI: 10.5120/10471-5200

[21] WANHAMMAR, L. *DSP Integrated Circuits*. Academic Press, 1999.

[22] STOICA, P., MOSES, R. L. *Spectral Analysis of Signals*. Upper Saddle River, NJ: Prentice-Hall, 2005.

About the Authors ...

K. RAJA RAJESWARI is graduated in Electronics and Communication Engineering from Andhra University during 1976. She obtained her M.E. and Ph.D. also from the same university. Presently, she is a Professor in the Dept. of ECE, College of Engg., Andhra University, India. She is a senior member of IEEE. Her research interests include radar/sonar signal processing and wireless mobile communications. She has more than 100 papers published to her credit in reputed international/national journals and conferences.

G.V.K. SHARMA obtained his Ph.D. and Master's degree from Andhra University and Indian Institute of Science, Bangalore respectively. Presently, he is an Associate Professor in the Dept. of ECE, GITAM University, Visakhapatnam, India. His research interests include radar signal processing, VLSI signal processing and multicarrier communications.

14

Level Crossing Rate of Macrodiversity System in the Presence of Multipath Fading and Shadowing

Branimir JAKSIC [1], *Dusan STEFANOVIC* [2], *Mihajlo STEFANOVIC* [1], *Petar SPALEVIC* [3], *Vladeta MILENKOVIC* [1]

[1] Faculty of Electrical Engineering, University of Nis, Aleksandra Medvedeva 14, 18000 Nis, Serbia
[2] College of Applied Technical Sciences, Aleksandra Medvedeva 20, 18000 Nis, Serbia
[3] Faculty of Technical Sciences, University of Pristina, Knjaza Milosa 7, 38220 Kosovska Mitrovica, Serbia

branimir.jaksic@pr.ac.rs, dusan.stefanovic@itcentar.rs, misa.profesor@gmail.com, petarspalevic@yahoo.com, vladeta.milenkovic@elfak.ni.ac.rs

Abstract. *Macrodiversity system including macrodiversity SC receiver and two microdiversity SC receivers is considered in this paper. Received signal experiences, simultaneously, both, long term fading and short term fading. Microdiversity SC receivers reduce Rayleigh fading effects on system performance and macrodiversity SC receivers mitigate Gamma shadowing effects on system performance. Closed form expressions for level crossing rate of microdiversity SC receivers output signals envelopes are calculated. This expression is used for evaluation of level crossing rate of macrodiversity SC receiver output signal envelope. Numerical expressions are illustrated to show the influence of Gamma shadowing severity on level crossing rate.*

Keywords

Macrodiversity selection combining (SC) receiver, Rayleigh multipath fading, Gamma shadowing, level crossing rate, correlation

1. Introduction

Small scale fading and large scale fading degrade system performance and limit quality of service (QoS). Received signal experiences multipath fading resulting in signal envelope variation and shadowing resulting in signal envelope power variation. Reflection and refraction of radio wave cause short term fading and large obstacles between transmitter and receiver cause long term fading. It is important to determine how small scale fading and large scale fading affect performance of wireless communication system as outage probability and bit error probability. There are more distributions which can be applied to describe short term signal envelope variation depending on propagation environment. The most used statistical models are Rayleigh, Rician, Nakagami-*m*, Weibull and α-μ distributions [1]. Rayleigh and Nakagami-*m* distributions can be applied to describe signal envelope variation in linear non line-of-sight (LOS) multipath fading channel. In line-of-sight multipath fading environment, small scale signal envelope variation can be described by using Rician distribution. Weibull and α-μ distributions can be used to describe small scale signal envelope variation in nonlinear multipath fading channel dependent on the number of clusters in propagation environment [2].

There are two distributions which can be applied to describe large scale signal envelope power variation in shadowing fading channels. These statistical models are log-normal distribution and Gamma distribution. Log-normal distribution well describes large scale signal envelope power variation but it does not lead to closed form expression for probability density function of output signal envelope. The expression for outage probability of wireless communication system subjected to long term fading has a closed form, when Gamma distribution describes signal envelope power variation.

Macrodiversity system has one macrodiversity receiver and two or more microdiversity receivers. Macrodiversity receiver reduces large scale fading effects on outage probability and microdiversity receivers reduce small scale fading effects on outage probability. Signal envelope at output of macrodiversity receiver is equal to signal envelope at output of microdiversity receiver with greater signal envelope power at inputs [3].

In this paper macrodiversity selection combining (SC) receiver with two microdiversity SC receivers is considered. There are several combining techniques that can be used to mitigate the influence long term fading and short term fading on system performance. The most frequently combining techniques are maximal ratio combining (MRC), equal gain combining (EGC) and selection combining (SC). MRC receiver enables the best performance and it has the highest implementation complexity. The MRC receiver requires channel state information on each diversity branch and need provide receiver train on each diversity branch. The EGC provides performance comparable to MRC and it has lower implementation complexity than the MRC receiver. The SC receiver has the least implementation complexity due to the processing being performed only on one diversity branch. The SC receiver selects and outputs branch with the strongest signal envelope [4–5].

The first order performances of wireless communication systems are outage probability, bit error probability and channel capacity. The second order performances of wireless systems are level crossing rate and average fade duration. Level crossing rate can be calculated as the average value of the first derivative of output signal envelope. Average fade duration can be calculated as a ratio of outage probability and level crossing rate. Outage probability is defined as probability that signal envelope falls below the threshold.

There are more works considering macrodiversity system with correlated branches [6] and level crossing rate of macrodiversity systems [7–9]. In [10], macrodiversity system with macrodiversity SC receiver and two microdiversity MRC receivers are analyzed. Received signal is subjected simultaneously to Nakagami-*m* multipath fading and Gamma shadowing. Macrodiversity SC receiver is applied to reduce Gamma shadowing and microdiversity MRC receivers are used to reduce Nakagami-*m* multipath fading. Closed form expressions for average level crossing rate and average fade duration are calculated. In [11], level crossing rate and average fade duration of macrodiversity system with macrodiversity SC receiver and two microdiversity MRC receivers operating over Gamma shadowed Rician multipath fading environment are evaluated.

Macrodiversity system with macrodiversity SC receiver and two microdiversity SC receivers is considered. Received signal experiences short term fading and long term fading resulting in system performance degradation. Microdiversity SC receivers are used to combat Rayleigh short term fading and macrodiversity SC receiver is used to combat long Gamma term fading. Closed form expression for level crossing rate considering macrodiversity system is calculated. To the best author's knowledge, wireless communication system with macro and micro structures operating over Gamma shadowed Rayleigh multipath fading channels is not reported in open technical literature. Obtained results in this paper can be used in performance analysis and designing macrodiversity systems with macrodiversity SC receiver and two or more microdiversity SC receivers in the presence of Gamma shadowing and Rayleigh multipath fading.

2. Rayleigh Random Variable Level Crossing Rate

Squared Rayleigh random variable can be expressed as a sum of two independent, zero mean Gaussian random variables x_1 and x_2 with the same variance:

$$x^2 = x_1^2 + x_2^2 \tag{1}$$

where x is Rayleigh random variable. The first derivative of x is

$$\dot{x} = \frac{1}{x}\left(x_1\dot{x}_1 + x_2\dot{x}_2\right) \tag{2}$$

where $\dot{x}_1$ and $\dot{x}_2$ are independent, zero mean Gaussian random variables. Therefore, $\dot{x}$ is Gaussian random variable as linear transformation of Gaussian random variables. The average value of $\dot{x}$ is:

$$\overline{\dot{x}} = \frac{1}{x}\left(x_1\overline{\dot{x}_1} + x_2\overline{\dot{x}_2}\right). \tag{3}$$

The variance of the first derivative of Rayleigh random variable is

$$\sigma_{\dot{x}}^2 = \frac{1}{x^2}\left(x_1^2\sigma_{\dot{x}_1}^2 + x_2^2\sigma_{\dot{x}_2}^2\right) \tag{4}$$

where

$$\sigma_{\dot{x}_1}^2 = \sigma_{\dot{x}_2}^2 = \pi^2 f_m^2 \sigma^2 = f_1^2 \tag{5}$$

where f_m is maximal Doppler frequency and f_1 is normalized Doppler frequency. After substituting (5) in (4), the expression for variance becomes:

$$\sigma_{\dot{x}}^2 = \frac{f_1^2}{x^2}\left(x_1^2 + x_2^2\right) = f_1^2. \tag{6}$$

The probability density function of $\dot{x}$ is

$$p_{\dot{x}}(\dot{x}) = \frac{1}{\sqrt{2\pi} f_1} e^{-\frac{\dot{x}^2}{2f_1^2}}. \tag{7}$$

Rayleigh random variable and the first derivative of Rayleigh random variable are independent. Therefore, joint probability density function of Rayleigh random variable and the first derivative of Rayleigh random variable is

$$p_{x\dot{x}}(x\dot{x}) = p_x(x)p_{\dot{x}}(\dot{x}) = \frac{2x}{\Omega} e^{-\frac{x^2}{\Omega}} \frac{1}{\sqrt{2\pi} f_1} e^{-\frac{\dot{x}^2}{2f_1^2}}. \tag{8}$$

The level crossing rate of Rayleigh random variable can be calculated as average value of the first derivative of Rayleigh random variable:

$$\begin{aligned} N_x &= \int_0^\infty p_{x\dot{x}}(x\dot{x}) \cdot \dot{x} \cdot d\dot{x} \\ &= \frac{2x}{\Omega} e^{-\frac{x^2}{\Omega}} \int_0^\infty \dot{x} \frac{1}{\sqrt{2\pi} f_1} e^{-\frac{\dot{x}^2}{2f_1^2}} d\dot{x} = \frac{2x}{\Omega} e^{-\frac{x^2}{\Omega}} f_1 \end{aligned} \tag{9}$$

The expression for level crossing rate of Rayleigh random variable can be used in performance analysis of wireless communication system operating over Rayleigh multipath fading environment.

3. Level Crossing Rate of SC Receiver Output Signal Envelope

The average level crossing rate (LCR) is a measure [12] that clearly reflects the performances of fading affected system and is used for modeling of wireless communication systems. LCR is related to criterion used to assess error probability of packets of distinct length [4], and to determinate parameters of equivalent channel, modeled by a Markov chain with defined number of states. LCR is used for determining of the rate at which the envelope of the received signal crosses a specified defined level.

Double SC receiver operating over Rayleigh multipath fading channel is considered. Signal envelopes at inputs of SC receiver are denoted with x_1 and x_2 and signal envelope at output of SC receiver is denoted with x. At inputs of SC receiver identical and independent Rayleigh multipath fading is present. Probability density function of SC receiver output signal is

$$\begin{aligned} p_x(x) &= p_{x_1}(x)F_{x_2}(x) + p_{x_2}(x)F_{x_1}(x) \\ &= 2p_{x_1}(x)F_{x_2}(x) \end{aligned} \quad (10)$$

where $F_{x2}(x)$ is cumulative distribution function of Rayleigh random variable:

$$F_{x_2}(x) = 1 - e^{-\frac{x^2}{\Omega}}. \quad (11)$$

After substituting (11) in (10), the expression for $p_x(x)$ becomes

$$p_x(x) = \frac{2x}{\Omega} e^{-\frac{x^2}{\Omega}} \left(1 - e^{-\frac{x^2}{\Omega}}\right) \quad (12)$$

where Ω is signal envelope power.

The joint probability density function of x and $\dot{x}$ is

$$\begin{aligned} p_{x\dot{x}}(x\dot{x}) &= p_{x_1\dot{x}_1}(x\dot{x})F_{x_2}(x) + p_{x_2\dot{x}_2}(x\dot{x})F_{x_1}(x) \\ &= 2p_{x_1\dot{x}_1}(x\dot{x})F_{x_2}(x) \\ &= \frac{2x}{\Omega} e^{-\frac{x^2}{\Omega}} \left(1 - e^{-\frac{x^2}{\Omega}}\right) \frac{1}{\sqrt{2\pi} f_1} e^{-\frac{\dot{x}^2}{2f_1^2}}. \end{aligned} \quad (13)$$

Level crossing rate of SC receiver output signal envelope is

$$\begin{aligned} N_x &= \int_0^\infty p_{x\dot{x}}(x\dot{x}) \cdot \dot{x} \cdot d\dot{x} \\ &= \frac{2x}{\Omega} e^{-\frac{x^2}{\Omega}} \left(1 - e^{-\frac{x^2}{\Omega}}\right) \int_0^\infty \dot{x} \frac{1}{\sqrt{2\pi} f_1} e^{-\frac{\dot{x}^2}{2f_1^2}} d\dot{x} \\ &= \frac{2xf_1}{\Omega} e^{-\frac{x^2}{\Omega}} \left(1 - e^{-\frac{x^2}{\Omega}}\right). \end{aligned} \quad (14)$$

The expression for level crossing rate can be applied for calculation of average fade duration of wireless communication system with dual SC receiver operating over Gamma shadowed Rayleigh multipath fading channels.

4. Level Crossing Rate of Macrodiversity SC Receiver Output Signal Envelope

Macrodiversity system with macrodiversity SC receiver and two microdiversity SC receivers operating over shadowed multipath fading environment is considered. Short term Rayleigh fading and Gamma correlated long term fading are presented at inputs of microdiversity SC receivers. Macrodiversity SC receiver reduces signal envelope power variation and microdiversity SC receivers reduce signal envelope variation on system performance. Signal envelopes at output of microdiversity receivers are denoted with x_1 and x_2 and macrodiversity SC receiver output signal envelope is denoted with x.

Signal envelopes powers at inputs in microdiversity receivers are correlated. Signal envelope powers Ω_1 and Ω_2 follow Gamma distribution [11]:

$$\begin{aligned} p_{\Omega_1\Omega_2}(\Omega_1\Omega_2) &= \frac{1}{\Gamma(c)(1-\rho^2)\rho^{\frac{c-1}{2}}\Omega_0^{c+1}} \times \\ &(\Omega_1\Omega_2)^{\frac{c-1}{2}} e^{-\frac{\Omega_1+\Omega_2}{\Omega_0(1-\rho^2)}} I_{c-1}\left(\frac{2\rho}{\Omega_0(1-\rho^2)}(\Omega_1\Omega_2)^{\frac{1}{2}}\right) \end{aligned} \quad (15)$$

where c is fading severity, ρ is correlation coefficient and Ω_o is average power of Ω_1 and Ω_2 ($\Omega_1 \geq 0$, $\Omega_2 \geq 0$). Level crossing rate of signal envelopes x_1 and x_2 can be calculated by using expression (14).

The level crossing rate of macrodiversity SC receiver output signal envelope is

$$\begin{aligned} N_x &= \int_0^\infty d\Omega_1 \int_0^{\Omega_1} d\Omega_2 N_{x_1/\Omega_1} p_{\Omega_1\Omega_2}(\Omega_1\Omega_2) \\ &+ \int_0^\infty d\Omega_2 \int_0^{\Omega_2} d\Omega_1 N_{x_2/\Omega_1} p_{\Omega_1\Omega_2}(\Omega_1\Omega_2) \\ &= 2\int_0^\infty d\Omega_1 \int_0^{\Omega_1} d\Omega_2 N_{x_1/\Omega_1} p_{\Omega_1\Omega_2}(\Omega_1\Omega_2) \\ &= \frac{4f_1 x}{\Gamma(c)(1-\rho^2)\rho^{\frac{c-1}{2}}\Omega_0^{c+1}} \sum_{i=0}^\infty \left(\frac{\rho}{\Omega_0(1-\rho^2)}\right)^{c+2i-1} \\ &\times \frac{(\Omega_0(1-\rho^2))^{c+i}}{\Gamma(c+i)i!} \int_0^\infty d\Omega_1 \Omega_1^{c+i-2} e^{-\frac{x^2}{\Omega_1}} e^{-\frac{\Omega_1}{\Omega_0(1-\rho^2)}} \\ &\times \left(1 - e^{-\frac{x^2}{\Omega_1}}\right) \gamma\left(c+i, \frac{\Omega_1}{\Omega_0(1-\rho^2)}\right). \end{aligned} \quad (16)$$

The incomplete Gamma function $\gamma(n,x)$ is [13]:

$$\begin{aligned}\gamma(n,x) &= \Gamma(n) - \frac{1}{n} x^n e^{-x} {}_1F_1(1, n+1, x) \\ &= \Gamma(n) - \frac{1}{n} x^n e^{-x} \sum_{i=0}^{\infty} \frac{x^i}{(n+1)_i}\end{aligned} \quad (17)$$

where $(a)_n$ denotes the Pocchammer symbol.

After substituting (17) in (16), the expression for level crossing rate of macrodiversity SC receiver output signal envelope becomes:

$$\begin{aligned}N_x = &\frac{4 f_1 x}{\Gamma(c)(1-\rho^2)\rho^{\frac{c-1}{2}}\Omega_0^{c+1}} \sum_{i=0}^{\infty} \left(\frac{\rho}{\Omega_0(1-\rho^2)} \right)^{c+2i-1} \\ &\times \frac{(\Omega_0(1-\rho^2))^{c+i}}{\Gamma(c+i) i!} (I_1 - I_2 - I_3 + I_4)\end{aligned} \quad (18)$$

where

$$I_1 = \int_0^{\infty} d\Omega_1 \Omega_1^{c+i-2} e^{-\frac{x^2}{\Omega_1} - \frac{\Omega_1}{\Omega_0(1-\rho^2)}} \Gamma(c+i), \quad (19)$$

$$\begin{aligned}I_2 = &\int_0^{\infty} d\Omega_1 \Omega_1^{c+i-2} e^{-\frac{x^2}{\Omega_1} - \frac{2\Omega_1}{\Omega_0(1-\rho^2)}} \frac{1}{c+i} \left(\frac{\Omega_1}{\Omega_0(1-\rho^2)} \right)^{c+i} \\ &\times \sum_{j=0}^{\infty} \left(\frac{\Omega_1}{\Omega_0(1-\rho^2)} \right)^j \frac{1}{(c+i+1)_j},\end{aligned} \quad (20)$$

$$I_3 = \int_0^{\infty} d\Omega_1 \Omega_1^{c+i-2} e^{-\frac{2x^2}{\Omega_1} - \frac{\Omega_1}{\Omega_0(1-\rho^2)}} \Gamma(c+i) \quad (21)$$

and

$$\begin{aligned}I_4 = &\int_0^{\infty} d\Omega_1 \Omega_1^{c+i-2} e^{-\frac{2x^2}{\Omega_1} - \frac{2\Omega_1}{\Omega_0(1-\rho^2)}} \frac{1}{c+i} \left(\frac{\Omega_1}{\Omega_0(1-\rho^2)} \right)^{c+i} \\ &\times \sum_{j=0}^{\infty} \left(\frac{\Omega_1}{\Omega_0(1-\rho^2)} \right)^j \frac{1}{(c+i+1)_j}.\end{aligned} \quad (22)$$

After processing and solving integrals (19), (20), (21) and (22), expression (18) becomes:

$$\begin{aligned}N_x = &\frac{4 f x_1}{\Gamma(c)(1-\rho^2)\rho^{\frac{c-1}{2}}\Omega_0^{c+1}} \\ &\times \sum_{i=0}^{\infty} \left(\frac{\rho}{\Omega_0(1-\rho^2)} \right)^{c+2i-1} \frac{(\Omega_0(1-\rho^2))^{c+i}}{\Gamma(c+i) i!} \\ &\times \Bigg[\Gamma(c+i)\left(x^2\Omega_0(1-\rho^2)\right)^{\frac{c+i-1}{2}} \\ &\times K_{c+i-1}\left(2\sqrt{\frac{x^2}{\Omega_0(1-\rho^2)}} \right) - \\ &- \frac{1}{c+i} \sum_{j=0}^{\infty} \frac{1}{(c+i+1)_j} \frac{1}{(\Omega_0(1-\rho^2))^{c+i+j}} \\ &\times \left(\frac{1}{2} x^2 \Omega_0 (1-\rho^2) \right)^{c+i+\frac{j}{2}-\frac{1}{2}} \\ &\times K_{2c+2i+j-1}\left(2\sqrt{\frac{2x^2}{\Omega_0(1-\rho^2)}} \right) - \\ &- \Gamma(c+i)\left(2x^2\Omega_0(1-\rho^2)\right)^{\frac{c+i-1}{2}} \\ &\times K_{c+i-1}\left(2\sqrt{\frac{2x^2}{\Omega_0(1-\rho^2)}} \right) + \\ &+ \frac{1}{c+i} \sum_{j=0}^{\infty} \frac{1}{(c+i+1)_j} \frac{1}{(\Omega_0(1-\rho^2))^{c+i+j}} \\ &\times \left(x^2 \Omega_0 (1-\rho^2) \right)^{c+i+\frac{j}{2}-\frac{1}{2}} \\ &\times K_{2c+2i+j-1}\left(2\sqrt{\frac{4x^2}{\Omega_0(1-\rho^2)}} \right) \Bigg]\end{aligned} . \quad (23)$$

where $K_n(x)$ is modified Bessel function of the second kind, order n and argument x.

5. Numerical Results

In Fig. 1 and Fig. 2, normalized average level crossing rate of macrodiversity SC receiver output signal envelope versus macrodiversity SC receiver output signal envelope is plotted. For lower values of SC receiver output signal envelope, average level crossing rate increases as signal envelope increases and for higher values of signal envelope, level crossing rate decreases as signal envelope increases. The influence of signal envelope on level crossing rate is greater for higher values of signal envelope. Level crossing rate decreases as Gamma fading severity increases. For higher values of Gamma shadowing severity, long term fading is of less severity. Graphical results are shown; average level crossing rate increases as correlation coefficient increases. Diversity gain decreases as correlation coefficient increases.

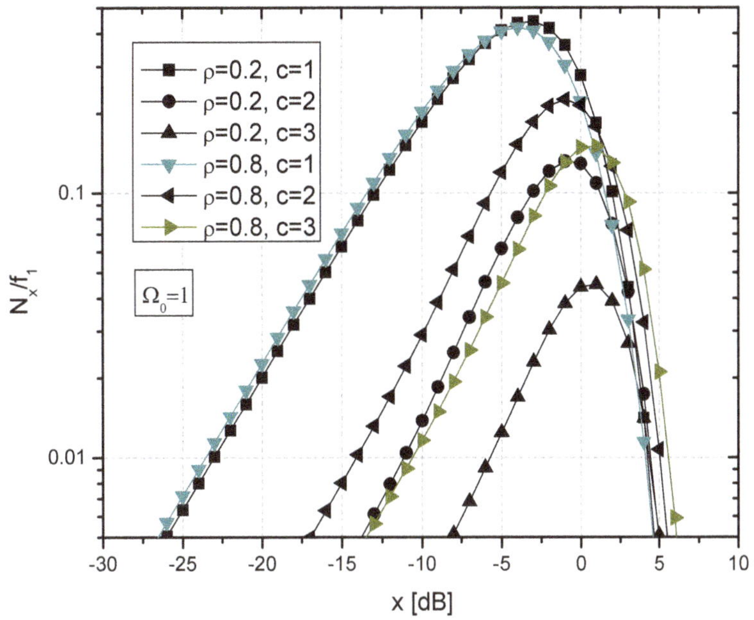

Fig. 1. Level crossing rate of macrodiversity SC receiver output signal envelope for different values of Gamma shadowing severity parameter c.

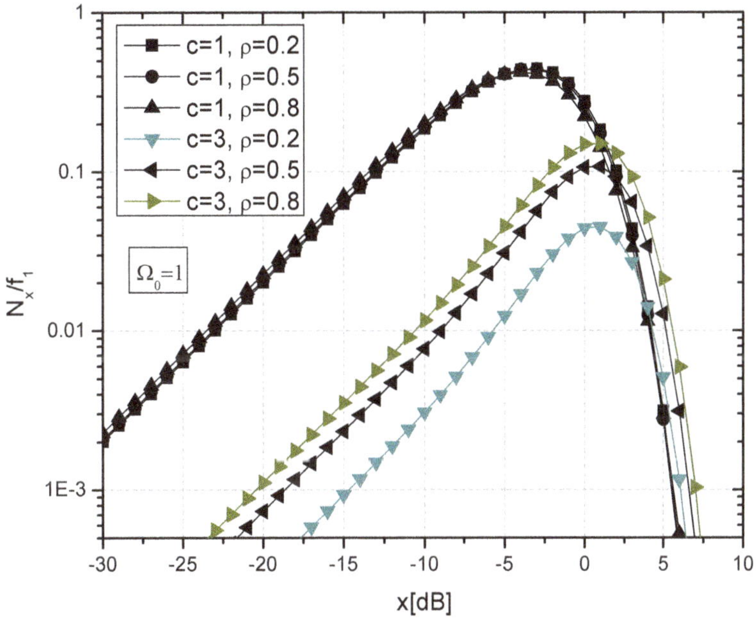

Fig. 2. Level crossing rate of macrodiversity SC receiver output signal envelope for different values of correlation coefficient ρ.

In Fig. 3, normalized average level crossing rate of macrodiversity SC receiver output signal envelope versus Gamma shadowing severity parameter c for several values of correlation coefficient is plotted. Average level crossing rate decreases as Gama fading severity increases. When Gamma fading severity parameter c goes to infinity composite Gamma shadowed, Rayleigh multipath channel goes to Rayleigh multipath fading channel.

Normalized average level crossing rate of macrodiversity SC receiver output signal envelope versus correlation coefficient ρ for several values of Gamma shadowing severity is illustrated in Fig. 4. Average level crossing rate increases as coefficient of correlation increases. Outage probability is better for lower values of average level crossing rate. When correlation coefficient goes to 1, level crossing rate has maximum and system performance is the worst. For the case, the least value of signal envelope occurs at both antennas.

6. Conclusion

Macrodiversity system with macrodiversity SC receiver and two microdiversity SC receivers subjected simultaneously to Rayleigh multipath fading and Gamma shadowing has been analyzed in this paper. Closed form

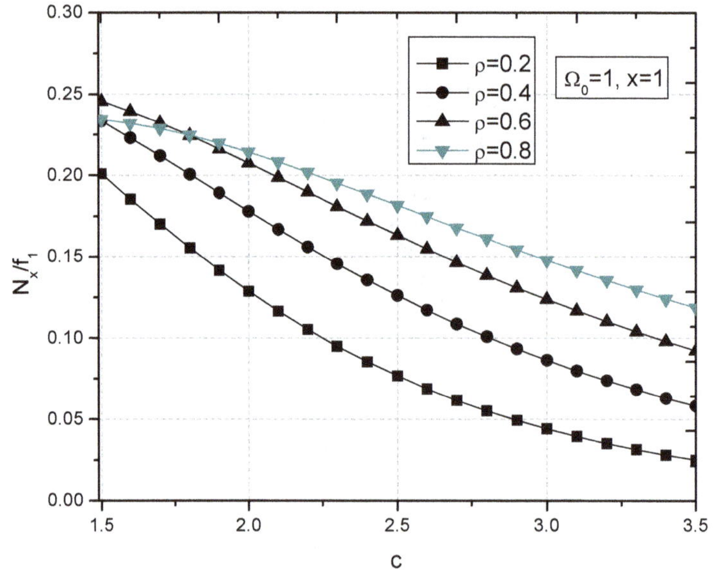

Fig. 3. Level crossing rate of macrodiversity SC receiver output signal envelope versus Gamma shadowing severity parameter c.

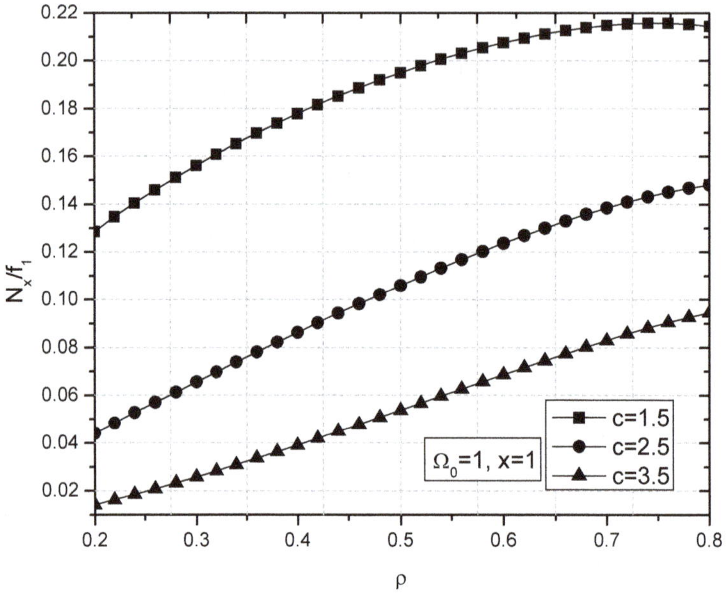

Fig. 4. Level crossing rate of macrodiversity SC receiver output signal envelope versus correlation coefficient ρ.

expressions for LCR of macrodiversity SC receiver output signal envelope are derived. Capitalizing on obtained expressions numerical results are presented graphically to show Gamma shadowing severity effects and correlation coefficient effects on LCR of macrodiversity system output signal envelope. LCR reduction has been analyzed as a function of Gamma fading severity decrease and correlation coefficient decrease.

Presented performance evaluation for proposed communication scenarios provides insight into the performance dependence on key system parameters. Presented LCR analysis allows system designers to perform trade-off studies among the various communication type/drawback combinations in order to determine the optimal choice in the presence of given constraints.

References

[1] SUBER, G. L. *Mobile Communication.* 2nd ed. Dordrecht: Kluwer Academic Publisher, 2003.

[2] PROAKIS, J. *Digital Communications.* 4th ed. New York: McGraw-Hill, 2001.

[3] MUKHERJEE, S., AVIDOR, D. Effect of microdiversity and correlated macrodiversity on outages in a cellular system. *IEEE Transactions on Wireless Technology*, 2003, vol. 2, no. 1, p. 50 to 58. DOI: 10.1109/TWC.2002.806363

[4] PANIC, S., STEFANOVIC, M., ANASTASOV, J., SPALEVIC, P. *Fading and Interference Mitigation in Wireless Communications.* USA: CRC Press, 2013.

[5] SIMON, M. K., ALOUINI, M. S. *Digital Communication over Fading Channels.* USA: John Wiley & Sons, 2000.

[6] SHANKAR, P. M. Analysis of microdiversity and dual channel macrodiversity in shadowed fading channels using a compound fading model. *International Journal of Electronics and Communications (AEUE)*, 2008, vol. 62, no. 6, p. 445–449. DOI: 10.1016/j.aeue.2007.06.008

[7] KLINGENBRUNN, T., MOGENSEN, P. Modelling cross-correlated shadowing in network simulations. In *Proceedings of the 50th IEEE Vehicular Technology Conf. (VTC 1999 Fall)*. Amsterdam (The Netherlands), 1999, vol. 3, p. 1407–1411. DOI: 10.1109/VETECF.1999.801494

[8] ZHANG, J., AALO, V. Effect of macrodiversity on average-error probabilities in a Rician fading channel with correlated lognormal shadowing. *IEEE Transactions on Communications*, 2001, vol. 49, no. 1, p. 14–18. DOI: 10.1109/26.898244

[9] SAFAK, A., PRASAD, R. Effects of correlated shadowing signals on channel reuse in mobile radio systems. *IEEE Transactions on Vehicular Technology*, 1991, vol. 40, no. 4, p. 708–713. DOI: 10.1109/25.108381

[10] STEFANOVIC, D., PANIC, S., SPALEVIC, P. Second-order statistics of SC macrodiversity system operating over Gamma shadowed Nakagami-m fading channels. *AEU - International Journal of Electronics and Communications,* 2011, vol. 65, no. 5, p. 413–418. DOI: 10.1016/j.aeue.2010.05.001

[11] SEKULOVIC, N., STEFANOVIC, M. Performance analysis of system with micro- and macrodiversity reception in correlated Gamma shadowed Rician fading channels. *Wireless Personal Communications,* 2012, vol. 65, no. 1, p. 143–156. DOI: 10.1007/s11277-011-0232-8

[12] ISKANDER, C. D., MATHIOPOULOS, P. T. Analytical level crossing rate and average fade duration in Nakagami fading channels. *IEEE Transactions on Communications*, 2002, vol. 50, no. 8, p 1301–1309. DOI: 10.1109/TCOMM.2002.801465

[13] GRADSHTEYN, I., RYZHIK, I. *Tables of Integrals, Series, and Products.* New York: Academic Press, 1994.

About the Authors ...

Branimir JAKSIC was born in Kosovska Mitrovica, Serbia, in 1984. He received B.Sc. and M.Sc. degrees in Electrical Engineering from the Faculty of Technical Sciences in Kosovska Mitrovica, University of Pristina, Serbia. He is Ph.D. candidate in the Faculty of Electronic Engineering, University of Nis, Serbia. Areas of research include statistical communication theory and optical telecommunications. He has authored several scientific papers on the above subject.

Dusan STEFANOVIC was born in Nis, Serbia, in 1979. He received the M.Sc. in Electrical Engineering from the Faculty of Electronic Engineering (Dept. of Telecommunications), University of Nis, Serbia, in 2005 and Ph.D. in the same department in 2012. He is interested in optical and wireless communication. Now he is working in College of Applied Technical Sciences as professor on the following subjects: Database administration, computer networks, administration of computer networks and opto-laser technologies.

Mihajlo STEFANOVIC was born in Nis, Serbia in 1947. He received B.Sc., M.Sc. and Ph.D. degrees in Electrical Engineering from the Faculty of Electronic Engineering (Dept. of Telecommunications), University of Nis, Serbia, in 1971, 1976 and 1979, respectively. His primary research interests are statistical communication theory, optical and satellite communications. He has written or co-authored a great number of journal publications. Dr. Stefanovic is a full-time professor with the Dept. of Telecommunications, Faculty of Electronic Engineering, University of Nis, Serbia.

Petar SPALEVIC was born in Kraljevo, Serbia, in 1973. He received the B.S. degree from the Faculty of Electronic Engineering, University of Pristina, in 1997, and M.S. and Ph.D. degrees from the Faculty of Electronic Engineering, University of Nis in 1999 and 2003, respectively. He is a Professor with the Faculty of Technical Sciences in Kosovska Mitrovica. His primary research interests are statistical communications theory, wireless communications, applied probability theory and optimal receiver design.

Vladeta MILENKOVIC was born in Nis, Serbia. He received B. Sc. and M. Sc. degrees in Electrical Engineering from the Faculty of Electronic Engineering, University of Nis, Serbia. He is Ph.D. candidate in the same faculty. He has authored several scientific papers on wireless communications, metrology and measurement techniques.

15

Optimization of Excitation in FDTD Method and Corresponding Source Modeling

Bojan DIMITRIJEVIC, Bojana NIKOLIC, Slavoljub ALEKSIC, Nebojsa RAICEVIC

Faculty of Electronic Engineering, University of Nis, A. Medvedeva 14, 18 000 Nis, Serbia

{bojan.dimitrijevic, bojana.nikolic, slavoljub.aleksic, nebojsa.raicevic}@elfak.ni.ac.rs

Abstract. *Source and excitation modeling in FDTD formulation has a significant impact on the method performance and the required simulation time. Since the abrupt source introduction yields intensive numerical variations in whole computational domain, a generally accepted solution is to slowly introduce the source, using appropriate shaping functions in time. The main goal of the optimization presented in this paper is to find balance between two opposite demands: minimal required computation time and acceptable degradation of simulation performance. Reducing the time necessary for source activation and deactivation is an important issue, especially in design of microwave structures, when the simulation is intensively repeated in the process of device parameter optimization. Here proposed optimized source models are realized and tested within an own developed FDTD simulation environment.*

Keywords

Finite difference time domain, source modeling, excitation, optimization

1. Introduction

The finite difference time domain (FDTD) method currently draws significant scientific attention as one of the most efficient methods for analysis and characterization of wide range of electromagnetic problems [1]. The proper excitation and source modeling in the FDTD computational domain is especially important issue in every application of FDTD simulation.

Introduction of discrete internal sources is usually done by applying either hard source or soft source excitation. The hard source excitation is consisted in assigning specific value of certain electric (or magnetic) field component at a single or several grid points in every time step through an appropriate time function [2]. The soft source excitation is introduced by adding the appropriate time function to the field value obtained in regular update equation [1].

Although it is not a physical reality, the plane wave excitation has enormously large significance in many theoretical and analytical considerations. For this reason it is very important to introduce the same excitation in simulation environments and to enable the comparative analysis of the results. The necessity of plane-wave source arose originally with the first FDTD modeling in the field of defense and bioelectromagnetics [1]. Considering scattering problems, where the particular structure of interest is far away from the radiation source and the incident wave can be considered as a plane wave, Yee was the first to introduce the initial-condition approach [3]. However, today mainly accepted approach for plane-wave excitation is total-field/ scattered-field (TF/SF) formulation [4], [5]. The TF/SF technique showed very good performance in FDTD modeling of long-duration pulsed or continuous wave excitation and it is widely used in guided-wave simulations [1]. The TF/SF technique has been extensively studied in the literature and many modifications and improvements of this basic method can be found [6–9].

However, another way of plane wave excitation modeling includes adding or assigning of an electric (or magnetic) field value at specific positions in one plane, unlike the commonly used TF/SF technique, where corrections are made in both electric and magnetic field components (displaced in time and space for a half time step) on the boundary surface. The advantage of this direct approach is its simplicity. Its main difficulty is, however, the existence of wave propagation in undesirable direction. But since very effective boundary conditions like the convolutional perfectly matched layer (CPML) [10] are available, this is no longer an obstacle to its application. The considerations regarding this approach can be found in [11].

The excitation modeling in FDTD formulation significantly affects the simulation performance. A sudden excitation of the domain causes undesirable numerical variations in whole computational domain. This problem is usually resolved by slow introduction of the source excitation, using the appropriate shaping functions in time. A number of time functions for slow introduction of source excitation are available in the literature [12]. However, a gradual raise of the excitation signal is time consuming and can be a significant difficulty in applications where the intensive repetition of simulations is required. For this reason, a certain compromise between the required time and satisfactory simulation performance must be achieved.

The optimization presented in this paper is conducted in order to minimize the propagation of the undesirable energy through the computational domain. As a result of the optimization process, very simple and closed-form optimal function is obtained. Numerically obtained optimal shaping function is compared with the solutions that can be found in the literature and some general remarks are derived.

2. Source Modeling

In hard source modeling, instead of calculating field value using FDTD update equations, the field value in the specific grid points is assigned using a time function. If the source is at position (*i*, *j*, *k*), it is then

$$E^n_{\nu(i,j,k)} = E^n{}_{source},\ \nu = (x, y, z). \quad (1)$$

In the same manner as in the case of the point source, the plane wave source can be also modeled as a hard source, applying (1) on a group of points belonging to a specific surface or even volume. Regardless of the case, hard source model is equivalent to the ideal voltage source and for this reason acts like an electric wall, causing reflections of any wave arrived to the source location [2], [13]. This means that such a source, more specifically in the case of plane wave excitation separates the computational domain in two independent regions without mutual interaction. Hard source modeling is commonly applied in excitation of guided-wave structures. In [14] the analytical solution for the FDTD hard source has been derived, so its application is broaden to the validation of FDTD codes or FDTD schemes.

One possible solution to the reflective behavior of the hard sources is to combine it with the regular FDTD update equations [1]. Namely, the hard source can be turned on only while there is a signal excitation. The simulation setup in that case should provide that the signal excitation ends before any scattered or reflected wave returns to the source position. After this period, the usual FDTD update of the field components can be applied. The main difficulty of this approach is that it can't be used for continuous excitation.

In soft source modeling, source excitation is added to the value obtained in applied FDTD update equations. If the source is at position (*i*, *j*, *k*), it is then

$$E^n_{\nu(i,j,k)} = E^n_{\nu(i,j,k)FDTDupdate} + E^n_{\nu\,source},\ \nu = (x, y, z). \quad (2)$$

The main advantage of the soft source is the fact that it is transparent to the incoming waves and it allows the different incident fields to interact [1], [15]. For soft source modeling of the plane wave it can be either TF/SF method or method of direct adding/assigning of electric (magnetic) field value [11] applied. When point soft source modeling is concerned, the additional source component is usually introduced through the current density $J^{t+1/2}_{\nu(i,j,k)}$, which is defined at the same spatial position as the resulting electric field component but the time step is the same as the one of the magnetic field. If the source is at position (*i*, *j*, *k*), it is then

$$\frac{\Delta E^{t+1}_{\nu(i,j,k)}}{\Delta t} = \frac{1}{\varepsilon_0}\frac{\Delta H^{t+1/2}_{w(i,j,k)}}{\Delta u} - \frac{1}{\varepsilon_0}\frac{\Delta H^{t+1/2}_{u(i,j,k)}}{\Delta w} - \frac{1}{\varepsilon_0} J^{t+1/2}_{\nu(i,j,k)} \quad (3)$$

$$\nu = (x, y, z),\ u = (z, x, y),\ w = (y, z, x),$$

$$i = \{0,\ldots,N_x - 1\},\ j = \{0,\ldots,N_y - 1\},\ k = \{0,\ldots,N_z - 1\},$$

$$n = \{0,\ldots,N_t - 1\}$$

where Δt is the time step, Δx, Δy and Δz are spatial steps along *x*, *y* and *z* axis, respectively. Total numbers of elementary cells along *x*, *y* and *z* axis are denoted by N_x, N_y and N_z, respectively. Total number of elementary time steps is denoted by N_t. The electric magnetic permittivity in vacuum is denoted by ε_0.

Since the FDTD cell is usually much shorter than one-tenth of the main wavelengths of interest, physically the soft source current acts as a Hertzian dipole antenna [14]. The point soft source physically corresponds to real current (or voltage) source.

One of the main difficulties regarding the soft source modeling is constant deposit of charges and generation of the charge-associated fields [16], [1]. This can be circumvented by using matched voltage or current sources. By applying the source resistance R_S, voltage source at position (*i*, *j*, *k*) can be introduced in update equation of the corresponding field component as

$$E^{t+1}_{\nu(i,j,k)} = \frac{1-b}{1+b} E^t_{\nu(i,j,k)} + $$
$$+ \frac{\Delta t/\varepsilon_0}{1+b}\left(\frac{\Delta H^{t+1/2}_{w(i,j,k)}}{\Delta u} - \frac{\Delta H^{t+1/2}_{u(i,j,k)}}{\Delta w} - \frac{V^{t+1/2}_{S\nu(i,j,k)}}{R_S \Delta u \Delta w}\right) \quad (4)$$

$$b = \frac{\sigma \Delta t}{2\varepsilon_0} + \frac{\Delta t \Delta \nu}{2 R_S \varepsilon_0 \Delta u \Delta w}$$

where σ is the electric conductivity.

It is shown in [14] that the fields radiated by hard and soft source models are identical and the relation that connects excitation $E^n_{\nu\,source}$ from (1) and $J^{t+1/2}_{\nu(i,j,k)}$ from (2) can be expressed as

$$E^n_{\nu\,source} = -\frac{1}{3\varepsilon_0}\int_{-\infty}^{t_n} J_S(t)dt, \quad (5)$$

$$J_S(t_{n+1/2}) = J^{n+1/2}_{\nu(i,j,k)}.$$

3. Excitation Modeling

A proper excitation of the FDTD computational domain is important in every FDTD application. The abrupt source introduction yields intensive numerical variations at high frequencies which propagate through the whole computational domain. A generally accepted solution is to slowly introduce the source. In order to fulfill this, it is necessary for the excitation function to satisfy following conditions: at zero time step the excitation function value must be zero (if fields are initiated at zero) and the excitation function must be smooth. The most commonly used time functions for slow introducing of pulsed source excitation available in the literature are given in Tab. 1.

Broad band Gaussian with DC component	$f(t)=\exp\left(-(t-t_0)^2/t_w^2\right),\ 0<t<2t_0$
Broad band Gaussian without DC component – Gaussian derivative	$f(t)=\frac{-2}{t_w}(t-t_0)\exp\left(-(t-t_0)^2/t_w^2\right)$
Blackman – Harris window	$b(t)=\sum_{n=0}^{3} a_n\cos\left(\frac{2\pi nt}{T}\right),\quad 0<t<T$ $a=\begin{bmatrix}0.353222222\\-0.488\\0.145\\-0.010222222\end{bmatrix},\ T=\frac{1.55}{f_{bw}}$ f_{bw} - half bandwidth of the pulse
Differentiated Blackman – Harris window	$d_b(t)=-\sum_{n=0}^{3} a_n n\sin\left(\frac{2\pi nt}{T}\right),\quad 0<t<T$
Raised cosine [17]	$z(t)=0.5\left(1-\cos\left(\frac{2\pi t}{T}\right)\right),\ 0<t<T_r/2$ T_r - period of ramped cosine ($T_r\approx 3T$)

Tab. 1. Time functions used for pulsed excitation [12].

The raised cosine [17] is considered to be the most suitable for excitation of FDTD domain and it is the preferred choice, especially compared to linear and exponential ramps [15]. However, the presented excitation functions (Tab. 1) are not designed to meet the specific requirements of FDTD formulation. In order to get a better insight in FDTD nature, we shell start from the case of z-polarized plane wave, propagating along y axis. Ampere's and Faraday's law in that case have the form

$$\frac{\partial H_x}{\partial t}=\frac{1}{\mu_0}\frac{\partial E_z}{\partial y}, \tag{6}$$

$$\frac{\partial E_z}{\partial t}=\frac{1}{\varepsilon_0}\frac{\partial H_x}{\partial y}-\frac{J_S}{\varepsilon_0}. \tag{7}$$

If one differentiates (1) over space variable y and (2) over time t, it yields

$$\frac{\partial^2 H_x}{\partial y\partial t}=\frac{1}{\mu_0}\frac{\partial^2 E_z}{\partial y^2}, \tag{8}$$

$$\frac{\partial^2 E_z}{\partial t^2}=\frac{1}{\varepsilon_0}\frac{\partial^2 H_x}{\partial t\partial y}-\frac{1}{\varepsilon_0}\frac{\partial J_S}{\partial t}. \tag{9}$$

Substituting (8) in (9), it is obtained

$$\frac{\partial^2 E_z}{\partial t^2}=\frac{1}{\varepsilon_0}\frac{1}{\mu_0}\frac{\partial^2 E_z}{\partial y^2}-\frac{1}{\varepsilon_0}\frac{\partial J_S}{\partial t}. \tag{10}$$

Equation (10) indicates that the second time derivative of the excitation function should be the one to investigate, since it causes the propagation through the computational domain.

4. Numerical Optimization of FDTD Domain Excitation Function

In order to minimize the propagation of the undesirable energy through the FDTD computational domain that consequently appears during its excitation, the minimization of the second time derivative of the excitation function should be performed.

Without loss of generality, the excitation function will be analyzed in its normalized form. Considering (10) and [17], in order to have the desirable properties, the excitation function $f(x)$ should satisfy the following criteria

1. $f(0) = 0$ and $f(1) = 1$;
2. $f(x)$ is an odd function with respect to the point (1/2, 1/2) in Cartesian coordinate system;
3. the first derivative of $f(x)$ is continuous function and $f'(0) = 0$ (in order to avoid large values in the second derivative).

Two possible solutions are considered.

4.1 Polynomial Optimization

For the purpose of optimization, the excitation function will be presented as a linear combination of basic functions which satisfy conditions 1, 2 and 3. Considered basic functions are in polynomial form and given as

$$f_n(x)=\begin{cases}0, & x<0\\ 2^{n-1}x^n, & 0\le x\le \frac{1}{2},\ n=2,3,\ldots\\ 1-2^{n-1}(1-x)^n, & \frac{1}{2}\le x\le 1\\ 1, & x>1\end{cases} \tag{11}$$

If functions $f_n(x)$ fulfill conditions 1, 2 and 3, then their linear combination

$$f(x)=C_2f_2(x)+C_3f_3(x)+\ldots \tag{12}$$

also fulfills the same condition 3. However, there is an additional requirement for $f(x)$, in order to satisfy the conditions 1 and 2

$$\sum_{i=2}^{+\infty} C_i=1. \tag{13}$$

Applying numerical iterative minimization of the mean square value of the second derivative $\partial^2 f(x)/\partial x^2$ ($\min_{C_i,i=2,3,4,\ldots}\overline{f''^2(x)}$), it is obtained

$$C_2 = \frac{3}{2},\ C_3 = \frac{1}{2},\ C_4 = C_5 = C_6 = \ldots = 0. \tag{14}$$

Thus, the optimal excitation function has the form

$$f_{opt}(x) = \begin{cases} 0, & x < 0 \\ 3x^2 - 2x^3, & 0 \le x \le 1 \\ 1, & x > 1 \end{cases} \tag{15}$$

4.2 Trigonometric Optimization

If the basic functions in series expansion of the excitation function are

$$f_m^{(M)}(x) = \begin{cases} 1, & x < 0 \\ \cos(2m+1)\pi x, & 0 \le x \le 1 \\ -1, & x > 1 \end{cases} \tag{16}$$

the excitation function is then

$$f^{(M)}(x) = \frac{1}{2} - \sum_{m=0}^{M} a_{2m+1} f_{2m+1}^{(M)}(x) \tag{17}$$

where a_{2m+1} are series coefficients, which should fulfill the criterion

$$\sum_{m=0}^{M} a_{2m+1} = \frac{1}{2}. \tag{18}$$

Applying numerical iterative minimization of the mean square value of the second derivative $\partial^2 f^{(M)}(x)/\partial x^2$ in this case, a function with expansion coefficients presented in Tab. 2 for different M values is obtained. The mean square values of the second derivative of function $f^{(M)}(x)$ are also given.

5. Optimization Results

In Fig. 1 excitation function obtained in the optimization process using polynomial basic functions (denoted as Opt), as well as the ones using trigonometric basic functions for different values of M are presented. The curve that corresponds to the value $M = 0$ is actually the excitation function that is widely used in the literature and known as raised cosine.

In Fig. 2 one can observe the second derivatives of the functions from Fig. 1. It can be seen from Fig. 2 that the second derivative of the excitation function $M = 0$ (raised cosine) significantly deviates from the second derivative of the optimal function obtained using polynomial expansion. It can be also observed that with the increase of M the second derivative of the function with trigonometric expansion converges to the one of the optimal function with polynomial expansion. This confirms that the same optimal result is obtained regardless of the applied type of basic functions in optimization process. Since the solution

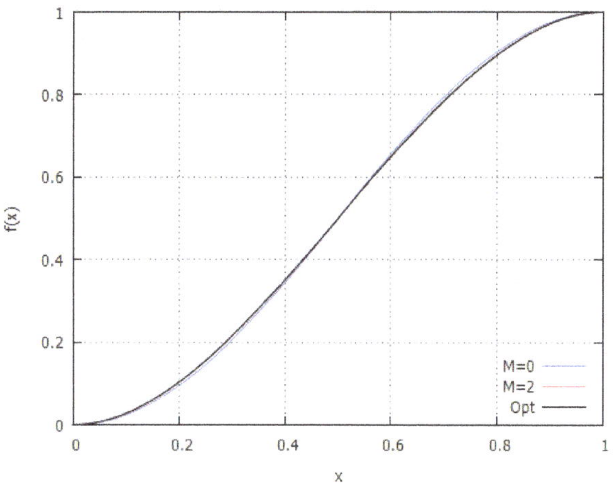

Fig. 1. The proposed optimized excitation functions.

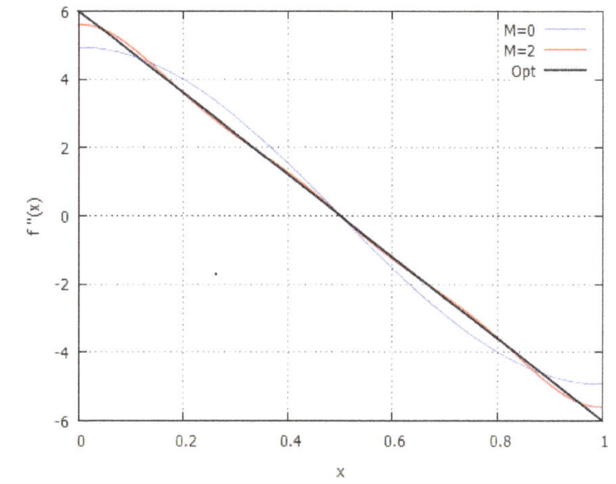

Fig. 2. Second derivative of the proposed optimized excitation functions.

obtained in polynomial form is simple and in a closed form, we propose function (15) as the excitation function for FDTD domain.

Optimality of the obtained function shape is verified in an own developed FDTD simulation environment [18].

6. Optimal Excitation Signals

Using the obtained optimal excitation function (Fig. 3), two pulsed signals are proposed for efficient excitation of FDTD domain. The total pulsed signal retains the optimal properties only if it is formed using the proposed optimal function as segments that are appropriately symmetrically extended or scaled in time and amplitude. Thus, the proposed pulsed function has the form

$$f_p(x) = \begin{cases} 0, & x < 0 \\ 3x^2 - 2x^3, & 0 \le x \le 1 \\ 3(2-x)^2 - 2(2-x)^3, & 1 \le x \le 2 \\ 0, & x > 2 \end{cases} \tag{19}$$

M	$\overline{f''^2(x)}$	a_1	a_3	a_5	a_7	a_9	…
0	12.1761	0.5	-	-	-	-	-
1	12.0276	0.493902	0.00609756	-	-	-	-
2	12.0086	0.493123	0.00608793	7.88996×10^{-4}	-	-	-
3	12.0037	0.492920	0.00608543	7.88672×10^{-4}	2.05298×10^{-4}	-	-
4	12.0019	0.492846	0.00608452	7.88554×10^{-4}	2.05267×10^{-4}	7.51175×10^{-5}	-
$\to \infty$	12.0000	0.492767	0.00608354	7.88427×10^{-4}	2.05234×10^{-4}	7.51054×10^{-5}	…

Tab. 2. Expansion coefficients in $f^{(M)}(x)$.

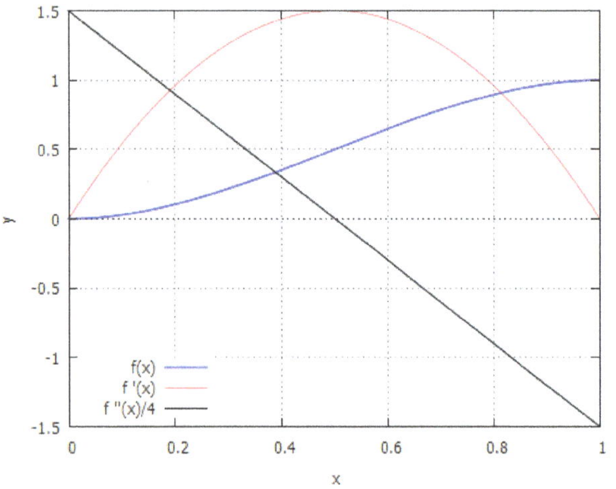

Fig. 3. Proposed optimal excitation function and its first and second derivative.

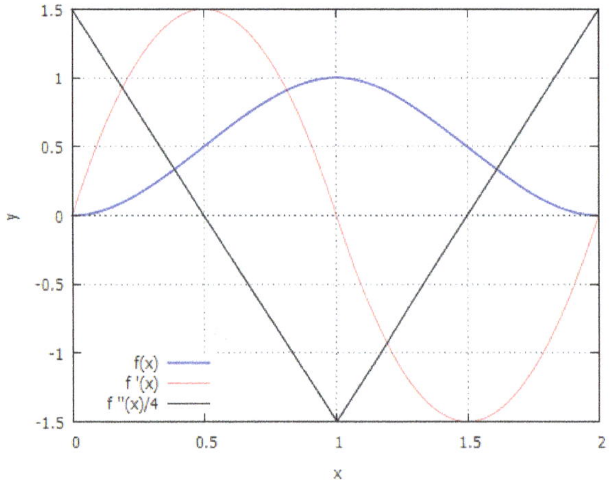

Fig. 4. Proposed optimal pulsed excitation function with DC component and its first and second derivative.

Function (19) along with its first and second derivative is presented in Fig. 4.

In case the pulsed signal with no DC component is required, it should be also obtained as symmetrically extended or adequately scaled optimal function (15). The pulsed signal with no DC component shouldn't be formed as the first derivative of the optimal pulsed function (Fig. 4), because in that case the resulting function would change its nature and wouldn't have optimal properties any more. Thus, we propose the pulsed function with no DC component in the form

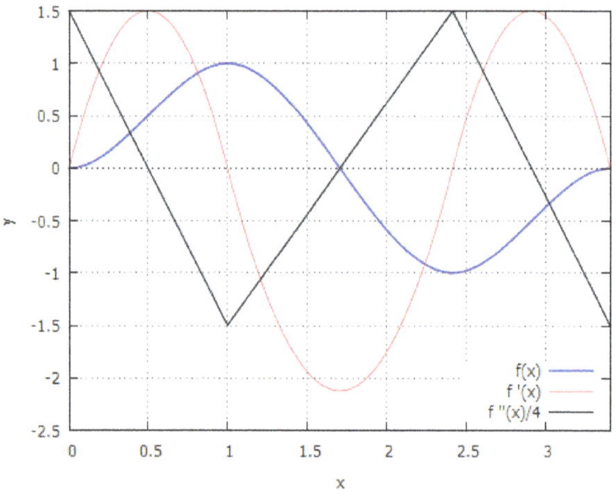

Fig. 5. Proposed optimal pulsed excitation function without DC component and its first and second derivative.

$$f_d(x) = \begin{cases} 0, & x < 0 \\ 3x^2 - 2x^3, & 0 \le x \le 1 \\ 3\left(1+\sqrt{2}-x\right)^2 - \sqrt{2}\left(1+\sqrt{2}-x\right)^3 - 1, & 1 \le x \le 1+\sqrt{2} \\ -3\left(2+\sqrt{2}-x\right)^2 + \sqrt{2}\left(2+\sqrt{2}-x\right)^3, & 1+\sqrt{2} \le x \le 2+\sqrt{2} \\ 0, & x > 2+\sqrt{2} \end{cases} \tag{20}$$

It is interesting to mention that in (20) in the function segment $1 \le x \le 1+\sqrt{2}$ the required transition is from 1 to -1, thus the optimal function (15) should be scaled by $\sqrt{2}$ over time and by 2 over amplitude in order to keep its optimal characteristics. Function (20), along with its first and second derivative, is presented in Fig. 5.

7. Conclusion

The first part of this paper contains the overview of the relevant principles in source modeling in FDTD, with special focus on differences between hard and soft sources and on different source geometry (plane wave sources and point sources). The second part of the paper is dedicated to the source optimization, more specifically to the optimization of the excitation time function, which has a significant influence on the behavior of the generator, regardless of its type. Generally accepted and the most frequently used excitation functions are listed. However, neither of them is designed primarily for FDTD application, taking into account specificities of FDTD method.

The optimization problem in this work is defined in order to minimize the propagation of the undesirable energy through the computational domain. This is accomplished by minimizing the mean square value of the second time derivative of the excitation function. As a result of the optimization process, very simple and closed-form optimal function is obtained.

In addition, two functions for pulsed excitation of FDTD domain, with and without DC component, are proposed.

Optimality of the obtained function shape is verified in the own developed FDTD simulation environment.

References

[1] TAFLOVE, A., HAGNESS, S. B. *Computational Electrodynamics: The Finite-Difference Time-Domain.* 3rd ed. Boston: Artech House, 2005.

[2] BUECHLER, D. N., ROPER, D., DURNEY, C. H., CHRISTENSEN, D. A. Modeling sources in the FDTD formulation and their use in quantifying source and boundary condition errors. *IEEE Transactions on Microwave Theory and Techniques*, 1995, vol. 43, p. 810–814. DOI: 10.1109/22.375228

[3] YEE, K. S. Numerical solution of initial boundary value problems involving Maxwell's equations in isotropic media. *IEEE Transactions on Antennas and Propagation*, 1966, vol. 14, no. 3, p. 302–307. DOI: 10.1109/TAP.1966.1138693

[4] MEREWETHER, D. E., FISHER, R., SMITH, F. W. On implementing a numeric Huygen's source scheme in a finite difference program to illuminate scattering bodies. *IEEE Transactions Nuclear Science*, 1980, vol. 27, no. 6, p. 1829–1833. DOI: 10.1109/TNS.1980.4331114

[5] UMASHANKAR, K. R., TAFLOVE, A. A novel method to analyze electromagnetic scattering of complex objects. *IEEE Transactions on Electromagnetic Compatibility*, 1982, vol. 24, no. 4, p. 397–405. DOI: 10.1109/TEMC.1982.304054

[6] WATTS, M. E., DIAZ, R. E. Perfect plane-wave injection into a finite FDTD domain through teleportation of fields. *Electromagnetics*, 2003, vol. 23, p. 187–201. DOI: 10.1080/02726340390159504

[7] OGUZ, U., GUREL, L., ARIKAN, O. An efficient and accurate technique for the incident-wave excitations in the FDTD method. *IEEE Microwave and Guided Wave Letters*, 1998, vol. 46, no. 6, p. 869–882. DOI: 10.1109/22.681215

[8] TAN, T., POTTER, M. Optimized analytic field propagator (O-AFP) for plane wave injection in FDTD simulations. *IEEE Transactions on Antennas and Propagation*, 2010, vol. 58, no. 3, p. 824–831. DOI: 10.1109/TAP.2009.2039310

[9] GUIFFAUT, C., MAHDJOUBI, K. A perfect wideband plane wave injector for FDTD method. In *Proceedings of the IEEE International Symposium on Antennas and Propagation.* Salt Lake City (UT, USA), 2000, vol. 1, p. 236–239. DOI: 10.1109/APS.2000.873752

[10] RODEN, J. A., GEDNEY, S. D. Convolution PML (CPML): An efficient FDTD implementation of the CFS–PML for arbitrary media. *Microwave and Optical Technology Letters*, 2000, vol. 27, no. 5, p. 334–339.

[11] MANSOURABADI, M., POURKAZEMI, A. FDTD hard source and soft source reviews and modifications. *Progress In Electromagnetics Research C*, 2008, vol. 3, p. 143–160. DOI: 10.2528/PIERC08032302

[12] GEDNEY, S. D. *Introduction to the Finite-Difference Time-Domain (FDTD) Method for Electromagnetics.* Course text, Morgan and Claypool Publishing, 2011. [Online] Cited 2015-03-03. Available at: http://www.engr.uky.edu/~gedney/courses/ee624.

[13] INAN, I. M., MARSHALL, R. A. *Numerical Electromagnetics – The FDTD Method.* Cambridge: Cambridge University Press, 2011.

[14] COSTEN, F., BERENGER, J.-P., BROWN, A. Comparison of FDTD hard source with FDTD soft source and accuracy assessment in Debye media. *IEEE Transactions on Antennas and Propagation*, 2009, vol. 57, no. 7, p. 2014–2022. DOI: 10.1109/TAP.2009.2021882

[15] KALIALAKIS, C. *Finite Difference Time Domain Analysis of Microstrip Antenna-Circuit Modules.* PhD Thesis. School of Electronic and Electrical Engineering, The University of Birmingham, July 1999.

[16] WAGNER, C. L., SCHNEIDER, J. B. Divergent fields, charge and capacitance in FDTD simulations. *IEEE Transactions on Microwave Theory and Techniques*, 1998, vol. 46, no. 12, p. 2131–2136. DOI: 10.1109/22.739294

[17] ROPER, D. H., BAIRD, J. M. Analysis of overmoded waveguides using the finite difference time domain method. In *Digest Proceedings of the IEEE Microwave Theory and Techniques Society Symposium.* Albuquerque (USA), 1992, p. 401–404. DOI: 10.1109/MWSYM.1992.187997

[18] NIKOLIC, B., DIMITRIJEVIC, B., RAICEVIC, N, ALEKSIC, S. Implementation of FDTD based simulation environment. *Facta Universitatis. Ser.: Electronics and Energetics.* 2013, vol. 26, p. 121–132. DOI: 10.2298/FUEE1302121N

About the Authors ...

Bojan DIMITRIJEVIĆ was born in Leskovac, Serbia, in 1972. He received the B.E.E., M.Sc., and Ph.D. degrees from the University of Niš in 1998, 2002, and 2006, respectively. His research interests include digital signal processing in telecommunications with special focus on interference suppression, adaptive filtering and synchronization, numerical methods in electromagnetics with special focus on FDTD method and signal, material and component modeling.

Bojana NIKOLIĆ was born in Niš, Serbia in 1982. She received the Dipl. – Ing. and Ph.D. degrees in Telecommunications from the Faculty of Electronic Engineering in Niš in 2007 and 2012, respectively. Her research interests include FDTD numerical modeling in electromagnetics and wireless communications.

Slavoljub ALEKSIĆ was born in Berčinac, Serbia in 1951. He received Dipl. – Ing., M. Sc. and Ph.D. degrees in Theoretical Electrical Engineering from the Faculty of Electronic Engineering, University of Niš, Serbia in 1975, 1979 and 1997, respectively. His researching areas are: electromagnetic field theory, numerical methods in electromagnetics, lightning protection systems, low-frequency EM fields, microstrip transmission lines with isotropic, anisotropic and bianisotropic media, cable joints and cable terminations, permanent magnets analysis, power lines.

Nebojša RAIČEVIĆ was born in Niš, Serbia in 1965. He received his the Dipl. – Ing., M.Sc. and Ph.D. degrees from the Faculty of Electronic Engineering of Niš, Serbia, in 1989, 1998 and 2010, respectively. His research interests include: cable terminations and joints, numerical methods for electromagnetic problems solving, microstrip transmission lines with isotropic, anisotropic and bianisotropic media, analysis of metamaterial structures, electromagnetic compatibility, nonlinear electrostatic problems, magnetic field calculation of coils and permanent magnets.

16

Signal Detection for QPSK Based Cognitive Radio Systems using Support Vector Machines

M. Tahir MUSHTAQ[1], *Inayatullah KHAN*[2], *M. S. KHAN*[3], *Otto KOUDELKA*[1]

[1]Inst. of Communications Networks and Satellite Communications, TU Graz, Austria
[2]Centers of Excellence in Science and Applied Technologies, Islamabad
[3] Dept. of Electrical Engineering, University of Gujrat, Gujrat, Pakistan

tahirmushtaq2006@yahoo.com, inayatkh@gmail.com, msaeedbaloch@gmail.com

Abstract. *Cognitive radio based network enables opportunistic dynamic spectrum access by sensing, adopting and utilizing the unused portion of licensed spectrum bands. Cognitive radio is intelligent enough to adapt the communication parameters of the unused licensed spectrum. Spectrum sensing is one of the most important tasks of the cognitive radio cycle. In this paper, the auto-correlation function kernel based Support Vector Machine (SVM) classifier along with Welch's Periodogram detector is successfully implemented for the detection of four QPSK (Quadrature Phase Shift Keying) based signals propagating through an AWGN (Additive White Gaussian Noise) channel. It is shown that the combination of statistical signal processing and machine learning concepts improve the spectrum sensing process and spectrum sensing is possible even at low Signal to Noise Ratio (SNR) values up to* -50 *dB.*

Keywords

Cognitive radio, auto-correlation function, machine learning, Support Vector Machine, spectrum sensing, statistical signal processing, opportunistic dynamic spectrum access

1. Introduction

The concept of cognitive radio raised a question on the effectiveness of fixed spectrum access and indicate the need to change the spectrum assignment policy. Radio spectrum is assigned and regulated by the government and international bodies following long-term fixed spectrum assignment policy [1].The unlicensed spectrum is overcrowded [2] due to tremendous increase in the usage of wireless devices and new services. Various studies show under-utilization of licensed spectrum [2, 3, 4] that leads to wastage of precious radio resources. According to Federal Communications Commission (FCC) 2002 report, the licensed bands are underutilized and the ISM (Industrial, scientific and Medical) radio bands are over crowded [2]. The report also stated that the average utilization of licensed bands is 15–85% [2].

The temporarily unused band is known as spectrum hole. The idea of cognitive radio introduced by J. Mitola [6] is to utilize these spectrum holes without affecting the communication priority of primary (licensed) users. It will provide more bandwidth to the un-licensed (cognitive) users. It will compensate the under-utilization of valuable licensed radio resources [5, 6] and provide more bandwidth to cognitive users. Cognitive radio is the enhanced version of SDR (Software Defined Radio). SDR transceiver is designed to work at various frequencies and modulation techniques with most of its functional implementation in software rather than hardware. SDR allows cognitive user to intelligently adapt communication parameters (carrier frequency, bandwidth, power, coding schemes, modulation scheme etc.) for the unused licensed spectrum, thus making dynamic spectrum access possible [7]. In the United States FCC allowed the dynamic access of the UHF-TV bands by the cognitive radio devices [8].

In order to utilize a licensed spectrum in an opportunistic way, first spectrum sensing is applied to detect the potential unused spectrum. Spectrum sensing improves the spectrum efficiency and reduces the interference [13, 14]. Welch's periodogram is a widely used energy based detection technique. Welch's Periodogram improved the Bartlett method and it reduces the noise in the estimated power spectra with the help of FFT (Fast Fourier Transform) [26]. Using this technique, spectrum estimation performance is improved by applying the arbitrary window on the data segments. It can be used as an easy and simple solution for detection of the spectrum holes and mitigation of noise in cognitive radio systems, and it requires less prior information of the source signal and channel. This technique can detect successfully the narrow-band signals up to -25 dB. However, it will be more interesting to evaluate its performance at very low Signal to Noise Ratio (SNR) values.

In this paper, the system design approach to meet the challenges of detecting very weak signals with the help of machine learning technique, is presented. Our goal is to present a practical system design for spectrum sensing functionality to improve the spectrum sensing performance [9]. Support Vector Machine (SVM) based classifier is used for detection of very weak signals. According to statistical machine learning theory, SVM works based on the structural risk minimization principle [10]. SVM has an advantage that can provide better generalization and improve performance for a small number of training samples. Recently spectral correlation analysis and feature detection have been used for the communication parameter detection using SVM [11, 12]. A mapping function known as a kernel is used by SVM to map the input space to high-dimensional feature space [15, 16]. SVM performance mainly depends on the chosen kernel function that makes the kernel a key part of SVM.

In this work, we have implemented two signal detectors. Firstly, the auto-correlation kernel based Support Vector Machine (SVM) classifier and secondly, Welch's Periodogram detector. For simulation purpose, we have used four QPSK (Quadrature Phase Shift Keying) based signals. These singles are propagating through AWGN (Additive White Gaussian Noise) channel with the help of SSB (Single Side Band) AM modulation technique with four different carrier frequencies $f_1,\ldots,f_4$ respectively. The AWGN noise is added in the four signals. In the simulations the signals are propagated with different SNR values in different cases.

The scalar SNR is a fundamental and important measure in communication systems. Generally, it is the ratio of the signal power to the noise power and is used as a device parameter for describing the noise performance of the detector. The SNR is a unit less quantity, often expressed in decibels (dB). A higher value indicates more signal than noise. In our simulation, SNR is taken as the ratio of signal (symbol) energy to the AWGN spectrum density. The results indicate high performance at low very SNR values. The signal detection is done with the help of a new SVM kernel known as the *Auto Correlation kernel Function (ACF)* [10]. The results show that the proposed detector detects the spectrum holes up to 89% accuracy for communication signals at very low SNR values in AWGN Channel for QPSK based four signals.

The rest of the paper is organized as follows: Section 2 contains brief introduction of Support Vector Machine for pattern classification. Section 3 describes auto-correlation SVM kernel function. Section 4 explains the system model and gives the implementation details. Simulation results are presented in Sec. 5 and, finally, the paper is concluded in Sec. 6.

2. Support Vector Machines for Pattern Classification

In this section, a brief introduction of Support Vector Machine (SVM) is presented, for further details see [17] and [16]. SVM belongs to the category of supervised learning algorithms that requires a supervisor, to reveal the true interpretation of the data, such interpretation involves class labeling of the data. The goal is to learn a mapping from the input data to the output, given by a supervisor. Depending on the nature of the output, the algorithms can be subdivided into a classification problem (where the output space is represented in a discrete form) or a regression problem (where the output space is a continuous form) [18].

The task of learning by support vector machines with general kernel functions for a binary classification problem is formulated as follows. Given training samples or data-set in the form of set of auto-correlation functions (the auto-correlation kernel will be discussed in the following section shortly):

$$\{(r_i,y_i)\}_{i=1}^{l} \subset \mathbb{R}^n \times \{-1,+1\}, \tag{1}$$

where l is the total number of training samples, $y_i \in \{-1,+1\}$ is the class label of signal $\mathbf{x}$. The SVMs learn a discriminant function of the form

$$d(f) = \operatorname{sgn}\left(\sum_{i=1}^{n} \alpha_i y_i \boldsymbol{K}(r, r_{si}) + b\right), \tag{2}$$

where the data $r_{si}(x)(1 \le i \le n)$ are the support vectors and the coefficients α_i are the solutions of the following quadratic programming problem,

$$\text{maximize} \sum_{i=1}^{l} \alpha_i - \frac{1}{2}\sum_{i,j=1}^{l} \alpha_i\alpha_j y_i y_j \boldsymbol{K}(r_i,r_j) \tag{3}$$

$$\text{subject to} \sum_{i=1}^{l} \alpha_i y_i = 0,\ 0 \le \alpha_i < C, i = 1,\ldots,l \tag{4}$$

and the bias b is given by

$$b = \frac{1}{n}\sum_{j=1}^{n}(y_j - \sum_{i=1}^{n} \alpha_i y_i \boldsymbol{K}(r, r_{si})). \tag{5}$$

Here, C is a user-defined constant and $\boldsymbol{K}(r_i,r_j)$ is the kernel function that must satisfy the Mercer theorem [22]. Generally, for the kernel functions either a linear inner-product, defined as

$$\boldsymbol{K}(r_i,r_j) = r_i \cdot r_j, \tag{6}$$

is used linear decision boundaries or the Gaussian kernel function, defined as

$$\begin{aligned}\boldsymbol{K}(r_i,r_j) &= \exp\left(-\gamma \parallel r_i - r_j \parallel^2\right) \\ &= \exp\left(-\gamma(r_i \cdot r_i + r_j \cdot r_j - 2r_i \cdot r_j)\right),\end{aligned} \tag{7}$$

is used for non-linear decision boundaries in the feature space. It follows that all kernels function that can be expressed in terms of the inner products of data and that satisfy the Mercer theorem can be used for computing the kernel matrix. We use the auto-correlation function for computing the kernel matrix corresponding to an SVM trained in auto-correlations feature space.

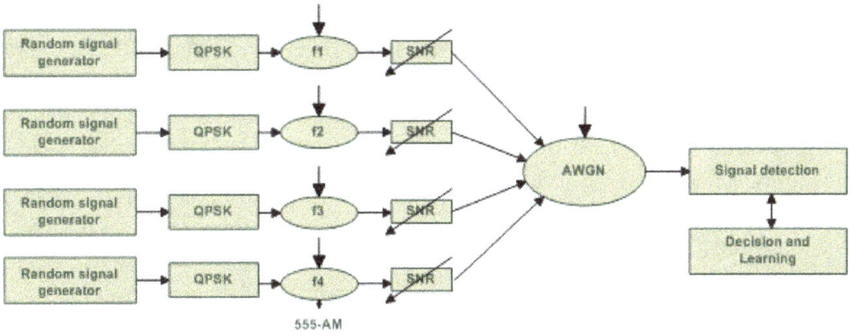

Fig. 1. System model: Experimental setup for QPSK AWGN pass-band signal detection.

3. Autocorrelation Kernel for SVM

Auto-correlation function is one of the important functions used in signal processing. It is the correlation between two samples x_k and x_l separated by time lag $(i-j)$. It is actually the measure of dependence of successive samples on the previous samples. Auto-correlation function is symmetric in time lag $(i-j)$. Definition of auto-correlation function is represented by (8),

$$r(i-j) = E[x_i x_j]. \quad (8)$$

An auto-correlation function is also an important feature of random signals. The auto-correlation function is an acceptable SVM kernel as it is symmetric function (satisfies Mercer condition), and the discrete-time Fourier transform of auto-correlation function is the power spectrum of the random signal x_i. So auto-correlation function satisfies both conditions of being an acceptable SVM kernel [10].

4. System Model and Implementation

This section describes the system model and gives its implementation details. In this work, we have implemented two signal detectors. Firstly, an auto-correlation kernel based SVM detector and secondly, Welch's Periodogram detector. For simulation purpose, we have used four QPSK based signals. These singles are propagating through AWGN channel with the help of signals sideband AM modulation technique with four different carrier frequencies $f_1,\ldots,f_4$ respectively. The AWGN noise is added in each of the four signals. In the simulations the signals were propagating with different SNR values in different cases.

The system is depicted in Fig. 1. It consists of four transmitters acting as primary users. At the receiver end, the signals are sampled and processed. These processed signals are then classified or detected by the detectors.

The signal detection problem can be formulated into a binary hypothesis form as

$$\mathcal{H}_0 \;: x[n] = w(n) \quad (9)$$

$$\mathcal{H}_1 \;: x[n] = s(n) + w(n) \quad (10)$$

where $\mathcal{H}_0$ is a *null hypothesis* which represents the presence of only noise or absence of a signal and $\mathcal{H}_1$ is the *alternative hypothesis* which shows the presence of the information signal. Here, $s(n)$ is the received signal $(n = 0,1,\ldots,N-1)$ and $w(n)$ represents the additive white Gaussian noise. We performed this signal classification and detection with two methods, Welch's Periodogram detector and auto-correlation kernel function based on SVM.

The Welch's Periodogram [26] is an energy detector worked on the basis of overlapping window used for spectral power density estimation (PSD). Welch's Periodogram reduces the noise in the estimated power spectra with the help of FFT. In this technique, Spectrum estimation performance is improved by applying the arbitrary window on the signal segments. The presence of the signal is shown by lobes. The comprehensive detail of this method is out of the scope of this paper. However, we briefly describe the steps involved in calculating the Welch's Periodogram as:

- The signal, x, is split into L overlapping segments and the length of each segment is M. The overlap is defined by a factor D. The overlap is 50% for $D = M/2$, and it is equal to 0 for $D = 0$.
- The overlapping segments are then windowed. For each segment a periodogram is calculated by computing discrete Fourier transform and then its square magnitude is computed. Then we calculate an average of the individual periodograms.
- The end result is shown by frequency bins versus the PSD graph. The presence of the signal is shown by lobes.

In the case of the Welch's Periodogram detector, we followed the Neyman-Pearson criterion. The criterion maximizes the probability of detection P_D for a given probability of the false alarm (P_{FA}) [19, 20].

In case of the SVM detector P_{FA} represents the 'false positive rate' and 'true positive rate' is represented by P_D. We use an auto correlation kernel function based SVMs for the detection of four signals propagating through AWGN channel. The receiver and the transmitters are both in fixed positions, so the Doppler shift is $f_{ds} = 0$. The signals are transmitted with adjustable SNR values range e.g., for a range of SNR values = [−80 : 10 : 20] dB where 10 is the step size we have a total of 11 SNR values. The transmission channel is modeled as a non-line of sight AWGN channel. These signals, generated using a simulator developed in MATLAB, are used as input for training the four binary SVM based classifiers. We have used the Chang et al implementation called libSVM for training SVMs [23].

For the experimentation purpose, we trained four binary SVMs using the auto correlation kernel on training data set. This class labeled data set is generated using our AWGN pass-band signal generator program. The number of the samples used for training the SVM varies and depends on the given SNR values range. For example, for a given SNR range [−40 : 2 : 20] we have a total of 31 SNR values and for each SNR value, we generate 1500 training data set of signals consisting of an equal number of 15 combinations of the four frequencies signals. This makes a total of 46500 training samples. To evaluate the performance of the detectors, a test data set is generated for the same SNR range of values.

5. Simulation Results

We performed a series of simulation experiments for the evaluation of the proposed detector. We use the Area Under ROC (Receiver Operating Characteristics) curve, or simply AUC as a measure of detector performance. In machine learning community, the AUC is widely used for the performance evaluation of machine learning algorithms. Interested reader can find more details about AUC in [24], and [25], here we describe it briefly. The ROC is a plot of the probability of classifying correctly the positive examples (True positive rate or true detection) against the rate of incorrectly classifying true negative examples (False positive rate or false detection). Fig. 2 depicts an example of the ROC curve. The diagonal line corresponds to the ROC curve of a classifier that predicts the class at random, and the performance improves the further the curve is near to the upper left corner of the plot. AUC is the performance measure extracted from the ROC curve. When AUC is equal to 1, the classifier achieves perfect accuracy, and a classifier that predicts the class at random has an associated AUC of 0.5. Another interesting point of the AUC is that it depicts a general behavior of the classifier since it is independent of the decision threshold used for obtaining a class label [25]. The AUC is computed by calculating the area using the following equation:

$$\mathbf{AUC} = \frac{\sum_{i=1}^{n^+}\sum_{j=1}^{n^-} 1_{f(x_i^+)>f(x_j^-)}}{n^+ n^-}, \tag{11}$$

where $f(.)$ is the decision or scoring function of a classifier, which is defined by (2), x^+ and x^- denote the positive class labeled and negative class labeled data samples, and n^+ and n^- are the number of positive and negative examples. The 1_π is equal to 1 if the predicate $\pi = f(x_i^+) > f(x_j^-)$ holds and 0 otherwise. The AUC of a classifier is maximal when $f(x_i^+) > f(x_j^-)$, $\forall_i = 1,\ldots,n^+$, $\forall j = 1,\ldots,n^-$. The AUC decreases when ever a negative data sample is ranked higher than a positive class labeled data sample.

The performance of the detector depends upon the number of training samples, SNR values range and frequency of a signal to be detected. At low SNR values the performance of detector is decreased. The low SNR can be dealt by decreasing the step size [21] or increasing the number of training samples. The simulation's results show that the signal detection is possible at very low SNR values.

The decision threshold value is a minimum confidence level for signal detection or for obtaining the signal class label. In the current scenario, this decision threshold value is selected with the help of standard cross validation procedure. In this work confidence level or decision threshold > 20% shows the presence of the signal. This value is fixed through experiments using the 10-fold cross validation procedure in order to make a balance between false detection and true detection. The maximum performance achieved by the SVM detector is measured with AUC equal to 89%). Figs. 2, 3, 4, and 5 illustrate the performance of the simulation results at different SNR values ranges.

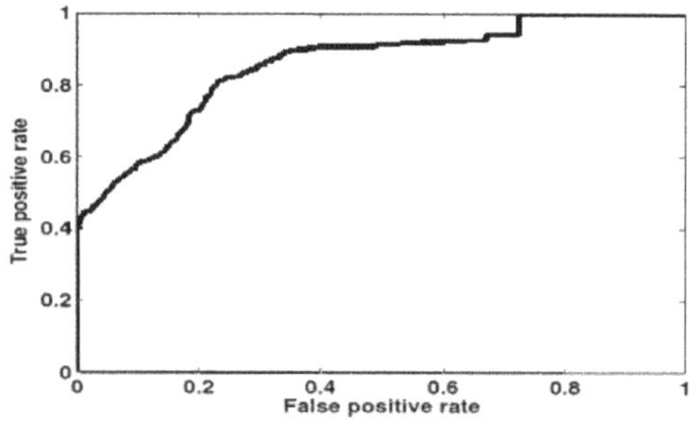

Fig. 2. ROC performance curve with **AUC = 0.86** for the auto-correlation kernel based SVM detector. The detector was trained and tested with data-set containing 20, 30, 40 and 60 kHz AWGN pass-band signals and SNR values in the range [−40 : 10 : 40].

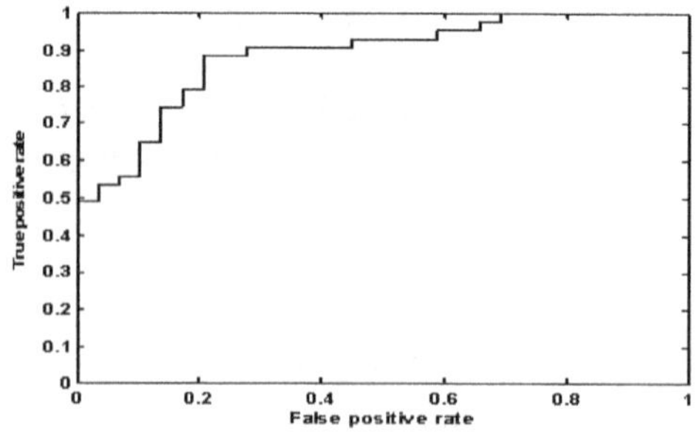

Fig. 3. ROC performance curve with **AUC = 0.89** for the auto-correlation kernel based SVM detector. The detector was trained and evaluated with data-set containing 20, 30, 40 and 60 kHz AWGN pass-band signals and SNR values in the range [−40 : 2 : 40].

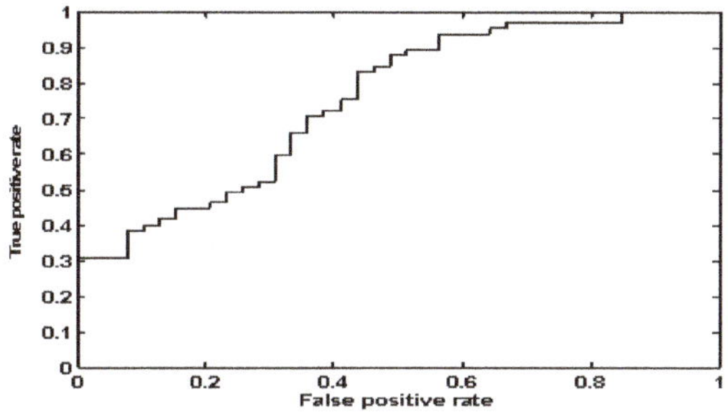

Fig. 4. ROC performance curve with **AUC= 0.75** for the auto-correlation kernel based SVM detector. The detector was trained and evaluated with data-set containing 20, 30, 40 and 60 kHz AWGN pass-band signals and SNR values in the range $[-80:10:40]$

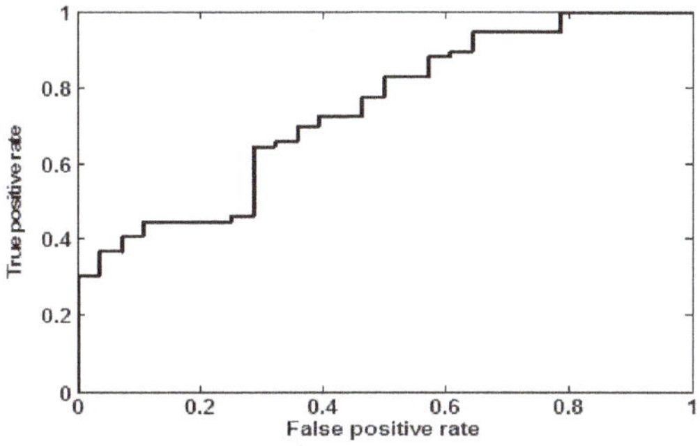

Fig. 5. ROC performance curve with **AUC = 0.74** for the auto-correlation kernel based SVM detector. The detector was trained and evaluated with data-set containing 200, 300, 400 and 500 kHz AWGN pass-band signals and SNR values in the range $[-80:10:20]$.

Our simulation results show that the SVM detector's performance depends on the step size and can be improved by decreasing step size, which will increase the size of training data-set. This can be easily observed in Fig. 2 and Fig. 3. In Fig. 2 AUC is 0.86 for the step size of 10 while in Fig. 3 AUC is improved to 0.89 as step size is decreased to 2 for the same SNR range. We can see in Fig. 4 that the performance decreases when SNR values range is increased.

The simulation results shown in Fig. 5 confirms that the performance of SVM detector depends upon the frequency range of the signal. It is shown that the increase in the frequency range of the signals will decrease the performance of the detector for the same SNR range of values and step size.

For the purpose of performance evaluation, we performed a number of experiments to compare the Welch's periodogram and auto-correlation kernel based SVM based techniques. For these experiments we trained four auto-correlation kernel based SVM detectors with a training data set containing class labeled of four 20, 30, 40 and 60 kHz carrier frequencies signals which were transmitted with adjustable SNR values range equal to $[-80:10:40]$.

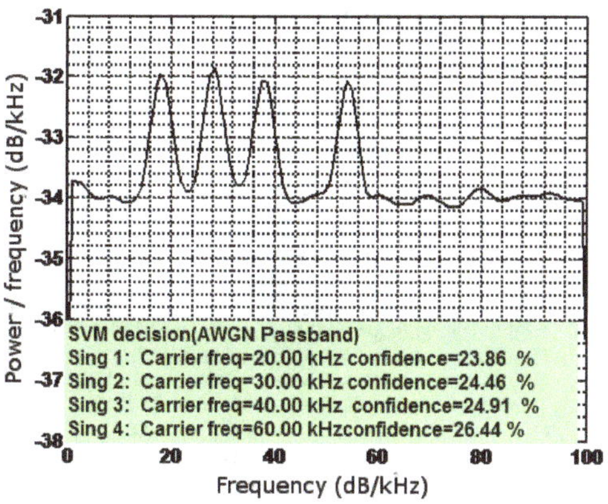

Fig. 6. Four test signals at $SNR = -10$ dB with four 20, 30, 40 and 60 kHz carrier AWGN Pass band frequencies. Both Welch's periodogram (shown by the upper graph) and the auto-correlation kernel based SVM detectors detect (shown in the lower green box) correctly all the four signals, the confidence > 20%.

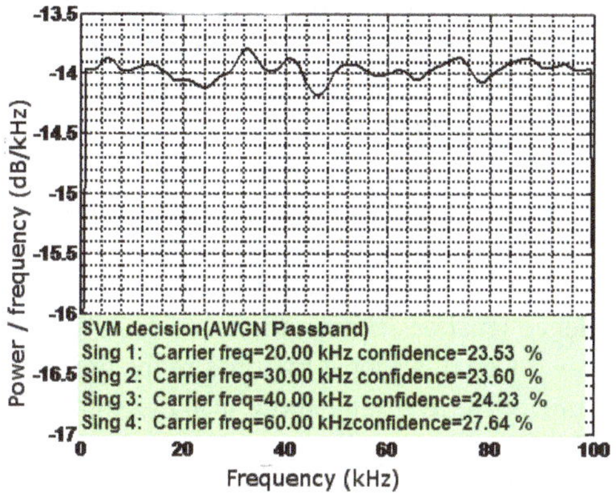

Fig. 7. Four test signals at −30 dB with four 20, 30, 40 and 60 kHz carrier AWGN Pass band frequencies. Only the auto-correlation kernel based SVM detectors detect (shown in the lower green box) correctly all the four frequencies of the signals, the confidence > 20%.

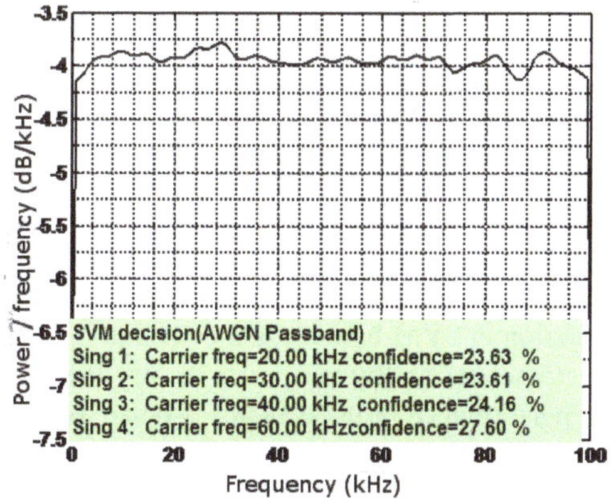

Fig. 8. Four test signals at −40 dB with four 20, 30, 40 and 60 kHz carrier AWGN Pass band frequencies. Only the auto-correlation kernel based SVM detectors detect (shown in the lower green box) correctly all the four frequencies of the signals, the confidence > 20%.

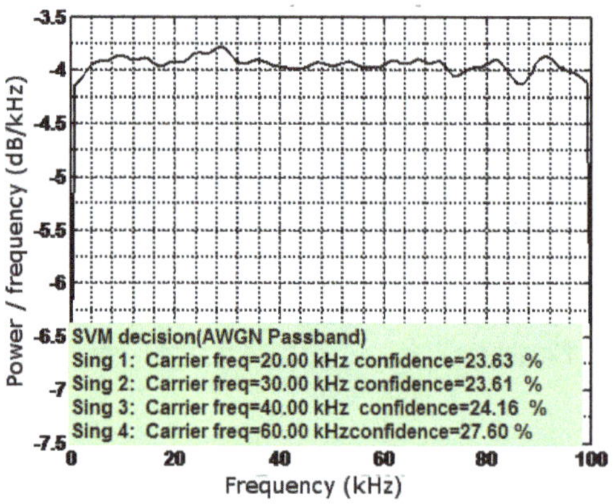

Fig. 9. Four test signals at −50 dB with four 20, 30, 40 and 60 kHz carrier AWGN Pass band frequencies. Only the auto-correlation kernel based SVM detectors detect (shown in the lower green box) correctly all the four frequencies of the signals, the confidence > 20%.

The results after passing test signals with four carrier frequencies at −10 dB, −30 dB, −40 dB and −50 dB SNR values are shown in Fig. 6, Fig. 7, Fig. 8 and Fig. 9 respectively. In these figures, the upper graph are the Welch's power spectral density estimates that show the results for detection of signals by Welch's periodogram while the lower green box shows the signal detection with the help of auto-correlation function kernel based SVM detector. It is worth reminding that the SVM detector detects a signal correctly if its confidence level or decision threshold > 20%, while the presence of a signal is shown by lobes in the graph Welch's periodogram. These results show that the signals are not detected by Welch's periodogram at SNR ≤ -30 dB while SVM detects the signals at SNR ≤ -30 dB.

The results suggest that the SVM detector identifies very weak signals. The combination of Welch's periodogram detector and an SVM based detector could provide means to detect signals at low SNR levels more reliably. However further investigation is required to obtain more accurate results.

6. Conclusions

In this paper, the auto-correlation function kernel based SVM classifier along with Welch's periodogram detector is successfully implemented for the detection of four QPSK based signals propagating through an AWGN channel. The main attraction in SVM-based signal detection is its flexibility and its potential ability to operate at very low SNR values. The performance of the detector is better as compared to conventional energy detectors.

Our results suggest the SVM detector identifies very weak signals. We have shown that increase in the frequency range decreases the performance of the detector. Furthermore, larger SNR range of values results in low AUC (ROC). Similarly, performance of the detector is improved by decreasing the SNR step size. By reducing the SNR step size the number of training samples will increase that will eventually increase the accuracy of the signal detection. Higher the number of training samples the better will be the performance of the detector. Additionally, we showed that the Welch's periodogram detector does not perform well at SNR ≤ -30 dB for an AWGN channel. On the other hand, signal detection could be done at SNR ≤ -30 dB with the help of the auto-correlation kernel based SVM detector.

Along with various advantages of using SVM for signal classification, it can detect the signals at very lower SNR values. In this paper, we presented a spectrum sensing technique with better accuracy at very low SNR values. Our results showed that by using the combination of statistical signal processing and machine learning concepts, spectrum sensing could be improved. With the use of machine learning techniques, cognitive radio networks could become self-aware and behave in a really adaptive way. In future work, we will use the auto correlation kernel function based signal detection technique for the detection of weak earthquake and star signals.

Acknowledgments

This work is sponsored by Higher Education Commission of Pakistan under overseas scholarship in selected areas phase II. We also offer special thanks to ICG TU Graz Austria for providing computational and technical expertise.

References

[1] TRAGOS, E., ZEADALLY, S., FRAQKIADAKIS, A., SIRIS , V. Spectrum assignment in cognitive radio networks: A comprehensive survey. *IEEE Communications Surveys and Tutorials*, 2013, vol. 15, no. 3, p. 1108–1135. DOI: 10.1109/SURV.2012.121112.00047

[2] FCC. *Report of the Spectrum Efficiency Working Group Spectrum Policy Task Force Report ET Docekt No 02-135.* 2002

[3] AKYILDIZ, I. F., LEE, W. Y., VURAN, M., MOHANTY, S. NeXt generation/dynamic spectrum access/cognitive radio wireless networks: a survey. *Computer Networks*, 2006, vol. 50, no. 13, p. 2127–2159. DOI: 10.1016/j.comnet.2006.05.001

[4] CABRIC, D., MISHRA, S., WILLKOMM, D., BRODERSEN, R., WOLISZ, A. A Cognitive radio approach for usage of virtual unlicensed spectrum. In *14th IST Mobile and Wireless Communications Summit*. 2005.

[5] DA SILVA, C. R. C., CHOI, B., KIM, K. Distributed spectrum sensing for cognitive radio systems. In *Proceedings of 2007 Information Theory and Applications Workshop*. 2007, p. 120–123.

[6] MITOLA, J. *Cognitive Radio: An Integrated Agent Architecture for Software Defined Radio.* PhD Thesis. Stockholm (Sweden): Royal Institute of Technology (KTH), 2000.

[7] KIM, S. J., GIANNAKIS, G. B. Rate-optimal and reduced-complexity sequential sensing algorithms for cognitive OFDM radios. *EURASIP Journal on Advances in Signal Processing*, 2009. DOI: 10.1155/2009/421540

[8] MOLISCH, A., GREENSTEIN, L., SHAFI, M. Propagation issues for cognitive radio. *Proceedings of the IEEE*, 2009, vol. 97, no. 5, p. 787–804. DOI: 10.1109/JPROC.2009.2015704

[9] KAMIL, N. H., KADHIM, D. J., LIU, W., CHENG, W. Signal processing techniques for robust spectrum sensing. In *ETP International Conference on Future Computer and Communication*. Wuhan (China), 2009, p. 120–123. DOI: 10.1109/FCC.2009.14

[10] KONG, R., ZHANG, B. Auto-correlation kernel functions for support vector machines. In *International Conference of Natural Computation*. 2007, vol. 1, p. 512–516.

[11] SADEGHI, H., AZMI, P. A novel primary user detection method for multiple antenna cognitive radio. In *International Symposium on Telecommunications (IST)*. Tehran (Iran), 2008, p. 188–192. DOI: 10.1109/ISTEL.2008.4651297

[12] HU, H., SONG, J. , WANG, Y. Signal classification based on spectral correlation analysis and svm in cognitive radio. In *22nd International Conference on Advanced Information Networking and Applications (AINA)*. Okinawa (Japan), 2008, p. 883–887.

[13] MITOLA, J., MAQUIRE, G. Q. Cognitive radio: making software radio more personal. *IEEE Personal Communications*, 1999, vol. 6, no. 4, p. 13–18. DOI: 10.1109/98.788210

[14] DIGHAM, F. F., ALOUINI, M.-S., SIMOK, M. L. On the energy detection of unknown signals over fading channels. In *IEEE International Conference on Communications (ICC)*. 2003, vol. 5, p. 3575–3579. DOI: 10.1109/ICC.2003.1204119

[15] CHRISTOPHER, J. A Tutorial on support vector machine for pattern recognition. *Data Mining and Knowledge Discovery*, 1998, vol. 2, no. 2, p. 121–167.

[16] CRISTIANINI, N., TAYOR, J. S. *An Introduction to Support Vector Machines and Other Kernel-based Learning Methods.* Cambridge University Press, 2000. DOI: 10.1017/CBO9780511801389

[17] VAPNIK, V. V. *The Nature of Statistical Learning Theory.* Springer, 1995. DOI: 10.1007/978-1-4757-2440-0

[18] SAFARI, A. *Multi-Class Semi-Supervised and Online Boosting Institute for Computer Graphics and Vision.* Ph.D Thesis. Graz (Austria): Graz University of Technology, 2010.

[19] CHEN, K. C., CHEN, P. Y., PRASAD, N., LIANG, Y. C., SUN, S. Trusted cognitive radio networking. *Wireless Communications and Mobile Computing*, 2010, vol. 10, no. 4, p. 467–485. DOI: 10.1002/wcm.777

[20] ZOU, Y., YAO, Y. D., ZHENG, B. Outage probability analysis of cognitive transmissions: Impact of spectrum sensing overhead. *IEEE Transactions on Wireless Communications*, 2010, vol. 9, no. 8, p. 2676–2688. DOI: 10.1109/TWC.2010.061710.100108

[21] KAY, S. M. *Fundamentals of Statistical Signal Processing: Detection Theory.* Prentice Hall, 1993

[22] SCHOLKOPF, B., SMOLA, A. *Learning with Kernels: Support Vector Machines, Regularization, Optimization and Beyond.* Cambridge (MA, USA): MIT Press, 2001.

[23] CHANG, C.-C., LIN, C.-J. A library for support vector machines. *ACM Transactions on Intelligent Systems and Technology*, 2011, vol. 2, no. 3, p. 1–27.

[24] BRADLEY, A. P. The use of the area under the ROC curve in the evaluation of machine learning algorithms. *Pattern Recognition*, 1997, vol. 30, no. 7, p. 1145–1159. DOI: 10.1016/S0031-3203(96)00142-2

[25] RAKOTOMAMONJY, A. *Support Vector Machines and Area Under ROC Curves*. 2004.

[26] WELCH, P. D. the use of Fast Fourier Transform for the estimation of power spectra: A method based on time averaging over short, modified periodograms. *IEEE Transactions on Audio and Electroacoustics*, 1967, vol. 15, no. 2, p. 70–73. DOI: 10.1109/TAU.1967.1161901

Design of High Performance Microstrip Dual-Band Bandpass Filter

Nafiseh KHAJAVI[1], Seyed Vahab AL-Din MAKKI[2], Sohrab MAJIDIFAR[3]

[1]Dept. of Electrical Engineering, Dezful Branch, Islamic Azad University, Dezful, Iran
[2]Dept. of Electrical Engineering, Razi University, Kermanshah, Iran
[3]Dept. of Electrical Engineering, Kermanshah University of Technology, Kermanshah, Iran

n_khajavi89@yahoo.com, v.makki@razi.ac.ir, s.majidifar@razi.ac.ir

Abstract. *This paper presents a new design of dual-band bandpass filters using coupled stepped-impedance resonators for wireless systems. This architecture uses multiple couple stubs to tune the passband frequencies and the filter characteristics are improved using defected ground structure (DGS) technique. Measurement results show insertion losses of 0.93 dB and 1.13 dB for the central frequencies of 2.35 GHz and 3.61 GHz, respectively. This filter is designed, fabricated and measured and the results of the simulation and measurement are in good agreement.*

Keywords

Dual-band bandpass filters, stepped-impedance resonators, defected ground structure (DGS), insertion loss

1. Introduction

Nowadays microwave systems have different applications in the society including satellite televisions, mobile phones, and civilian and military satellite systems. Given the growth in daily demand of communication systems like satellite communication and mobile phones, a contemporary trend in microwave technology can be seen towards compact and small-sized circuits and lower prices. To achieve this objective, active and passive microwave circuits are designed in a compact, multi-band and frequency-adjustable manner. After developing portable communication systems, multi-band systems have attracted great attention. Using this characteristic, results in smaller sizes and lower prices. Designing dual-band filters using loaded-stub open-loop resonators is a common practice [1]. This structure has return loss of more than 3 dB in two bands. Dual-band bandpass filters have also been provided using spiral stepped-impedance resonators [2]. In [3] dual-band filters have been designed using defected ground structure resonator and a dual-mode open-stub loaded stepped impedance resonator. A dual mode microstrip fractal resonator is proposed in [4] and optimized perimeter of the proposed resonator by using fourth iteration T-square fractal shape. L. C. Tsai [5] combined a wide-band bandpass filter and a band-stop filter and designed a dual-band filter. In another work, X. Guan designed dual-band filters using transmission lines and open-ended stubs [6]. A non-degenerate dual-band microstrip filter has been developed using non-degenerate resonator loads and open-ended stubs in [7]. A dual-band bandpass microstrip filter has been designed in [8] using microstrip periodic stepped-impedance ring resonators. For better description, design process of a LC model is presented in [9]. High selectivity and good suppression is obtained in [10] using embedded scheme resonator. The parallel LC resonant circuit can be represented for equivalent circuit of the DGS [11]. In other case a bandpass filter is designed using coupled DGS open loop resonators [12]. By applying the fractal theory and defected ground structure, filter dimensions have been reduced in [13].

In this paper a microstrip dual-band bandpass filter is designed using coupled stepped-impedance resonators. Designing filter is performed in three stages. To adjust the frequency of pass bands, we add a low impedance section to the middle of the basic structure. For fixed frequencies according to WLAN standards, multiple coupling structures is added to the proposed resonator and filter characteristics like return loss, insertion loss and bandwidth were enhanced using defected ground structure technique. In design process of the proposed filter, the exact combination of several known methods (resonator design base on the λ/2-stepped impedance structure, multiple coupling- defected ground structure) is used instead of a complex dual band structure. In this method each parts of filter response is influenced by a part of the design process and complexity of the final structure is replaced with the multiplicity of design process.

2. Filter Design

Figure 1 shows basic structure of open-ended stepped-impedance resonator with the half-wave of λ/2. This structure consists of a high impedance section Z_1 with the electric length of θ_1 and two low impedance sections Z_2 with the electric length of θ_2 as in [14].

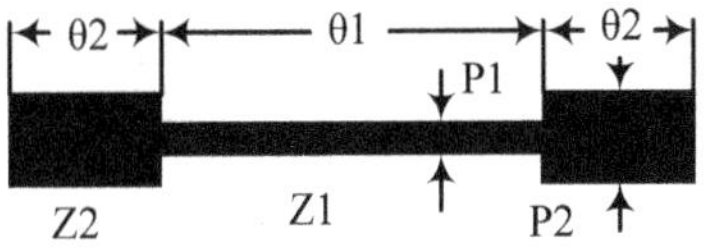

Fig. 1. Basic structure stepped-impedance resonator with the half-wave of λ/2.

The resonance condition of the resonator can be derived by:

$$R_Z = Z_2 / Z_1 = \tan\theta_1 \tan \theta_2 \qquad (1)$$

where R_Z is the impedance ratio of the stepped impedance resonator (SIR). The fundamental frequency f_0 and first spurious frequency f_{sb1} of the resonator are related by:

$$\frac{f_{sb1}}{f_0} = \frac{\pi}{2\tan^{-1}\sqrt{R_Z}}. \qquad (2)$$

Figure 2 shows the normalization of the first neutral frequency as a function of R_Z. By selecting a lower impedance ratio, the first neutralization mode shifts towards the upper frequency range.

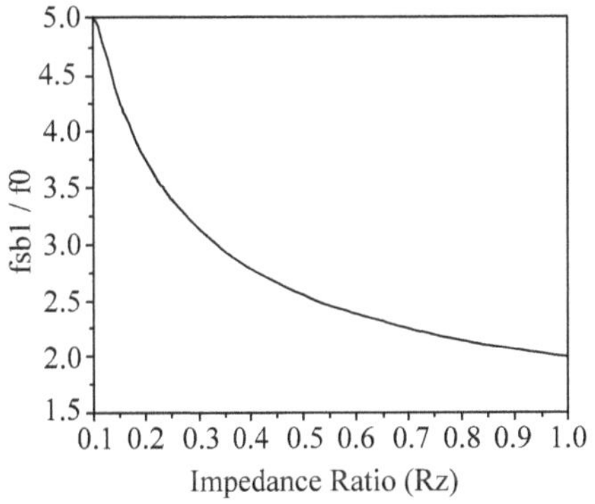

Fig. 2. Normalization of the first neutral frequency as a function of R_Z.

Designing of the filter is performed in three stages:

- Resonator design.
- Adding multiple couplings to the proposed resonator for neutralize neutral harmonics in stop band.
- Using defected ground structure (DGS) technique to improve the filter characteristics.

2.1 Resonator Design

2.1.1 Basic Resonator

According to (2) and f_0 / f_{sb1} = 1.85, we will have R_Z =1.3. Now, if Z_1 = 100 Ω, given (1), Z_2 = 130 Ω. Given the values of Z_1 and Z_2, P1 = 0.289 mm and L2 = 0.13 mm are calculated. As shown in Fig. 3(a), the basic resonator is designed using structure of Fig. 1. Dimension of Fig. 3(a) are as follow: L1 = 30.4 mm, P1 = 0.298 mm, L2 = 0.13 mm, P2 = 2.16 mm and G = 0.1 mm.

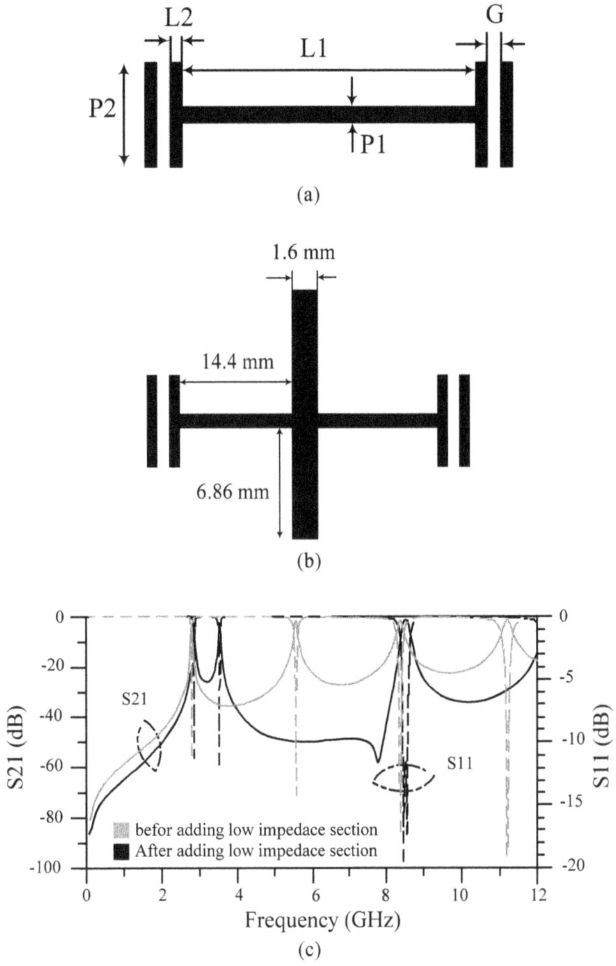

Fig. 3. Basic resonator. (a) The layout of the basic resonator. (b) The layout of the basic resonator with added low impedance section. (c) Simulation result of the part (a) and (b).

2.1.2 Tuning of the Passband Frequencies

In the basic resonator, the first passband frequency reaches 2.7 GHz and the second one reaches 5.5 GHz. With adding a low impedance stub to the basic resonator (Fig. 3b), the second passband frequency shifts to the first one. Passband frequencies can be adjusted in two stages:

- By changing L1, first passband frequency is placed at a desired value.
- By changing the dimensions of the low impedance stub, while the first pass band frequency is fix, the second one tunes at the desired value.

2.1.3 Physical Size Optimizations

To optimize the dimensions of the filter, we change the 30.4 mm long middle transmission line as shown in Fig. 4(a). Dimensions of the proposed resonator are as follow: W1 = 0.13 mm, L1 = 0.16 mm, W2 = 0.13 mm, L2 = 17.57 mm, W3 = 0.186 mm, L3 = 1.86 mm, L4 = 7.852 mm, L5 = 14.412 mm, W4 = 0.4 mm, L6 = 2.4 mm, W5 = 0.43 mm, L7=4.065 mm, W6 = 0.289 mm, G = 0.2 mm.

2.1.4 Proposed LC Model of the Resonator

In order to better describe the designed resonator, its LC model, as it is shown in Fig. 4(b) is presented. In this model, Cg represents the coupling capacitance between the open stubs and the transmission lines are modeled by a T-junction circuit including two series inductances and a central ground ended capacitor. Lt1 and Ct1 are the inductance and capacitance of L7 and Cop1, Cop2 are the capacitances of the open stubs with respect to ground. Lt2 and Ct2 are the inductance and capacitance of T-junction part of the middle loop. The values of circuit parameters are calculated using equations (3)-(8) [15].

$$\varepsilon_{ref} = \frac{\varepsilon_r + 1}{2} + \left\{ \left(1 + 12\frac{H}{W}\right)^{-0.5} + 0.04\left(1 - \frac{W}{H}\right)^2 \right\} \qquad \frac{W}{H} \le 1 \tag{3}$$

$$\varepsilon_{ref} = \frac{\varepsilon_r + 1}{2} + \frac{\varepsilon_r - 1}{2}\left(1 + 12\frac{H}{W}\right)^{-0.5} \qquad \frac{W}{H} \ge 1$$

$$Z_C = \frac{120\pi}{\frac{C_a}{\varepsilon_r}\sqrt{\varepsilon_{ref}}} \tag{4}$$

where W is the microstrip line width, ε_r is the relative dielectric constant of the substrate, H is the substrate thickness, Z_C is the characteristic impedance, ε_{ref} is the effective dielectric constant, C_a is the capacitance per unit length with the dielectric substrate replaced by air and c is the velocity of electromagnetic waves in free space (c = 3.0×10^8 m/s). C_a can be determined by (5).

$$C_a = \frac{2\pi\varepsilon_r}{l_n\left(\frac{8H}{W} + \frac{W}{4H}\right)} \qquad \frac{W}{H} \le 1 \tag{5}$$

$$C_a = \varepsilon_r\left(\frac{W}{H} + 1.393 + 0.66 l_n\left(\frac{W}{H} + 1.444\right)\right) \qquad \frac{W}{H} \ge 1$$

While l is the length of the microstrip line, L and C as inductance and capacitance of microstrip line can be determined by (6) and (7).

$$L = \frac{Z_C l}{V_P}, \tag{6}$$

$$C = \frac{l}{Z_C V_P} \tag{7}$$

V_p is the phase velocity and can be determined by (8).

$$V_p = \frac{c}{\sqrt{\varepsilon_{ref}}} \tag{8}$$

The calculated values of the circuit parameters are as follow: L = 0.08 nH, Cop1 = 0.346 pF, Cg = 0.3 pF, Cop2 = 0.36 pF, Lt1 = 1.09 nH, Ct1 = 0.17 pF, Lt2 = 3.916 nH and Ct2 = 0.67 pF. The electromagnetic (EM) and LC simulated results of this structure are shown in Fig. 4(c). As it is shown, good agreement between the EM and LC simulated results is achieved. This resonator is designed on Ro 4003 substrate with 3.38 dielectric constant, 20 mil heights and 0.0021 loss tangent.

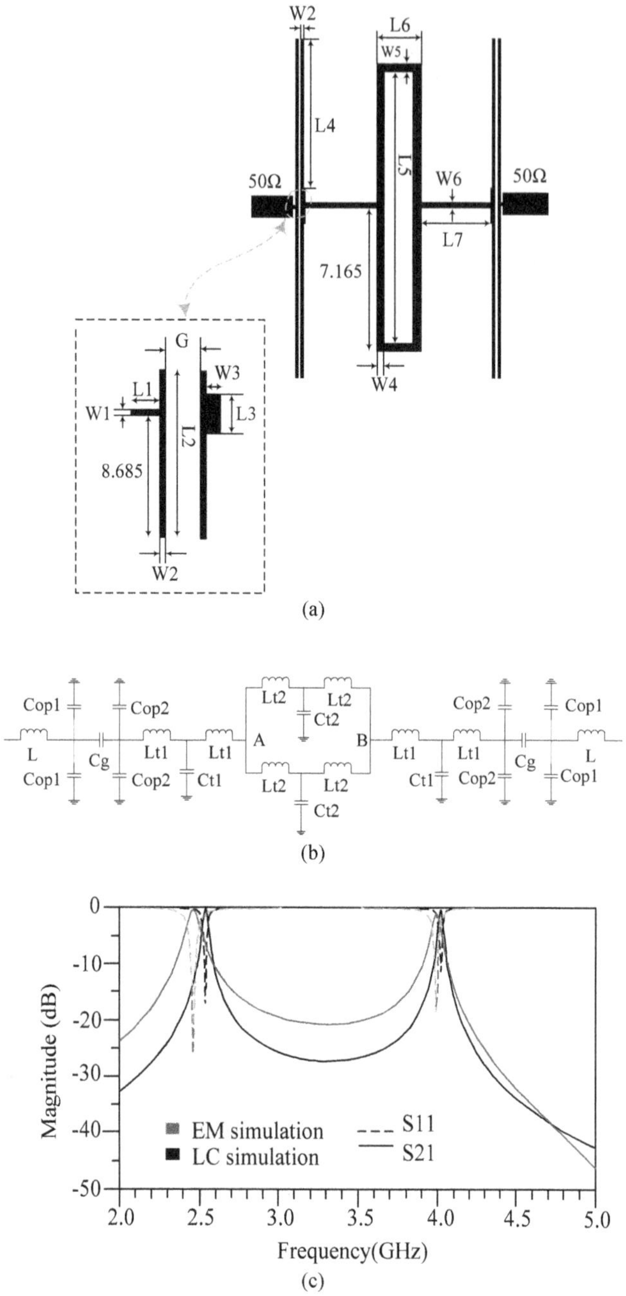

Fig. 4. Proposed resonator. (a) The layout of the proposed resonator. (b) LC equivalent circuit of the proposed resonator. (c) EM and LC circuit simulated results.

The frequency response of the proposed resonator in a wide span is shown in Fig. 5. As can be seen in this figure, this resonator has two bands with the central frequencies of f_1= 2.40 GHz and f_2= 4.0 GHz. The objective is to design a dual-band bandpass filter with the same central frequencies as those of WLAN. Central frequency of the first band is fixed on 2.4 GHz, while in the second band our goal is to achieve central frequency of 3.7 GHz. These bands have insertion losses of 0.478 dB and 1.137 dB, return losses of 26.05 dB and 18.549 dB and bandwidths of 76 MHz and 49 MHz, respectively. In order to achieve the better response characteristics of the resonator, the following changes should be considered: balancing the insertion loss of two bands, decreasing the return loss of two bands and increasing the passbands' bandwidth. Furthermore, the neutral harmonic at the 6.866 GHz decreases the stopband bandwidth.

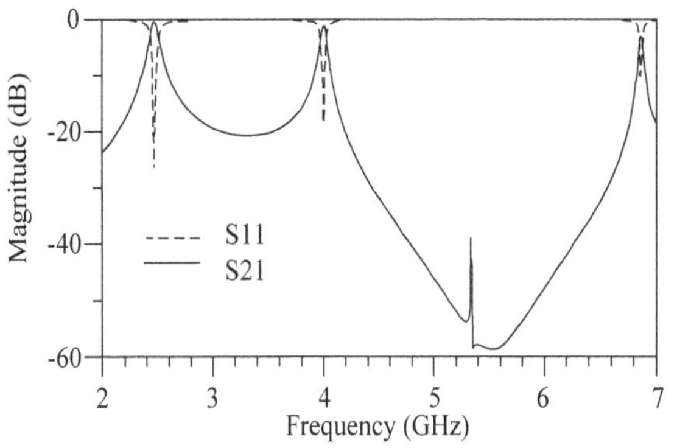

Fig. 5. Frequency response of the proposed resonator in a wide span.

2.2 Adding Multiple Couplings to the Proposed Resonator for Neutralizing Neutral Harmonics in the Stopband

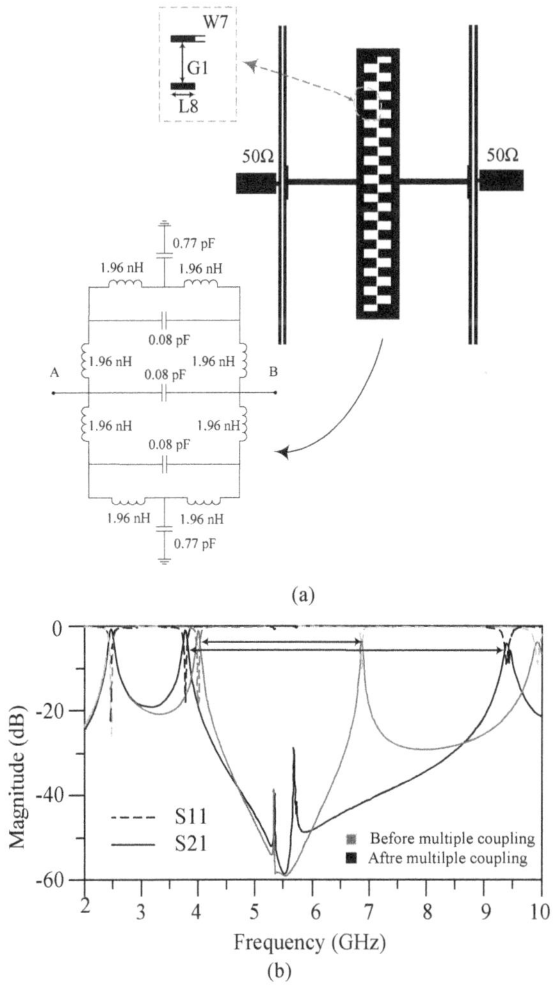

Fig. 6. The proposed filter (a) Layout and LC model (b) Compares between frequency responses of the proposed resonator and the proposed filter.

One of the problems in the last stage is the limited stopband bandwidth. This problem can be solved using multiple couplings. Figure 6(a) shows the proposed filter. In this filter the added multiple coupling are located inside the middle lope of the resonator. The dimensions of these coupling stubs are as follow: W7 = 0.524 mm, L8 = 0.762 mm, G1 = 0.524 mm. The size and number of the coupling stubs and its LC model are depicted in Fig. 6(a). In this model the Lt2 is divided into two parts and the effects of the multiple couplings are depicted with added coupling capacitors and increased grounded capacitors. Simulation results of the proposed filter are shown in Fig. 6(b).

As it is shown in Fig. 6(b), this method suppresses the harmonics from 4.2 GHz to 9.377 GHz, increases the stopband bandwidth and sets the second passband at the 3.7 GHz. In the proposed filter, the first and second passbands have insertion losses of 0.521 dB and 0.842 dB, return losses of 25.3 dB and 17.914 dB and bandwidths of 77 MHz and 61 MHz, respectively. This results show that return loss is increased but passband bandwidth is improved in two bands. In the next stage we try to increase the passbands bandwidth and achieve modified return loss in two bands.

2.3 Using Defected Ground Structure (DGS) Technique to Improve the Filter Characteristics

Figure 7 shows the applied defected ground structure and its LC model to improve the filter characteristics. In this filter, improvement of the passband performance and also increasing the bandwidth of each band are the reasons of DGS utilization. Dumbbell and rectangular shape were used in defected ground structure technique. As it is shown in Fig. 7(a) the dumbbell shaped DGS is introduced by a parallel LC resonator and this part is added to the LC model of the multiple couplings. The rectangular shape DGS is applied to moderate the effect of the bends (on both sides of the central line).

The dimensions of the dumbbell parts and rectangular planes are as follow: W8 = 0.9 mm, L9 = 2.79 mm, W9 = 0.198 mm, L10 = 3.33 mm, L11 = 0.89 mm, L12 = 1.69 mm, L13 = 0.34 mm. A picture of the fabricated filter and the results of the simulation and measurement of the final filter are depicted in Fig. 7(b) and Fig. 7(c), respectively. EM simulations are performed in ADS and measurement performed using the Agilent network analyzer N5230A.

Figure 8 shows the current density distribution at the surface of the filter, with/without DGS. As it is shown in Fig. 8(a) and Fig. 8(b), dumbbell shaped DGS is utilized because of the balanced increasing at the current density distribution of the filter and this has improved the filter response in the passbands (in terms of insertion loss, bandwidth and return loss).From Fig. 8(c) it is apparent that the

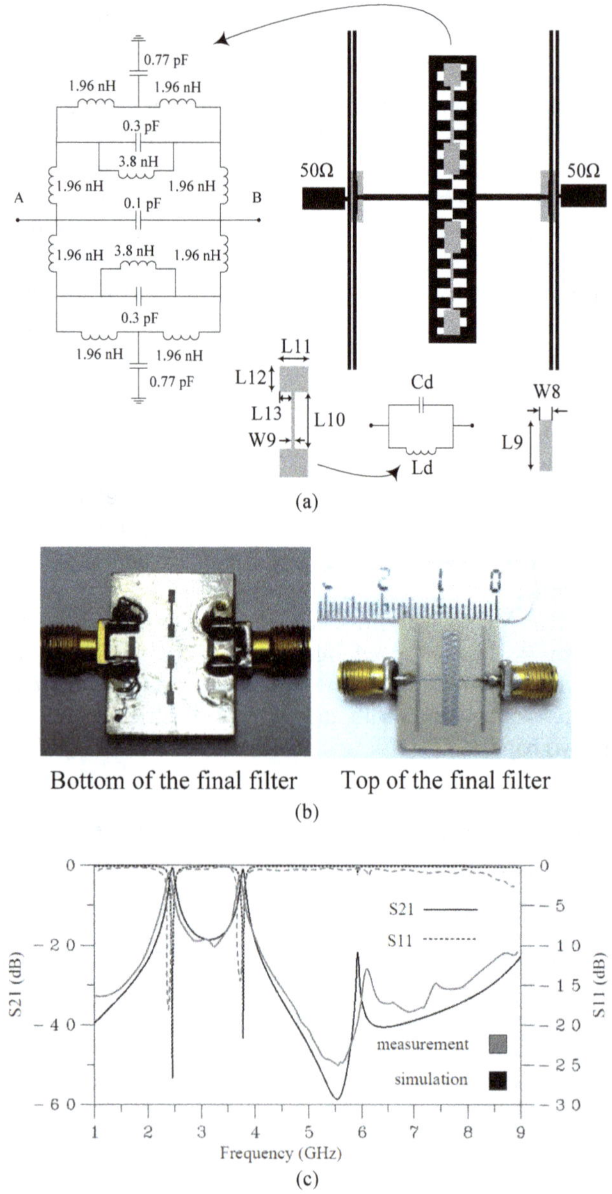

Fig. 7. Final filter. (a) Layout and LC model of the proposed filter. (b) A picture of the fabricated filter. (c) Simulation and measurement results of the filter.

Characteristics of proposed resonator and proposed filter	S_{21} in the first and second passband respectively (dB)	S_{11} in the first and second passband respectively (dB)	Bandwidth in the first and second passband respectively (MHz)
Proposed resonator (without multiple coupling)	- 0.478 & - 1.137	- 26.05 & - 18.549	76 & 49
Proposed filter (using multiple coupling)	- 0.521 & - 0.842	- 25.3 & - 17.914	77 & 61
Final structure of the proposed filter (using DGS technique)	- 0.457 & - 0.682	- 23.57 & - 17.2	95 & 87

Tab. 1. Compares between response characteristics of proposed structures in passbands.

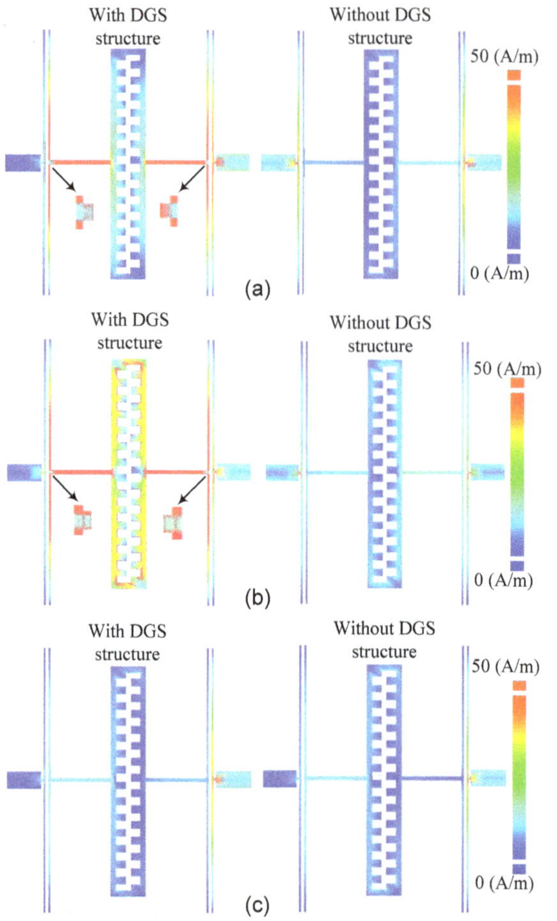

Fig. 8. Current density distribution at the surface of the filter: (a) at 2.45 GHz; (b) at 3.7 GHz; (c) at 3.07 GHz.

current density distribution of the filter with/ without DGS has no obvious change at 3.07 GHz (as the isolation frequency between two bands). So dumbbell shaped DGS has improved the passband performance of the filter. Also, the isolation between passbands and stopband characteristics has not been changed. The bends affect on the high frequency harmonics. In order to decrease the effects of these bends, the rectangular shaped DGS is utilized. As Figures 8(a) and 8(b) indicate the current density distribution of the bends is fixed using rectangular shaped DGS (while the current density distribution is increased in other lines) and the effects of them are limited. A comparison between the response characteristics of the proposed structures (in term of passbands) is shown in Tab. 1.

3. Conclusion

In this paper, at the first, using specified relations, a basic structure is designed. In order to shift the closest harmonic as the second passband towards the first passband, and also, to attenuate other harmonics, we added a low impedance stub to the middle of the basic structure. Using this structure, the proposed resonator for optimization of dimensions is provided. In order to further suppress the harmonics, and also to fix frequencies according to WLAN standards, multiple coupling structures are added to the proposed resonator. Finally, DGS technique was applied to improve filter response specifications. In final structure of the filter, frequency of the first and second passbands are 2.35 GHz and 3.61 GHz respectively, which have return losses of 0.93 dB and 1.13 dB and insertion losses of 17.56 dB and 14.08 dB and bandwidths of 89 MHz and 81 MHz, respectively.

References

[1] MONDAL, P., MANDAL, M. K. Design of dual-band bandpass filters using stub-loaded open-loop resonators. *IEEE Transactions on Microwave Theory and Techniques*, 2008, vol. 56, no. 1, p. 150–155. DOI: 10.1109/TMTT.2007.912204

[2] GUO, L., YU, Z., ZHANG, Y. A dual-band band-pass filter using stepped impedance resonator. *Microwave and Optical Technology Letters*, 2011, vol. 53, no. 1, p. 123–125. DOI: 10.1002/mop.25648

[3] WANG, L., GUAN, R. Novel compact and high selectivity dual-band BPF with wide stopband. *Radioengineering,* 2012, vol. 21, no. 1, p. 492–495.

[4] AHMED, E. S. Dual-mode dual-band microstrip bandpass filter based on fourth iteration T-square fractal and shorting pin. *Radioengineering,* 2012, vol. 21, no. 2, p. 617–623.

[5] YUN, T. S., NOH, S. K., KIM, OH, E. K., SON, H. M., LEE, J. C. Compact dualband bandpass filter with two transmission zeros using dual-mode microstrip resonator and tapped-line geometry. *Microwave and Optical Technology Letters*, 2011, vol. 53, no. 1, p. 108–111. DOI: 10.1002/mop.25627

[6] TSAI, L. C., HSUE, C. W. Dual-band bandpass filters using equal-length coupled-serial-shunted line and Z-transform technique. *IEEE Transactions on Microwave Theory and Techniques,* 2004, vol. 52, no. 4, p. 1111–1117. DOI: 10.1109/TMTT.2004.825680

[7] GUAN, X., MA, Z., CAI, P., KOBAYASHI, Y., ANADA, T., HAGIWARA, G. Synthesis of dual-band bandpass filters using successive frequency transformation and circuit conversions. *IEEE Transactions on Microwave and Wireless Components Letters,* 2006, vol. 16, no. 3, p. 110–112. DOI: 10.1109/LMWC.2006.869868

[8] SHETA, A. F. Narrow band compact non-degenerate dual-mode microstrip filter. In *Proceedings of the 25th National Radio Science Conference.* Egypt, 2008.

[9] MAKKI, S. V. AL-DIN, AHMADI, A., MAJIDIFAR, S., SARIRI, H., RAHMANI, Z. Sharp resonator microstrip LPF using folded stepped impedance open stubs. *Radioengineering,* 2013, vol. 22, no. 1, p. 328–329.

[10] ZONG, B. F., WANG, G. M., ZENG, H. Y., WANG, Y. W. Compact and high performance dual-band bandpass filter using resonator-embedded scheme for WLANs. *Radioengineering,* 2012, vol. 21, no. 4, p. 1050–1053.

[11] PARUI, S. K., DAS S. A new defected ground structure for different microstrip circuit applications. *Radioengineering,* 2007, vol. 16, no. 1, p. 16–22.

[12] VAGNER, P., KASAL, M. A novel bandpass filter using a combination of open-loop defected ground structure and half-wavelength microstrip resonators. *Radioengineering,* 2010, vol. 19, no. 3, p. 392–396.

[13] KUFA, M., RAIDA, Z. Comparison of planar fractal filters on defected ground substrate. *Radioengineering,* 2012, vol. 21, no. 4, p. 1019–1024.

[14] SARKAR, P., GHATAK, R., PODDAR, D. R. A dual-band bandpass filter using SIR suitable for WiMAX band. In *International Conference on Information and Electronics Engineering IPCSIT.* (Malaysia), 2011.

[15] HONG, J. S., LANCASTER, M. J. *Microstrip Filters for RF/ Microwave Applications.* A Wiley-Interscience Publication. John Wiley & Sons, 2001. ISBN 0-471-38877-7

About the Authors...

Nafiseh KHAJAVI received the B.Sc. degree in Medical Engineering from Islamic Azad University Dezful Branch, 2008, and the M.Sc. degree in Electrical Engineering from Kermanshah Science and Research Branch, Islamic Azad University, Kermanshah, 2012. She has been with the Department of Engineering, Islamic Azad University Dezful Branch. Her research interests include microstrip filter, the analysis and design of high-frequency electronics and microwave passive circuits.

Seyed Vahab Al-Din MAKKI was born in Kermanshah. He received his PhD in Electrical Engineering-Waves from Khaje Nasir Toosi University in 2008. He is with the Electrical Engineering Department of Razi University in Kermanshah and Islamic Azad University, Kermanshah Branch, since 2008. His current research interests include modern digital radio propagation systems, microwave devices and radio transmitters.

Sohrab MAJIDIFAR received his B.Sc. and M.Sc. in Electrical Engineering from Razi University in 2009 and 2011, respectively. He joined Kermanshah University of Technology in 2011 as a lecturer. His research interests include microwave passive circuits and RFIC.

18

On the Design and Performance Analysis of Low-Correlation Compact Space-Multimode Diversity Stacked Microstrip Antenna Arrays for MIMO-OFDM WLANs over Statistically-Clustered Indoor Radio Channels

Asuman SAVASCIHABES [1], *Ozgur ERTUG* [1], *Erdem YAZGAN* [2]

[1] Telecommunications and Signal Processing Laboratory, Dept. of Electrical and Electronics Engineering, University of Gazi, Ankara, Turkey
[2] Dept. of Electrical and Electronics Engineering, University of Hacettepe, Ankara, Turkey

ahabes@nny.edu.tr, ertug@gazi.edu.tr, yazgan@hacettepe.edu.tr

Abstract. *The support of high spectral efficiency MIMO spatial-multiplexing communication in OFDM-based WLAN systems conforming to IEEE 802.11n standard requires the design and use of compact antennas and arrays with low correlation ports. For this purpose, compact space-multimode diversity provisioning stacked circular multimode microstrip patch antenna arrays (SCP-ULA) are proposed in this paper and their performance in terms of spatial and modal correlations, ergodic spectral efficiencies as well as compactness with respect to antenna arrays formed of vertically-oriented center-fed dipole elements (DP-ULA) and dominant-mode operating circular microstrip patch antennas (CP-ULA) are presented. The lower spatial and modal correlations and the consequent higher spectral efficiency of SCP-ULA with ML detection over statistically-clustered Kronecker-based spatially-correlated NLOS Ricean fading channels with respect to DP-ULA and CP-ULA at significantly lower antenna and array sizes represents SCP-ULA as a promising solution for deployment in terminals, modems and access points of next-generation high-speed 802.11n MIMO-OFDM WLAN systems.*

Keywords

IEEE 802.11n MIMO-OFDM WLAN, spectral efficiency, spatial correlation, multimode antenna, spatial-multiplexing, Kronecker channel model, NLOS Ricean fading

1. Introduction

In the concurrent and next-generation communication systems, the spectral efficiency and transmission quality can be vastly enhanced by multiple-input multiple-output (MIMO) communication techniques [1]. In communication systems employing MIMO spatial-multiplexing, higher data rates can be achieved when there are a large number of scatterers between the transmit and receive antennas i.e. rich-scattering environment. However, the spatial correlation between transmit and receive antenna ports that is dependent on antenna-specific parameters such as the radiation patterns, the distance between the antenna elements as well as the channel characteristics such as unfavorable spatial distribution of scatterers and angular spread severely degrades the capacity and quality achievable by MIMO spatial-multiplexing systems.

The space consumption of MIMO antennas is especially vital in applications such as access points, modems and end-user terminal equipments (laptops, PDAs etc.) of WLAN and WIMAX systems. When regularly spaced antenna elements are used in MIMO systems, the correlation between the antenna elements in a space diversity system and hence the channel capacity and transmission quality are dependent on the distance between antenna array elements, the number of antenna elements and the array geometry. However, due to the physical constraints and the concerns on ergonomics and aesthetics, the distance between antenna elements in practice cannot be extended beyond a certain level which limits the use of space-only diversity MIMO spatial-multiplexing systems to achieve the desired spectral efficiencies and transmission qualities. As an alternative solution to achieve compactness in MIMO systems, the use of pattern diversity [2], [3], multimode diversity [4], [5], and polarization diversity [6], [7] techniques in conjunction with space diversity are proposed in the literature.

Besides polarization diversity that is well-known, multimode and pattern diversity techniques that are less addressed in antenna engineering community are achieved by using higher-order mode generation in antenna structures and in general microstrip, biconical, helical, spiral, sinuous and log-periodic antenna structures are amenable to higher-order mode generation. In this manner, the higher-order modes generated in a single antenna structure with directional radiation patterns resulting in low spatial correlation in angle space are used as diversity ports in a MIMO system within a compact space. In pattern diversity on the other hand that is slightly different than multimode diversity, orthogonal radiation patterns generated on distinct antennas that are co-located at the phase-centers are generated and used as diversity ports.

In this work, a multimode stacked circular microstrip patch antenna used in a uniform linear array structure (SCP-ULA) for MIMO-OFDM WLAN systems conforming to IEEE 802.11n standard is designed and the associated spatial power correlation, ergodic spectral efficiency and compactness with respect to omnidirectional dipole (DP-ULA) and circular microstrip uniform linear arrays (CP-ULA) operating in the dominant isotropic TM01 mode are analyzed. Section 2 represents the 802.11n MIMO-OFDM WLAN details as well as the associated system and statistically-clustered Kronecker-based correlated channel models for MIMO spatial-multiplexing. Section 3 is dedicated to the design procedure of stacked circular microstrip antenna for IEEE 802.11n MIMO-OFDM WLAN communication in HFSSv.11@TM CAD program and the analysis of the marginal and superimposition radiation patterns as well as S-parameters and VSWR variations versus frequency respectively. In Section 4, the gains of SCP-ULA with respect to CP-ULA and DP-ULA in terms of spatial/modal correlations, ergodic spectral efficiency as well as compactness are presented. Finally, Section 5 concludes the paper.

2. IEEE 802.11N WLANS and Associated System/Channel Models

The wireless local area network (WLAN) technology for medium-range indoor/outdoor wireless communications standardized by IEEE P802.11 working group has emerged from pre-802.11 standards towards spectrally-efficient and multipath-robust OFDM modulation based 802.11b and 802.11a/g with data rates increasing up to a maximum of 54 Mbps for 802.11a/g. These standards are limited to the use of single transmit and receive antenna at the access points and modems as well as laptops/PDAs in WLANs for end-users forming a SISO-OFDM (single-input single-output OFDM) channel. In the last decade, with the proliferation of MIMO spatial-multiplexing technology using multiple transmit and receive antennas achieving much higher data rates without sacrificing either bandwidth or transmit power with respect to SISO systems, IEEE 802.11 standard is later extended with the version 802.11n incorporating MIMO capability with the first amendment published in 2009 [8] proposing operation at lower and upper ISM bands of 2.4 GHz and 5.8 GHz with the corresponding bandwidths of 20 MHz and 40 MHz respectively. By the use of MIMO spatial-multiplexing technology, higher-order 64/128/256-QAM modulations, a 40 MHz channel bandwidth at 5.8 GHz that is double that of legacy IEEE 802.11 a/b/g systems, a cyclic prefix (CP) of 400 ns that is half of legacy systems which reduces symbol time and hence increases data rates, more efficient OFDM structure with 52 subcarriers with respect to 48 subcarriers in legacy systems, and frame-aggregation/block-acknowledgement protocol with higher packet sizes at the MAC layer [9], [10], IEEE 802.11n standard sets forth the basis for multimedia-enabling high-throughput next-generation WI-FI networks.

The transmitter and receiver block diagrams of an IEEE 802.11n based MIMO-OFDM WLAN system is illustrated in Fig. 1. The binary data is first encoded by channel encoder after which the encoded bits are multiplexed into sub-streams, modulated, and transmitted from each antenna. At the receiver side, after a digital representation of N received signals is obtained by ADCs, the CP is removed and N-point DFT is performed per receiver branch. Since the MIMO-OFDM system turns into a narrow band flat-fading MIMO channel per sub-carrier over multipath fading channels, the received signal vector per sub-carrier is given by:

$$\mathbf{y}(i,k) = \mathbf{H}(i)\mathbf{s}(i,k) + \mathbf{n}(i,k) \qquad (1)$$

where i and k are sub-carrier and symbol indices respectively. Here, $\mathbf{H}(i)$ represents $N \times M$ dimensional channel matrix for the ith sub-carrier, $\mathbf{s}(i, k)$ represents modulated transmit symbol vectors for the ith sub-carrier and $\mathbf{n}(i, k)$ represents additive white Gaussian noise at the ith subcarrier and kth symbol index with N independent and identically distributed (i.i.d) zero-mean complex elements with variance of σ_n^2.

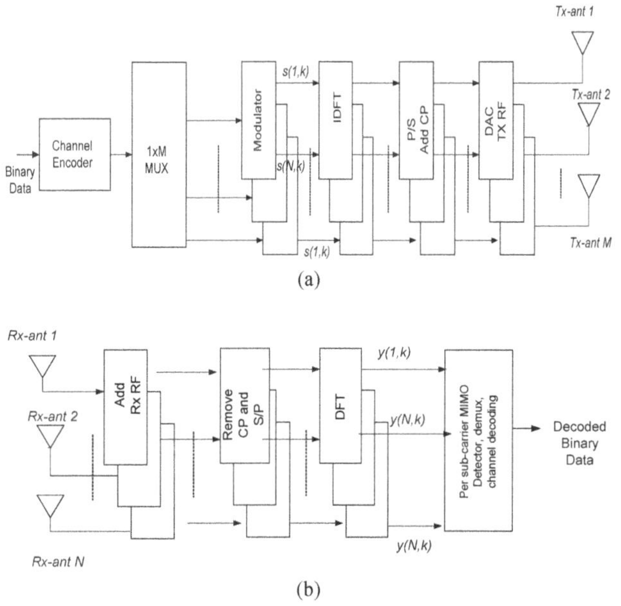

Fig. 1. Transmitter and receiver diagrams of a MIMO-OFDM WLAN system: (a) Transmitter. (b) Receiver.

Amongst the MIMO channel models proposed in the literature such as deterministic models (ray-tracing, recorded impulse response etc.) or stochastic models (geometric ring, parametric, correlation-based etc.), we deploy as the channel model the statistically-clustered Kronecker channel model for the characterization of the spatially-correlated MIMO channel in this work which has also been standardized for MIMO-OFDM IEEE 802.11n WLAN systems in [9]. Kronecker model assumes seperability between transmit/receive spatial correlations and the same model is also used for the performance analysis of indoor MIMO systems with stacked circular microstrip patch antennas with pattern diversity in [3].

The general geometry of the statistically-clustered Kronecker model we employ throughout the sequel representing the clusters and the transmission paths is represented in Fig. 2. For simplicity, only reflection from a single cluster is assumed in this work for the evaluation of the correlation and consequently spectral efficiencies, and the extension of the model to include a higher number of clusters is straightforward due to the additivity of spatial correlations.

The spatially-correlated channel matrix $\mathbf{H}_c$ in Kronecker model is formulated as:

$$\mathbf{H}_c = \sqrt{\frac{K}{1+K}}\mathbf{H}_{LOS} + \sqrt{\frac{1}{1+K}}\mathbf{H}_{NLOS} \tag{2}$$

and includes Rayleigh fading for $K = 0$, Ricean fading for higher K values and also AWGN channel for $K = \infty$ where $\mathbf{H}_{LOS}$ and $\mathbf{H}_{NLOS}$ are LOS (line-of-sight) and NLOS (nonline-of-sight) components respectively, K is the Ricean K-factor given by the ratio of the LOS component's power over NLOS component's power [1] and the NLOS channel matrix is defined as [9]:

$$\mathbf{H}_{NLOS} = \mathbf{R}_r^{1/2}\mathbf{H}_w\mathbf{R}_t^{1/2} \tag{3}$$

where $\mathbf{H}_w$ denotes Wishart-type random matrix, $\mathbf{R}_r$ and $\mathbf{R}_t$ denote the transmit and receive spatial power correlation matrices respectively. The LOS component of the channel is also assumed to be rank one, and defined as [11]:

$$\mathbf{H}_{LOS} = \mathbf{a}(\Omega_{\mathrm{LOS,r}})\cdot\mathbf{a}(\Omega_{\mathrm{LOS,t}})^H \tag{4}$$

where $\mathbf{a}(\Omega)$ is the array response as a function of the solid angle $\Omega = (\phi,\theta)$, while $\Omega_{\mathrm{LOS,t}}$ and $\Omega_{\mathrm{LOS,r}}$ are AoA/AoD corresponding to the LOS component at the transmitter and receiver respectively.

3. Design of Multimode Stacked Circular Patch (SCP) Microstrip Antenna

The multimode stacked circular microstrip patch antenna (SCP) proposed for IEEE 802.11n MIMO-OFDM WLAN application in this work has the upper antenna in the stack excited at the TM11 mode and the bottom antenna excited at the TM21 mode to meet the compactness requirements since the radius of a circular microstrip patch antenna scales up with the mode number m excited given by the formula [12]:

$$a = \frac{\chi_m \lambda}{2\pi\sqrt{\varepsilon_r}} \tag{5}$$

where χ_m indicates the first zero of the derivative of the first kind Bessel function of order m $J_m(x)$ as presented in Tab. 1.

	TM_{01}	TM_{11}	TM_{21}	TM_{31}	TM_{41}	TM_{51}	TM_{61}
χ_m	3.82	1.84	3.04	4.18	5.29	6.38	7.46

Tab. 1. χ_m for different modal orders.

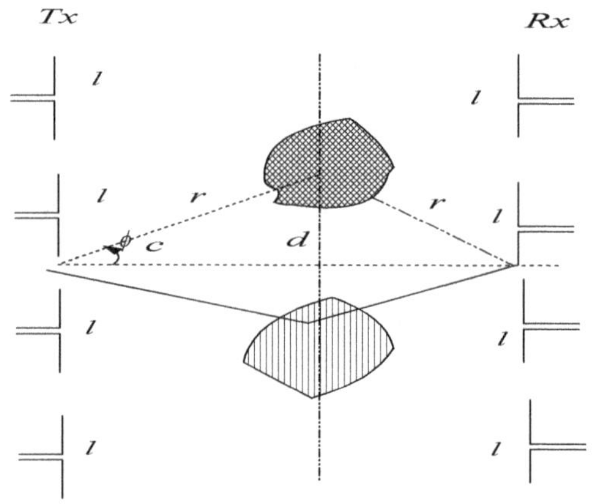

Fig. 2. Geometry of the clustered channel model representing clusters and transmit/receive antenna arrays (4×4 array) where ϕ_c is the mean A.O.D. of the cluster.

The far-field radiation patterns for a circular microstrip patch antenna excited at the m-th mode is further given by [13]:

$$\vec{E}_m = \frac{e^{-jk_f r}}{r}\left(E_{m,\theta}\vec{\theta} + E_{m,\phi}\vec{\phi}\right), \tag{6}$$

$$E_{m,\theta} = \frac{j^m V_m^0 k_f a}{2}\left[J_{m+1}(z) - J_{m-1}(z)\right]\cos(m(\phi-\phi_0)), \tag{7}$$

$$E_{m,\phi} = -\frac{j^m V_m^0 k_f a}{2}\left[J_{m+1}(z) + J_{m-1}(z)\right]\cos(\theta)\sin(m(\phi-\phi_0)) \tag{8}$$

where $k_f = 2\pi/\lambda$ is the wavenumber, $J_m(x)$ is the Bessel function of first kind of order m, a is the patch radius, V_m^0 is the peak input voltage of the mth mode, ϕ_0 is the reference azimuth angle for the feed of the circular patch, and $z = k_f\, a \sin(\theta)$. Via these far-field radiation patterns, neglecting elevation spread that is acceptable for indoor propagation environments and assuming the look-direction coincident with broadside is $\theta = \pi/2$, only θ components of the far-field radiation pattern dependent on the azimuth angle ϕ in (6) remains. Azimuth plane radiation patterns for mode orders $m = 1, 2, 3$ are presented in Fig. 3.

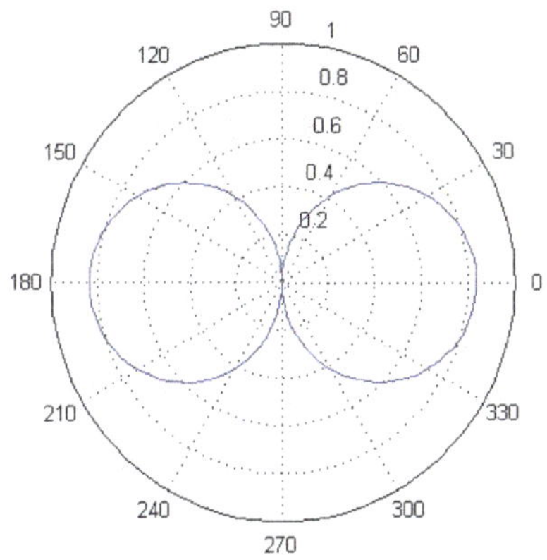

(a)

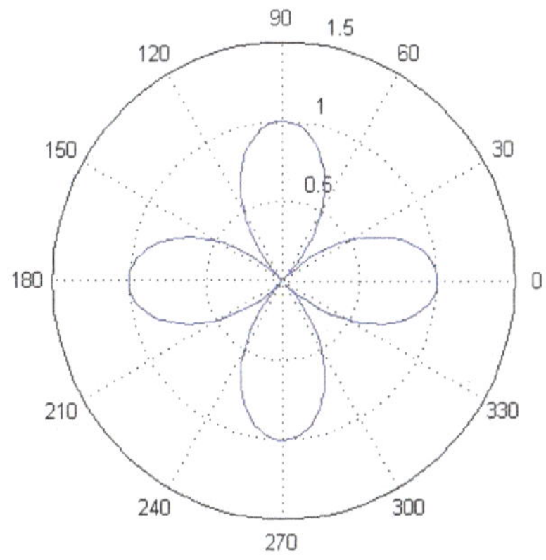

(b)

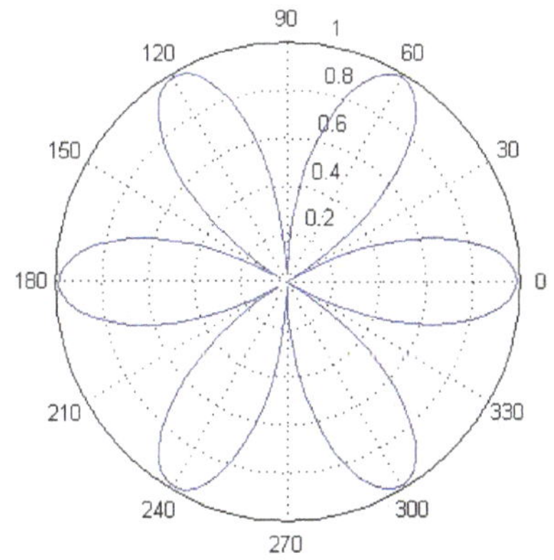

(c)

Fig. 3. Radiation patterns of multimode circular patch antennas: (a) TM11 mode, (b)TM21 mode, (c)TM31 mode.

The SCP antenna fed by microstrip feed lines designed in HFSSv.11 3D EM design and analysis software is presented in Fig. 4. To stay at the proper point of the trade-off between antenna bandwidth improvement and the antenna efficiency decrease/pattern distortion as well as to keep mutual coupling low, the distances between ground and bottom antenna as well as between circular antennas are kept as 0.5 mm. To ensure high radiation efficiency, the dielectric constant of the substrate is further chosen low as $\varepsilon_r = 2.2$.

The theoretical radius of the top and bottom antennas in SCP antenna excited by TM11 and TM21 modes via (5)

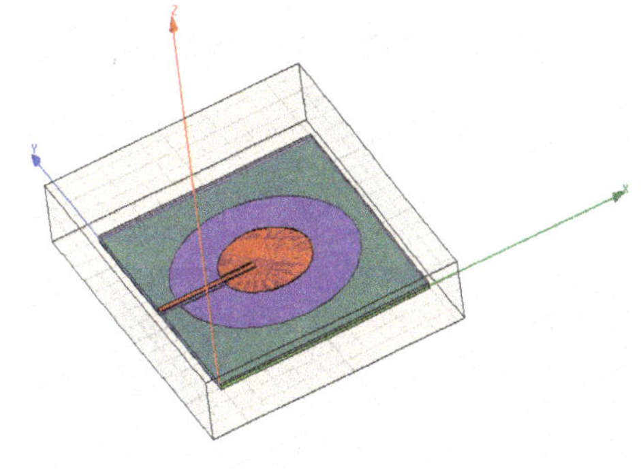

(a)

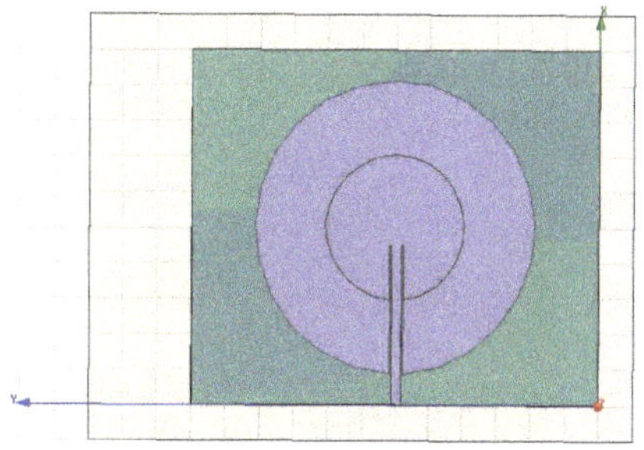

(b)

Fig. 4. The side (a) and the top (b) view of the SCP antenna with microstrip feed lines in HFSS.

are respectively given by 10.2 mm and 17 mm respectively. However, including the fringing effect [7], the actual effective radius of the circular patch antenna that is larger than estimated by (7) is obtained by the formula:

$$a_e = a\left(1 + \frac{2h}{\pi a \varepsilon_r}\left(\ln(\frac{a}{2h}) + (1{,}41\varepsilon_r + 1{,}77) + \frac{h}{a}(0{,}26\varepsilon_r + 1{,}65)\right)\right)^{1/2} \tag{9}$$

through which the effective radiuses of TM11 and TM21 modes are found to be 10.6 mm and 17.4 mm respectively. To operate both stack elements at the same resonant frequency, the radius of the antenna of TM21 mode is further modified and set as 20.67 mm in HFSSv.11@TM while the radius of the antenna operating on TM11 mode is used according to the theoretical value of 10.26 mm.

For the impedance matching of the SCP antenna at 802.11n upper ISM band WLAN operating frequency of 5.8 GHz for a 50 Ω load, the radiation resistances for both stack antennas have to be evaluated first for TM11 mode and TM21 modes. The real Poynting vector is given by [13]:

$$P + jQ = \frac{1}{2}\iint_{size}\left(\vec{E}_{ax}\vec{H}_a^{\,*}\right)\hat{z}\,dxdy \tag{10}$$

and integrating the real part of (10):

$$P_r = \frac{1}{2\eta_0}\int_0^{2\pi}\int_0^{\pi/2}\left(\left|E_\theta\right|^2 + \left|E_\varphi\right|^2\right) r^2 \sin\theta d\theta d\varphi \,. \tag{11}$$

The relation between (11) and the radiation resistance R_r is given by:

$$P_r = \frac{1}{2}G_r(E_0h)^2 = \frac{1}{2}G_rV_0^{\ 2}, \tag{12}$$

$$R_r = \frac{1}{G_r} \tag{13}$$

where G_r is the radiation conductance. In this manner, the radiation resistance of elements operating on TM11 mode and TM21 mode are obtained as 342.57 Ω and 156.13 Ω respectively. In the stack antenna the radiation resistance of TM21 mode is further obtained as 131.21 Ω and the radiation resistance of TM11 mode is found as 315.04 Ω in HFSSv.11@TM.

Antenna input impedances are described by the equations:

$$Z_{in} = R_{in} + jX_{in}, \tag{14}$$

$$R_{in} = \frac{1}{G_{in}}, \tag{15}$$

$$G_{in} = G_r + G_d + G_c \tag{16}$$

where Z_{in}, is input impedance, R_{in} is input resistance, X_{in} is input reactance, G_{in} is input conductance, G_d is dielectric conductance and G_c is the conductor conductance respectively. Since the antenna patch used in the design is selected as a zero-loss conductor, there exists no conductor resistance. On the other hand, the dielectric conductance value can be calculated via [14] as:

$$G_d = \frac{\varepsilon_{m0}\tan\delta}{4\mu_0 h f_r}\left[(ka_r)^2 - m^2\right] \tag{17}$$

where $\varepsilon_{m0} = 2$ for $m = 0$ and $\varepsilon_{m0} = 1$ for $m \neq 0$, and h is height of dielectric substrate of microstrip line. The effective loss tangent of the dielectric material with $\varepsilon_r = 2.2$ is determined as $\delta = 0.0009$. Using these values, the dielectric conductances for TM11 and TM21 modes are obtained as $G_d = 1.4925 \times 10^{-4}$ and $G_d = 6.0936 \times 10^{-4}$ respectively. These values are quite small so that it is rational not to take dielectric conductances into account for G_{in} calculations and to state that the input resistances of the antenna operating at TM11 mode is 342.57 Ω, and the input resistance of the antenna operating at TM21 mode is 156.13 Ω. To adapt these resistance values to 50 Ω, microstrip feed line widths for both elements in the stacked antenna are determined as 1.5 mm via the guideline equations in [15] such that both stack elements are matched to 50Ω input resistance.

In the design of SCP antenna, the antenna reactances are further calculated via

$$X_{in} = -jw\mu_0 h\left[\frac{1}{\pi a_e^2 k^2} + \sum_{m=2}^{\infty}\frac{j_0^2 k_{0m}\rho_0}{\pi a^2 j_0^2 k_{0m} a_e (k^2 - k_{0m}^2)} + \frac{2}{\pi}\sum_{m=1}^{\infty}\left(\frac{\sin(n\Delta)}{n\Delta}\right)^2\right] \tag{18}$$

and found as -j162.7 Ω and –j107.13 Ω for TM11 and TM21 modes respectively. In the stack antenna, the evaluation of the reactances of antenna with TM21 mode is obtained as j77.316 Ω while that of TM11 mode is found as j123.25 Ω that are quite high values for matching and can be minimized by optimizing the length of the microstrip feed lines.

The characteristic impedance of the feed line can be found depending on the size of the microstrip feed line by using:

$$Z_0 = \begin{cases} \dfrac{\eta}{2\pi\sqrt{\varepsilon_{re}}}\ln\left(\dfrac{8h}{L} + 0.25\dfrac{L}{h}\right), & (L/h \le 1) \\ \dfrac{\eta}{\sqrt{\varepsilon_{re}}}\left\{\dfrac{L}{h} + 1.393 + 0.667\ln\left(\dfrac{L}{h} + 1.444\right)\right\}^{-1} & (L/h \ge 1) \end{cases} \tag{19}$$

where L is microstrip feed line width. Effective dielectric constant is also calculated via:

$$\varepsilon_{re} = \frac{\varepsilon_{re}+1}{2} + \frac{\varepsilon_{re}-1}{2}F(L/h). \tag{20}$$

According to the desired characteristic impedance value, the ratio of the microstrip feedline width L to dielectric substrate height h is calculated via:

$$\frac{L}{h} = \begin{cases} \dfrac{8\exp(A)}{\exp(2A)-2}, & (L/h<2) \\ \left\{B-1-\ln(2B-1)+\dfrac{\varepsilon_r-1}{2\varepsilon_r}\left[\ln(B-1)+0.39-\dfrac{0.61}{\varepsilon_r}\right]\right\}, & (L/h \ge 1) \end{cases} \tag{21}$$

where A and B are:

$$A = \frac{Z_0}{60}\sqrt{\frac{\varepsilon_r+1}{2}} + \frac{\varepsilon_r-1}{\varepsilon_r+1}\left(0.23 + \frac{0.11}{\varepsilon_r}\right), \quad B = \frac{377\pi}{2Z_0\sqrt{\varepsilon_r}}. \tag{22}$$

With the height of the dielectric substrate $h = 0.5$ mm, the line width L is calculated as 1.55 mm via (22) for the microstrip feed lines and in this manner, the antenna reactances of TM11 mode and TM21 modes are minimized to j11.513 Ω and j4.5072 Ω respectively with the corresponding antenna input impedances finally obtained and used in S-parameter and VSWR analysis as $Z_{in1} = 45.507 + j11.513$ Ω and $Z_{in2} = 50.678 + j4.5072$ Ω for TM11 and TM21 modes respectively.

The S-parameters versus frequency of the designed SCP antenna are analyzed in Fig. 5. The resonance of both modes are achieved around 5.8 GHz that is suitable for IEEE 802.11n MIMO-OFDM WLANs operating in upper-ISM band and the operating bandwidth of SCP antenna is measured to be around 55 MHz via S11 and S22 return-loss S-parameters that is sufficiently adequate for IEEE 802.11n WLAN communication which requires 40 MHz

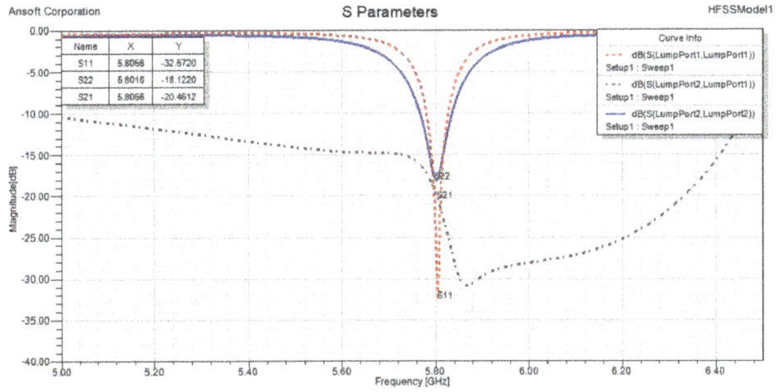

Fig. 5. S-parameters versus frequency of SCP antenna.

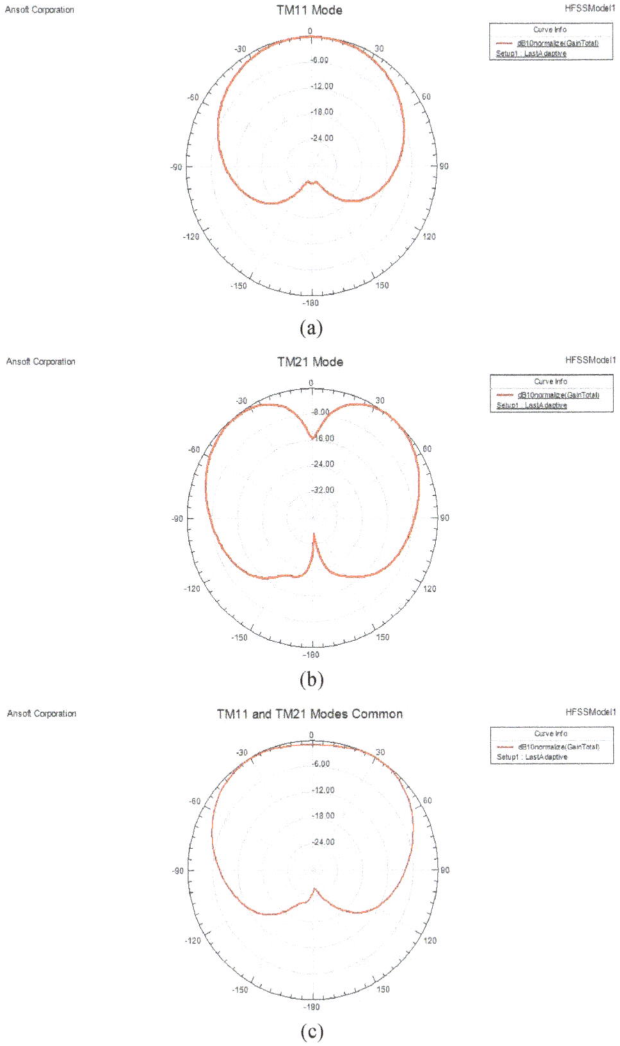

Fig. 6. Radiation patterns of SCP antenna: (a) TM11 alone, (b) TM21 alone, (c) TM11 andTM21 superimposed.

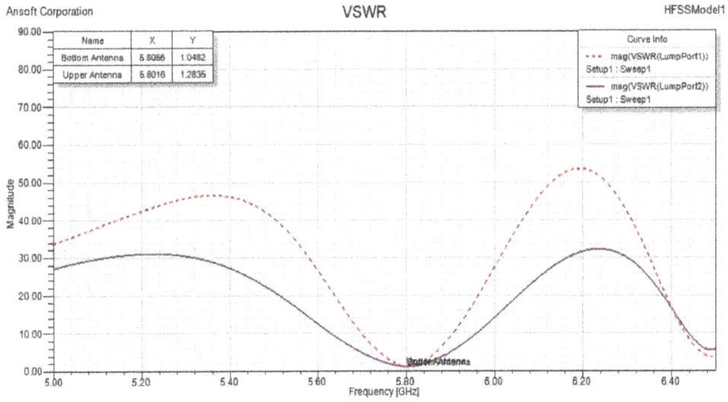

Fig. 7. VSWR plots versus frequency for SCP antenna.

bandwidth. Furthermore, the S21 mutual-coupling S-parameter in Fig. 4 is well below approximately -20 dB within the operating bandwidth around 5.8 GHz letting the omittance of pattern distortion due to the coupling effects between the collocated antennas. This fact is also well apparent in Fig. 6 where the radiation patterns of both modes are presented marginally and in superimposed form. Due to the low mutual coupling achieved, the radiation pattern distortion imposed by TM21 mode on TM11 mode is negligible in the superimposed radiation pattern in Fig. 6(c).

The VSWR plots of the SCP antenna for both TM11 and TM21 modes versus frequency are further presented in Fig. 7. The VSWR values for the TM11 and TM21 modes at the resonance frequency of 5.8 GHz is measured in HFSSv.11@TM as 1.0482 and 1.2835 respectively that are very close to the theoretical lower bound of unity validating the reliability of the impedance matching of SCP antenna.

4. Spatial/Modal Correlations and Spectral Efficiency Analysis of SCP-ULA

To evaluate the achievable gains with multimode SCP antenna used as a uniform linear array (SCP-ULA) for IEEE 802.11n MIMO-OFDM WLAN communications when employed at access points, modems or end-user terminal equipments, the spatial/modal power correlations of the SCP-ULA is compared with the spatial/modal power correlations of uniform linear arrays of center-fed dipole antennas of length $\lambda/2$ (DP-ULA) and the spatial/modal power correlations of uniform linear arrays of isotropic TM01 dominant-mode operating circular patch antennas in Fig. 8 for a statistically-clustered indoor propagation environment with a single-cluster of angular spread σ_c^2 and uniformly-distributed mean angle of arrival (AoA) and angle of departure (AoD) with respect to array broadsides. The spatial/modal normalized complex correlation coefficient of all three types of antennas is given by:

$$\rho = \frac{\int_{-\pi}^{\pi} E_1(\phi)E_2^*(\phi)PAS(\phi;\phi_c,\sigma_c^2)\exp\left(-jk_f d(n_2-n_1)\sin(\phi)\right)d\phi}{\sqrt{\int_{-\pi}^{\pi}|E_1(\phi)|^2 PAS(\phi;\phi_c,\sigma_c^2)d\phi}\sqrt{\int_{-\pi}^{\pi}|E_2(\phi)|^2 PAS(\phi;\phi_c,\sigma_c^2)d\phi}} \tag{23}$$

and the power azimuth spectrum PAS (ϕ; ϕ_c, σ_c^2) that defines the distribution of power over the sub-multipath components within a cluster is used as truncated-Laplacian density with mean azimuth angle ϕ_c and angular spread σ_c^2 the form of which is given by [16]:

$$PAS(\phi;\phi_c,\sigma_c^2) = \frac{e^{-\left|\frac{\sqrt{2}(\phi-\phi_c)}{\sigma_c}\right|}}{\sqrt{2}\sigma_c\left(1-e^{-\frac{\sqrt{2}\pi}{\sigma_c}}\right)}. \tag{24}$$

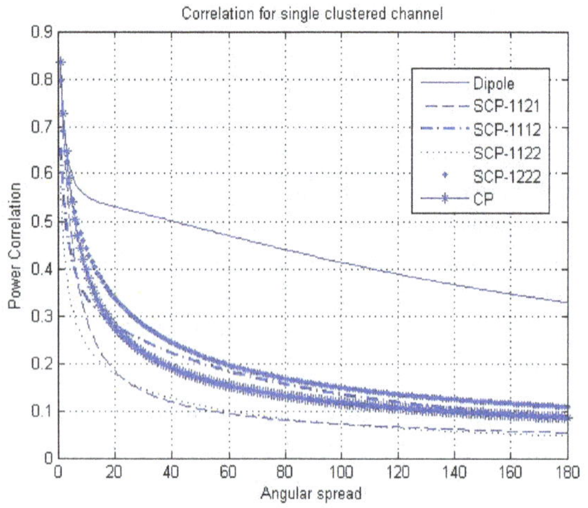

Fig. 8. Power correlation values of DP-ULA, CP-ULA and SCP-ULA averaged over azimuth angle ϕ for a 2×2 array.

The power correlation value that is strictly related to SNR scaling for MIMO channels is then given by $\rho_p = |\rho|^2$ and the power correlation values are averaged over uniformly-distributed mean azimuth angles over azimuth angle ϕ.

The power correlation values of the SCP antenna for all angular spread values and spatial/modal combinations in the densest configuration where stack antennas are nearby in the most compact scenario allowed by the physical radius of the bottom antenna for 2×2 configuration as presented in Fig. 8 are much lower than that of the two nearest antennas spaced at the patch radius of SCP bottom antenna in $\lambda/2$- length DP-ULA and CP-ULA which dictates spectral efficiency gains for MIMO spatial-multiplexing.

In Fig. 9, the power correlations of SCP-ULA versus mean azimuth angle for two different values of low (5°) and high (20°) angular spread are presented. For all angular spread values, DP-ULA has the maximum correlation values in comparison with CP and SCP at broadside, where the cluster is at the center of the antenna arrays and the oscillations of the patch antennas SCP-1122 (upper antenna radiated at TM11 and lower antenna radiated at TM22 mode), and SCP-1222 (upper antenna radiated at TM12 and lower antenna radiated at TM22 mode) decrease directly with the increase in angular spread. Besides, the maximum value of the correlations for all antennas is inversely proportional with angular spread.

The ergodic spectral efficiencies in bps/Hz of MIMO-OFDM WLAN system with DP-ULA, CP-ULA and SCP-ULA versus angular spread for 2×2, 4×4, 6×6 and 8×8 configurations are further presented in Fig. 10 over single-cluster Kronecker-model NLOS Rayleigh fading channel via [17]:

$$\eta_e = E\left\{\frac{1}{N_c}\sum_{i=1}^{N_c}\log_2\left(\det\left(\mathbf{I}_N + \frac{SNR}{M}\mathbf{H}_c\mathbf{H}_c^{\ H}\right)\right)\right\} \quad (25)$$

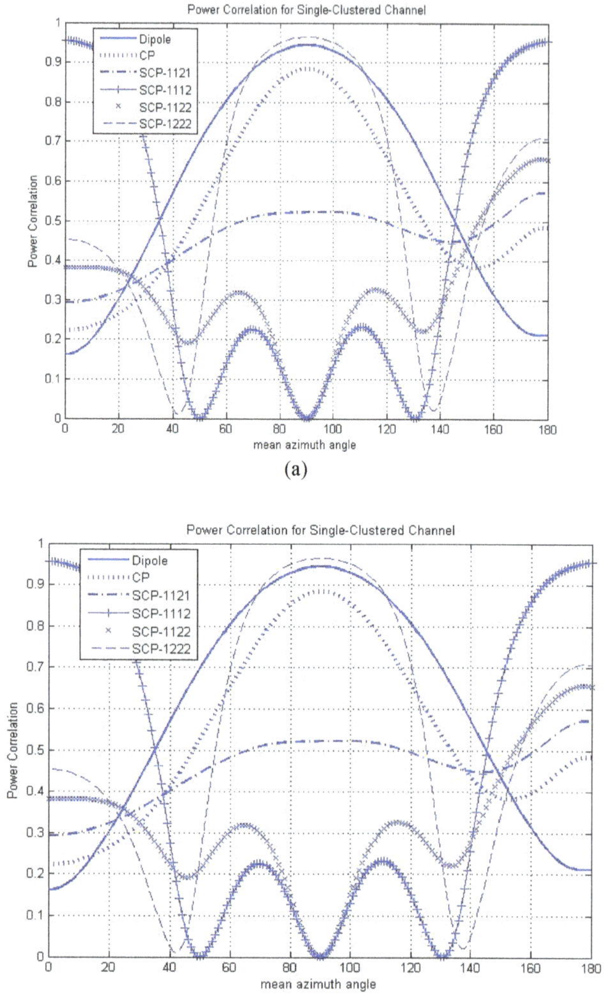

(b)

Fig. 9. Power correlation values of DP-ULA, CP-ULA and SCP-ULA versus mean azimuth angle (a): $\sigma_c^2 = 5°$, (b) $\sigma_c^2 = 20°$.

where the correlated channel matrix $\mathbf{H}_c$ has the Kronecker form [18-20]; $\mathbf{H}_c = \mathbf{R}_{RX}^{1/2}\mathbf{H}_w\mathbf{R}_{TX}^{1/2}$, in terms of the receive and transmit normalized power correlation matrices $\mathbf{R}_{RX}$ and $\mathbf{R}_{TX}$ and the elementwise-independent $N \times M$ Wishart-type random matrix $\mathbf{H}_w$ with elements distributed as $CN(0.1)$.

For 2×2 and 4×4 configurations, the SCP-ULA has much higher spectral efficiency than DP-ULA and slightly higher spectral efficiency than CP-ULA. On the other hand, for 6×6 and 8×8 configurations, CP-ULA and SCP-ULA have nearly the same spectral efficiencies which is also higher than DP-ULA. Furthermore, based on the spectral efficiency results presented in Fig. 10, data rates achievable with SCP-ULA conforming to IEEE 802.11n standards with 40 MHz bandwidth at 5.8 GHz for 2×2, 4×4, 6×6 and 8×8 array configurations are 164 Mbps, 324 Mbps, 480 Mbps, 644 Mbps respectively for high angular spreads that are much higher than standard 54 Mbps data rate achievable with SISO-OFDM 802.11 a/g WLAN systems.

The ergodic spectral efficiency of SCP-ULA, CP-ULA and DP-ULA versus mean azimuth angle of the cluster with respect to the antenna array boresights are further presented in Fig. 11 for low and high angular spreads of $\sigma_c^2 = 50$ and $\sigma_c^2 = 200$ respectively. In both cases, the ergodic spectral efficiency of SCP-ULA is the highest when the cluster is at the boresight; i.e. $\phi_c = 90°$ followed by CP-ULA and DP-ULA. On the other hand, as the cluster moves towards endfires with respect to the

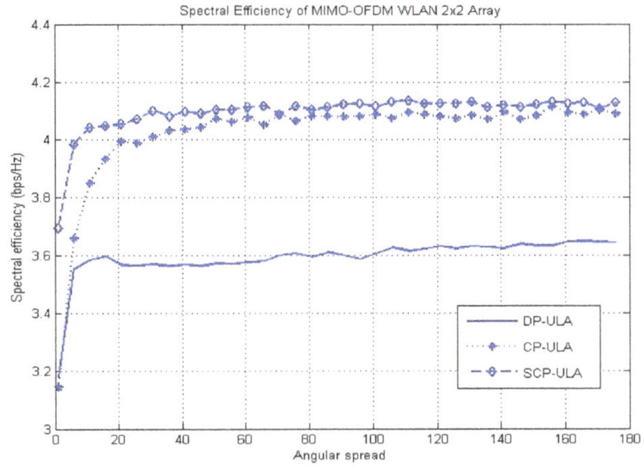

(a)

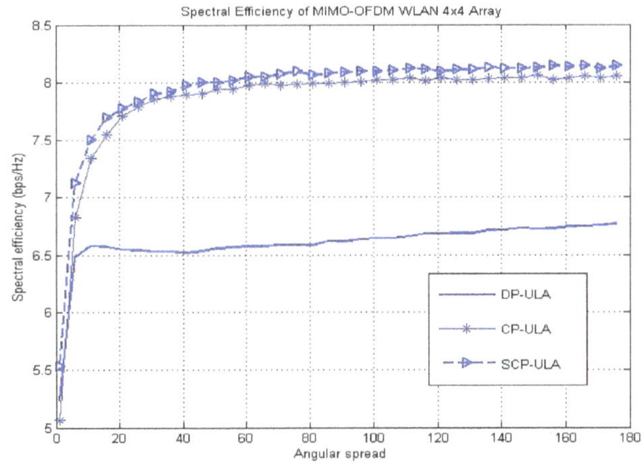

(b)

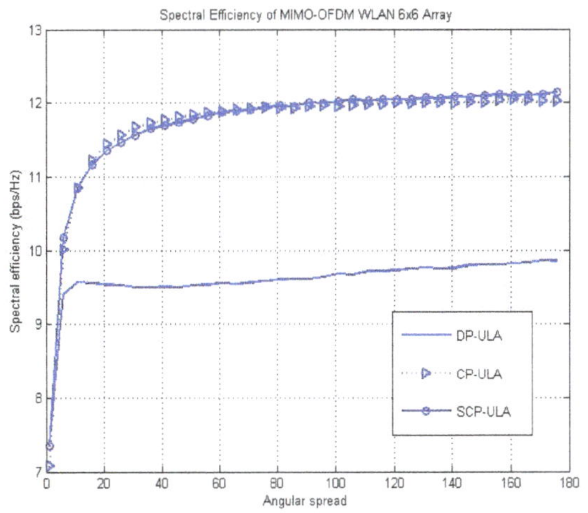

(c)

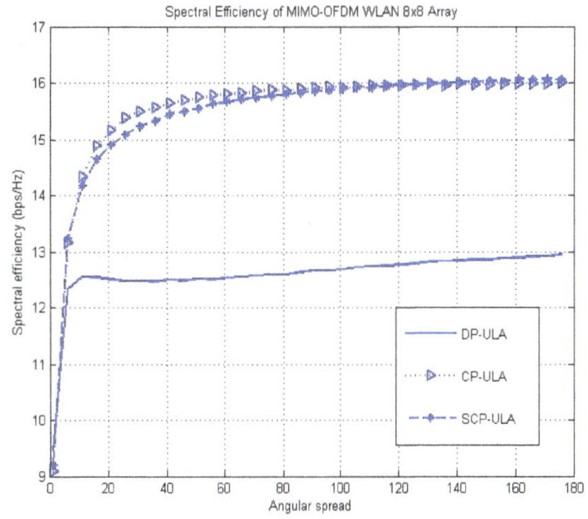

(d)

Fig. 10. Ergodic spectral efficiency of MIMO-OFDM WLAN over Kronecker-based statistically-clustered single-cluster NLOS Rayleigh fading channel with DP-ULA, CP-ULA and SCP-ULA versus angular spread at SNR = 10 dB: (a) 2×2, (b) 4×4, (c) 6×6, (d) 8×8 array.

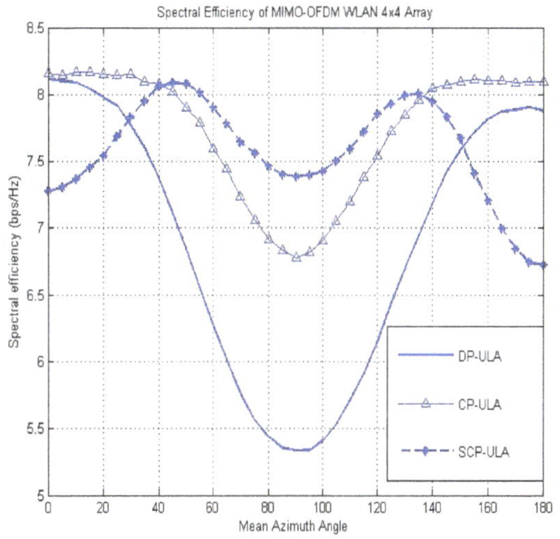

(a)

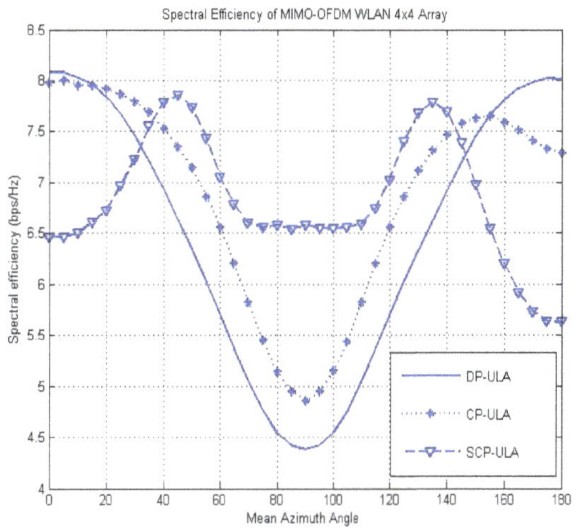

(b)

Fig. 11. Ergodic spectral efficiency of MIMO-OFDM WLAN over NLOS fading channel versus mean azimuth angle for 4×4 configuration of DP-ULA, CP-ULA and SCP-ULA at (a) $\sigma_c^2 = 5°$, (b) $\sigma_c^2 = 20°$.

antenna array boresights; i.e. $\phi_c = 0°$ and $\phi_c = 180°$, CP-ULA and DP-ULA respectively presents higher spectral efficiencies with respect to SCP-ULA. Despite this lower spectral efficiency of SCP-ULA with respect to CP-ULA and DP-ULA at endfires, the clusters in practical scenarios are mainly concentrated around boresight of antenna arrays in indoor propagation mediums and the range of mean azimuth angles around boresight over which SCP-ULA dominates CP-ULA and DP-ULA in terms of spectral efficiency is where the clusters reside realistically.

Most importantly, these lower correlation and higher spectral efficiency gains are achieved at 33.3% and 49.9%, 40% and 50%, and 42,9% and 50% denser space in compactness with SCP-ULA with respect to DP-ULA and CP-ULA for 4×4, 6×6 and 8×8 configurations respectively as tabulated in Tab. 2 and Tab. 3.

(Type)/(N×M)	2×2 (%)	4×4 (%)	6×6(%)	8×8(%)
DP-ULA	0	33.3	40.0	42.9
CP-ULA	49.7	49.9	50.0	50.0

Tab. 2. Compactness gain of SCP-ULA with respect to CP-ULA and DP-ULA.

(TYPE)/(N×M)	2×2 (mm)	4×4 (mm)	6×6 (mm)	8×8 (mm)
DP-ULA	21.2	63.6	106	148.4
CP-ULA	42.4	84.8	127.2	169.6
SCP-ULA	16.9	33.8	50.7	67.6

Tab. 3. Physical array length of SCP-ULA, CP-ULA and DP-ULA for minimum distance between elements.

5. Conclusions

In this paper, we proposed space-multimode stacked circular patch uniform linear antenna arrays (SCP-ULA) for deployment in size-constrained access points, modems and end-user terminal equipments of IEEE 802.11n MIMO-OFDM WLAN systems at 5.8 GHz. The performance of SCP-ULA in terms of spatio-modal power correlations, ergodic spectral efficiencies and compactness are compared with these of dominant-mode operating circular patch antenna arrays (CP-ULA) and center-fed dipole arrays (DP-ULA). The higher or comparable spectral efficiencies and lower spatio-modal correlations achieved by SCP-ULA over Kronecker-based statistically-clustered single-cluster NLOS Rayleigh fading channels in much more compact space requirements present space-multimode SCP-ULA antennas as a favorable solution for deployment in spatially-compact equipments of next-generation high-speed indoor IEEE 802.11n MIMO-OFDM WLAN systems.

References

[1] TELATAR, I. E. Capacity of multi-antenna Gaussian channels. *European Transactions on Telecommunications*, 1999, vol. 10, no. 6, p. 585–596. DOI: 10.1002/ett.4460100604

[2] FOSCHINI, G. J., GANS, M. On limits of wireless communications in a fading environment when using multiple antennas. *Wireless Personal Communications*, 1998, vol. 6, p. 311–355. DOI: 10.1023/A:1008889222784

[3] FORENZA, A., HEATH, R. W. Benefit of pattern diversity via two-element array of circular patch antennas in indoor clustered MIMO channels. *IEEE Transactions on Communications*, 2006, vol. 54, no. 5, p. 943–954. DOI: 10.1109/TCOMM.2006.873978

[4] SANCHEZ-FERNANDEZ, M., RAJO-IGLESIAS, E., QUEVEDO-TERUEL, O., PABLO-GONZALEZ, M. L. Spectral efficiency in MIMO systems using space and pattern diversities under compactness constraints. *IEEE Transactions on Vehicular Technology*, 2008, vol. 57, no. 3, p. 1637–1645. DOI: 10.1109/TVT.2007.909279

[5] SVANTESSON, T. Correlation and channel capacity of MIMO systems employing multimode antennas. *IEEE Transactions on Vehicular Technology*, 2002, vol. 51, no. 6, p. 1304–1312. DOI: 10.1109/TVT.2002.804856

[6] MUKHERJEE, A., KWON, H. M. Compact multi-user wideband MIMO system using multiple-mode microstrip antennas. In *Proceedings of 65th Vehicular Technology Conference VTC 2007 Spring*. Dublin (Ireland), April 2007, p. 584–588. DOI: 10.1109/VETECS.2007.131

[7] WALDSCHMIDT, C., KUHNERT, C., SCHULTEIS, S. WIESBECK, W. Compact MIMO-arrays based on polarization-diversity. In *Proceedings of IEEE Antennas and Propagation Symposium 2003*. Columbus (USA), 2003, vol. 2, p. 499–502. DOI: 10.1109/APS.2003.1219284

[8] *IEEE 802.11n-2009–Amendment 5: Enhancements for Higher Throughput*. IEEE-SA. 29 October 2009. DOI: 10.1109/IEEESTD.2009.5307322.

[9] *IEEE P802.11n Wireless LANs: TGn Channel Models*. IEEE-SA. May 10 2004. doc.: IEEE 802.11-03/940r4.

[10] Wi-Fi CERTIFIED n: Longer-Range, Faster-Throughput, Multimedia-Grade Wi-Fi® Networks, Wi-Fi Alliance. Sept. 2009.

[11] PAULRAJ, A., NABAR, R., GORE, D. *Introduction to Space-Time Wireless Communications*. New York: Cambridge University Press, 2003.

[12] SVANTESSON, T. On the capacity and correlation of multi-antenna systems employing multiple polarizations. In *Proc. of IEEE Antennas and Propagation Symposium 2002*. San Antonio (USA), 2002, vol. 3, p. 202–205. DOI: 10.1109/APS.2002.1018190

[13] VAUGHAN, R. G. Two-port higher mode circular microstrip antennas. *IEEE Transactions on Antennas and Propagation*, 1988, vol. 36, no. 3, p. 309–321. DOI: 10.1109/8.192112

[14] GARG, R., BHARTIA, P., BAHL, I. *Microstrip Antenna Design Handbook*. Artech House, 2001.

[15] BALANIS, C.A. *Antenna Theory: Analysis and Design*. 2nd ed. United States of America: John Wiley & Sons Inc., 1997.

[16] BAHL, J. *Lumped Elements for RF and Microwave Circuits*. Boston: Artech House, 2003.

[17] VAN ZELST, A., VAN NEE, R., AWATER, G. A. Space-division multiplexing for OFDM systems. In *IEEE Vehicular Technology Conference* 2000. Tokyo (Japan), 2000, vol. 2, p. 1070–1074. DOI: 10.1109/VETECS.2000.851289

[18] CHE-NEE CHUAH, KAHN, J. M., TSE, D. N. C. Capacity of multi-antenna array systems in indoor wireless environments. In *Proceedings of GLOBECOM'98*. Sidney (Australia), 1998, vol. 4, p. 1894–1899. DOI: 10.1109/GLOCOM.1998.775873

[19] GOLDSMITH, A. *MIMO Wireless Communications*. Cambridge University Press, 2007.

[20] VAN ZELST, A. Space-division multiplexing algorithms. In *10th Mediterranean Electrotechnical Conference* 2000, MELECON 2000. Cyprus, 2000, vol. 3, p. 1218–1221. DOI: 10.1109/MELCON.2000.879755

About the Authors ...

Asuman SAVAŞCIHABEŞ has received her Ph.D. degree in Electrical Engineering in 2013 specializing on Telecommunications and Signal Processing at Gazi University, Ankara, Turkey. She is currently working as an assistant professor in Electrical and Electronics Engineering Department of Nuh Naci Yazgan University, Kayseri, Turkey. Her research interests include wireless communication systems, MIMO communication and transmit/receive diversity of MIMO-OFDM.

Özgür ERTUG was born in Ankara, Turkey in 1975. He received his B.Sc. degree in 1997 from University of Southern California, USA, M.Sc. degree from Rice University in 1999 and Ph.D. degree from Middle East Technical University in 2005. He is currently working as an assistant professor in Electrical and Electronics Engineering Department of Gazi University. His main research interests lie in algorithm and architecture design as well as theoretical and simulation-based performance analysis of wireless communication systems especially in the physical layer.

Erdem YAZGAN received the B.S. and M.S. degrees from the Middle East Technical University, Ankara, Turkey, in 1971 and 1973, respectively, and the Ph.D. degree from Hacettepe University, Ankara, Turkey, in 1980, all in Electrical Engineering. Since 1990, she has been a Professor with the Department of Electrical Engineering, Hacettepe University. In 1989, she was a Visiting Professor with Essex University, Essex, U.K. In 1994, she was with the Electroscience Laboratory, Ohio State University, Columbus. Her research interests include HF propagation, low altitude radar systems, mobile communications, MICs, reflector and microstrip antennas, Gaussian beam solutions, conformal mapping, and medical electronics.

Uplink Multiuser MIMO Detection Scheme with Reduced Computational Complexity

Soobum CHO [1], *Sang Kyu PARK* [2]

[1] Dept. of Electrical Engineering, Stanford University, Stanford, CA 94305, USA
[2] Dept. of Electronics and Computer Engineering, Hanyang University, Seoul 133-791, South Korea

sbcho98@gmail.com, skpark@hanyang.ac.kr

Abstract. *The wireless communication systems with multiple antennas have recently received significant attention due to their higher capacity and better immunity to fading channels as compared to single antenna systems. A fast antenna selection scheme has been introduced for the uplink multiuser multiple-input multiple-output (MIMO) detection to achieve diversity gains, but the computational complexity of the fast antenna selection scheme in multiuser systems is very high due to repetitive pseudo-inversion computations. In this paper, a new uplink multiuser detection scheme is proposed adopting a switch-and-examine combining (SEC) scheme and the Cholesky decomposition to solve the computational complexity problem. K users are considered that each user is equipped with two transmit antennas for Alamouti space-time block code (STBC) over wireless Rayleigh fading channels. Simulation results show that the computational complexity of the proposed scheme is much lower than the systems with exhaustive and fast antenna selection, while the proposed scheme does not experience the degradations of bit error rate (BER) performances.*

Keywords

Cholesky decomposition, fast antenna selection, multiuser MIMO, SEC, STBC

1. Introduction

High system capacity, high spectral efficiency, and strong tolerance with respect to interference and multipath propagation are required in wireless communication systems. Multiple-input multiple-output (MIMO) systems satisfy these requirements by increasing the transmission rate and improving the reliability over the fading channel [1–3]. Even though the MIMO systems theoretically have several advantages, it may be difficult to equip over two antennas into mobile devices such as smart phone, tablet PC, and laptop computer. The installation of the multiple antennas into the small devices may lead to high hardware complexity, high cost, large size, and low battery life.

To solve the problem of the MIMO systems, multiuser MIMO approaches have been proposed [4–6]. The multiuser MIMO systems allow multiple users to deploy spatially separated transmit antennas and to deliver independent signal streams. Therefore, multiuser MIMO systems provide higher data rates and need a small number of antennas than the MIMO systems. However, multiuser MIMO channels suffer from co-channel interference (CCI), since they use the same frequency.

An uplink multiuser MIMO detection method for suppressing CCIs was introduced [7]. In this work, CCIs were suppressed using a beamforming process based on the Rayleigh-Ritz theorem, and a number of RF chains were required to achieve sufficient bit error rate (BER) performance. However, since adding a number of RF chains is costly, this is undesirable from a practical point of view. To solve the problem of the beamforming process based on the Rayleigh-Ritz theorem, exhaust and fast antenna selection schemes have been proposed in multiuser detection process [8]. The BER performance of the systems that use such selection schemes is higher than that of systems using the same number of RF chains without any selection. However, in the process of selecting the optimum antenna subset, repetitive matrix inversion computations occur, which is the main cause of the increase in computational complexity. Generally, matrix inversion can be computed via Gaussian-Jordan elimination. However, if the matrix is positive definite, this approach is computationally expensive and does not exploit the characteristics of Hermititan matrices. In this case, it is better to start with the Cholesky decomposition.

Furthermore, a multi-branch switched diversity system based on a fixed switching threshold was introduced [9]. It was also shown that a switch-and-examine combining (SEC) scheme improves performance along the additional diversity branches [10]. In a subset of the available diversity branches with limited resources, the SEC scheme is adequate for selecting branches based on the signal-to-interference plus noise ratio (SINR) with low complexity.

In this paper, the SEC scheme is adopted to multiuser detection in order to reduce the number of repetitive matrix inversion computations. Additionally, the matrices are positive definite so the Cholesky decomposition can be used, which has a lower computational complexity than the Gauss-Jordan method.

This paper is organized as follows. The uplink multiuser communication system model is explained in Sec. 2. In Sec. 3, a beamforming process based on the Rayleigh-Ritz theorem is performed to suppress CCIs. Section 4 describes proposed multiuser decoding method with reduced computational complexity. Section 5 describes adaptive switching thresholds and the BER according to the fading model. Finally, conclusions are given in Sec. 6.

2. System Model

Figure 1 shows an uplink multiuser MIMO system over a multiple access channel.

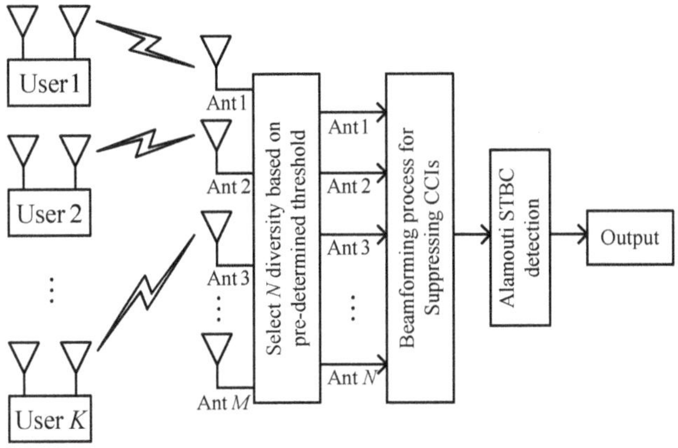

Fig. 1. Uplink multiuser MIMO system based on SEC and beamforming process.

K users are equipped with two transmit antennas, and the base station (BS) has M diversity branches with N available RF chains. It has been proven that when each user has two transmit antennas, using $2 \times (K-1)+r$ receiver antennas for any $r \geq 1$, the receiver can completely separate the signals of the K users, where r is the redundant receiver antenna for obtaining receiver diversity gains [11]. All users use the Alamouti transmission scheme [12] to achieve the transmit diversity gain. During the first time slot, $s_1^{(k)}$ and $s_2^{(k)}$ are transmitted at two transmit antennas, and during the second time slot, $-[s_2^{(k)}]^*$ and $[s_1^{(k)}]^*$ are transmitted, where the superscripts $[\cdot]^*$ denote complex conjugation and $k = 1, 2, \ldots, K$. $s_1^{(k)}$ and $s_2^{(k)}$ have unit symbol energy. It is assumed that the K users transmit simultaneously in the same frequency band, and their transmissions are perfectly synchronized in time. The channel matrix for user k can be described as

$$\mathbf{H}^{(k)} = \begin{bmatrix} h_{11}^{(k)} & h_{12}^{(k)} \\ h_{21}^{(k)} & h_{22}^{(k)} \\ \vdots & \vdots \\ h_{N1}^{(k)} & h_{N2}^{(k)} \end{bmatrix} \qquad (1)$$

where $h_{ij}^{(k)}$ is the Rayleigh fading channel gain of user k from the jth transmit antenna to the ith receiver antenna. Each entry in the channel matrix is modeled as a statistically independent and identically distributed (i.i.d.) zero-mean complex Gaussian variable with a variance of 1/2. It is assumed that none of the users have channel state information (CSI), whereas CSI is known at the BS.

In reality, the channel between transmit and receive antennas is not ideally the i.i.d. Rayleigh fading because of several factors [13]. One of them is spatial correlation caused by reflection and diffraction wave propagation [14]. If the main scattering appears close to the antenna arrays, the usage of Kronecker model is reasonable [15].

3. Beamforming Process for CCI Cancellation

An Alamouti space-time block code (STBC) word sent over the two transmit antennas of user k during two symbol periods is represented as

$$\mathbf{S}^{(k)} = \begin{bmatrix} s_1^{(k)} & -\left[s_2^{(k)}\right]^* \\ s_2^{(k)} & \left[s_1^{(k)}\right]^* \end{bmatrix}, \; k = 1, 2, \ldots, K. \qquad (2)$$

The correlation matrix of the Alamouti STBC is expressed as

$$\mathrm{E}\left[\mathbf{S}^{(k)}\left\{\mathbf{S}^{(k)}\right\}^H\right] = 2E_s\mathbf{I}_2 \qquad (3)$$

where E_s denotes symbol energy, $\mathbf{I}_2$ is a 2×2 identity matrix, and E[·] is the expected value operator. If user 1 is the desired user, the signal received at the BS can be represented as

$$\mathbf{Y} = \mathbf{H}^{(1)}\mathbf{S}^{(1)} + \sum_{k=2}^{K}\mathbf{H}^{(k)}\mathbf{S}^{(k)} + \mathbf{n}_0 \qquad (4)$$

where $\mathbf{n}_0$ is an $N \times 2$ matrix that is additive white complex valued Gaussian noise (AWGN), whose elements are i.i.d. Gaussian random variables with zero mean and a variance σ^2.

The main purpose of the beamforming process is to suppress CCIs. In other words, the beamforming process plays a role in minimizing the power of unwanted signals. To suppress CCIs, an $N \times 1$ beamforming weight vector $\mathbf{x}$ for user 1 is assumed and then the received signal $\mathbf{Y}$ is multiplied by $\mathbf{x}^H$ at the beamformer. The signal after multiplying $\mathbf{Y}$ by $\mathbf{x}^H$ is represented as

$$\mathbf{x}^H\mathbf{Y} = \mathbf{x}^H\mathbf{H}^{(1)}\mathbf{S}^{(1)} + \mathbf{x}^H\sum_{k=2}^{K}\mathbf{H}^{(k)}\mathbf{S}^{(k)} + \mathbf{x}^H\mathbf{n}_0 \qquad (5)$$

where the superscript $[\cdot]^H$ denotes a Hermitian conjugate. The signal correlation matrix $\mathbf{R}_s$ is denoted as

$$\mathbf{R}_s = E\left[\mathbf{x}^H\mathbf{H}^{(1)}\mathbf{S}^{(1)}\left(\mathbf{x}^H\mathbf{H}^{(1)}\mathbf{S}^{(1)}\right)^H\right] = 2E_s\,\mathbf{x}^H\mathbf{H}^{(1)}\left\{\mathbf{H}^{(1)}\right\}^H\mathbf{x} \quad (6)$$

and the interference plus noise correlation matrix $\mathbf{R}_{in}$ is expressed as

$$\begin{aligned}\mathbf{R}_{in} &= E\left[\left(\mathbf{x}^H\sum_{k=2}^{K}\mathbf{H}^{(k)}\mathbf{S}^{(k)} + \mathbf{x}^H\mathbf{n}_0\right)\left(\mathbf{x}^H\sum_{k=2}^{K}\mathbf{H}^{(k)}\mathbf{S}^{(k)} + \mathbf{x}^H\mathbf{n}_0\right)^H\right]\\ &= \mathbf{x}^H\left[\sum_{k=2}^{K}\mathbf{H}^{(k)}E\left[\mathbf{S}^{(k)}\left\{\mathbf{S}^{(k)}\right\}^H\right]\left\{\mathbf{H}^{(k)}\right\}^H + E\left[\mathbf{n}_0\left\{\mathbf{n}_0\right\}^H\right]\right]\mathbf{x}\\ &= \mathbf{x}^H\left(E_s\sum_{k=2}^{K}\mathbf{H}^{(k)}2E_s\mathbf{I}_2\left\{\mathbf{H}^{(k)}\right\}^H + 2\sigma^2\mathbf{I}_M\right)\mathbf{x}\end{aligned} \quad (7)$$

where it is assumed that E_s is unit symbol energy. Finally, the output SINR of the desired user 1 at the beamformer is defined as

$$\mathrm{SINR}^{(1)} = \frac{\mathbf{R}_s}{\mathbf{R}_{in}} = \frac{\mathbf{x}^H\mathbf{H}^{(1)}\left\{\mathbf{H}^{(1)}\right\}^H\mathbf{x}}{\mathbf{x}^H\left[\sum_{k=2}^{K}\mathbf{H}^{(K)}\left\{\mathbf{H}^{(K)}\right\}^H + \sigma^2\mathbf{I}_M\right]\mathbf{x}}. \quad (8)$$

The goal of beamforming process is to find the beamforming weight vector $\mathbf{x}$ that maximizes the output SINR of user 1. From (8), the problem of maximizing the output SINR can be considered as a special case of the more general problem of finding a critical point of the Rayleigh quotient, and its value is bounded by the maximum and minimum eigenvalues of $\left[\sum_{k=2}^{K}\mathbf{H}^{(k)}\left\{\mathbf{H}^{(k)}\right\}^H + \sigma^2\mathbf{I}_M\right]^{-1}\mathbf{H}^{(1)}\left\{\mathbf{H}^{(1)}\right\}^H$. To maximize the output SINR, $\mathbf{x}$ should be chosen as the eigenvector of $\left[\sum_{k=2}^{K}\mathbf{H}^{(k)}\left\{\mathbf{H}^{(k)}\right\}^H + \sigma^2\mathbf{I}_M\right]^{-1}\mathbf{H}^{(1)}\left\{\mathbf{H}^{(1)}\right\}^H$ associated with the maximum eigenvalue. By multiplying $\mathbf{x}^H$ by the received signal $\mathbf{Y}$, a 1×2 equivalent channel vector is obtained, which seems like a channel matrix when it is equipped with two transmit antennas and one receiver antenna. It enables to detect user 1 by using a general Alamouti decoding scheme.

4. Multiuser Detection with Reduced Computational Complexity

Fast antenna selection schemes improve system performance with a low hardware cost. If the BS has M diversity branches with N available RF chains, there are C_N^M possible antenna subsets, which are expressed as

$$\Theta = \left\{\Phi_1, \Phi_2, \ldots, \Phi_{C_N^M}\right\} \quad (9)$$

where C_N^M denotes the number of ways to choose N elements from a set of M elements and Φ_i denotes the ith antenna subset. In multiuser MIMO systems, the system performance is mainly determined by the minimum SINR among all users. Therefore, if $\mathrm{SINR}_{\min,i}$ denotes the minimum SINR among all users for the subset Φ_i, the best antenna subset can be selected by using a max-min criterion, represented as

$$\Phi_{opt} = \arg\max_{\Phi_i \in \Theta} SINR_{\min,i} \ , i = 1, 2, \ldots, C_N^M. \quad (10)$$

To search for the optimum antenna subset that satisfies (10), repetitive SINR computations are required between each user and the BS equipped with subset Φ_i. If there are C_N^M antenna subsets, $K \times C_N^M$ SINR computations are required for selecting the best antenna subset. Moreover, in the SINR calculation process, the matrix inversion of $\left[\sum_{k=2}^{K}\mathbf{H}^{(k)}\left\{\mathbf{H}^{(k)}\right\}^H + \sigma^2\mathbf{I}_M\right]$ should be calculated, which makes the computational complexity high. In short, the complexity of the fast antenna selection scheme in multiuser systems is very high due to the many pseudo-inverse computations.

In this paper, two alternative solutions are proposed based on the SEC scheme and the Cholesky decomposition to solve the problem mentioned above and to compare the systems in terms of computational complexity and BER performance. The SEC scheme is adopted in multiuser detection in order to reduce the number of repetitive matrix inversions. Additionally, since the matrix is positive definite, the Cholesky decomposition can be exploited, which is less computationally complex than the Gauss-Jordan method.

4.1 Switch-and-Examine Combining

In the case of a conventional SEC in a single user system, the receiver needs to monitor the signal quality of the currently used diversity branch and to compare it with the pre-determined switching threshold. If the signal quality falls below the threshold, the receiver switches to another diversity branch.

In this paper, the SEC scheme is adopted to the multiuser detection system instead of fast antenna selection schemes in order to reduce the number of repetitive matrix inversions while maintaining BER performance. In the proposed multiuser detection method, the receiver first calculates the switching decision criterion presented as the square of the Frobenius norm of the minimum SINR matrix from the currently in use antenna subset Φ_i and then compares it with the pre-determined switching threshold. If the switching decision criterion falls below the pre-determined threshold, the receiver switches to another antenna subset. For the SEC, special care must be taken in choosing the switching threshold for performance maximization, and the proper thresholds that improve BER performance can be obtained through repetitive simulations. The detailed algorithm is depicted in Fig. 2.

For $j = 1 : C_N^M$

$$\mathrm{SINR}_j = \left[\sum_{k=2}^{K} \mathbf{H}_{\mathrm{subj}^{(k)}} \left\{ \mathbf{H}_{\mathrm{subj}^{(k)}} \right\}^H + \sigma^2 \mathbf{I}_M \right]^{-1} \mathbf{H}_{\mathrm{subj}^{(1)}} \left\{ \mathbf{H}_{\mathrm{subj}^{(1)}} \right\}^H$$

$$\Lambda_j = \left| \mathrm{SINR}_j \right|_F^2 = Tr\left[\mathrm{SINR}_j \left(\mathrm{SINR}_j \right)^H \right]$$

if $\Lambda_j >=$ pre-determined switching threshold

for $i = 1 : P$

$\mathbf{H}^{(k)} = \mathbf{H}_{\mathrm{subj}^{(k)}}$

end

break;

end

end

Fig. 2. Algorithm for the proper threshold.

In the algorithm, $\mathbf{H}_{\mathrm{subj}^{(k)}}$ denotes the channel matrix between user k and the antenna subset Φ_i . If the minimum SINRs from all possible antenna subsets are smaller than the pre-determined switching threshold, the best antenna subset is selected by using the max-min criterion.

4.2 Cholesky Decomposition

Generally, Cholesky decomposition is used to solve linear equations for which the coefficient matrix is a special form, namely, positive definite. A Hermitian matrix is positive definite if it satisfies

$$\mathbf{a}^H \mathbf{A} \mathbf{a} > 0, \tag{11}$$

where **a** denotes all nonzero vectors. If **A** is positive definite, it can be decomposed in exactly one way into the following form:

$$\mathbf{A} = \mathbf{R}^H \mathbf{R}, \tag{12}$$

where **R** is upper triangular, and all main diagonal entries are positive. **R** is called the Cholesky factor of **A**. As it is already known, the matrix inversion of $\left[\sum_{k=2}^{K} \mathbf{H}^{(k)} \left\{ \mathbf{H}^{(k)} \right\}^H + \sigma^2 \mathbf{I}_M \right]$ should be calculated repetitively to search for the optimum antenna subset. One thing to note is that the matrix $\left[\sum_{k=2}^{K} \mathbf{H}^{(k)} \left\{ \mathbf{H}^{(k)} \right\}^H + \sigma^2 \mathbf{I}_M \right]$ is positive definite, so it is more efficient to start with the Cholesky decomposition and then invert the lower triangular matrix and compute its Gram. First, it is assumed that $\mathbf{A} = \left[\sum_{k=2}^{K} \mathbf{H}^{(k)} \left\{ \mathbf{H}^{(k)} \right\}^H + \sigma^2 \mathbf{I}_M \right]$ and **A** is decomposed as $\mathbf{A}=\mathbf{R}^H\mathbf{R}$ using Cholesky's method. The lower triangular matrix **R** can be inverted, and then the Gram $(\mathbf{R}^{-1})^{\mathrm{H}} \mathbf{R}^{-1}$ of $\mathbf{R}^{-1}$ can be built. As a result, $\frac{1}{2}N^3 + \frac{3}{2}N^2$ multiplications, $\frac{1}{2}N^3 - \frac{1}{2}N^2$ summations, and N square-root operations are required to calculate the inverse of **A**, which yields the FLOPs of Cholesky's method as

$$FLOPs \text{ of Cholesky's method} = N^3 + N^2 + N. \tag{13}$$

In contrast, the Gauss-Jordan elimination method is used to find inverse matrices in normal cases. In this case, the routine solves the N versions of the linear equation problem with different constant vectors. Therefore, the FLOPs for the Gauss-Jordan method is denoted as

$$FLOPs \text{ of the Gauss-Jordan method} = \frac{1}{2}N^4 + \frac{3}{2}N^3. \tag{14}$$

However, this method is computationally expensive and does not exploit the Hermitian structure of the matrix. In short, when matrix **A** is positive definite, it is more efficient to calculate its inverse with the Cholesky decomposition rather than Gauss-Jordan elimination.

5. Simulation Results

In this section, simulation results are provided that confirm the performance of the system model explained in the previous sections. The BER performance comparisons of various multiuser detection methods are shown over a Rayleigh fading channel. It is assumed that two and three users are equipped with two antennas, and each user transmits an Alamouti STBC. Quadrature phase shift keying (QPSK) modulation scheme is used, which provides a rate equal to two bits per channel user. To effectively suppress CCIs, more than $2 \times (K - 1) + r$ receiver antennas should be selected. In addition, the fast antenna selection and SEC schemes are used in the detection process to achieve receiver diversity gains. The threshold values for the simulation of the proposed multiuser detection method are given in Tab. 1. When the number of user K is 2 and the number of available RF chain N is 3, it is assumed that the number of diversity branch M is 4 or 5. Furthermore, it is assumed that M is 6 or 7, when K is 3 and N is 5.

Fig. 3 and Fig. 4 show the comparisons of BER performances of the proposed multiuser detection method, fast antenna selection scheme, and no selection scheme, when the number of user is two and three respectively. Fig. 3 and Fig. 4 demonstrate that the system with the fast antenna selection and the proposed multiuser detection method

E_b/N_0 (dB)	Proper threshold			
	K=2		K=3	
	M=4, N=3	M=5, N=3	M=6, N=5	M=7, N=5
0	1.7	3	1.7	1.4
2.5	6	10	9	9
5	22	30	3.5	35
7.5	42	80	80	80
10	120	180	160	250
12.5	240	400	310	400
15	500	900	500	980

Tab. 1. Optimum threshold values for the proposed multiuser detection method.

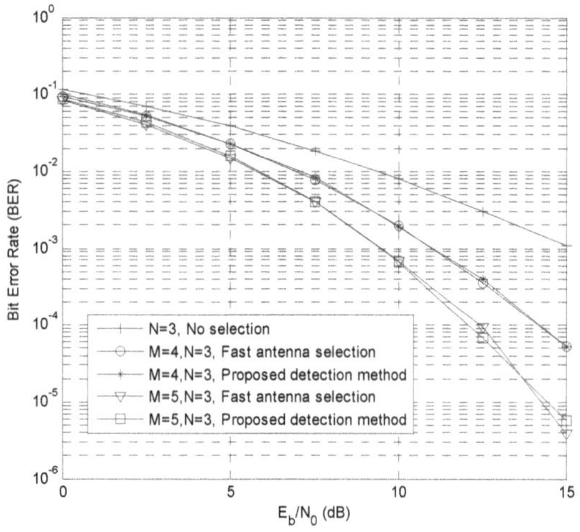

Fig. 3. Comparison of BER performance with two users.

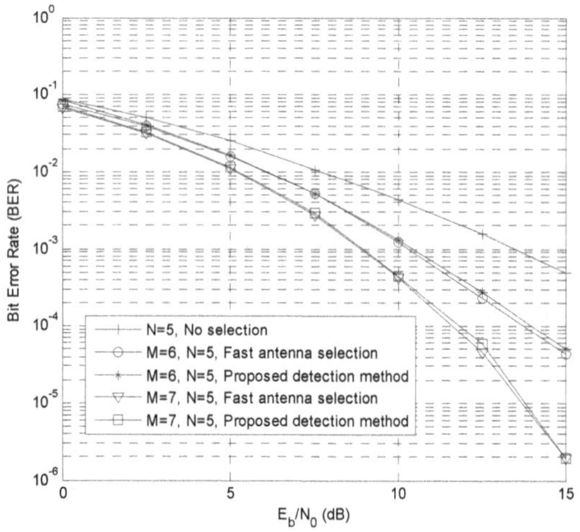

Fig. 4. Comparison of BER performance with three users.

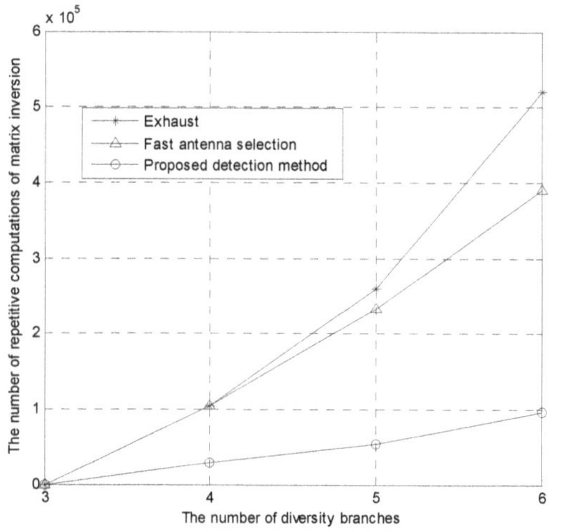

Fig. 5. Comparison of the number of repetitive computations of matrix inversion as a function of the number of available diversity branches with two users.

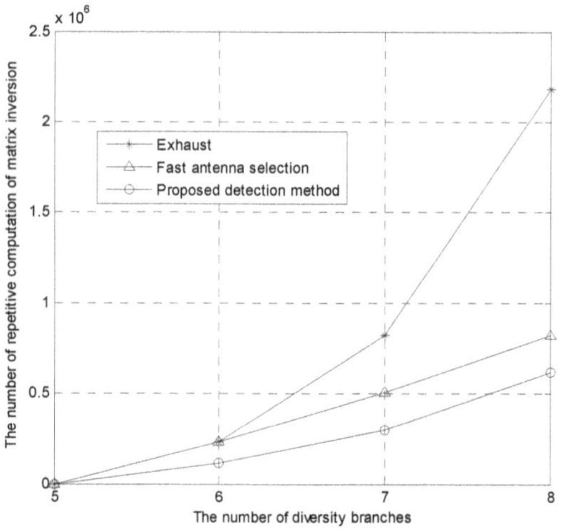

Fig. 6. Comparison of the number of repetitive computations of matrix inversion as a function of the number of available diversity branches with three users.

outperforms the system without the selection schemes in terms of BER performance. The existence of diversity gains can be recognized by confirming the slope of the BER curves between the system with the selection scheme and the system without it. Moreover, Fig. 3 and Fig. 4 also show that the BER performance of the system with the proposed multiuser detection method is very close to that of the fast antenna selection scheme. Therefore, it is clear that the proposed multiuser detection method can significantly reduce the FLOPs without the degradation of the BER performance.

As already mentioned in Sec. 4, in the process of searching for the optimal receiver antenna subset, the repetitive matrix inversion computations required to obtain the minimum SINR for each possible subset are the main cause of the high computational complexity. Therefore, reducing the number of computations in the matrix inversion can be a valuable method for lowering system complexity.

When BS has three RF chains, Fig. 5 and Fig. 6 represent the number of computations of matrix inversion that occur repetitively until final optimum antenna subset is selected. The results prove that the computational complexity of the proposed multiuser detection method is much lower than that of the systems with exhaustive and fast antenna selection. It is evident that the system adopting the fast antenna selection scheme is poor in terms of complexity performance, since it is efficient when $M >> N$. Therefore, if the number of available diversity branches M increases, the fast antenna selection scheme makes great performance difference as compared with exhaustive selection, whereas the proposed multiuser detection method still has good performance despite $M >> N$. Therefore, the proposed multiuser detection method has the best performance from a computational complexity point of view regardless of the difference between M and N.

We also compare the proposed method with other Exhaust and Fast antenna selection methods by means of computational time. To compare the time, we assume that the system uses 2.5 GHz CPU and one command occurs in each clock time. We also assume that BS has three RF chains and the numbers of diversity branches are 3, 4, 5, and 6 for two users and 5, 6, 7, and 8 for three users. Tab. 2 and Tab. 3 show the comparison of the time when the number of users is two and three respectively.

The number of diversity branches	Exhaust	Fast antenna selection	Proposed method
4	40 μs	40 μs	12 μs
5	104 μs	94 μs	22 μs
6	208 μs	156 μs	40 μs

Tab. 2. Comparison of the computational time as a function of the number of available diversity branches with 2 users.

The number of diversity branches	Exhaust	Fast antenna selection	Proposed method
6	0.1 ms	0.1 ms	0.052 ms
7	0.32 ms	0.2 ms	0.12 ms
8	0.88 ms	0.28 ms	0.25 ms

Tab. 3. Comparison of the computational time as a function of the number of available diversity branches with 3 users.

6. Conclusion

In this paper, an uplink multiuser MIMO detection scheme using the SEC and Cholesky decomposition was proposed for reducing system complexity performance. The proposed multiuser detection method was compared with the system adopting exhaustive and fast antenna selection with regard to BER and computational complexity over a Rayleigh fading channel employing QPSK modulation. In addition, the differences in the FLOP counts and computational time are compared for the matrix inversion calculation between the Cholesky decomposition and Gauss-Jordan elimination. As a result, it was shown that the three detection schemes have similar BER performance, whereas the proposed multiuser detection method has a much lower computational complexity when the difference between M and N is small. Furthermore, the FLOP counts and computational time for matrix inversion based on the Cholesky decomposition was much lower than those for Gauss-Jordan elimination.

For future work, we will analyze the effects between antenna arrays of point sources and virtual point sources to reduce the system computational complexity for real world environment. In this paper, we assumed that each user is equipped with two antennas. Therefore, we will extend our results when each user uses more antennas which cause degradations of the robustness because of the antenna characteristics such as the gain, the nonuniform distribution of the emitted power, the distance and the coupling between antennas. We will also apply the effect of channel correlation.

References

[1] GOLDSMITH, A. *Wireless Communications.* Cambridge: Cambridge University Press, 2005.

[2] WU, Q., YANG, X.-S., ZHANG, Y.-T. Research on 2 × 2 MIMO channel with truncated Laplacian azimuth power spectrum. *Radioengineering*, 2013, vol. 22, no. 2, p. 544–548.

[3] DIOUM, I., CLEMENTE, M., DIALLO, A., LUXEY, C., ROSSI, J. P., FARSSI, S. M. Meandered monopoles for 700 MHz LTE handsets and improved MIMO channel capacity performance. *Radioengineering*, 2011, vol. 20, no. 4, p. 726–732.

[4] LIM, C., YOO, T., CLERCKX, B., LEE, B., SHIM, B. Recent trend of multiuser MIMO in LTE-advanced. *IEEE Communications Magazine*, 2013, vol. 51, no. 3, p. 127–135. DOI: 10.1109/MCOM.2013.6476877

[5] PHROMPICHAI, S. SGD frequency-domain space-frequency semiblind multiuser receiver with an adaptive optimal mixing parameter. *Radioengineering*, 2013, vol. 22, no. 1, p. 400–410.

[6] CZINK, N., BANDEMER, B., OESTGES, C., ZEMEN, T., PAULRAJ, A. Analytical multi-user MIMO channel modeling: subspace alignment matters. *IEEE Transactions on Wireless Communications*, 2012, vol. 11, no. 1, p. 367–377. DOI: 10.1109/TWC.2012.010412.110980

[7] SUN, C., KARMAKAR, N. C., LIM, K. S., FENG, A. Combining beamforming with Alamouti scheme for multiuser MIMO communications. In *Proceedings of the 60th IEEE Vehicular Technology Conference (VTC 2004 Fall).* Los Angeles (CA, USA), 2004, vol. 2, p. 1415–1419. DOI: 10.1109/VETECF.2004.1400257

[8] CHEN, M.-X., XU, C.-Q., ZHANG, X.-G. Fast antenna selection for Alamouti multi-user MIMO detection. In *Proceedings of the 4th International Conference on Wireless Communications, Networking and Mobile Computing.* Dalian (China), 2008, p. 1–4. DOI: 10.1109/WiCom.2008.112

[9] YANG, H.-C., ALOUINI, M.-S. Performance analysis of multibranch switched diversity systems. *IEEE Transactions on Communications*, 2003, vol. 51, no. 5, p. 782–794. DOI: 10.1109/TCOMM.2003.811408

[10] NAM, H., ALOUINI, M.-S. Multi-branch switched diversity with adaptive switching thresholds. In *Proceedings of the International Symposium on Information Theory and its Applications.* Auckland (New Zealand), 2008, p. 1–6.

[11] TAROKH, V., NAGUIB, A., SESHADRI, N., CALDERBANK, A. R. Combined array processing and space-time coding. *IEEE Transactions on Information Theory*, 1999, vol. 45, no. 4, p. 1121 to 1128. DOI: 10.1109/18.761255

[12] ALAMOUTI, S. M. A simple transmit diversity technique for wireless communications. *IEEE Journal on Selected Areas in Communications*, 1998, vol. 16, no. 8, p. 1451–1458. DOI: 10.1109/49.730453

[13] GESBERT, D., BOLCSKEI, H., GORE, D. A., PAULRAJ, A. J. Outdoor MIMO wireless channels: models and performance prediction. *IEEE Transactions on Communications*, 2002, vol. 50, no. 12, p. 1926–1934. DOI: 10.1109/TCOMM.2002.806555

[14] LOYKA, S. L. Channel capacity of MIMO architecture using the exponential correlation matrix. *IEEE Communications Letters*, 2001, vol. 5, no. 9, p. 369–371. DOI: 10.1109/4234.951380

[15] YU, K., BENGTSSON, M., OTTERSTEN, B., MCNAMARA, D., KARLSSON, P., BEACH, M. Modeling of wide-band MIMO radio channels based on NLoS indoor measurements. *IEEE Transactions on Vehicular Technology*, 2004, vol. 53, no. 3, p. 655–665. DOI: 10.1109/TVT.2004.827164

About the Authors ...

Soobum CHO received the B.S. degree in Electronics Engineering from Dong-A University, Korea, in 2006, the

M.S. and Ph.D. degrees in Electronics and Computer Engineering from Hanyang University, Korea, in 2008 and 2012, respectively. He is currently a postdoctoral visiting scholar in Electrical Engineering at Stanford University, California. His research interests are in the areas of wireless communications theory, massive MIMO, wireless power transfer, and car connectivity technique.

Sang Kyu PARK received the B.S. degree from Seoul National University, Korea, in 1974, the M.S. degree from Duke University, U.S.A. in 1980, and the Ph.D. degree from the University of Michigan, U.S.A. in 1987, all in Electrical Engineering. From July 1976 to October 1978, he was a research Engineer at the Agency for Defense Development, Korea. From August 1990 to August 1991, he was a visiting scholar at the University of Southern California, U.S.A. Since March 1987, he has been with the Department of Electronics and Computer Engineering at Hanyang University, Korea, where he is currently a Professor. His research interests are in the areas of communications theory, wireless communications, mobile communications, spread spectrum communications, and secure communications.

An Electronically Reconfigurable Patch Antenna Design for Polarization Diversity with Fixed Resonant Frequency

Mohamed Nasrun OSMAN[1], *Mohamad Kamal A. RAHIM*[1], *Peter GARDNER*[2], *Mohamad Rijal HAMID*[1], *Mohd Fairus MOHD YUSOFF*[1], *Huda A. MAJID*[1]

[1]Communication Engineering Dept., Faculty of Electrical Engineering, Universiti Teknologi Malaysia, 81310 Skudai, Johor, Malaysia
[2]School of Electronic, Electrical and Computer Engineering, University of Birmingham, Edgbaston, Birmingham, B15 2TT, United Kingdom.

mnasrun2@live.utm.my, mkamal@fke.utm.my, p.gardner@bham.ac.uk, rijal@fke.utm.my, fairus@fke.utm.my, huda2@live.utm.my

Abstract. *In this paper, an electronically polarization reconfigurable circular patch antenna, with fixed resonant frequency and operating at Wireless Local Area Network (WLAN) frequency band (2.4–2.48 GHz,) is presented. The structure of the proposed design consists of a circular patch as a radiating element fed by a coaxial probe, cooperated with four equal-length slits etched on the edge along the x-axis and y-axis. A total of four switches were used and embedded across the slits at specific locations, thus controlling the length of the slits. By activating and deactivating the switches (ON and OFF) across the slits, the current on the patch is changed, thus modifying the electric field and polarization of the antenna. Consequently, the polarization excited by the proposed antenna can be switched into three types; linear polarization, left-hand circular polarization or right-hand circular polarization. This paper proposes a simple approach that enables switching the polarizations and excites at the same operating frequency. Simulated and measured results of the ideal case (using copper strip switches) and the real case (using PIN diode switches) are compared and presented to demonstrate the performance of the antenna.*

Keywords

Polarization reconfigurable antenna, circular patch, slit perturbations

1. Introduction

Antenna is a key and critical component in wireless telecommunication systems. The evolution of the antenna has rapidly grown and been extensively investigated. It leads to a change of system requirements, hence making it more complex. Adjustments of the operating system scenario are needed in order to meet current trends and demands from end users - such as light weight and compacted devices with enhanced performance. Conventional antenna may face restrictions and limitations to adapt to the new adjustments, due to their characteristics of being fixed and inflexible. One solution to overcome this restriction is the use of reconfigurable antenna.

Reconfigurable antennas, or multi-functional antennas, have recently become quite an active and important topic of research all over the globe. This is due to the fact that the characteristics and properties of the reconfigurable antenna - such as operating frequency/bandwidth, far-field radiation pattern and polarization [1] - can be altered dynamically using external control, hence providing additional functionality and versatility to the systems. Flexibility and reconfigurability features could result in the adaptability of the antenna to address complex system requirements and new specifications. Furthermore, through this concept, the number of components, sizes and hardware complexities could be reduced [2] - hence the cost would be more economically efficient. These benefits would render reconfigurable antennas to become a highly desired feature in modern radio-frequency (RF) systems for wireless and satellite communications.

Reconfigurability can be achieved in different ways, such as through deliberate modification of the antenna physical structure, radiating edge and feeding network. These changes will alter the current flows and distributions on the antenna structure, hence modifying the performance characteristics. To achieve this reconfigurability feature, mechanical [3], optical (photoconducting switches) [4] or electronic (RF switching devices) [5] switches can be used. Among these types of switches, RF PIN diodes and RF MEMS are typical components used in designing the reconfigurable antenna. The PIN diode needs less complicated biasing components, and the lower cost results in it becoming a more popular choice amongst researchers. Even though PIN diodes also have a few drawbacks, such as high insertion loss and lower efficiency, the simplicity in biasing is a key criterion for the selection. The equivalent circuit of the PIN diode for forward bias and reverse bias is shown in Fig. 1.

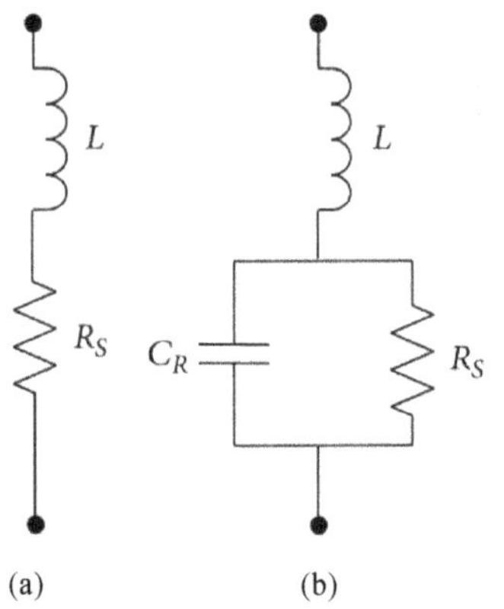

Fig. 1. PIN diode equivalent circuit. (a) forward bias; (b) reverse bias.

Recently, polarization reconfigurable antennas, which can incorporate and offer various polarizations from a single antenna, have attracted considerable attention. This is because the antenna has the ability to reduce the multipath fading [7], and the realization of frequency re-use [8]. In addition, this type of antenna is experimentally proven to improve the capacity offered in the multiple-input-multiple-output (MIMO) systems [9], [10]. For instance, the polarization reconfigurable antenna might take place between linear polarization (LP) and circular polarization (CP), between orthogonal LPs [11], [12], or between left-hand (LHCP) and right-hand circular polarization (RHCP) [13], [14]. Ideally, the main goals in designing reconfigurable antennas are to have the characteristics to be separable and independent. This is because the major challenge to achieve this kind of reconfigurability is to accomplish, without significant changes, on the frequency response and impedance matching. Therefore, careful design approaches are required in order to achieve this feature.

A few papers on the polarization reconfigurable antenna between LP and CP are reported in the literature. Reconfigurable polarization is achieved in [15], through the control of two crossing diagonal slots on the square patch. A similar technique is also used in [16]. In [17], the authors incorporated PIN diodes on the inset of the U-shaped slotline. However, the design required switchable matching stubs in order to obtain good reflection coefficient, which adds more size and complexity to the structure. Furthermore, reference [18] proposed an interesting design approach to achieve polarization reconfigurability features by creating loop slots in the ground plane. The switches were embedded on every slot of the ground plane, which altered the desired polarization sense. Even though the proposed solutions [16-18] managed to obtain a polarization reconfigurable antenna, however the changes of the polarization sense were excited at a different resonant frequency.

Hence, this paper proposes polarization reconfigurable antenna with fixed operating frequency. Details of the design, development and analysis of the polarization reconfigurable circular patch antenna with switchable slit length are presented. Two pairs of slits were located at the edge of the circular radiating patch, positioned on the x-axis and y-axis. The polarization reconfigurability feature of the antenna was executed by placing four switches at specific positions across the slits. The change of the state of the switches consequently alters the length of the slit, hence creating a length difference between the slit in the x-axis and the y-axis. This, accordingly, will determine the types of polarization excited by the antenna - either LHCP, RHCP or LP. Interestingly, all types of polarization senses produced are operated at the same resonant frequency.

Work done in [19] proposed quite a similar approach. However, the polarization was only able to be reconfigured between LHCP and RHCP, without consideration made for the LP. Despite that, the biasing technique applied in this paper is much simpler and requires smaller dimensions.

This manuscript is divided into two main sections based on the type of switch adopted: "Design A" (ideal diode) and "Design B" (PIN diode). The simulated and measured results for both designs are presented, compared and analyzed.

2. Stage 1: Proof of Concept (Copper Strips)

In this section, the initial stage of the design started with the employment of ideal switch. The ideal switch is represented by copper strips or metal pins, with dimensions of 1 mm × 1 mm. The presence of the copper strips indicates the switch in the 'ON' state, meanwhile the absence denotes the 'OFF' state condition.

2.1 Antenna Geometry and Operation

The schematic geometry of the proposed antenna is illustrated in Fig. 2. This consists of a circular patch with radius of r on the finite fully grounded Taconic RF-35 substrate (dimensions of $L \times W$ in mm) with a dielectric constant, ε_r of 3.52, thickness, h of 1.52 mm and tangent loss, $\tan\delta$ of 0.0018. Four slits, which have a length of L_s mm and width of W_s mm are etched at the corner of the circular patch, located along the x-axis and y-axis of the structure with 90° apart from each other. Each slit will have a switch located across it, thus altering the length of the slits. As shown in Fig. 2, the switches are placed at the distance of L_p from the center of the patch. The location of the switch is crucial as it determines the optimum result for the axial ratio. The feed location is positioned diagonally and is fed from the back of the structure through subminiature (SMA) probe. Parameter d is optimized to achieve good impedance matching.

The design procedure generally started with determining the value of unloaded Q_o of the patch, which depends on the radius r, thickness of the substrate h, and the dielectric constant ε_r. By using (1), the value of slit perturbation is determined [20].

$$\left|\frac{\Delta S}{S}\right| = \frac{1}{1.841 Q_0} \tag{1}$$

where S is area of patch and ΔS is the area of slit perturbation. Then, the dimension of the slit is optimized using parameter sweep in order to obtain good circular polarization.

The approach and mechanism of the proposed antenna can be explained using the cavity model method. Due to the diagonal feeding on the structure and perturbation segments, the two near degenerated orthogonal resonant modes TM_{01} and TM_{10} are excited simultaneously. The existence of the perturbation slits will drive the surface current at the edge of the structure to move along it. The lengthened path is only affected to the current travel in the perpendicular direction to the L_s. However, from the parallel direction, the small width, W_s of the perturbation slits will slightly affect the current that is coming from that route. For example, when the pair of slits is cut on the x-axis of the patch, only TM_{01} mode will be affected without giving much effect on the TM_{10} mode, and vice versa.

Due to the presence of the copper strips, the slit length will be shortened, thereby causing an effective surface current to flow across it (shortest distance) instead of travelling around the slits. Consequently, the length difference between the slit on the x-axis and on the y-axis will provide the phase delay between both orthogonal resonant modes. At a specific length difference, the two orthogonal degenerated resonant modes will have the same amplitude and in-phase quadrature, which the CP is excited. Meanwhile, LP is excited when two switches on the x-axis and y-axis are ON, which means no phase difference between both orthogonal resonant modes.

Overall, this proposed antenna works in three polarization modes; namely LHCP, RHCP and LP. The type of excited polarization depends on the configurations of four switches. The switching conditions of all switches, with the respected modes, are tabulated in Tab. 1. The other configurations are not presented, as they provide the same response and redundancy results due to the geometry symmetry. Fig. 3 shows the physical structure used in the simulation, and the photograph of the prototype is shown in Fig. 4.

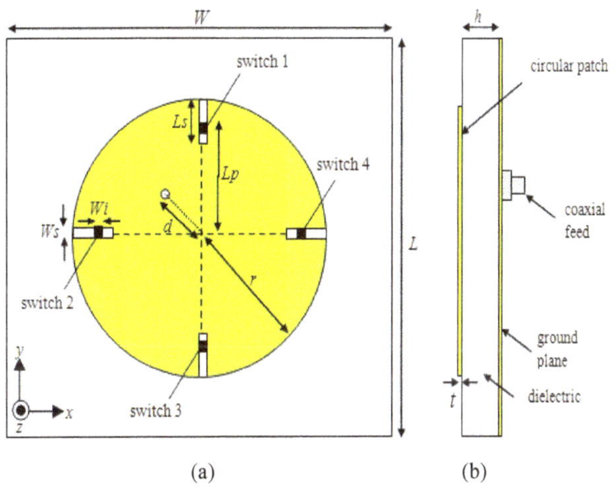

Fig. 2. Geometry of the proposed antenna: (a) front view; (b) side view.

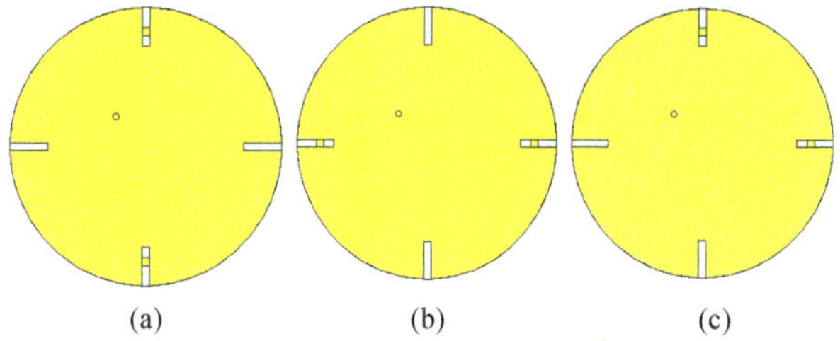

Fig. 3. The physical structure of the proposed antenna used in the simulation: (a) C1-LHCP; (b) C2-RHCP; (c) C3-LP.

Configuration	State of switch				Polarization mode
	SW1	SW2	SW3	SW4	
C1-LHCP	ON	OFF	ON	OFF	LHCP
C2-RHCP	OFF	ON	OFF	ON	RHCP
C3-LP	ON	ON	OFF	OFF	LP

Tab. 1. Switching configurations of the proposed antenna.

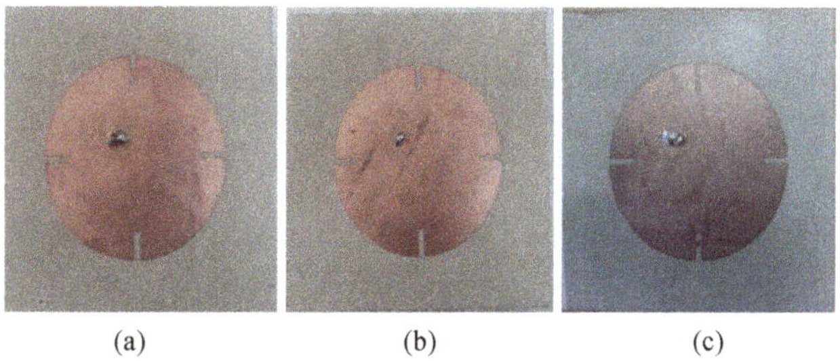

Fig. 4. Front view photograph of fabricated antenna prototypes. (a) C1-LHCP. (b) C2-RHCP. (c) C3-LP.

2.2 Result and Analysis of the Proposed Antenna

The prototype of the antenna is tested and analyzed. S-parameters and far-field test is measured using a Rohde & Schwartz ZVL network analyzer and in an anechoic chamber, respectively. The simulation is carried out using Computer Simulation Technology (CST) Microwave Studio full wave simulator.

Figure 5 shows the comparison between simulated and measured reflection coefficient of the proposed antenna for all switch configurations. It can clearly be seen that for C1-LHCP and C2-RHCP, the two degenerated resonant modes (TM_{01} and TM_{10}) are excited with close frequency at 2.45 GHz/2.49 GHz and 2.45 GHz/2.50 GHz, respectively. Meanwhile, the resonant frequency for C3-LP is 2.48 GHz. The measured -10 dB bandwidths (BW) of S_{11} are 71 MHz (2.438 ÷ 2.509 GHz), 84 MHz (2.437 GHz to 2.521 GHz) and 43 MHz (2.462 ÷ 2.505 GHz) for C1-LHCP, C2-RHCP and C3-LP, respectively. A good impedance matching is achieved for all configurations. It is observed that the measured center frequency is slightly higher than the simulated one. This frequency shift may be due to fabrication uncertainty.

The simulated and measured result of the axial ratio of two CP operations is shown in Fig. 6. For C1-LHCP, the

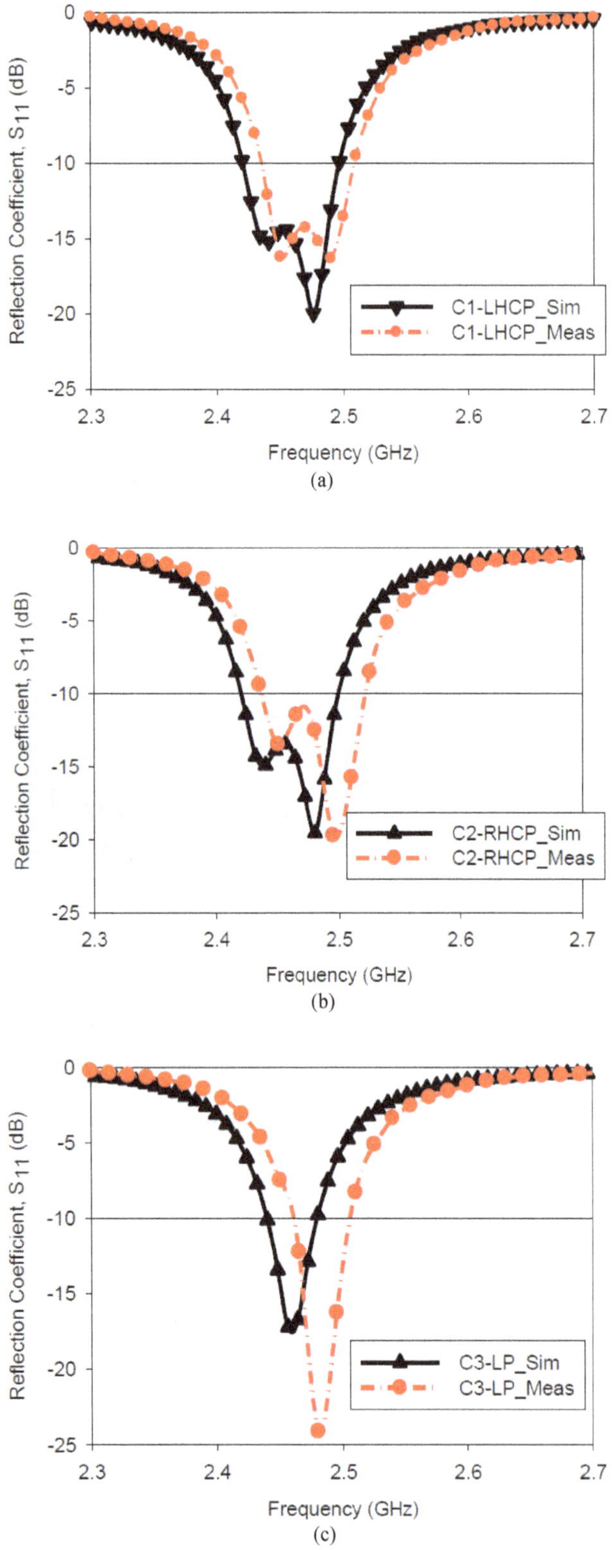

Fig. 5. Comparison of simulated and measured reflection coefficient of the proposed antenna: (a) C1-LHCP; (b) C2-RHCP; (c) C3-LP.

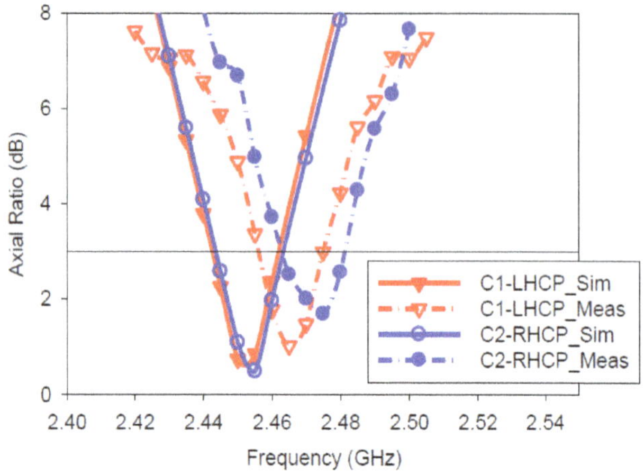

Fig. 6. Simulated and measured axial ratio.

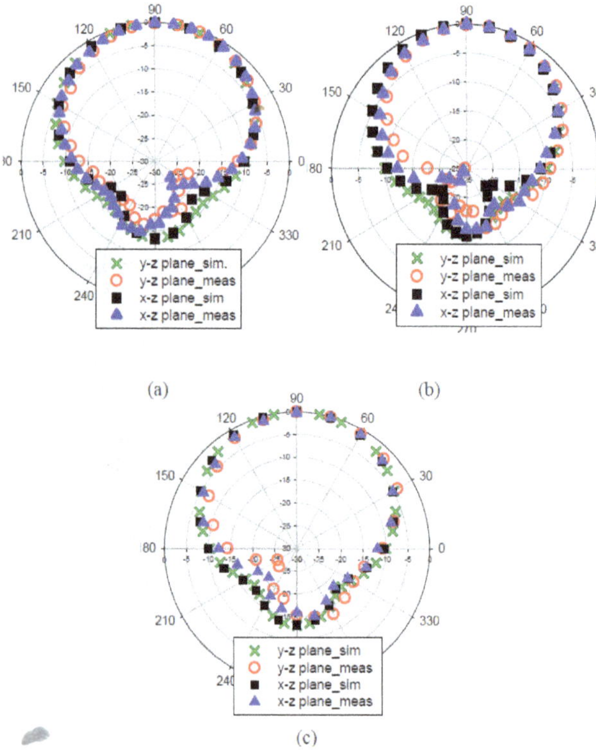

Fig. 7. Simulated and measured normalized far-field radiation pattern at frequency of 2.47 GHz: (a) C1-LHCP; (b) C2-RHCP; (c) C3-LP.

measured 3dB axial ratio bandwidth is 19 MHz (2.456 ÷ 2.475 GHz), or 0.77% with respect to the center frequency of 2.468 GHz. In addition, the BWCP for C2-RHCP is 18 MHz (2.463 ÷ 2.481 GHz), or 0.73% with respect to frequency of 2.472 GHz. The phenomenon variance of the reflection coefficient result leads to the axial ratio to be shifted as well. The far-field radiation of the proposed antenna is investigated. Figure 7 presents the simulated and measured normalized radiation pattern (*y-z* plane and *x-z* plane) at 2.47 GHz. Good radiation pattern performance, with broadside form, is achieved for all types of polarization modes at the respective center frequency.

3. Stage 2: Antenna Design and Operation; Realization of Active Switches (PIN Diodes)

In this section, the copper strips are replaced by RF PIN diode. In this design, the RF PIN diode being used is

Infineon BAR 50-02L. For activating and deactivating the diodes, the antenna structure is modified by integrating a biasing circuit into the structure. To activate the diode, forward bias current is supplied to the ON-state, and the diode is left unbiased for the OFF state condition.

3.1 Modified Structure of the Antenna with Integrated Biasing Circuit

Figure 8 illustrates the antenna structure with modifications, including biased line and lump elements for the purpose of DC biasing. To accommodate the biasing network, minor adjustment was made on the design and re-simulation was done using the CST software in order to obtain the optimum results. The comparison of the dimension dissimilarity, between without biasing network (design A) and with consideration of biasing network (design B), is tabulated in Tab. 2.

In this structure, L-shape narrow slots with width of W_b = 0.3 mm and a length of 9 mm (L_{b1} + L_{b2}) are created to separate between the positive and negative areas of the DC voltage supply. Additional lump elements, RF capacitors and RF inductors are introduced into the design for a specific purpose. Five RF chokes inductors are placed between the radiator and biased wire to improve the isolation between RF current and DC supply [7]. The value of all inductors is 27 nH. Furthermore, considering the effect of the narrow slots, twelve DC block capacitors of 100 pF are mounted across it in order to preserve the continuity of the RF current across the radiator.

The cathodes of the diodes are connected to the four small copper areas and supplied with negative DC voltage. Meanwhile, the anodes of the diodes, which are soldered to the remaining copper region on the circular patch, are connected to the positive terminal of the positive DC battery. The DC is directly supplied using copper wire to the radiator through a via-hole and small circle slots created on the ground plane. The circle slots, with a diameter of L_c are introduced to provide separation of DC voltage. This technique is taken to avoid the presence of the biased wire from

Parameter	Value (mm)		Parameter	Value (mm)	
	A	B		A	B
L	55	55	W_b	-	0.3
W	55	55	W_i	1	-
h	1.524	1.524	L_s	5	5
t	0.035	0.035	L_p	15.3	17
d	5.5	5.5	L_{b1}	-	5
r	17.9	17.9	L_{b2}	-	4
W_s	1	1	L_c	-	4

Tab. 2. Dimension of the designed antennas (Design A and Design B).

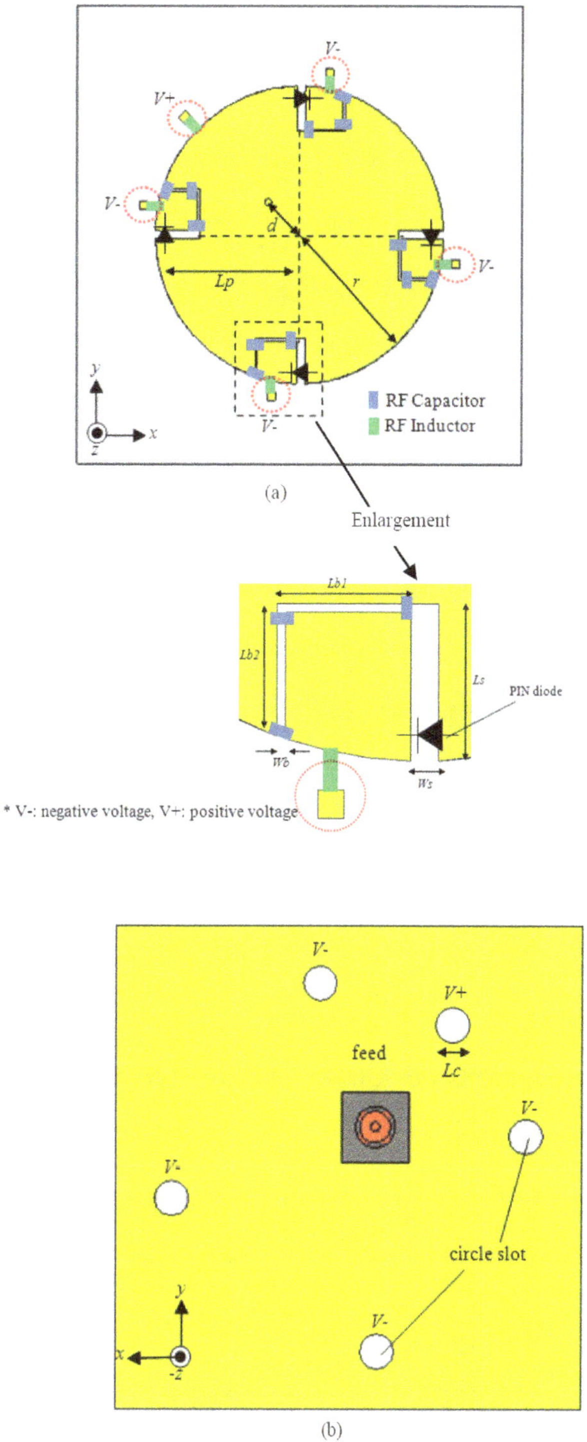

Fig. 8. Geometry of the proposed antenna with the integration of biasing mechanism: (a) front view, (b) back view.

interrupting the antennas far-field radiation pattern. In the meantime, the wire is terminated at the small square copper located outside the circular patch. For the activation of the diodes, the external switching control board is constructed. The switching board is composed of copper wire, 9V DC battery, a dip switch to control the biasing and the resistor for current limitation. The resistors with a value of 100 Ω are chosen so that the forward biased current 90 mA is obtained (ON-state) and left unbiased in the OFF-state condition. The photograph of the fabricated antenna is shown in Fig. 9.

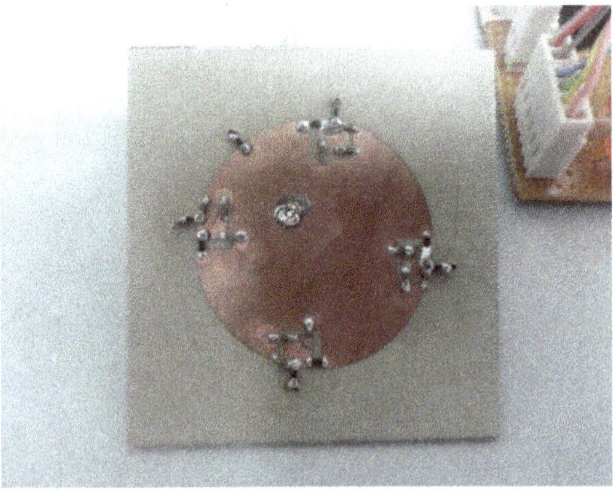

(a)

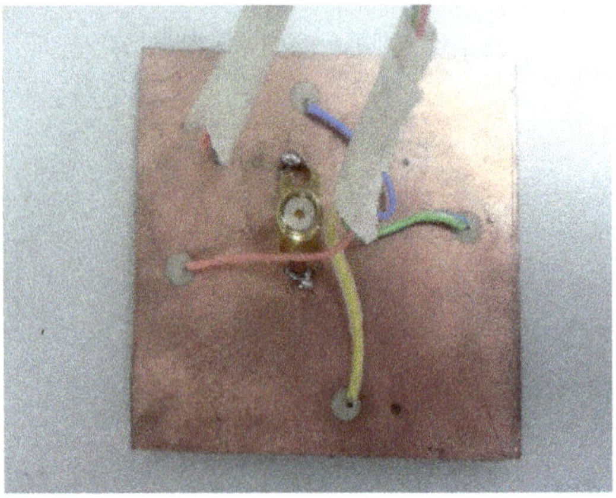

(b)

Fig. 9. Photograph of fabricated antenna prototypes: (a) front view, (b) back view.

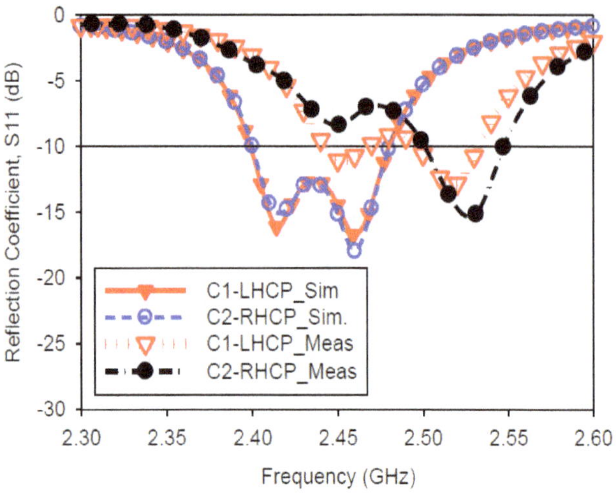

(a)

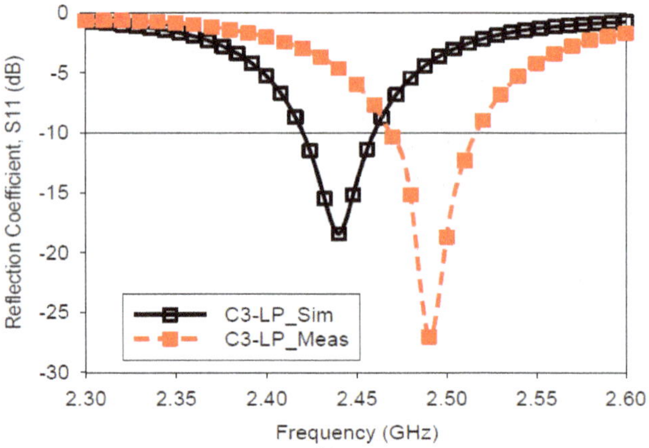

(b)

Fig. 10. Comparison between the simulated and measured reflection coefficient result of the modified structure. (a) Circular polarization. (b) C3-LP.

3.2 Results and Analysis of the Modified Antenna

For accurate simulation results, the PIN diode (Infineon BAR 50-02L) was included in the design and simulated using CST. This step was done by the utilization of touchstone block that contains s2p file which can be obtained from the manufacturer. The touchstone file consists of the information s-parameter of the diode for the ON and OFF state condition.

The optimized antenna was measured using the similar equipment and set up used for the measurement of the ideal diode. The simulated and measured reflection coefficients of the modified structure for all switch configurations are presented in Fig. 10. Good measured impedance matching was achieved for C3-LP, with -10 dB reflection coefficient bandwidth is in the frequency range of 2.469 to 2.515 GHz, or 46 MHz.

However, the reflection coefficient for CP operations is slightly above -10 dB. The impedance bandwidths, based on -10 dB reflection coefficient is 40 MHz (2.494 GHz to 2.534 GHz) and 48 MHz (2.5÷2.548 GHz) for C1-LHCP and C2-RHCP, respectively.

It can be seen that measurement is shifted to the higher frequency. It is also observed that the good impedance matching was obtained except for C2-RHCP configuration. The lowest value of the magnitude of S_{11} for C2-RHCP is -6.7 dB obtained at a frequency of 2.473 GHz. The slight discrepancies between simulation and measurement results could be due to fabrication tolerance, SMA port or parasitic effects from the PIN diode. As mentioned before, the position of the diode in the slits is highly crucial as it determines the axial ratio result and could affect the impedance of the antenna.

Figure 11 shows the simulated and measured results for axial ratio. The measured BW_{CP}, as referred to 3 dB axial ratio bandwidth, is 31 MHz (2.483 ÷ 2.452 GHz) and 22 MHz (2.474 ÷ 2.452 GHz) for C1-LHCP and C2-RHCP, respectively. It has to be noted that the BW_{CP} is totally covered by the BW for linear polarization. Since the proposed antenna performs polarization reconfiguration at a single operating frequency, thus the available working bandwidth is mainly determined by BW_{CP}. Due to symmetry structure, the performances for the both CP operations are almost identical for reflection coefficient and axial ratio result, like the case of ideal diode. The frequency shifted of the reflection coefficient cause the variance between simulated and measured result for the axial ratio. This is due to the shift of the two degenerated orthogonal modes which will consequently shift the frequency of the excited CP mode.

The measured far-field radiation patterns in the *y-z* plane and *x-z* plane for all operation modes are plotted in Fig. 12. Measured results show that the antenna has a broadside radiation pattern at the resonant frequency and 3-dB beamwidth of more than 90° is obtained. In addition, the graph depicts that the proposed antenna has almost the same pattern in both planes. It is also noticed that the gain for C3-LP is higher than the CP modes. The comparison of results between design A and design B is summarized in Tab. 3.

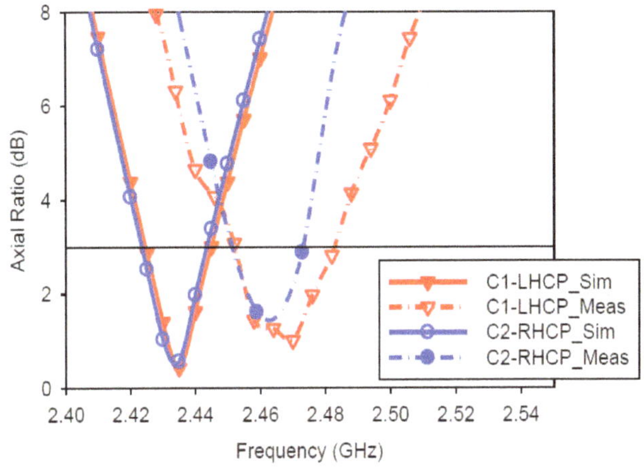

Fig. 11. Simulated and measured axial ratio of the modified structure.

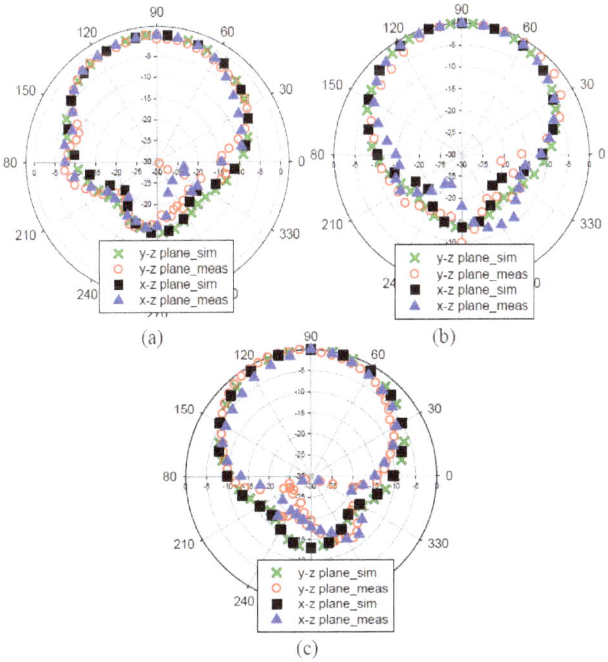

Fig. 12. Measured normalized radiation pattern of the modified structure: (a) C1-LHCP at 2.47 GHz; (b) C2-RHCP at 2.47 GHz; (c) C3-LP at 2.49 GHz.

Conf.	Polar.	Meas. S_{11} (dB)		Meas. Bandwidth (MHz)			
		Design A	Design B	-10dB BW		3-dB BW (Axial Ratio)	
				A	B	A	B
C1-LHCP	LHCP	-16.48	-12.60	71	40	19	31
C2-RHCP	RHCP	-19.83	-15.25	84	48	18	22
C3-LP	LP	-24.06	-26.62	43	46	NA	

*A: Design A (Ideal diode) *B: Design B (PIN diode) *NA: Not applicable

Tab. 3. Comparison between Design A and Design B.

4. Conclusions

This paper presents a simple technique to achieve polarization reconfigurable antenna, with a fixed resonant frequency, that operates within WLAN frequency band. The study begins with the use of copper strips as a switch for proof of the concept. Then, minor modifications were made on the structure in order to accommodate the employment of real switches (RF PIN diode) and switching network. Using four switches embedded across the slits, the slit length is altered, hence producing a phase difference between two degenerated orthogonal resonant modes. Consequently, depending on the switching state, the polarization excited by the proposed antenna is capable to be reconfigured between three types - LHCP, RHCP or LP - that functioned at the same operating frequency. The results of reflection coefficient, axial ratio and radiation pattern are presented and comparisons are made between all different bias conditions. The results demonstrate good impedance matching, and broadside radiation pattern. The discrepancies of the results might be due to parasitic effects of the PIN diode, or fabrication tolerance. Small in dimension, and with unique features, the proposed antenna is potentially suitable in wireless applications for limited space and multipath rich environments; such as to serve as an access point in an indoor scenario for WIFI application.

Acknowledgements

The authors thank the Ministry of Higher Education (MOHE) for supporting the research work, Research Management Centre (RMC), School of Postgraduate (SPS) and Communication Engineering Dept., Universiti Teknologi Malaysia (UTM) for the support of the research under grant no QJ13000.7123.04H38, 4L811, 05H35 and 4F360.

References

[1] BYUN, S.-B., LEE, J.-A., LIM, J.-H., YUN, T.-Y. Reconfigurable ground-slotted patch antenna using PIN diode switching. *ETRI Journal*, Dec. 2007, vol. 29, no. 6, p. 832–834.

[2] KUMAR, S., GOLI, H., BASKARAN, P., ANGELA, P. P. Novel reconfigurable microstrip antenna. In *IEEE 3rd International Conference on Industrial and Information Systems*. Kharagpur (India), Dec. 2008, p. 1–5, DOI: 10.1109/ICIINFS.2008.4798448

[3] RUVIO, G., AMMANN, M. J., CHEN, Z. N. Wideband reconfigurable rolled planar monopole antenna. *IEEE Transactions on Antennas and Propagation*, 2007, vol. 55, no. 6, p. 1760–1767. DOI: 10.1109/TAP.2007.898575

[4] PANAGAMUWA, C. J., CHAURAYA, A. C., VARDAXOGLOU, J. C. Frequency and beam reconfigurable antenna using photoconducting switches. *IEEE Transactions on Antennas and Propagation*, Feb. 2006, vol. 54, no. 2, p. 449–454. DOI: 10.1109/TAP.2005.863393

[5] NIKOLAOU, S., KIM, B., VRYONIDES, P. Reconfiguring antenna characteristics using PIN diodes. In *The 3rd European Conference on Antennas and Propagation EuCap2009*. Berlin (Germany), 2009, p. 3748–3752.

[6] ISMAIL, M. F., A. RAHIM, M. K., MAJID, H. A. The investigation of PIN diode switch on reconfigurable antenna. In *IEEE International RF and Microwave Conference RFM 2011*. Seremban (Negeri Sembilan, Malaysia), December 2011, p. 234–237. DOI: 10.1109/RFM.2011.6168737

[7] FERRERO, F., LUXEY, C., JACQUEMOD, G., STARAJ, R., FUSCO, V. Polarisation-reconfigurable patch antenna. In *International Workshop on Antenna Technology: Small and Smart Antennas Metamaterials and Applications IWAT 2007*. Cambridge (UK), 2007, no. 1, p. 73–76. DOI: 10.1109/IWAT.2007.370083

[8] LIM, J.-H., BACK, G.-T., YUN, T.-Y. Polarization-diversity cross-shaped patch antenna for satellite-DMB systems. *ETRI Journal*, Apr. 2010, vol. 32, no. 2, p. 312–318. DOI: 10.4218/etrij.10.0109.0280

[9] PIAZZA, D., MOOKIAH, P., D'AMICO, M., DANDEKAR, K. R. Pattern and polarization reconfigurable circular patch for MIMO systems. In *European Conference on Antennas and Propagation EuCap 2009*. Berlin (Germany), 2009, p. 1047–1051.

[10] PIAZZA, D., MOOKIAH, P., D'AMICO, M., DANDEKAR, K. R. Experimental analysis of pattern and polarization reconfigurable circular patch antennas for MIMO systems. *IEEE Transactions on Vehicular Technology*, 2010, vol. 59, no. 5, p. 2352–2362. DOI: 10.1109/TVT.2010.2043275

[11] LI, Y., ZHANG, Z., CHEN, W., FENG, Z. Polarization reconfigurable slot antenna with a novel compact CPW-to-slotline transition for WLAN application. *IEEE Antennas and Wireless Propagation Letters*, 2010, vol. 9, p. 252–255. DOI: 10.1109/LAWP.2010.2046006

[12] AMINI, M. H., HASSANI, H. R., NEZHAD, S. M. A. A single feed reconfigurable polarization printed monopole antenna. In *European Conference on Antennas and Propagation EuCap 2012*. Prague (Czech Rep.), 2012, p. 1–4. DOI: 10.1109/EuCAP.2012.6205882

[13] LAI, C.-H., HAN, T.-Y., CHEN, T.-R. Circularly-polarized reconfigurable microstrip antenna. *Journal of Electromagnetic Waves and Applications*, 2009, vol. 23, no. 2-3, p. 195–201. DOI: 10.1163/156939309787604373

[14] WANG, C.-C., CHEN, L.-T., ROW, J.-S. Reconfigurable slot antennas with circular polarization. *Progress in Electromagnetics Research Letters*, 2012, vol. 34, p. 101–110. DOI:10.2528/PIERL12072410

[15] MONTI, G., CORCHIA, L., TARRICONE, L. Patch antenna with reconfigurable polarization. *Progress in Electromagnetics Research C*, 2009, vol. 9, p. 13–23.

[16] HYUN, D.-H., BAIK, J.-W., LEE, S. H., KIM, Y.-S. Reconfigurable microstrip antenna with polarisation diversity. *Electronics Letters*, 2008, vol. 44, no. 8, p. 509–511. DOI: 10.1049/el:20080125

[17] KIM, B., PAN, B., NIKOLAOU, S., KIM, Y.-S., PAPAPOLYMEROU, J., TENTZERIS, M. M. A novel single-feed circular microstrip antenna with reconfigurable polarization capability. *IEEE Transactions on Antennas and Propagation*, 2008, vol. 56, no. 3, p. 630–638. DOI: 10.1109/TAP.2008.916894

[18] YANG, X.-X., SHAO, B.-C., YANG, F., ELSHERBENI, A. Z., GONG, B. A polarization reconfigurable patch antenna with loop slots on the ground plane. *IEEE Antennas and Wireless Propagation Letters*, 2012, vol. 11, p. 69–72. DOI: 10.1109/LAWP.2011.2182595

[19] CHEN, Y. B., CHEN, T. B., JIAO, Y. C., ZHANG, F. S. A reconfigurable microstrip antenna for switchable polarization. *Journal of Electromagnetic Waves and Applications*, 2006, vol. 20, no. 10, p. 1391–1398. DOI: 10.1163/156939306779276820

[20] GARG, R., BHARTIA, P., BAHL, I., ITTIPIBOON, A. *Microstrip Antenna Design Handbook*, Artech House Publishers, January 2001.

About the Authors ...

Mohamed Nasrun OSMAN was born in November 1987. He received his Electrical Engineering degree major telecommunication from Universiti Teknologi Malaysia in 2010. He is currently working towards his PhD degree in Electrical Engineering at the same university. His research interests include reconfigurable antenna design, RF design and wireless MIMO system.

Mohamad Kamal A RAHIM was born in Alor Star Kedah Malaysia on 3rd November, 1964. He received the B Eng degree in Electrical and Electronic Engineering from University of Strathclyde, UK in 1987. He obtained his Master Engineering from University of New South Wales, Australia in 1992. He graduated his PhD in 2003 from University of Birmingham, U.K., in the field of Wideband Active Antenna. From 1992 to 1999, he was a lecturer at the Faculty of Electrical Engineering, Universiti Teknologi Malaysia. From 2005 to 2007, he was a senior lecturer at the Department of Communication Engineering, Faculty of Electrical Engineering, Universiti Teknologi Malaysia. He is now a Professor at Universiti Teknologi, Malaysia. His research interest includes the design of active and passive antennas, dielectric resonator antennas, microstrip antennas, reflect-array antennas, electromagnetic band gap (EBG), artificial magnetic conductors (AMC), left-handed metamaterials and computer aided design for antennas.

Peter GARDNER graduated in Physics from the University of Oxford in 1980 with first class honors. He gained eight years of industrial experience in research and development of active microwave components for Ferranti International, Poynton, Cheshire and later as an independent consultant. He obtained his MSc in 1990 and PhD in 1992. He joined the School of Electronic and Electrical Engineering at Birmingham in 1994, as a lecturer, and currently is a Head of School for School of Electronic, Electrical and Computer Engineering. His research interests include active and adaptive antenna, microwave amplifier, transmitter linearization and micromachine antennas.

Mohamad Rijal HAMID received the M.Sc. degrees in Communication Engineering from the Universiti Teknologi

Malaysia, Johor Bahru, Malaysia, in 2001 and the PhD Degree at the University of Birmingham, Birmingham, United Kingdom. He has been with Universiti Teknologi Malaysia at the Faculty of Electrical Engineering, UTM since 1999. Currently his position is a Senior Lecturer. His major research interest is reconfigurable antenna design for multimode wireless applications.

Mohd Fairus MOHD YUSOFF is a graduate faculty member of the Faculty of Electrical Engineering, UTM. He joined UTM in 2002 as a Tutor. He received his Bachelor in Engineering (Electrical-Telecommunication) in 2002 and Master of Electrical Engineering (Electrical-Electronics and Telecommunications) in 2005 from Universiti Teknologi Malaysia. He obtained his PhD in 2012 from University of Rennes 1, France in area of Signal Processing and Telecommunication. His main research interest and areas is antenna design, millimeter waves and microwave devices.

Huda A. MAJID obtained his first degree in Electrical Engineering majoring Telecommunication in 2007 and his M.Sc in 2010 from Universiti Teknologi Malaysia, Johor Bahru. He received his PhD degree from the same institution in 2013. His research interest includes design of metamaterials, reconfigurable antennas, RF and microwave. He is currently working as a Postdoctoral Research Fellow since January 2014.

Fuzzy Chance-Constrained Programming Based Security Information Optimization for Low Probability of Identification Enhancement in Radar Network Systems

Chenguang SHI, Fei WANG, Jianjiang ZHOU, Jun CHEN

Key Laboratory of Radar Imaging and Microwave Photonics, Ministry of Education, Nanjing University of Aeronautics and Astronautics, No. 29, Yudao Street, Qinhuai District, Nanjing city, Jiangsu Province, Nanjing 210016, P. R. China

zjjee@nuaa.edu.cn

Abstract. *In this paper, the problem of low probability of identification (LPID) improvement for radar network systems is investigated. Firstly, the security information is derived to evaluate the LPID performance for radar network. Then, without any prior knowledge of hostile intercept receiver, a novel fuzzy chance-constrained programming (FCCP) based security information optimization scheme is presented to achieve enhanced LPID performance in radar network systems, which focuses on minimizing the achievable mutual information (MI) at interceptor, while the attainable MI outage probability at radar network is enforced to be greater than a specified confidence level. Regarding to the complexity and uncertainty of electromagnetic environment in the modern battlefield, the trapezoidal fuzzy number is used to describe the threshold of achievable MI at radar network based on the credibility theory. Finally, the FCCP model is transformed to a crisp equivalent form with the property of trapezoidal fuzzy number. Numerical simulation results demonstrating the performance of the proposed strategy are provided.*

Keywords

Security information, low probability of identification (LPID), power allocation, fuzzy chance-constrained programming (FCCP), radar network systems

1. Introduction

Radar network architecture, which often refers to distributed multiple-input multiple-output (MIMO) radar [1], is pioneered by Fisher in [2] and has drawn considerable attentions due to its advantage of signal and spatial diversities. Moreover, radar network outperforms traditional monostatic radar in target detection, localization accuracy and information extraction [3].

Recent years have witnessed an increasing interest on the radar network configuration which has been extensively studied from various perspectives. The authors in [4] consider the optimal waveform design for MIMO radar in colored noise based on maximization of mutual information (MI) and relative entropy. Yang and Blum present two radar waveform design schemes with constraints on waveform power [5]: the one is maximization of the MI between the target impulse response and the reflected waveform, the other is minimization of the minimum mean-square error (MMSE) in estimating the target impulse response. A novel two-stage waveform optimization algorithm for distributed MIMO radar is proposed in [6], where it is demonstrated that this method can provide great performance improvement in target information extraction. In [7], the authors present three power allocation criteria integrating propagation losses into distributed MIMO radar signal model: maximizing the MI, minimizing the MMSE and maximizing the echo energy. Shi et al. in [8], [9] investigate the low probability of intercept (LPI) optimization strategies in radar network configurations for the first time, which are shown to be effective to enhance the LPI performance for radar networks.

In recent years, pursing high physical-layer (PHY) security is becoming a central issue in wireless communications, in which secrecy capacity is utilized as a metric for secrecy communication performance [10]. In [11], the authors study the use of artificial interference in maximizing secrecy capacity, where a portion of the transmitting power is allocated to broadcast the information signal with enough power to guarantee a certain signal-to-interference-plus-noise ratio (SINR) for the intended receiver, while the rest of the power is utilized to broadcast artificial interference to jam the passive eavesdroppers. Zhou et al. in [12] investigate the problem of secure communication in fading channels. While [13] proposes an optimization strategy for achieving security over multiple-input single-output (MISO) channels by beamforming and artificial interference combined with the "protected zone". Mukherjee and Swindlehurst model the interactions between the legitimate transmitter and active eavesdropper as a two-person zero-sum game [14]. The authors in [15] present a multiuser scheduling algorithm to improve the cognitive transmission security. Further, Wang et al. in [16] propose security information factor to evaluate radar radio frequency (RF) stealth, where it is illustrated airborne radar RF stealth effects based on security information factor concept under some conditions. Shi et al. extend the work in [16] and provide a security information based optimal power allocation scheme for LPID performance in radar networks [17], [18].

However, most researches on secrecy capacity are mainly towards maximizing the secrecy rate for communication with guaranteeing system requirements. The use of security information for LPID performance in radar network systems has rarely been studied previously, which motivates us to consider this problem. In addition, the modern electromagnetic environment is becoming more and more complicated, large difficulties for radar mission are caused by amounts of uncertain factors in electronic warfare, which cannot be completely solved by stochastic theory. The theory of fuzzy set has drawn considerable attentions since this concept was initiated by Zadeh [19] in 1965. In 2002, Liu [20] proposed the concept of credibility measure, and established the theory of fuzzy chance-constrained programming (FCCP) [21], which is a branch of mathematics for studying fuzzy phenomena.

This paper will investigate the FCCP based security information optimization for LPID enhancement in radar networks. The main contributions of this paper are summarized as follows. Firstly, we derive an analytical closed-form expression of security information. Secondly, when the prior knowledge of intercept receiver is unavailable, a novel FCCP based security information optimization algorithm is formulated to minimize the achievable MI at intercept receiver, while the achievable MI outage probability at radar network is enforced to be greater than a specified confidence level. Regarding to the complexity and uncertainty of electromagnetic environment in the modern electronic warfare, the trapezoidal fuzzy number is utilized to describe the threshold of achievable MI at radar network. Finally, the FCCP model is transformed to a crisp equivalent form with the property of credibility theory. Numerical simulations are provided to demonstrate that our proposed algorithm can improve the LPID performance for radar networks to defend against passive intercept receivers. To the best of authors' knowledge, no literature discussing FCCP based security information optimization for improved LPID performance in radar network systems was conducted prior to this work.

The remainder of this paper is organized as follows. Section 2 introduces the basic concepts of credibility theory and the system model for radar network. We first derive the analytical closed-form expression of security information with cooperative jamming (CJ) for radar network in Sec. 3 and formulate the FCCP based security information optimization algorithm for radar network system. Section 4 provides some numerical simulation results. Finally, conclusion remarks are drawn in Sec. 5.

2. Preliminaries and System Model

2.1 Credibility Theory

The theory of fuzzy set has received close attention by the scientific community over the last several decades, which was pioneered by Zadeh via membership function in 1965. In 1978, Zadeh presented the concept of possibility measure, which is utilized to measure a fuzzy set. Although possibility measure has been widely used in both theory and practice, it has no self-duality property. In 2002, Liu proposed the concept of credibility measure to define a self-dual measure. After that, Liu established the credibility theory in 2004, which is a branch of mathematics for studying fuzzy phenomena. Some basic concepts of credibility theory are provided in the following.

Definition 2.1: (Liu & Liu [20]) Let Θ be a nonempty set, and P the power set of Θ. The set function Cr is called a credibility measure if it satisfies the following four axioms:

Axiom 1: $Cr\{\Theta\}=1$.

Axiom 2: $Cr\{A\}\le Cr\{B\}$, whenever $A\subset B$.

Axiom 3: $Cr\{A\}+Cr\{A^c\}=1$ for any event $A\in P$.

Axiom 4: $Cr\{\cup_i A_i\}=\sup_i Cr\{A_i\}$ for any events $\{A_i\}$ with $\sup_i Cr\{A_i\}<0.5$.

Then, the triplet (Θ,P,Cr) is called a credibility space.

Definition 2.2: (Liu [21]) A fuzzy variable is a measurable function from a credibility space (Θ,P,Cr) to the set of real numbers $\Re$.

Definition 2.3: (Liu [21]) Let ξ be a fuzzy variable defined on the credibility space (Θ,P,Cr). Then its membership function is derived from the credibility measure by:

$$\mu(x)=\left(2Cr\{\xi=x\}\right)\wedge 1,\quad x\in\Re. \tag{1}$$

Theorem 2.1 (Credibility Inversion Theorem): (Liu [21]) Let ξ be a fuzzy variable with membership function $\mu(x)$. Then for any set B of real numbers, we have:

$$Cr\{\xi\in B\}=\frac{1}{2}\left(\sup_{x\in B}\mu(x)+1-\sup_{x\in B^c}\mu(x)\right). \tag{2}$$

Definition 2.4: (Liu [21]) The credibility distribution $\Phi:\Re\to[0,1]$ of a fuzzy variable ξ is defined by:

$$\Phi(x)=Cr\{\theta\in\Theta\,|\,\xi(\theta)\le x\}. \tag{3}$$

That is, $\Phi(x)$ is the credibility that the fuzzy variable ξ takes a value less than or equal to x.

2.2 Radar Network SNR Equation

Let us consider a radar network system with N_t transmitters and N_r receivers, which can be broken down into $N_t \times N_r$ transmitter-receiver pairs each with a bistatic component contributing to the entirety of the radar network signal-to-noise ratio (SNR), as depicted in Fig. 1. The radar network system has a common precise knowledge of space and time. In addition, it is worth pointing out that orthogonal polyphase codes are utilized in radar network system, which have a large main lobe-to-side lobe ratio. These codes have a more complicated signal structure making it harder to intercept and identify by a noncooperative intercept receiver.

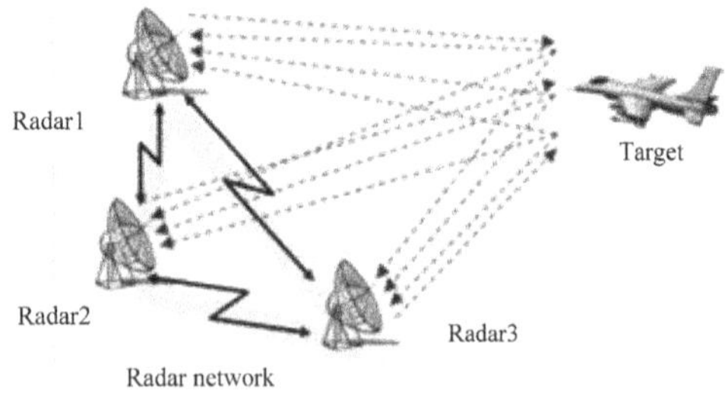

Fig. 1. Example of a radar network.

The radar network SNR can be calculated by summing up the SNR of each transmit-receive pair as in [1]:

$$SNR_{net} = \sum_{m=1}^{N_t}\sum_{n=1}^{N_r} \frac{P_{tm} G_{tm} G_{rn} \sigma_{tmn} \lambda_m^2}{(4\pi)^3 k T_{omn} B_{rm} F_{rn} R_{tm}^2 R_{rn}^2 L_{mn}} \quad (4)$$

where the P_{tm} is the mth transmitter power, G_{tm} is the mth transmit antenna gain, G_{rn} is the nth receive antenna gain, σ_{tmn} is the radar cross section (RCS) of the target for the mth transmitter and nth receiver, λ_m is the mth transmitted wavelength, k is Boltzmann's constant, T_{omn} is the receiving system noise temperature at the nth receiver, B_{rm} is the bandwidth of the matched filter for the mth transmitted waveform, F_{rn} is the noise factor for the nth receiver, L_{mn} is the system loss between the mth transmitter and nth receiver, R_{tm} is the distance from the mth transmitter to the target and R_{rn} is the distance from the target to the nth receiver.

2.3 Radar Network Signal Model

Let K denote the discrete time index, then we can express the radar network signal model as:

$$\boldsymbol{Y}_r = \boldsymbol{X}\boldsymbol{H}_r + \boldsymbol{W}_r \quad (5)$$

where $\boldsymbol{X} = [\boldsymbol{x}_1, \boldsymbol{x}_2, \ldots, \boldsymbol{x}_{N_t}] \in \mathbb{C}^{K \times N_t}$, is the set of transmission sequences, $\boldsymbol{H}_r = [\boldsymbol{h}_{r1}, \boldsymbol{h}_{r2}, \ldots, \boldsymbol{h}_{rN_t}] \in \mathbb{C}^{N_t \times N_r}$ refers to the path gain matrix for radar network system, $\boldsymbol{W}_r = [\boldsymbol{w}_{r1}, \boldsymbol{w}_{r2}, \ldots, \boldsymbol{w}_{rN_r}] \in \mathbb{C}^{K \times N_r}$ represents the system noise, and the received signal matrix can be written as $\boldsymbol{Y}_r = [\boldsymbol{y}_{r1}, \boldsymbol{y}_{r2}, \ldots, \boldsymbol{y}_{rN_t}] \in \mathbb{C}^{K \times N_r}$. For convenience, it is assumed that the noise matrix $\boldsymbol{W}_r$ does not depend on the transmitted waveform $\boldsymbol{X}$, and $\boldsymbol{H}_r$ and $\boldsymbol{W}_r$ are mutually independent.

According to the discussions in [7], the path gain $\boldsymbol{h}_{rn}$ contains the target reflection coefficient g_{mn} and the propagation loss factor p_{mn}. Based on the central limit theorem, $g_{mn} \sim CN(0, \sigma_g^2)$, where g_{mn} denotes the target reflection gain between the radar m and radar n. The propagation loss factor p_{mn} can be expressed as:

$$p_{mn} = \frac{\sqrt{G_{tm} G_{rn}}}{R_{tm} R_{rn}}. \quad (6)$$

Furthermore, with the consideration of propagation losses and target scattered matrix, the radar network signal model (5) can be rewritten as:

$$\boldsymbol{Y}_r = \boldsymbol{X}(\boldsymbol{G} \odot \boldsymbol{P}) + \boldsymbol{W}_r \quad (7)$$

where $\boldsymbol{G} = [\boldsymbol{g}_1, \boldsymbol{g}_2, \ldots, \boldsymbol{g}_{N_r}]$, $\boldsymbol{P} = [\boldsymbol{p}_1, \boldsymbol{p}_2, \ldots, \boldsymbol{p}_{N_r}]$, $\boldsymbol{w}_{rn} \sim \boldsymbol{CN}(0, \sigma_{w_r}^2 \boldsymbol{I}_K)$, and $\odot$ denotes the Hadamard product.

3. Problem Formulation

3.1 Security Information for Radar Network Systems

With the definition of MI in [17], [18], we can obtain the MI between the transmitting signal of radar network X and the backscatter signal $\boldsymbol{Y}_r$ as follows:

$$\begin{aligned} \boldsymbol{I}(\mathrm{X}, \mathrm{Y}_r) &= \boldsymbol{H}(\mathrm{Y}_r) - \boldsymbol{H}(\mathrm{Y}_r \mid \mathrm{X}) \\ &= \boldsymbol{H}(\mathrm{Y}_r) - \boldsymbol{H}(\mathrm{W}_r) \\ &= \sum_{m=1}^{N_t}\sum_{n=1}^{N_r} \ln\left(1 + \frac{P_{tm}\sigma_g^2 p_{mn}^2}{\sigma_{w_r}^2}\right) \\ &= \sum_{m=1}^{N_t}\sum_{n=1}^{N_r} \ln\left(1 + \frac{P_{tm}\sigma_g^2 G_{tm} G_{rn}}{\sigma_{w_r}^2 R_{tm}^2 R_{rn}^2}\right) \triangleq I_{net}(P_r) \end{aligned} \quad (8)$$

where $\boldsymbol{I}(\boldsymbol{X}, \boldsymbol{Y}_r)$ is the MI between $\boldsymbol{Y}_r$ and $\boldsymbol{X}$, $\boldsymbol{H}(\boldsymbol{Y}_r)$ is the entropy of backscatter signal, and $\boldsymbol{H}(\boldsymbol{W}_r)$ is the entropy of Gaussian white noise.

Similarly, we can express the MI between the transmitting signal of radar network X and the received signal of intercept receiver $\boldsymbol{Y}_i$ as:

$$\begin{aligned} I(\boldsymbol{X}, \boldsymbol{Y}_i) &= H(\boldsymbol{Y}_i) - H(\boldsymbol{Y}_i \mid \boldsymbol{X}) \\ &= \sum_{m=1}^{N_t} \ln\left(1 + \frac{P_{tm} G_{tm} G_{int}}{\sigma_{w_i}^2 R_{tm}^2}\right) \end{aligned} \quad (9)$$

where G_{int} is the antenna gain of intercept receiver, $\sigma_{w_i}^2$ denotes the noise covariance of intercept receiver.

As introduced in [17], [18], in modern electronic warfare, cooperative jammer is indispensable to keep the radar network in LPID state. This means that CJ is to jam the hostile intercept receiver so that the achievable MI at interceptor can be degraded by the CJ signals while the radar network system is unaffected. With the consideration of CJ, we can modify (12) as follows:

$$I(X,Y_i)=\sum_{m=1}^{N_t}\ln\left[1+\frac{P_{tm}G_{tm}G_{\text{int}}}{\left(\sigma_{w_i}^2+\frac{P_jG_jG_{\text{int}}}{R_j^2}\right)R_{tm}^2}\right]\triangleq I_{\text{int}}(P_r,P_j)\quad(10)$$

where P_j is the total transmitting power for CJ signal, G_j is the antenna gain of cooperative jammer, R_j is the distance from the target to cooperative jammer.

For convenience, we assume that the radar network system can simultaneously transmit radar modulating signal to track target and CJ signal to interfere passive intercept receiver for simplicity of discussion, while the CJ signal is designed to be completely orthogonal to radar modulating signal and generated to jam the intercept receiver without affecting the radar network.

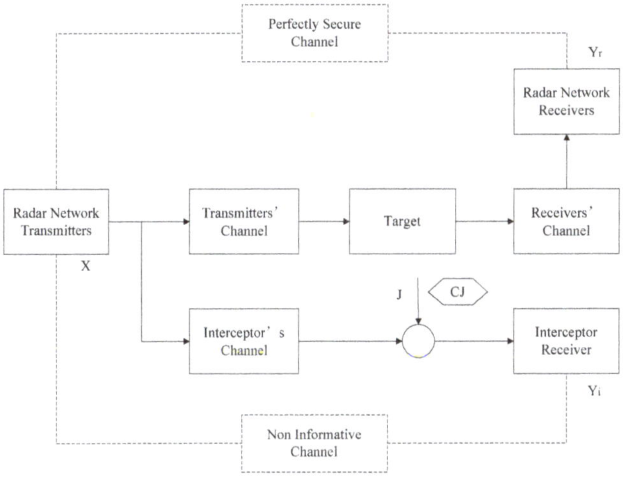

Fig. 2. The notional sketch of our proposed perfectly secure radar network system.

For simplicity of derivation, it is supposed that:

$$R_{net}^2\approx R_{ti}R_{rj}\left(\forall m=1,\ldots,N_t,n=1,\ldots,N_r\right),\quad(11)$$

$$P_{tm}=\frac{P_r}{N_t}\left(\forall m=1,\ldots,N_t\right)\quad(12)$$

where R_{net} is approximately the distance from target to radar network system, P_r is the total transmitting power for radar modulating signal. It is also assumed that each netted radar node in the network is the same. Therefore, originating from the secrecy capacity in wireless communications, we define security information to measure the LPID performance for radar network system [17], [18]:

$$\begin{aligned}&I_{\text{sec}}(P_r,P_j)\\&\triangleq\left[I_{net}(P_r)-I_{\text{int}}(P_r,P_j)\right]^+\\&\doteq\left\{\begin{aligned}&N_tN_r\ln\left(1+\frac{P_r\sigma_g^2G_tG_r}{N_t\sigma_{w_r}^2R_{net}^4}\right)\\&-N_t\ln\left[1+\frac{P_rG_tG_{\text{int}}}{N_t\left(\sigma_{w_i}^2+\frac{P_jG_jG_{\text{int}}}{R_j^2}\right)R_{net}^2}\right]\end{aligned}\right\}^+\end{aligned}\quad(13)$$

where $[x]^+=\max(0,x)$. It has been pointed out in [17], [18] that $I_{\text{sec}}>0$ means that radar network is in completely secure state while tracking target, and that the larger the achievable security information I_{sec} obtained, the better LPID performance to finish the system mission.

The notional sketch of our proposed perfectly secure radar network system is illustrated in Fig. 2. This amounts to say that, if the radar network system experiences an SNR higher than that of the noncooperative intercept receiver, a positive rate can be sustained, while the intercept receiver gets maximally confused based on the utilized secrecy criterion in (13).

3.2 FCCP Based Security Information Optimization

In practical applications, it would be impossible to suppose that any prior information about the hostile intercept receiver is available, such as the sensitivity of interceptor, the processing gain, et al. The system model is illustrated in Fig. 3, where the target and the intercept receiver are separated at different places.

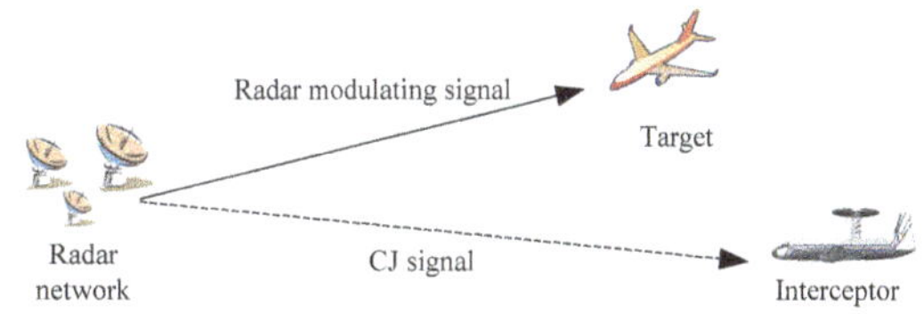

Fig. 3. The geometry of radar network, target and interceptor.

With the derivation of security information as (13), we can observe that security information is based on satisfying MI constraints both at radar network and at intercept receiver. With a proper choice of the power allocation, a perfectly securing channel can be devised such that:

$$\left.\begin{aligned}&I_{net}(P_r)\geq\delta^{th}\\&I_{\text{int}}(P_r,P_j)\xrightarrow{P_j\to\infty}0\end{aligned}\right\}\quad(14)$$

where δ^{th} is the predefined threshold of MI at radar network. Equation (14) means asymptotically perfect security in radar network system.

In this paper, we focus on minimizing the achievable MI at intercept receiver $I_{\text{int}}(P_r, P_j)$ to guarantee a predefined threshold of MI at radar network δ^{th}. To be more specific, minimization of the achievable MI at noncooperative intercept receiver can significantly defend against interceptor, showing LPID performance enhancement of exploiting CJ when any prior knowledge about the intercept receiver is unavailable.

The security information optimization strategy can be summarized as follows:

1) Specify the desired MI threshold δ^{th} for radar network system, which is utilized as a metric for target detection performance.
2) Allocate some transmission power to achieve the desired MI threshold δ^{th} for target detection.
3) Minimize the achievable MI at intercept receiver $I_{\text{int}}(P_r, P_j)$ by distributing the remaining transmission power to yield as much interference as possible, while guaranteeing that the CJ signal is designed to be completely orthogonal to radar modulating signal and generated to jam the intercept receiver without affecting the radar network.

Hence, the security information optimization for enhanced LPID performance can be formulated as:

$$\left.\begin{aligned} &\min_{P_r,P_j} \; I_{\text{int}}(P_r,P_j) \\ &s.t.:\; I_{net}(P_r) \geq \delta^{th} \\ &\qquad P_r + P_j \leq P_{tot}^{\max} \\ &\qquad P_r \in \left(0, P_r^{\max}\right] \end{aligned}\right\} \tag{15}$$

where $P_{tot}^{\max}$ is the maximum transmitting power for radar network, $P_r^{\max}$ is the maximum transmitting power for radar modulating signal. While regarding to the complexity and uncertainty of electromagnetic environment in the modern electronic warfare, the predefined threshold of MI at radar network δ^{th} would be uncertain. Herein, a fuzzy variable δ^{th}_{fuzzy} is utilized to evaluate the predefined threshold of MI at radar network. Based on the concepts of credibility theory, the achievable MI outage probability at radar network is enforced to be greater than a specified confidence level α, that is:

$$Cr\left\{I_{net}(P_r) \geq \delta^{th}_{fuzzy}\right\} \geq \alpha \tag{16}$$

where $Cr\{\cdot\}$ indicates the credibility that $\{\cdot\}$ will occur. Therefore, we have the FCCP based security information optimization for LPID enhancement in radar network as:

$$\left.\begin{aligned} &\min_{P_r,P_j} \; I_{\text{int}}(P_r,P_j) \\ &s.t.:\; Cr\left\{I_{net}(P_r) \geq \delta^{th}_{fuzzy}\right\} \geq \alpha \\ &\qquad P_r + P_j \leq P_{tot}^{\max} \\ &\qquad P_r \in \left(0, P_r^{\max}\right] \end{aligned}\right\} \tag{17}$$

With the FCCP model (17), we can observe that increasing the confidence level leads to enlarging the feasible set of the true problem, which in turn may result in decreasing of the optimal value of the true problem [22]. It is also worth pointing out that there exists a restrictive relationship between the confidence level and the achievable MI at intercept receiver.

3.3 The Crisp Equivalent Form of FCCP Model

The FCCP model (17) is a fuzzy linear programming, which can be transformed into the crisp equivalent form. In this paper, we set $\delta^{th}_{fuzzy} = (a, b, c, d)$ $(a < b \leq c < d)$ to be a trapezoidal fuzzy variable.

Definition 3.1: (Liu, Zhao & Wang [23]) By a trapezoidal fuzzy variable, we mean that the fuzzy variable fully determined by the quadruplet (r_1, r_2, r_3, r_4) of crisp numbers with $r_1 < r_2 \leq r_3 < r_4$, whose membership function is given by:

$$\mu(x) = \begin{cases} \dfrac{x-r_1}{r_2-r_1}, & \text{if } r_1 \leq x \leq r_2 \\ 1, & \text{if } r_2 \leq x \leq r_3 \\ \dfrac{x-r_4}{r_3-r_4}, & \text{if } r_3 \leq x \leq r_4 \\ 0, & \text{else} \end{cases} \tag{18}$$

Definition 3.2: (Liu, Zhao & Wang [23]) The credibility distribution of a trapezoidal fuzzy variable (r_1, r_2, r_3, r_4) is:

$$\Phi(x) = \begin{cases} 0, & \text{if } x \leq r_1 \\ \dfrac{x-r_1}{2(r_2-r_1)}, & \text{if } r_1 \leq x \leq r_2 \\ \dfrac{1}{2}, & \text{if } r_2 \leq x \leq r_3 \\ \dfrac{x-2r_3+r_4}{2(r_4-r_3)}, & \text{if } r_3 \leq x \leq r_4 \\ 1, & \text{if } r_4 \leq x \end{cases} \tag{19}$$

Theorem 3.1: (Liu [24]) If ξ is a trapezoidal fuzzy number $\xi = (r_1, r_2, r_3, r_4)(r_1 < r_2 \leq r_3 < r_4)$, for the given confidence level $\alpha \in (0.5, 1]$,the following equivalent transformation can be derived as:

$$Cr\{\xi \geq x\} \geq \alpha \Leftrightarrow x \leq (2\alpha-1)r_1 + (2-2\alpha)r_2, \tag{20a}$$

$$Cr\{\xi \leq x\} \geq \alpha \Leftrightarrow x \geq (2-2\alpha)r_3 + (2\alpha-1)r_4. \tag{20b}$$

Based on the properties of trapezoidal fuzzy number, we have that:

$$Cr\left\{I_{net}(P_r) \geq \delta^{th}_{fuzzy}\right\} \geq \alpha \Leftrightarrow I_{net}(P_r) \geq 2(1-\alpha)c + (2\alpha-1)d \tag{21}$$

The proposed FCCP based security information algorithm (17) could be transformed into the following crisp equivalent form:

$$\left.\begin{aligned} &\min_{P_r, P_j} \; I_{\text{int}}\left(P_r, P_j\right) \\ &s.t.: \; I_{net}\left(P_r\right) \geq 2(1-\alpha)c + (2\alpha - 1)d \\ &P_r + P_j \leq P_{tot}^{\max} \\ &P_r \in \left(0, P_r^{\max}\right] \end{aligned}\right\} \quad (22)$$

Problem (17) takes radar network mission δ^{th}_{fuzzy} into consideration because radar network system must accomplish its mission in modern battlefield. To be specific, for the predetermined system detection probability P_d^{net} and false alarm probability P_{fa}^{net}, if $Cr\{I_{\text{net}}(P_r) \geq \delta^{th}_{fuzzy}\} \geq \alpha$ and $I_{\text{sec}}(P_r, P_j) > 0$, the detection probability of intercept receiver P_d^{int} would be significantly less than 0.5, which means that the interceptor could not intercept and identify radar modulating signal, and that the radar network system is in completely LPID state [16].

Based on the above derivations as (13) and (22), it is worth pointing out the simplicity of the proposed algorithm which relies on basic mathematical calculations and does not require highly complex computation. Moreover, our proposed algorithm is significantly simple to implement.

So far, we have completed the achievable security information derivation and the FCCP based security information optimization for LPID enhancement in radar network systems. In what follows, some numerical simulations are provided to show the feasibility and effectiveness of our presented algorithm.

4. Numerical Simulations and Analysis

In this section, we evaluate the proposed algorithm through some numerical simulations. Let us consider a 4×4 radar network architecture ($N_t = N_r = 4$), which is depicted in Fig. 4 that the netted radars in the network are spatially distributed in the surveillance area.

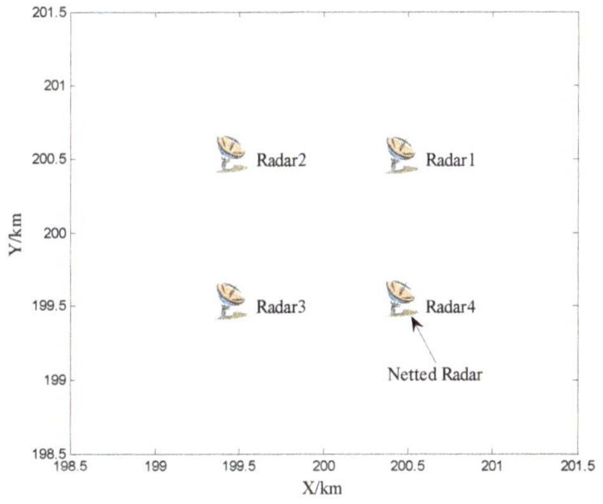

Fig. 4. The radar network system configuration in two dimensions.

Herein, we set the simulation parameters $P_{tot}^{\max} = P_r + P_j = 25$ kW, $G_t = G_r = G_j = 30$ dB, G_i=0 dB, $\sigma_{w_r}^2 = 4.57 \times 10^{-12}$ W, $\sigma_{w_i}^2 = 8.77 \times 10^{-8}$ W and $\sigma_g^2 = 1$. The radar network can detect the target whose RCS is 1 m^2 in the distance 180 km by transmitting the maximum power $P_r^{\max} = 24$ kW. The sensitivity of intercept receiver $S_{i\,\min}$ is set to be –80 dBmW. Based on some experimental data, the trapezoidal fuzzy number is set to be δ^{th}_{fuzzy} = (4.4, 8.1, 13.0, 25.6)nats, which equals to the fuzzy SNR (7.0, 10.2, 13.0, 18.0) dB. This is because that the radar network system can track the target steadily when the SNR is between 10.0 dB and 13.0 dB, and the value of the membership function is set to be 1.

4.1 Security Information Analysis

As shown in Fig. 5, for all the cases $N_r = 1$, $N_r = 2$ and $N_r = 4$, with an increasing N_t, the achievable security information is increased correspondingly. As the number of transmitters N_t continues increasing beyond a certain value, the achievable security information of (16) leads to approximate constant. Fig. 5 also demonstrates that as the number of receivers increases from $N_r = 1$ to $N_r = 4$, the achievable security information for radar network can be significantly increased. To be specific, increasing the number of radars can effectively improve security information for radar network. This is because that radar network can offer great transmit and receive diversities in terms of the achievable security information, which confirms the LPID benefits by exploiting radar network system to defend against passive intercept receiver attacks.

Figure 6 shows the achievable security information versus R_{net} with $P_r = 25$ kW, $P_j = 5$ kW and different R_j. It can be seen from Fig. 6 that as R_{net} and R_j decrease, the achievable security information is increased as theoretically proved in (16). Furthermore, it is depicted that with the same R_{net}, the available security information can be increased as R_j decreases, which shows the advantage of exploiting CJ to defend against intercept receiver.

Figure 7 illustrates achievable MI at intercept receiver versus the confidence level α at radar network for different sensitivities of intercept receiver $S_{i\min}$ with R_{net} = 100 km and $R_j = 300$ km. In Fig. 7, for all the cases, one can observe that the achievable MI at intercept receiver is

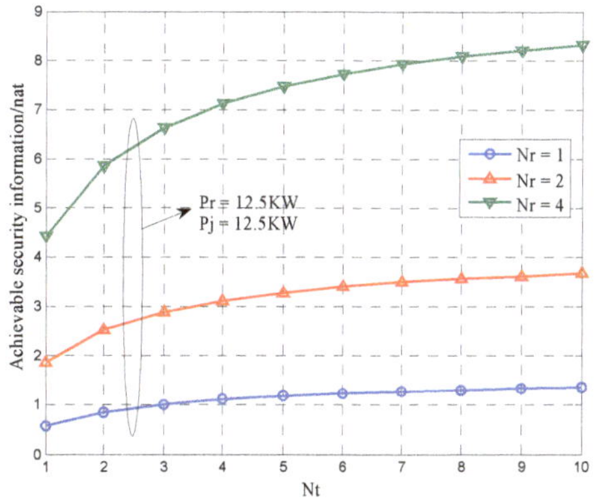

Fig. 5. Achievable security information versus the number of transmitters N_t for different number of receivers N_r with R_{net} = 150 km, R_j = 250 km, and $P_r = P_j$=12.5 kW.

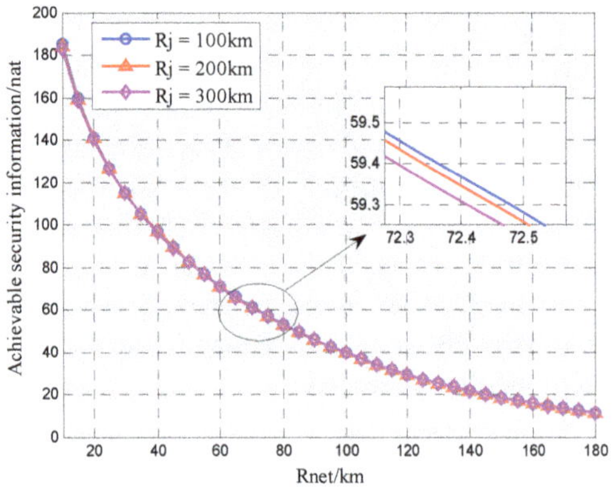

Fig. 6. Achievable security information versus R_{net} with P_r =20 kW, P_j =5 kW and different R_j .

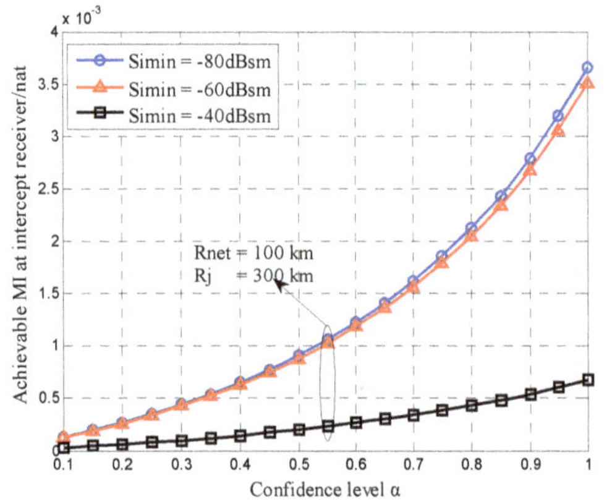

Fig. 7. Achievable MI at intercept receiver versus the confidence level α at radar network for different sensitivities of intercept receiver $S_{i\min}$ with R_{net} = 100 km and R_j = 300 km.

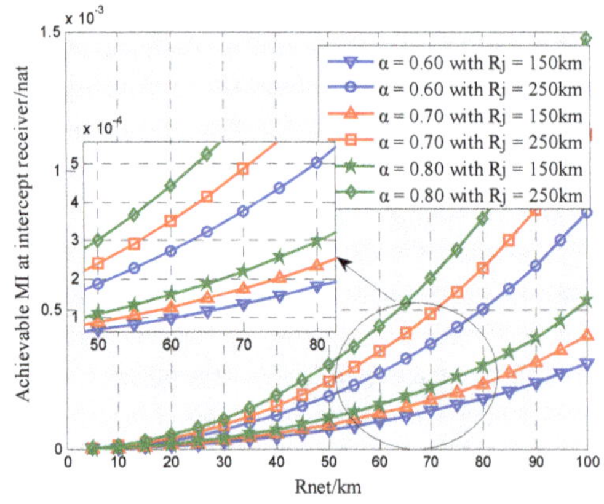

Fig. 8. Achievable MI at intercept receiver versus R_{net} for different confidence levels α and different R_j.

increased as the confidence level α at radar network increases, which shows that there exists a restrictive relationship between the confidence level α and the achievable MI at intercept receiver. This is due to the fact that more radar modulating signal would be transmitted to satisfy the system requirement with the increase of confidence level at radar network. The radar modulating signal could be intercepted by interceptor easily, and the achievable MI at intercept receiver will increase subsequently. Moreover, the achievable MI at intercept receiver is reduced as the sensitivity of intercept receiver $S_{i\min}$ decreases.

In Fig. 8, we depict the achievable MI at intercept receiver versus the distance between radar network and target R_{net} for different confidence levels at radar network α and different distances between radar network and intercept receiver R_j. One can observe from Fig. that for all the cases, the achievable MI at intercept receiver is increased as the distance between radar network and target increases from R_{net} = 5 km to R_{net} = 100 km. As mentioned before, this is because that more radar modulating signal is transmitted to satisfy the requirement for target detection as R_{net} increases, so less power is remained to generate CJ signal to jam the intercept receiver. Moreover, one can observe that as the distance between radar network and intercept receiver R_j decreases, the MI at interceptor is reduced, showing the LPID performance enhancement of exploiting CJ to defend against interceptors in radar network system.

4.2 Target Tracking with FCCP Based Security Information Optimization

This subsection presents the numerical results of our proposed security information optimization scheme in target tracking scenario. We track a single target by employing particle filtering (PF) method. For simplicity, it is assumed that the passive intercept receiver is carried by the target. Figure 9 shows one realization of the target trajectory for 50 s. Figure 10 illustrates the distance changing curve between radar network and target.

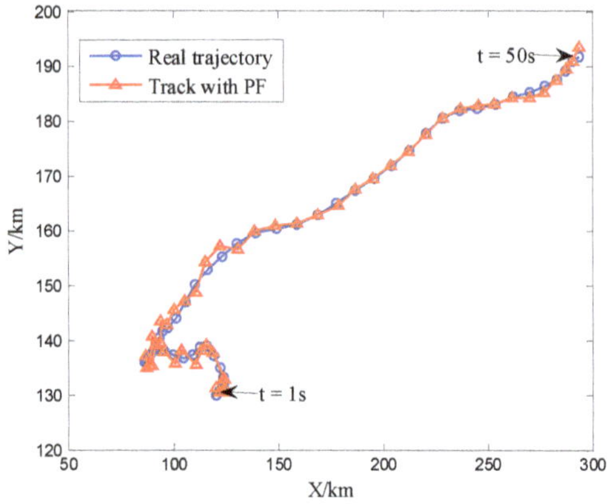

Fig. 9. Target tracking scenario.

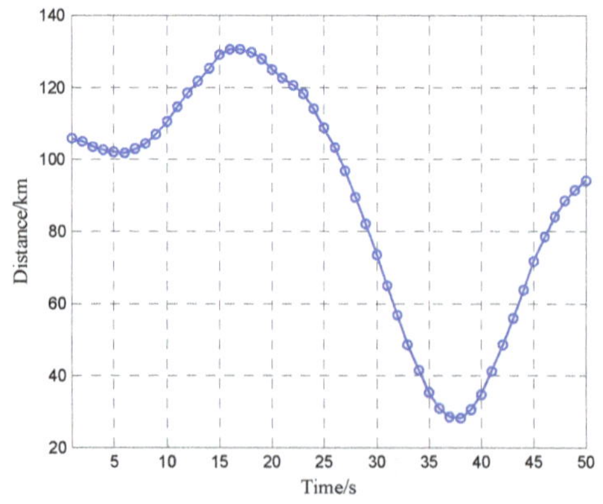

Fig. 10. The distance between the radar network and the target.

Figure 11 illustrates the achievable MI at intercept receiver with different confidence levels. We can see from Fig. 11 that the achievable MI at intercept receiver is increased as the confidence level increases. This is due to the fact that more radar modulating signal will be allocated to satisfy the system requirement with the increase of confidence level at radar network, in which way the radar signal would be intercepted and identified by hostile interceptor easily. It is also worth pointing out that the achievable MI at interceptor changes accordingly with the distance between radar network and target. To be specific, when the target is far away from the radar network, the network system would allocate more power for modulating signal to obtain a better capability of radar network to estimate the target. In contrast, as the distance decreases, more power for CJ signal could be transmitted to defend against the passive intercept receiver, which makes the achievable security information robust in terms of LPID performance.

In Fig. 12, we compare the performance of the proposed FCCP based security information optimization algorithm, the energy-efficiency based algorithm [17] and the adaptive security information optimization algorithm [18]. The algorithm provided in [17] aims at optimizing the energy-efficiency and will terminate to maintain certain security information, while the algorithm proposed in [18] aims at the optimization of the overall security information by optimizing the transmission power allocation between radar modulating signal and CJ signal. As Fig. 12 shows, the proposed algorithm significantly outperforms the algorithms proposed in [17] and [18], which is due to the fact that the achievable MI at interceptor is remarkable lower than that of the compared algorithms across the whole region. In addition, the proposed algorithm is more practical than the algorithm proposed in [18] because of the former's lower complexity.

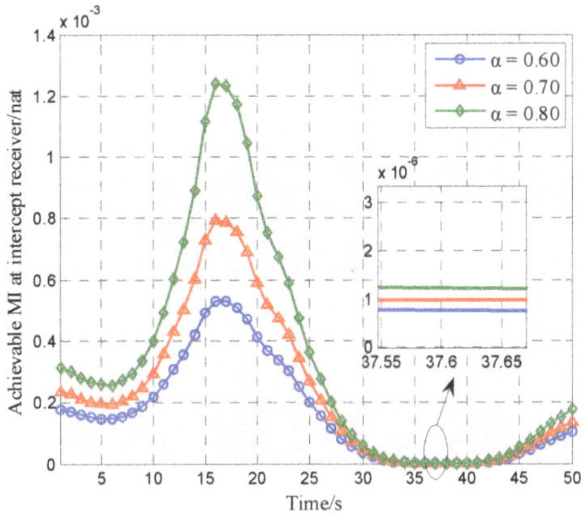

Fig. 11. Achievable MI at intercept receiver in the tracking process.

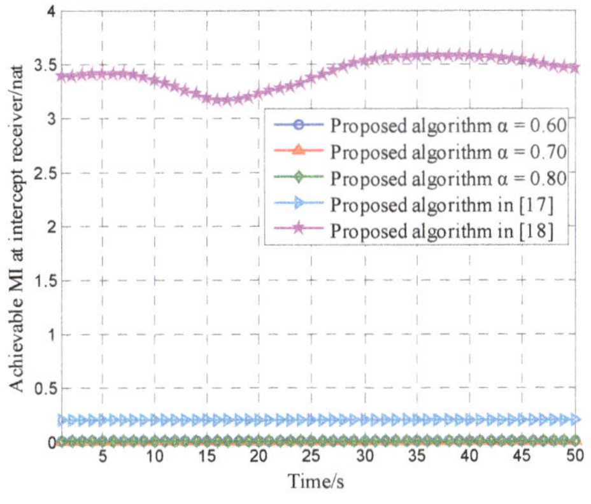

Fig. 12. Achievable MI at intercept receiver of various algorithms.

4. Conclusions

This paper has proposed a novel FCCP based security information optimization algorithm to achieve improved LPID performance in radar network systems without any prior knowledge of noncooperative intercept receiver, whose purpose is to minimize the achievable MI at interceptor, while the achievable MI outage probability at radar network is enforced to be greater than a specified confidence level. It is worth pointing out that our proposed algorithm is presented by simple analytical closed-form expression. Simulation results demonstrate that our proposed algorithm is effective to enhance LPID performance for radar network to defend against passive interceptor attacks. For future research, other optimization criteria need to be addressed to improve LPID performance for radar network systems.

Acknowledgements

The authors would like to thank the anonymous reviewers for their comments that help to improve the quality of this article. The support provided by the National Natural Science Foundation of China (Grant No. 61371170), the Fundamental Research Funds for the Central Universities (Grant No. NJ20140010), Funding of Jiangsu Innovation Program for Graduate Education (CXLX13_154), the Fundamental Research Funds for the Central Universities, the Priority Academic Program Development of Jiangsu Higher Education Institutions (PADA) and Key Laboratory of Radar Imaging and Microwave Photonics (Nanjing Univ. Aeronaut. Astronaut.), Ministry of Education, Nanjing University of Aeronautics and Astronautics, Nanjing, 210016, China are gratefully acknowledged.

References

[5] PACE, P. E. *Detecting and Classifying Low Probability of Intercept Radar*. Boston: Artech House, 2009, p. 342–352.

[6] FISHER, E., HAIMOVICH, A., BLUM, R. S., CIMINI, L. J., CHIZHIK, D., VALENZUELA, R. A. Spatial diversity in radars: models and detection performance. *IEEE Transactions on Signal Processing*, 2006, vol. 54, no. 3, p. 823–838. DOI: 10.1109/TSP.2005.862813

[3] HAIMOVICH, A. M., BLUM, R. S., CIMINI, L. J. JR. MIMO radar with widely separated antennas. *IEEE Signal Processing Magazine*, 2008, vol. 25, no. 1, p. 116–129. DOI: 10.1109/MSP.2008.4408448

[4] TANG, B., TANG, J., PENG, Y. N. MIMO radar waveform design in colored noise based on information theory. *IEEE Transactions on Signal Processing*, 2010, vol. 58, no. 9, p. 4684–4697. DOI: 10.1109/TSP.2010.2050885

[5] YANG, Y., BLUM, R. S. MIMO radar waveform design based on mutual information and minimum mean-square error estimation. *IEEE Transactions on Aerospace and Electronic System*, 2007, vol. 43, no. 1, p. 330–343. DOI: 10.1109/TAES.2007.357137

[6] CHEN, Y. F., NIJSURE, Y., YUEN, C., CHEW, Y. H., DING, Z. G. Adaptive distributed MIMO radar waveform optimization based on mutual information. *IEEE Transactions on Aerospace and Electronic System*, 2013, vol. 49, no. 2, p. 1374–1385. DOI: 10.1109/TAES.2013.6494422

[7] SONG, X. F., WILLETT, P., ZHOU, S. L. Optimal power allocation for MIMO radars with heterogeneous propagation losses. In *IEEE International Conference on Acoustics, Speech and Signal Processing ICASSP 2012.* Kyoto (Japan), 2012, p. 2465–2468. DOI: 10.1109/ICASSP.2012.6288415

[8] SHI, C. G., ZHOU, J. J., WANG, F. Low probability of intercept optimization for radar network based on mutual information. In *2014 2nd IEEE China Summit & International Conference on Signal and Information Processing (ChinaSIP).* Xi'an (China), 2014, p. 683–687. DOI: 10.1109/ChinaSIP.2014.6889331

[9] SHI, C. G., WANG, F., SELLATHURAI, M., ZHOU, J. J. LPI optimization framework for target tracking in radar network architectures using information-theoretic criteria. *International Journal of Antennas and Propagation*, 2014, 10 p. DOI: 10.1155/2014/654561

[10] WYNER, A. X. The wiretap channel. *The Bell System Technical Journal*, 1975, vol. 54, no. 8, p. 1355–1387. DOI: 10.1002/j.1538-7305.1975.tb02040.x

[11] SWINDLEHURST, A. L. Fixed SINR solutions for the MIMO wiretap channel. In *IEEE International Conference on Acoustics, Speech and Signal Processing ICASSP 2009.* Taipei, 2009, p. 2437–2440. DOI: 10.1109/ICASSP.2009.4960114

[12] ZHOU, X. Y., MCKAY, M. R. Secure transmission with artificial noise over fading channels: achievable rate and optimal power allocation. *IEEE Transactions on Vehicular Technology*, 2010, vol. 59, no. 8, p. 3831–3842. DOI: 10.1109/TVT.2010.2059057

[13] ROMERO-ZURITA, N., MCLERNON, D., GHOGHO, M., SWAMI, A. PHY layer security based on protected zone and artificial noise. *IEEE Signal Processing Letters*, 2013, vol. 20, no. 5, p. 487–490. DOI: 10.1109/LSP.2013.2252898

[14] MUKHERJEE, A., SWINDLEHURST, A. L. Jamming games in the MIMO wiretap channel with an active eavesdropper. *IEEE Transactions on Signal Processing*, 2013, vol. 61, no. 1, p. 82–91. DOI: 10.1109/TSP.2012.2222386

[15] ZOU, Y. L., WANG, X. B., SHEN, W. M. Physical-layer security with multiuser scheduling in cognitive radio networks. *IEEE Transactions on Communications*, 2013, vol. 61, no. 12, p. 5103 to 5113.

[16] WANG, F., SELLATHURAI, M., LIU, W. G., ZHOU, J. J. Security information factor based airborne radar RF stealth. *Journal of Systems Engineering and Electronics*, 2014. (Unpublished)

[17] SHI, C. G., ZHOU, J. J., WANG, F., CHEN, J. Optimal power allocation for low probability of identification in radar network based on security information with cooperative jamming. *ICIC Express Letters*, 2014, vol. 8, no. 12, p. 3401–3406.

[18] SHI, C. G., ZHOU, J. J., WANG, F., CHEN, J. LPID optimization with security information in radar network. *Industrial Electronics and Engineering*, 2014, vol. 93, p. 255–263. DOI: 10.2495/ICIEE140291

[19] ZADEH, L. A. Fuzzy sets. *Information and Control*, 1965, vol. 8, no. 3, p. 338–353. DOI: 10.1016/S0019-9958(65)90241-X

[20] LIU, B. D., LIU, Y. K. Expected value of fuzzy variable and fuzzy expected value models. *IEEE Transactions on Fuzzy Systems*, 2002, vol. 10, no. 4, p. 445–450. DOI: 10.1109/TFUZZ.2002.800692

[21] LIU, B. D. *Uncertainty Theory: An Introduction to Its Axiomatic Foundations.* Berlin: Springer-Verlag, 2004, p. 109–128.

[22] PAGNONCELLI, B. K., AHMED, S., SHAPIRO, A. Sample average approximation method for chance constrained programming: theory and application. *Journal of Optimization Theory and Application*, 2009, vol. 142, no. 2, p. 399–416. DOI: 10.1007/s10957-009-9523-6

[23] LIU, B. D., ZHAO, R. Q., WANG, G. *Uncertainty programming with Application.* Beijing: Tsinghua University Press, 2003, p. 138-187. (In Chinese)

[24] LIU, B. D. *Theory and Practice of Uncertainty Programming.* Heidelberg: Physical-Verlag, 2002, p. 53–74.

About the Authors ...

Chenguang SHI was born in 1989. He received B.S. from Nanjing University of Aeronautics and Astronautics (NUAA) in 2012, and he is currently working toward his Ph.D. degree in NUAA. His main research interest is aircraft radio frequency stealth, radar target tracking.

Jianjiang ZHOU (corresponding author) was born in 1962. He received M.S. and Ph.D. from Nanjing University of Aeronautics and Astronautics (NUAA) in 1988 and 2001 respectively, and then became a professor there. His main research interest is aircraft radio frequency stealth, radar signal processing.

22

Pipelined Two-Operand Modular Adders

Maciej CZYŻAK, Jacek HORISZNY, Robert SMYK

Faculty of Electrical and Control Engineering, Gdansk University of Technology, G Narutowicza 11/12, 80-233

mczyzak@ely.pg.gda.pl, jhor@ely.pg.gda.pl, rsmyk@ely.pg.gda.pl

Abstract. *Pipelined two-operand modular adder (TOMA) is one of basic components used in digital signal processing (DSP) systems that use the residue number system (RNS). Such modular adders are used in binary/residue and residue/binary converters, residue multipliers and scalers as well as within residue processing channels. The structure of pipelined TOMAs is usually obtained by inserting an appropriate number of pipeline register layers within a nonpipelined TOMA structure. Hence the area of pipelined TOMAs is determined by the nonpipelined TOMA structure and by the total number of pipeline registers. In this paper we propose a new pipelined TOMA, that has a considerably smaller area and the attainable pipelining frequency comparable with other known pipelined TOMA structures. We perform comparisons of the area and pipelining frequency with TOMAs based on ripple carry adder (RCA), Hiasat TOMA and parallel-prefix adder (PPA) using the data from the very large scale of integration (VLSI) standard cell library.*

Keywords

Carry-lookahead adder, FPGA, modular adder, parallel-prefix adder, residue number system (RNS), ripple-carry adder, VLSI design

1. Introduction

Modular addition plays an important role in the implementation of digital signal processing systems that use the residue number system [1–4] as well as its derivatives like the quadratic residue number system (QRNS) [5] and modified quadratic residue number system (MQRNS) [6] for processing of complex signals. The RNS is a non-weighted integer number system that is determined by its base $\boldsymbol{B}=\{m_1, m_2, ..., m_n\}$ being the set of positive pairwise prime integers m_i , i = 1, 2,.., n. Each integer $X \in \boldsymbol{Z}_M$, $M = \prod_{i=1}^{n} m_i$ and can be represented as $X \leftrightarrow (x_1,x_2,...,x_n) = (|X|_{m_1}, |X|_{m_2}, ..., |X|_{m_n})$ with $x_i \in \boldsymbol{Z}_{m_i}$. This mapping is the bijection and for $X, Y \in \boldsymbol{Z}_M$ and for $x_i, y_i \in \boldsymbol{Z}_{m_i}$, we have $z_i = |x_i \otimes y_i|_{m_i}$,where $\otimes$ denotes addition, subtraction or multiplication.

The reverse conversion from the RNS to a weighted system can be performed using the Chinese remainder theorem (CRT) [1], [2] or the mixed-radix system (MRS) [1], [2]. The main advantage of the RNS comes from the fact that addition, subtraction and multiplication are carry-free and can be performed without carries between individual positions of the number. The principal advantage of the RNS with respect to the high-speed DSP is due to the replacement of large multipliers that limit the pipelining frequency, by small multipliers modulo m_i. If their binary size $l = \lceil(\log_2 m_i)\rceil$, where $\lceil\bullet\rceil$ denotes rounding off to an integer, does not exceed six bits, multiplications by a constant can be performed by look-up with small ROMs or using combinatorial networks. General multiplications are also easier to perform because their standard realizations are small or segmentation of operands can be used for the combinatorial realization. It is worth mentioning that moduli with $l < 7$ may provide for the dynamic ranges over 90 bits [7]. The additional advantage of the RNS is the possibility of reducing power dissipation in CMOS circuits which is due to the lower switching activity and reduction of supply voltages [9]. The RNS has found numerous applications in the DSP, for example, in FIR filters [8–11], FFT processors [12], digital downconversion [13] and image processing [14], [15].

Generally TOMAs can be divided into two main categories determined by the type of the modulus. TOMAs for moduli akin to 2^n represent the first category and those for generic moduli the other. There are several works in the literature that consider the TOMA design.

Banerji [16] presented a look-up approach, Agrawal and Rao [17] proposed a TOMA for moduli of the form $(2^n + 1)$ based on binary adders. Soderstrand [18] introduced a hybrid approach based on look-up table along with the binary adder. Bayoumi and Jullien [19] described TOMAs using the table approach and binary adders approach. Dugdale [20] demonstrated an implementation of TOMAs that used binary adders, Piestrak [21] proposed a TOMA based on the carry-save adder (CSA) and two binary adders. Zimmermann [22] introduced modulo $(2^n \pm 1)$ adders based on parallel prefix-architecture (PPA). Hiasat [23] proposed a TOMA with the reduced area based on the carry-look-ahead (CLA) adder. Also a novel delay-power-area-efficient approach to the TOMA design was given by Patel et al. [24]. Their TOMA structure was based on the cascaded connection of the modified carry-save adder

(CSA) and reduced carry-propagate adder (CPA). The used CPA designs included ELM [25], Kogge-Stone [26] and Ladner Fischer [27] PPA.

In this paper we propose a new TOMA based on a modified CLA adder. This TOMA has the smaller area than other considered TOMAs and allows to derive a new pipelined TOMA that is better than other known pipelined TOMAs in terms of the area and the number of stages of pipeline registers. We shall show the structure of the new pipelined TOMA and, for comparison, TOMAs based on the RCA, PPA in the Brent-Kung form [28] and Hiasat TOMA [23]. Comparisons are made using the data from the VLSI standard cell library. We shall compare structures of individual TOMAs in terms of area, delay and pipelining frequency with the use of the additive method. The method uses summation of areas of individual components expressed in gate equivalents (GE), where 1 GE is the area of the NAND with the fan-out = 1 for the given standard cell library. The propagation delay of an individual element is taken as the worst case delay for all possible inputs. The analysis relies upon the established 130 nm Samsung standard cell library STDH150 [29]. Calculations of areas and delays of individual components are practically technology independent and they can be scaled down for VLSI technologies such as 28 nm or 22 nm. Therefore we may therefore suppose that for comparison of individual digital structures, the assumed technology will give sufficient and dependable information. The paper has the following structure: in Sec. 2 we review the basic TOMA structures, in Sec. 3 we consider the TOMA-RCA, and in Sec. 4 Hiasat TOMA, in Sec. 5 we present the TOMA based on the PPA adder and finally in Sec. 6 a new TOMA. In each section we analyze a nonpipelined and pipelined form.

2. Basic TOMA Structures Based on Binary Adders

In this section we shall shortly describe the basic known TOMA structures that use exclusively binary adders in series and which therefore may be the most suitable for transformation to the pipelined form and not those that use two parallel adders as in [21]. Two-operand modular addition for small m, $\lceil \log m \rceil \leq 6$ can be implemented by using the ROM ($2^{2\cdot\lceil \log m \rceil} \times \lceil \log m \rceil$), but such approach remarkably reduces the attainable pipelining frequency.

The TOMA computes $r_m = |X+Y|_m$, where r_m is the least nonnegative remainder from the division $X+Y$ by the modulus m. Assuming $Z = 2^{\lceil \log m \rceil} - m$, the computation can be also expressed as

$$r = \begin{cases} |X+Y+Z|_{2^{\lceil \log m \rceil}} & \text{if } X+Y+Z \geq 2^{\lceil \log m \rceil} \\ X+Y & \text{otherwise} \end{cases} \quad (1a)$$

In Fig. 1 to 3 three basic TOMA structures are shown Bayoumi-Jullien, Hiasat, and Piestrak.

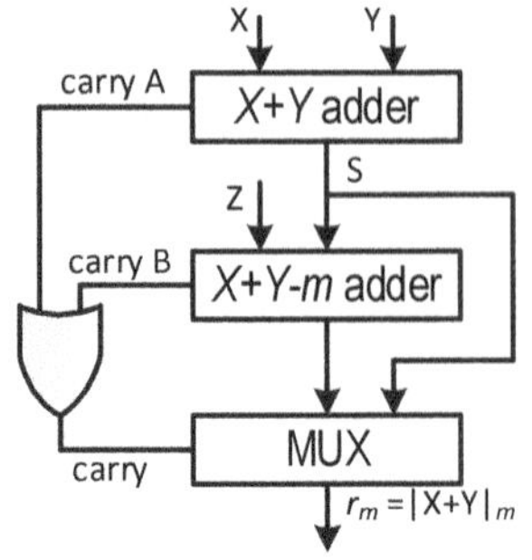

Fig. 1. Bayoumi-Jullien TOMA [19].

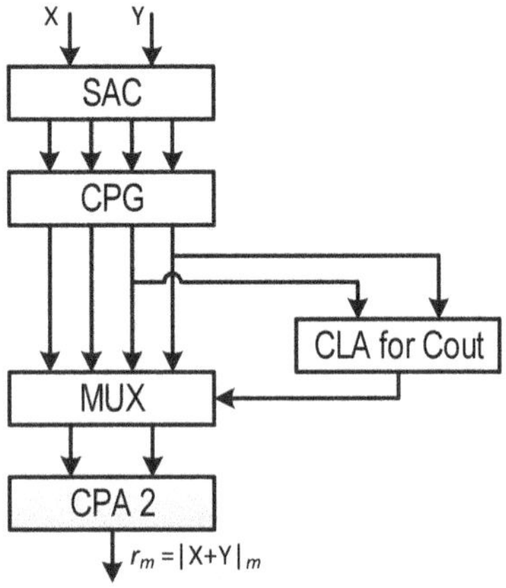

Fig. 2. Hiasat TOMA [23].

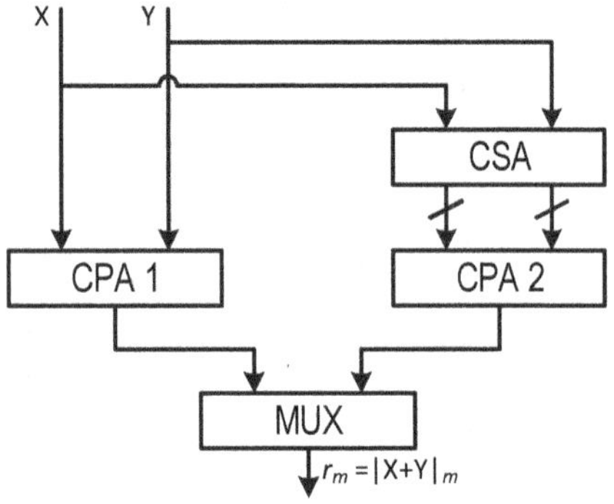

Fig. 3. Piestrak TOMA [21].

We shall shortly analyze the operation of the Bayoumi- Jullien TOMA (Fig. 1) because this structure will be the basis for the design of selected TOMAs. The binary adder in the first stage of this TOMA computes $X+Y$, whereas the second adder $X+Y-m$. The output of the TOMA is selected using $carry = carryA \vee carryB$. For $X+Y<m$, $carry = 0$ and $r_m = X+Y$, whereas for $X+Y \geq m$, $carry = 1$ and $r_m = X+Y-m$.

3. TOMA-RCA

By way of introduction we shall consider the realization of the Bayoumi-Jullien TOMA based on the RCA. In order to obtain a pipelined structure, layers of pipeline registers consisting of flip-flops (FFs) have to be inserted between individual adders as shown in Fig. 5.

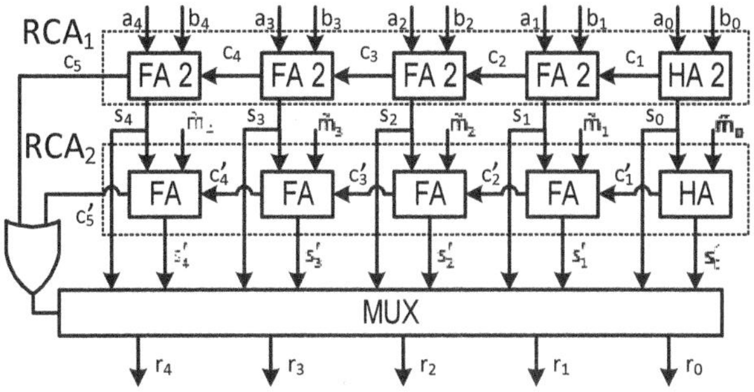

Fig. 4. Bayoumi-Jullien TOMA based on the RCA.

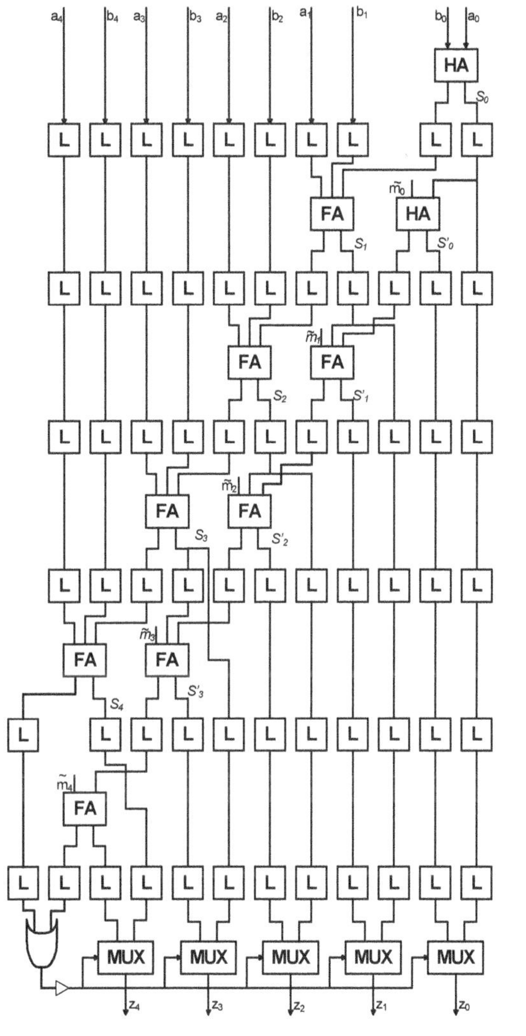

Fig. 5. Pipelined TOMA based on Bayoumi-Jullien TOMA and the RCAs.

In the following we shall analyze the area of the TOMA-RCA expressed in GE, the delay and the maximum attainable pipelining frequency. The area will be estimated using the areas of the individual components from STDH150, the delay for a nonpipelined structure will be evaluated by using the maximum delays for the individual components. In order to estimate the pipelining frequency a structure is divided into balanced layers with respect to the delay and the maximum pipelining frequency is obtained as the inverse of the sum of the delay of the slowest layer and the FF delay.

A. Nonpipelined 5-bit TOMA-RCA area

This area of 5-bit TOMA-RCA can be expressed in the following manner:

$$A_{TOMA\_RCA} = A_{HAd2} + 4\cdot A_{FAd2} + A_{HAd1} + 4\cdot A_{FAd1} + A_{OR2d1} + A_{NID6} + 5\cdot A_{MX2d1}. \quad (1)$$

The indices of the individual components come from STDH150. The data of individual components is given in Appendix A. After inserting these data into (1) we obtain $A_{TOMA\_RCA} = 98.68\text{GE}$. The area given by (1) does not depend upon the form the of the two's complement system (TCS) representation of $-m$, $\tilde{m} = (1,\ldots,\tilde{m}_4,\tilde{m}_3,\tilde{m}_2,\tilde{m}_1,\tilde{m}_0)$. The particular form of this representation allows to reduce the area for the given modulus. For example, if $m_i = 0$, the HA reduces to single connection and for $m_i = 1$ to one connection and to one inverter. For the FA and $m_i = 0$, we have one XOR gate and a single AND gate, and for $m_i = 1$ one OR gate and exclusive NOR. For $m = 29$ and $\tilde{m} = (1,\ldots,0,0,0,1,1)$, we obtain $A_{TOMA\text{-}RCA} = 81.68GE$.

B. Nonpipelined 5-bit TOMA-RCA delay

We shall estimate the delay of the structure of Fig. 4 taking into consideration individual delays of signals inside individual HAs and FAs.

The delay of the 5-bit TOMA-RCA can be expressed as

$$t_{TOMA-RCA} = \max(t_{s_4}, t_{s'_4}, \max(t_{c_5}, t_{c'_5})) + t_{OR2d1} + t_{MX2d1}. \quad (2)$$

The delay for s_4 and c_5 bits can be calculated as

$$t_{s_4} = \max(t_{HAd2\_ACO}, t_{HAd2\_BCO}) + 3\cdot t_{FAd2\_CICO} + t_{FAd2\_CIS},$$

$$t_{c_5} = \max(t_{HAd2\_ACO}, t_{HA\_BCO}) + 4\cdot t_{FAd2\_CICO}.$$

In order to compute c'_5, we shall first calculate t_{c_i} and $t_{c'_i}$, $i = 1, 2, 3, 4$. We have

$$t_{c_1} = \max(t_{HAd2\_ACO}, t_{HAd2\_BCO}), \quad (3a)$$

$$t_{c_i} = t_{c_{i-1}} + t_{FAd2\_CICO}, \; i = 2,3,4,5. \quad (3b)$$

Consequently

$$t_{c'_1} = \max(t_{HAd2\_AS}, t_{HAd2\_BS}) + t_{HAd1\_BCO}, \quad (4a)$$

$$t_{c'_i} = \max(t_{c_{i-1}} + t_{FAd2\_CIS} + t_{FAd2\_BCO}, t_{c'_{i-1}} + t_{FAd1\_CICO}),$$
$$i = 2,3,4,5, \quad (4b)$$

and $t_{s'_4} = t_{c'_4} + t_{FAd1\_CIS}$. (4c)

Example 1. Computation of 5-bit TOMA-RCA delay for components from the STDH150.

We shall first compute t_{s_4} and t_{c_5} as

$$t_{s_4} = \max(0.092\,\text{ns}, 0.074\,\text{ns}) + 3 \cdot 0.089\ \text{ns} + 0.150\,\text{ns} = = 0.509\,\text{ns},$$

$$t_{c_5} = \max(0.092\,\text{ns}, 0.074\,\text{ns}) + 4 \cdot 0.089\,\text{ns} = 0.448\,\text{ns}.$$

Before we can compute c_5', we have to determine c_i, $i = 1, 2, 3, 4$. We have

$$t_{c_1} = \max(0.092\ \text{ns}, 0.074\ \text{ns}) = 0.092\ \text{ns},$$

$$t_{c_2} = 0.181\ \text{ns},$$

$$t_{c_3} = 0.270\ \text{ns},$$

$$t_{c_4} = 0.359\ \text{ns}.$$

Subsequently we obtain

$$t_{c_1'} = \max(0.092\,\text{ns}, 0.074\,\text{ns}) + 0.055\,\text{ns} = 0.147\,\text{ns},$$

$$t_{c_2'} = \max(0.092\,\text{ns} + 0.102\,\text{ns} + 0.152\,\text{ns}, 0.147\,\text{ns} + 0.083\,\text{ns}) = \max(0.346\,\text{ns}, 0.230\,\text{ns}) = 0.346\,\text{ns},$$

$$t_{c_3'} = \max(0.181\,\text{ns} + 0.102\,\text{ns} + 0.143\,\text{ns}, 0.346\,\text{ns} + 0.083\,\text{ns}) = \max(0.426\,\text{ns}, 0.429\,\text{ns}) = 0.429\,\text{ns},$$

$$t_{c_4'} = \max(0.270\,\text{ns} + 0.102\,\text{ns} + 0.143\,\text{ns}, 0.429\,\text{ns} + 0.083\,\text{ns}) = \max(0.515\,\text{ns}, 0.512\,\text{ns}) = 0.429\,\text{ns},$$

$$t_{c_5'} = \max(0.359\,\text{ns} + 0.102\,\text{ns} + 0.143\,\text{ns}, 0.515\,\text{ns} + 0.083\,\text{ns}) = \max(0.604\,\text{ns}, 0.598\,\text{ns}) = 0.604\,\text{ns},$$

$$t_{s_4'} = 0.515\,\text{ns} + 0.095\,\text{ns} = 0.601\,\text{ns}.$$

Finally, we may determine the TOMA-RCA delay as

$$t_D^{TOMA-RCA} = \max(0.509\ \text{ns}, 0.601\ \text{ns}, 0.604\ \text{ns} + 0.065\ \text{ns}) + 0.078\ \text{ns} = 0.747\ \text{ns}.$$

C. The area of pipelined 5-bit TOMA-RCA

In Fig. 5 a pipelined form of the RCA-TOMA is presented. Six flip-flops stages are used with 66 flip-flops. The area is the sum of the nonpipelined 5-bit TOMA-RCA area and the area of pipeline registers. In this case these registers require $n_s = 66$ FFs. Thus the area can be expressed as

$$A_{TOMA\_RCA\_p} = A_{TOMA\_RCA} + n_s \cdot A_{FF}\,. \quad (5)$$

As A_{FF} we shall use the area of the flip-flop FD1Q, A_{FD1Q} from STDH150. For the structure from Fig. 5 we receive $A_{TOMA\text{-}RCAp} = 472.9$ GE.

D. Pipelined 5-bit RCA-TOMA pipelining rate

In order to design a pipelined structure of a TOMA, we have to decompose its nonpipelined structure into a certain number of layers and place pipeline registers between them. The decomposition is, to certain extent, arbitrary. The lower limit of the number of layers is two and the upper limit is determined by a delay of the component that we treat as indivisible. The minimum pipelining rate is approximately the sum of the delay of the layer with the maximum delay and the delay of the pipeline register. In this case we have assumed that after each FA or HA a register layer is placed and the OR gate and the MUXs are in the same layer. Hence we may evaluate the maximum delay of the layer as

$$t_{LD}^{TOMA\_RCA} = \max(t_{FAd1}, t_{OR2d1} + t_{MX2d1}) + t_{FD1Q} \quad (6)$$

where t_{FD1Q} is the maximum delay of the flip-flop.

Using the data from the STDH150, we may evaluate a theoretical maximum pipelining frequency as

$$f_{PF\_\max}^{TOMA\_RCA} = 1/(\max(0.143\,\text{ns}, 0.065\,\text{ns} + 0.078\,\text{ns}) + 0.094\,\text{ns}) = 1/0.237\,\text{ns} = 4.22\ \text{GHz}.$$

4. Hiasat TOMA

In the following we shall examine the results of transforming the Hiasat TOMA which requires the smallest hardware amount among known TOMAs. This TOMA consists of the serial connection of five units: the sum-and-carry (SAC), the carry propagate and generate (CPG), CLA for c_{OUT}, multiplexer (MUX), CLA and Summation (CLAS). The SAC is composed of HAs and HALs (the modified HAs in [23]). The SAC performs

$$s_i = x_i \oplus y_i \oplus z_i, \quad (7)$$

$$c_{i+1} = x_i \cdot y_i + x_i \cdot z_i + y_i \cdot z_i, \quad (8)$$

for the individual bits of $X + Y$, and $X + Y - m$, with the assumption that TCS representation of $-m$ without the sign bit is $(z_{n-1}, \ldots, z_0)$ with $n = 5$. Regarding that $z_i = 0$ or $z_i = 1$, the HAL is obtained that implements

$$A_i = x_i \oplus y_i, \quad (9a)$$

$$\hat{A}_i = \overline{x_i \oplus y_i}, \quad (9b)$$

$$B_{i+1} = x_i \cdot y_i, \quad (9c)$$

$$\hat{B}_{i+1} = x_i + y_i\,. \quad (9d)$$

As $(z_{n-1}, \ldots, z_0)$ may have w bits for which $z_i = 0$ and $n - w$ bits for which $z_i = 1$. Hence the SAC has w HAs and $n - w$ HAL cells. The CFG computes the carry generate and carry propagate vectors as in the standard CLA $P_i = A_i \oplus B_i$, $G_i = A_i\ \ B_i$ and $p_i = \hat{A}_i \oplus \hat{B}_i$, $g_i = \hat{A}_i \cdot \hat{B}_i$. This unit has at most $2k - 2$ HAs. In the CLAS p_i and g_i are used to compute c_{OUT}, that controls the selection of $X + Y$ or $X + Y - m$. Regarding that $c_0 = 0$, $g_0 = 0$, c_{OUT} can be computed for the five-bit Hiasat adder as

$$c_{OUT} = B_5 + g_4 + g_3 \cdot p_4 + g_2 \cdot p_3 \cdot p_4 + g_1 \cdot p_2 \cdot P_3 \cdot P_4 . \quad (10)$$

The following stage, MUX selects using c_{OUT} the carry's and generate's $p_i^{'} = p_i$ or $p_i^{'} = P_i$ and $g_i^{'} = g_i$ or $g_i^{'} = G_i$, $i = 0, 1, ..., 4$.

The final stage, the five-bit CLA adder computes the carries

$$c_1 = g_0^{'}, \quad (11)$$

$$c_2 = g_1^{'} + g_0^{'} \cdot p_1^{'}, \quad (12)$$

$$c_3 = g_2^{'} + g_1^{'} \cdot p_2^{'} + g_0^{'} \cdot p_1^{'} \cdot p_2^{'}, \quad (13)$$

$$c_4 = g_3^{'} + g_2^{'} \cdot p_3^{'} + g_1^{'} \cdot p_2^{'} \cdot p_3^{'} + g_0^{'} \cdot p_1^{'} \cdot p_2^{'} \cdot p_3^{'} . \quad (14)$$

In the next step the sum bits are calculated as

$$s_i^{'} = c_i \oplus p_i^{'}, \; i = 0, 1, 2, 3, 4. \quad (15)$$

First we shall determine the area for components of the Hiasat five-bit TOMA and then the area for $m = 29$.

A. 5-bit Hiasat TOMA area

The area of the five-bit Hiasat TOMA can be computed as follows

$$A_{TOMA\_Hiasat} = A_{SAC\_5} + A_{CFG\_5} + A_{CLA\_Cout\_5} + A_{MUX\_5} + A_{CLAS\_5} . \quad (16)$$

The areas of the individual blocks from (16) can be expressed as:

$$A_{SAC\_5} = 2 \cdot A_{HAd1} + A_{HAd2} + A_{HAL}, \quad (17)$$

with

$$A_{HAL} = A_{OR2d1} + A_{AND2d1} + A_{XOR2d1} + A_{IVd1} . \quad (18)$$

In general, the area of the CFG_5 can be expressed as

$$A_{CFG\_5} = 5 \cdot A_{HAd2} + A_{HAd1}, \quad (19)$$

$$A_{CLA\_out\_5} = A_{AND2d1} + A_{AND3d1} + A_{AND4d1} + A_{OR5d1} + A_{NID6}, \quad (20)$$

$$A_{MUX\_5} = A_{MX2d1} + A_{MX2d2} + 3 \cdot A_{MX4d1} . \quad (21)$$

The CLAS block consists of the five-bit Propagate-Generate Unit (PGU_5), Carry-Generate Unit (CGU_5) and Summation Unit (SU_5). Its hardware amount can be estimated as

$$A_{CLAS\_5} = A_{CGU\_5} + A_{SU\_5}, \quad (22)$$

with the fan-outs 1, 3, 3, 4, 2. We get

$$A_{CGU\_5} = A_{AND2d1} + A_{OR2d1} + A_{AND2d2} + A_{AND3d1} + A_{OR3d1}, \quad (23)$$

and

$$A_{SU\_5} = 5 A_{XOR2d1} = 15.0 \text{ GE} . \quad (24)$$

Example 2. Area of the five-bit Hiasat TOMA for $m = 29$.

The TCS representation of $(-m)$ is equal to 100011, hence $w = 3$, and $k - w = 2$ (the sign bit is excluded). Thus we obtain

$$A_{SAC\_5} = 2A_{HAd1} + A_{HAd2} + 2A_{HAL}$$
$$= 2 \; 4.67 \text{ GE} + 5.67 \text{ GE} + 7.34 \text{ GE}$$
$$= 22.350 \text{ GE},$$

$$A_{CFG\_5} = 5 \cdot A_{HAd2} + A_{HAd2} = 5 \cdot 5.67 \text{GE} + 4.67 \text{GE}$$
$$= 33.02 \text{GE},$$

$$A_{CLA\_Cout\_5} = 1.67 \text{ GE} + 2 \text{ GE} + 2.33 \text{ GE} + 3.33 \text{ GE} + 3.67 \text{ GE} = 13 \text{ GE},$$

$$A_{MUX\_5} = 3 \text{ GE} + 3.33 \text{ GE} + 3 \cdot 6.33 \text{ GE} = 25.32 \text{ GE},$$

$$A_{CGU\_5} = 1.67 \text{GE} + 1.67 \text{GE} + 2 \text{GE} + 2 \text{GE} + 2 \text{GE} + 1.67 \text{GE} + 2 \text{GE} + 2.33 \text{GE} + 3 \text{GE}$$
$$= 18.34 \text{GE},$$

$$A_{SU\_5} = 5 \cdot 3 \text{GE} = 15 \text{GE},$$

$$A_{CLAS\_5} = 18.34 \text{GE} + 15 \text{GE} = 33.34 \text{GE}.$$

In effect we obtain the area of the five-bit Hiasat TOMA as

$$A_{TOMA\_Hiasat\_5} = 22.35 \text{GE} + 33.02 \text{GE} + 13 \text{GE} + 25.32 \text{GE} + 33.34 \text{GE} = 127.03 \text{GE}.$$

B. 5-bit Hiasat TOMA delay

The Hiasat five-bit TOMA delay, t_{H} can be expressed as

$$t_D^{Hiasat-TOMA} = t_{HAL} + t_{HAd2} + t_{AND4d1} + t_{OR5d1} + t_{NID6} + t_{MX2d4} + t_{AND4d1} + t_{OR2d1} + t_{OR4d1} + t_{XORd1} =$$
$$0.119 \text{ns} + 0.092 \text{ns} + 0.082 \text{ns} + 0.094 \text{ns} + 0.054 \text{ns} + 0.092 \text{ns} + 0.082 \text{ns} + 0.090 \text{ns} + 0.076 \text{ns} + 0.090 \text{ns} = 0.871 \text{ns},$$

with

$$t_{HAL} = t_{XOR2d1} + t_{IVd1} = 0.090 \text{ ns} + 0.029 \text{ ns} = 0.119 \text{ ns} .$$

C. Pipelined 5-bit Hiasat TOMA area

The area of the Hiasat pipelined 5-bit TOMA can be expressed as

$$A_{TOMA\_Hiasat\_p} = A_{TOMA\_Hiasat} + n_h A_{FF} \quad (25)$$

where n_h is a number of flip-flops used in pipeline registers. For example, for the structure from Fig. 6 we obtain

$$A_{TOMA\_Hiasat\_p} = 127.03 \text{GE} + 64 \cdot 5.67 \text{GE}$$
$$= 489.91 \text{GE}.$$

D. Pipelining frequency of pipelined 5-bit Hiasat TOMA

In Fig. 6, a pipelined form of the Hiasat TOMA is presented. Five pipeline register stages are used with 58 flip-flops.

In this case we have adopted a decomposition into six layers that leads to a balanced structure. In order to evaluate the maximum pipelining frequency we shall calculate delays of the adopted individual layers. The maximum pipelining frequency will depend on the delay of the layer with the maximum delay and the delay of the assumed pipeline register. These layers have the following delays:

layer 1 $t_D^{L1,H}$: $t_{HAL} = 0.119\,\text{ns}$,

layer 2 $t_D^{L2,H}$: $t_{HAd1} = 0.088\,\text{ns}$,

layer 3 $t_D^{L3,H}$: $t_{AND4d1} + t_{OR5d1} = 0.176\,\text{ns}$,

layer 4 $t_D^{4,H}$: $t_{MX2d1} + t_{NID6} = 0.132\,\text{ns}$,

layer 5 $t_D^{L5,H}$: $t_{AND4d1} + t_{XOR2d1} = 0.172\,\text{ns}$,

layer 6 $t_D^{L6,H}$: $t_{XOR2d1} + t_{OR4d1} = 0.166\,\text{ns}$.

Using $t_D^{L3,H}$ as the maximum layer delay, we may evaluate the maximum pipelining frequency as

$$f_{PF\_\max}^{TOMA\_Hiasat} = 1/(0.176\,\text{ns} + 0.094\,\text{ns}) = 1/0.27\,\text{ns} = 3.7\,\text{GHz}.$$

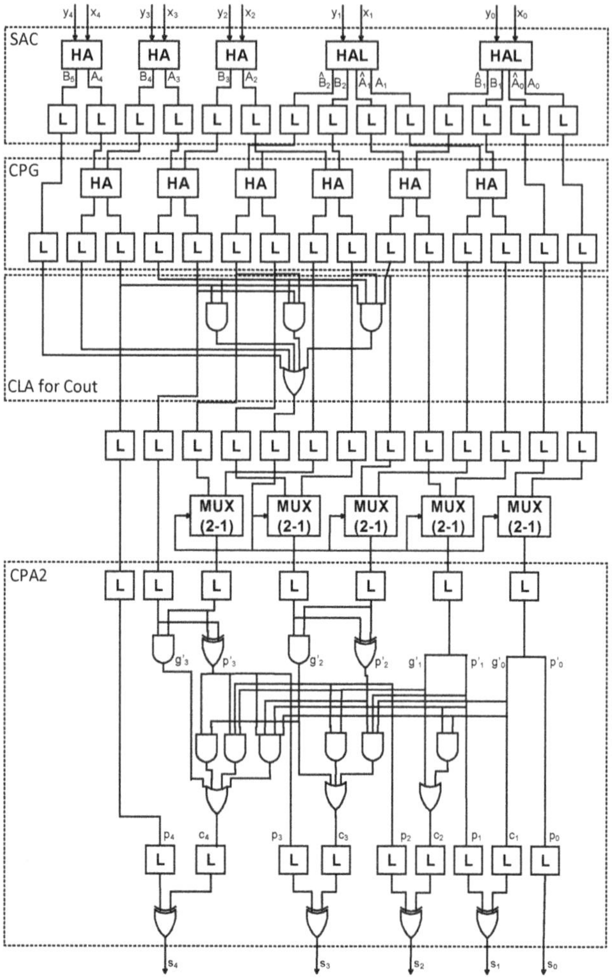

Fig. 6. Pipelined TOMA based on Hiasat TOMA.

5. PPA-based TOMA

As the next structure we shall consider the TOMA based on a PPA. As the PPA the Brent-Kung (BK) [28] adder has been selected. The Brent-Kung TOMA can be relatively easy transformed to the pipelined form, moreover the use of the Brent-Kung PPA allows one to simplify the adder used in the second stage when one of addends is a constant. The prefix operator ϕ is defined as

$$(g, p) = (g', p')\,\phi\,(g'', p''), \tag{26}$$

where

$$g = g'' + g' \cdot p'', \tag{27a}$$

$$p = p' \cdot p''. \tag{27b}$$

The block that implements (27a-b) will be denoted as BK_i. Subsequently we shall analyze the area and delay of the TOMA based on two BK adders.

The area of the TOMA BK A_{TOMA_BK} can be expressed as

$$A_{TOMA\_BK} = A_{BK} + A_{BK-m} \tag{28}$$

where A_{BK}, $A_{BK\text{-}m}$ represent the area of the *BK* adder and the modified *BK-m* adder that subtracts *m*, respectively.

A. The area of BK adder

A_{BK} can be calculated as

$$A_{BK} = 4 \cdot A_{HAd2} + A_{HAd1} + A_{BK0} + A_{BK1} + A_{BK2} + A_{BK3} + \quad + A_{BK4} + 4 \cdot A_{XOR2d1}. \tag{29}$$

The area of the first two terms is

$$4 \cdot A_{HAd2} + A_{HAd1} = 25.99\ \text{GE}. \tag{30}$$

After transforming the logic functions used for the realization of individual adders in (29), we receive the following areas

$$A_{BK_0} = A_{IVd1} + A_{NAND2d1} + A_{NAND2d2} + A_{AND2d2} = 6\ \text{GE}, \tag{31a}$$

$$A_{BK_1} = A_{IVd1} + A_{NAND2d1} + A_{NAND2d1} + A_{AND2d1} = = 4.67\,\text{GE}, \tag{31b}$$

$$A_{BK_2} = A_{IVd1} + A_{NAND2d1} + A_{NAND2d1} = 3\ \text{GE}, \tag{31c}$$

$$A_{BK_4} = A_{IVd1} + A_{NAND2d1} + A_{NAND2d2} + A_{AND2d2} = = 6\ \text{GE}, \tag{31d}$$

$$A_{BK_4} = A_{BK_2} = 3\ \text{GE}. \tag{31e}$$

Using (29), (30) and (31a-e) we obtain

$$A_{BK} = 62.02\ \text{GE}. \tag{31f}$$

B. The delay of BK adder

The BK adder delay can be expressed as

$$t_{BK} = t_{HA2} + \max(t_{BK_0}, t_{BK_1}) + \max(t_{BK_2}, t_{BK_3}) + t_{BK_4} + t_{XOR2d1}, \quad (32)$$

where

$t_{BK_0} = 2 \cdot t_{NAND2d1}$, $t_{BK_1} = t_{NAND2d1} + t_{NAND2d2}$, $t_{BK_2} = t_{BK_1}$, $t_{BK_3} = t_{BK_1}$, $t_{BK_4} = t_{BK_0}$.

Using the data from the STDH150, we have t_{BK_0} = 0.074 ns and t_{BK_1} = 0.068 ns.

Finally we obtain

$$t_{BK} = 0.092 \text{ ns} + 0.074 \text{ ns} + 0.068 \text{ ns} + 0.074 \text{ ns} + 0.09 \text{ ns} = 0.398 \text{ ns.}$$

C. The area of BK-m adder

The form of the first layer of the BK-*m* adder depends on the TCS representation of $-m$, $\tilde{m}$. We shall analyze the prefix operator computation for a pair of bits ($\tilde{m}_i$, $\tilde{m}_{i+1}$).

(27a-b) can be expressed as

$$g_{\tilde{m}} = s_{i+1} \cdot \tilde{m}_{i+1} + s_i \cdot \tilde{m}_i \cdot (s_{i+1} \oplus \tilde{m}_{i+1}), \quad (33a)$$

$$p_{\tilde{m}} = (s_{i+1} \oplus \tilde{m}_{i+1}) \cdot (s_i \oplus \tilde{m}_{i1}). \quad (33b)$$

For individual combinations of $(\tilde{m}_i, \tilde{m}_{i+1})$ we get

$(\tilde{m}_i, \tilde{m}_{i+1}) = (0,0)$ $g_{\tilde{m}} = 0$ and $p_{\tilde{m}} = s_i \cdot s_{i+1}$,
$(\tilde{m}_i, \tilde{m}_{i+1}) = (0,1)$ $g_{\tilde{m}} = \bar{s}_i \cdot s_{i+1}$ and $p_{\tilde{m}} = s_i \cdot s_{i+1}$,
$(\tilde{m}_i, \tilde{m}_{i+1}) = (1,0)$ $g_{\tilde{m}} = s_i$ and $p_{\tilde{m}} = s_i \cdot \bar{s}_{i+1}$,
$(\tilde{m}_i, \tilde{m}_{i+1}) = (1,1)$ $g_{\tilde{m}} = s_{i+1} + s_i \cdot \bar{s}_{i+1}$ and $p_{\tilde{m}} = \bar{s}_0 \cdot \bar{s}_1$.

The HA's become reduced, for we have $g_i = 0$, and the XOR gate that computes p_i, is reduced to the direct connection, i.e. $p_i = s_i$. For $\tilde{m}_i = 1$, $g_i = s_i$, the XOR gate that computes p_i .becomes an inverter, i.e. $p_i = \bar{s}_i$. The form of $g_{\tilde{m}}$ and $p_{\tilde{m}}$ influences the form of BK_0 and BK_1.

Next we shall analyze the BK-*m* adder for $m = 29$ in order to have a comparison with the adder presented by Hiasat [23]. The TCS representation of $m = 29$ has the form 100011, then for HA_0, g_0 - connection, p_0 - inversion, for HA_1 g_1 - connection, p_1 - inversion, for HA_2 $g_2 = 0$, p_2- connection, for HA_3 $g_3 = 0$, p_3 - connection, for HA_4 $g_4 = 0$, p_4 - connection.

Moreover, regarding that $\tilde{m}_0 = 1$ and $\tilde{m}_1 = 1$, we may transform BK_0, to obtain $BK_{0\text{-}m}$ as

$$g_{\tilde{m},BK_0} = s_{i+1} + s_i \cdot \bar{s}_{i+1} \quad (34a)$$

and

$$p_{\tilde{m},BK_0} = \bar{s}_i \cdot \bar{s}_{i+1} = \overline{s_i + s_{i+1}}, \quad (34b)$$

and the $A_{BK_{0-m}}$ can be calculated as

$$A_{BK_{0-m}} = A_{IVd1} + A_{AND2d1} + A_{OR2d1} + A_{NOR2d2} = 6.34\,\text{GE} \quad (35)$$

and the delay

$$t_{BK_{0-m}} = t_{IVd1} + t_{ANDd1} + t_{ORd1} = 0.05 \text{ ns} + 0.105 \text{ ns} + 0.111 \text{ ns} = 0.266 \text{ ns.} \quad (36)$$

For BK_1 $\tilde{m}_2 = 0$, $\tilde{m}_3 = 0$, hence $g_{\tilde{m}} = 0$ and $p_{\tilde{m}} = s_i \cdot s_{i+1}$.

Assuming the direct realization we receive

$$A_{BK_{1-m}} = A_{AND2d1}, \quad (37)$$

$$t_{BK_{1-m}} = t_{AND2d1}. \quad (38)$$

For other blocks we have

$$A_{BK_{2-m}} = A_{BK_2}, \; A_{BK_{3-m}} = A_{BK_3}, \; A_{BK_{4-m}} = A_{BK_4}. \quad (39)$$

We finally receive for the BK-*m* adder

$$A_{BK-m} = 2A_{IV1d2} + A_{BK0m} + A_{BK1m} + A_{BK2} + A_{BK3} + A_{BK4} + 4A_{XOR2d1} = 32.34\,\text{GE}, \quad (40)$$

and for TOMA for $m = 29$ based on BK adders

$$A_{TOMA\ BK} = A_{BK} + A_{BK\ m} = 59.99\,\text{GE} + 31.68\,\text{GE} = 91.67\,\text{GE}. \quad (41)$$

The *BK-m* delay can be calculated as

$$t_{BK-m} = t_{IV1} + t_{BK0\text{-}m} + \max(t_{BK2\text{-}m}, t_{BK3\text{-}m}) + t_{BK4m} + t_{XOR2d1} + t_{MX2d1}. \quad (42)$$

Hence

$$t_{BK-m} = 0.03 \text{ ns} + 0.15 \text{ ns} + 0.07 \text{ ns} + 0.07 \text{ ns} + 0.09 \text{ ns} + 0.08 \text{ ns} = 0.49 \text{ ns.}$$

Finally we obtain

$$t_{TOMA\_BK} = t_{BK} + t_{BK\text{-}m} = 0.398 \text{ ns} + 0.490 \text{ ns} = 0.888 \text{ ns}.$$

D. The area of the pipelined TOMA BK

This area can be evaluated as

$$A_{TOMA\_BK\_p} = A_{BK} + A_{BK\ m} + n_{BK} \cdot A_{FF} \quad (43)$$

where n_{BK} is the number of flip-flops in pipeline registers. For the structure from Fig. 7 with $n_{BK} = 51$ and $A_{FF} = A_{FDIQ} = 5.67$ GE, we get $A_{TOMA\_BK\_p} = 380.84$ GE.

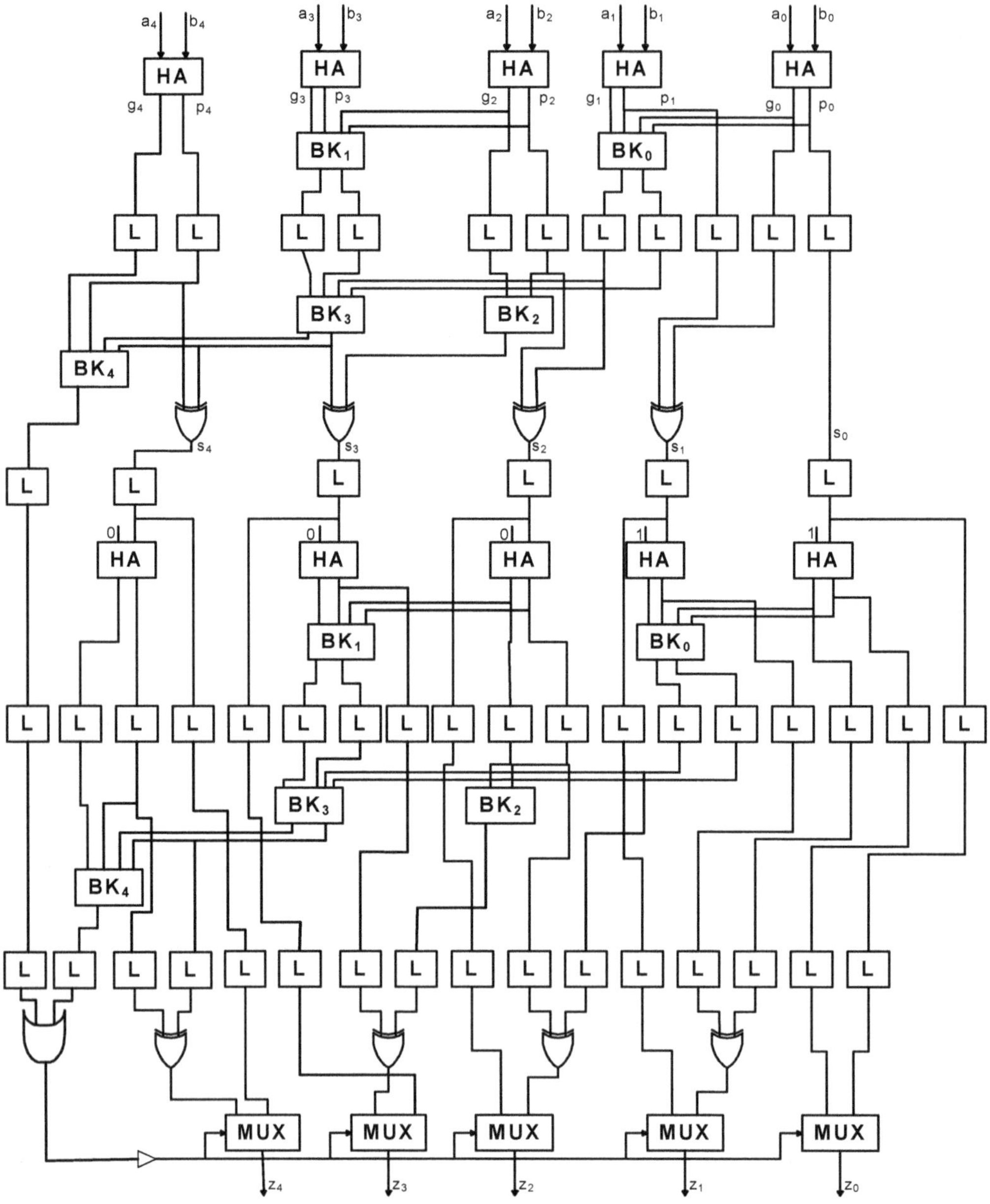

Fig. 7. Pipelined TOMA based on Brent-Kung adder.

E. Pipelining frequency of pipelined TOMA BK:

layer 1

$t_D^{L1,BK}$: $t_{HAd2} + t_{BK1} = 0.092 \text{ ns} + 0.068 \text{ ns} = 0.160 \text{ ns}$,

layer 2

$t_D^{L2,BK}$:

$t_{BK3} + t_{BK4} + t_{XOR2d1} =$

$= 0.068 \text{ ns} + 0.074 \text{ ns} + 0.090 \text{ ns} = 0.232 \text{ ns}$.

layer 3

$t_D^{L3,BK}$: $t_{HAd1} + t_{BK1} = 0.088 \text{ ns} + 0.074 \text{ ns} = 0.162 \text{ ns}$,

layer 4

$t_D^{4,BK}$: $t_{BK3} + t_{BK4} = 0.152 \text{ ns}$,

layer5

$t_D^{L5,BK} = t_{XOR2d1} + t_{NID6} + t_{MX2d1} = 0.222 \text{ ns}$.

Using the $t_D^{L2,BK}$ as the maximum layer delay, we receive the maximum pipelining frequency

$$f_{PF\_max}^{TOMA\_BK} = 1/(t_D^{L2,BK} + t_{FD1Q}) = 1/(0.232 \text{ ns} + 0.094 \text{ ns})$$
$$= 1/0.326 \text{ ns} = 3.06 \text{ GHz}.$$

6. New Five-bit TOMA

In this section we shall show a new TOMA structure and its pipelined form that requires smaller area than other TOMA structures. The TOMA is configured as a serial connection $X + Y$ adder and $X + Y - m$ adder that are designed in such a manner that leads to a substantial simplification and thus to a smaller delay or a smaller number of pipeline levels. Both adders are modifications of the standard CLA adder. In the first stage of the proposed structure the propagate's and generate's and transfer functions [30] $t_i = a_i + b_i$ are used. The first three carries c_1, c_2 and c_3 are computed simultaneously, and c_3 is used to generate c_4 and c_5.

Generally, the computation of the carry c_i can be expressed, assuming $c_0 = 0$, as

$$c_1 = g_0, \qquad (44a)$$

$$c_2 = g_1 + c_1 \cdot t_1, \qquad (44b)$$

$$c_3 = g_2 + c_2 \cdot t_2, \qquad (44c)$$

$$c_4 = g_3 + c_3 \cdot t_3, \qquad (44d)$$

$$c_5 = g_4 + c_4 \cdot t_4. \qquad (44e)$$

In the above formulas instead of p_i, the transfer function $t_i = a_i + b_i$ is used, which is justified as follows

$$c_{i+1} = g_i + c_i \cdot p_i, \qquad (45)$$

$$c_{i+1} = g_i + c_i \cdot p_i + c_i \cdot g_i = g_i + c_i \cdot (g_i + p_i) = g_i + c_i \cdot t_i, \qquad (46)$$

with $t_i = a_i + b_i$, $g_i = a_i b_i$ and $p_i = a_i \oplus b_i$.

We may express c_2 and c_3 as the functions of g_i and t_i as

$$c_2 = g_1 + c_1 \cdot t_1, \qquad (47)$$

$$c_3 = g_2 + g_1 \cdot t_2 + g_0 \cdot t_1 \cdot t_2. \qquad (48)$$

Consequently, we receive

$$c_4 = g_1 + c_3 \cdot t_3, \qquad (49)$$

and

$$c_5 = g_4 + g_3 \cdot t_4 + c_3 \cdot t_3 \cdot t_4. \qquad (50)$$

In the adder realization the above equations are transformed to the NAND form. The sum bits are generated using $s_i = p_{i-1} \oplus c_i$., 1, 2, 3, 4 with $s_0 = p_0$. The second stage of the TOMA implements the subtraction of $-m$ making use of the TCS representation of $-m$, $\widetilde{m} = (1, \widetilde{m}_4, \widetilde{m}_3, \widetilde{m}_2, \widetilde{m}_1, \widetilde{m}_0)$.

Regarding that the second operand of the $X + Y - m$ adder is $\widetilde{m}$, we can write

$$c_1 = s_0 \cdot \widetilde{m}_0, \qquad (51a)$$

$$c_2 = s_1 \cdot \widetilde{m}_1 + s_0 \cdot s_1 \cdot \widetilde{m}_0 + s_0 \cdot \widetilde{m}_0 \cdot \widetilde{m}_1, \qquad (51b)$$

$$\begin{aligned} c_3 = {} & s_2 \cdot \widetilde{m}_2 + s_1 \cdot s_2 \cdot \widetilde{m}_1 + s_0 \cdot s_1 \cdot s_2 \cdot \widetilde{m}_0 + \\ & + s_0 \cdot s_2 \cdot \widetilde{m}_0 \cdot \widetilde{m}_1 + + s_1 \cdot s_2 \cdot \widetilde{m}_1 \cdot \widetilde{m}_2 \\ & + s_0 \cdot s_1 \cdot \widetilde{m}_0 \cdot \widetilde{m}_2 + s_0 \cdot \widetilde{m}_0 \cdot \widetilde{m}_1 \cdot \widetilde{m}_2 \end{aligned} \qquad (51c)$$

$$c_4 = s_3 \cdot \widetilde{m}_3 + c_3 \cdot s_3 + c_3 \cdot \widetilde{m}_3, \qquad (51d)$$

$$\begin{aligned} c_5 = {} & s_4 \cdot \widetilde{m}_4 + s_3 \cdot s_4 \cdot \widetilde{m}_3 + c_3 \cdot s_3 \cdot s_4 + c_3 \cdot s_4 \cdot \widetilde{m}_3 + \\ & + s_3 \cdot \widetilde{m}_3 \cdot \widetilde{m}_4 + c_3 \cdot s_3 \cdot \widetilde{m}_4 + c_3 \cdot \widetilde{m}_3 \cdot \widetilde{m}_4 \end{aligned} \qquad (51e)$$

We may simplify the above equations by substituting $\widetilde{m}$ values of the individual five-bit moduli. The results of this simplification are given in Tab. 1.

m	c_1	c_2	c_5	c_4	c_5
17	s_0	$s_1 + s_0$	$s_2 + s_1 + s_0$	$s_3 + c_3$	$s_4 \cdot (s_3 + c_3)$
19	0	s_1	$s_2 + s_1$	$s_3 + c_3$	$s_4 \cdot (s_3 + c_3)$
21	s_0	$s_1 + s_0$	$s_2 \cdot (s_1 + s_0)$	$s_3 + c_3$	$s_4 \cdot (s_3 + c_3)$
23	s_0	$s_1 \cdot s_0$	$s_2 \cdot s_1 \cdot s_0$	$s_3 + c_3$	$s_4 \cdot (s_3 + c_3)$
25	s_0	$s_1 + s_0$	$s_2 + s_1 + s_0$	$s_3 \cdot c_3$	$c_3 \cdot s_4 \cdot s_3 \cdot$
27	s_0	$s_1 \cdot s_0$	$s_2 + s_1 s_0$	$s_3 \cdot c_3$	$c_3 \cdot s_4 \cdot s_3 \cdot$
29	s_0	$s_1 + s_0$	$s_2 \cdot (s_1 + s_0)$	$s_3 \cdot c_3$	$c_3 \cdot s_4 \cdot s_3 \cdot$
31	s_0	$s_1 \cdot s_0$	$s_2 \cdot s_1 \cdot s_0$	$s_3 \cdot c_3$	$c_3 \cdot s_4 \cdot s_3 \cdot$

Tab. 1. Logical functions for realizations of the carries of $X + Y - m$ adder.

In Fig. 8, the TOMA based on the new principle for $m = 29$ is depicted.

A. 5-bit new TOMA area

We shall analyze the area and delay of the new TOMA for $m = 29$.The area of the new TOMA can be computed as

$$A_{TOMA\_New} = A_{X+Y} + A_{X+Y-m}. \qquad (52)$$

The hardware amount of the $X + Y$ adder can be expressed as

$$A_{X+Y} = A_{HA-stage} + A_{t_i} + A_{c_2} + A_{c_3} + A_{c_4} + A_{c_5} + A_{SU} \qquad (53)$$

where $A_{HA-stage}$ is the area of the input summation stage (HAs and ORs), A_{c_i} are the areas of circuits generating the individual carries c_i.

Subsequently we have

$$A_{HA\_stage} = 2 \cdot A_{HAd1} + 3 \cdot A_{HAd2} + 4 A_{OR2d1} = 27.35 \text{ GE},$$

$$A_{c_2} = A_{IVd1} + 2 \cdot A_{NAND2d1} = 3 \text{ GE},$$

$$A_{c_3} = A_{IVd1} + 2 \cdot A_{NAND2d2} + A_{NAND3d1} + A_{NAND3d2} = 9.67 \text{ GE}$$

$$A_{c_4} = A_{IVd1} + 2 \cdot A_{NAND2d1} = 3 \text{ GE},$$

$$A_{c_5} = A_{IVd1} + A_{NAND2d1} + 2 A_{NAND3d1} = 4.67 \text{ GE},$$

$$A_{SU} = 5 \cdot A_{XOR2d1} = 15 \text{ GE}.$$

In effect, we receive

$$A_{X+Y} = 55.69 \text{ GE}.$$

For the $X + Y - m$ adder we have

$$A_{X+Y-m} = A_{c_2-m} + A_{c_3-m} + A_{c_4-m} + A_{c_5-m} + A_{SU-m},$$

where

$A_{c_1-m} = 0$ (direct connection),

$A_{c_2-m} = A_{OR2d2} = 1.67 \text{ GE},$

$A_{c_3-m} = A_{AND2d2} = 2 \text{ GE},$

$A_{c_4-m} = A_{AND2d1} = 1.67\text{GE},$

$A_{c_5-m} = A_{AND3d1} = 2 \text{ GE},$

$A_{SU-m} = A_{OR2d1} + A_{NID6} + 5 \cdot A_{MX2d1} = 20.34 \text{ GE}.$

We receive $A_{X+Y-m} = 27.68 \text{ GE}$.

The total hardware amount is $A_{TOMA\_New} = 110.06\text{GE}$.

B. 5-bit New TOMA delay

The delay of the new TOMA can be written as

$$t_{TOMA\_New} = t_{X+Y} + t_{X+Y-m}, \quad (54)$$

where

$t_{X+Y} = t_{HA\_Stage} + t_{c_1} + \max(t_{c_2}, t_{c_3}) + \max(t_{c_4}, t_{c_5}) + t_{SU}$ and t_{c_i}, $i = 1, 2,\ldots,5$, denote the individual carry generator delays

$t_{c_1} = \max(t_{HAd2\_ACO}, t_{HAd2\_BCO}),$

$t_{c_2} = \max(t_{NAND2d1}, t_{IVd1}) + t_{NAND2d1},$

$t_{c_3} = \max(t_{NAND2d1}, t_{NAND3d1}, t_{IVd1}) + t_{NAND3d1},$

$t_{c_4} = \max(t_{NAND2d1}, t_{IVd1}) + t_{NAND2d1},$

$t_{c_5} = \max(t_{NAND2d1}, t_{NAND3d1}, t_{IVd1}) + t_{NAND3d1},$

$t_{X+Y} = t_{HAd2} + t_{NAND3d1} + t_{NAND3d2} + 2 \cdot t_{NAND2d1} + t_{XOR2d1},$

$t_{X+Y} = 0.092\text{ns} + 0.052\text{ns} + 0.044\text{ns} + 2 \cdot 0.037\text{ns} + 0.09\text{ns}$
$= 0.352\text{ns}$

and for $X + Y - m$ adder we have

$t_{X+Y-m} = t_{c_5'} + \max(t_{OR2d1} + t_{NID6}, t_{XOR2d1})$, where

$t_{c_5'} = t_{OR2d1} + t_{AND2d1} + t_{AND3d1},$

$t_{c_4\_c_5} = \max(t_{c_4}, t_{c_5}),$

$t_{X+Y-m} = t_{NAND2d1} + t_{NAND3d2} + t_{AND3d1} + t_{XOR2d1} + t_{MX2d1},$

$t_{X+Y-m} = 0.037 \text{ ns} + 0.044 \text{ ns} + 0.066 \text{ ns} +$
$+ 0.09 \text{ ns} + 0.078 \text{ ns} = 0.315 \text{ ns},$

$t_{TOMA\_New} = 0.352 \text{ ns} + 0.315 \text{ ns} = 0.667 \text{ ns}.$

D. The area of the pipelined new TOMA

This area is expressed as

$$A_{TOMA\_New\_p} = A_{X+Y} + A_{X+Y-m} + n_N \cdot A_{FF}$$

where n_N is the number of flip-flops in pipeline registers. For the structure from Fig. 8 with $n_N = 30$ and $A_{FF}=A_{FDIQ}=5.67\text{GE}$, we get . $A_{TOMA\_New} = 280.82\text{GE}$

E. Pipelining frequency of the pipelined new TOMA

For the individual layers in the pipelined structure of the new TOMA, shown in Fig. 8, we have the following delays:

layer 1:
$t_D^{L1,N} = t_{HAd1} + 2 \cdot t_{NAND3d1} = 0.192 \text{ ns},$

layer 2:
$t_D^{L2,N} = 2 \cdot t_{NAND3d1} + t_{XOR2d1} = 0.194 \text{ ns},$

layer 3:
$t_D^{L3,N} = t_{OR2d1} + t_{AND2d1} + t_{AND3d1} = 0.185 \text{ ns},$

layer 4:
$t_D^{L4,N} = t_{XOR2d1} + t_{NID6} + t_{MX2d1} = 0.222 \text{ ns}.$

The design of the pipelined structure aimed at the minimization of the number of pipeline stages while preserving possibly high pipelining frequency. The structure allows one to employ only three pipeline register stages with 30 flip-flops with the maximum pipelining frequency equal to

$$f_{PF\_\max}^{new\_TOMA} = 1/(0.222 \text{ ns} + .094 \text{ ns}) = 1/0.316 \text{ ns} = 3.16 \text{ GHz}.$$

In Tab. 2 the summary of the obtained TOMA parameters is given.

	TOMA-RCA	TOMA-BK	TOMA-Hiasat	New TOMA I
Area [GE] (nonpipelined)	81.68	99.01	127.03	110.72
Delay[ns]	0.747	0.888	0.886	0.667
Area x delay	61.01	87.64	112.55	73.41
Number of pipeline layers	6	4	5	3
Number of FFs	66	58	64	30
Area [GE] (pipelined)	472.90	380.84	489.91	280.82
Pipelining frequency max [GHz]	4.22	3.06	3.7	3.16

Tab. 2. TOMA parameters for $m = 29$.

It is seen that the area-delay product has the best values for the TOMA-RCA and the new TOMA, moreover the new TOMA requires the smallest area for the pipelined structure but at the cost of the reduced maximum pipelining frequency. In general the new pipelined TOMA calls for about 35% less area than the TOMA-BK, the best of three other considered structures.

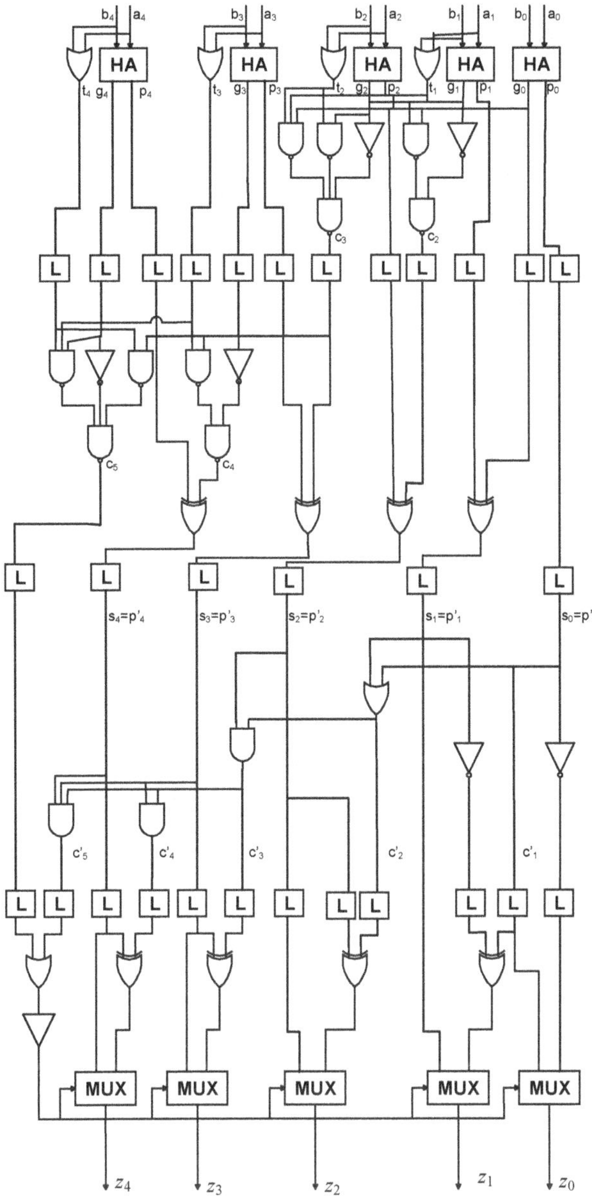

Fig. 8. New five-bit TOMA for $m = 29$.

7. Conclusions

The structures of pipelined two-operand modular adders for five-bit moduli based on ripple carry-adder, Brent-Kung adder and Hiasat adder have been presented and analyzed with respect to the area, number of layers and attainable pipelining frequency. Also a new structure of the two-operand modular adder based on the modified carry-look ahead adder has been proposed. It has been shown that the new pipelined adder has the smallest number of pipeline layers as well as the area smaller by about 35% than the best of other considered structures.

References

[1] SZABO, N. S., TANAKA, R. I. *Residue Arithmetic and its Applications to Computer Technology*. McGraw-Hill Inc., 1967.

[2] SODERSTRAND, M. A., JENKINS, M., JULLIEN, G. A., TAYLOR, F. J. *Residue Number System Arithmetic: Modern Applications in Digital Signal Processing*. IEEE Press, 1986.

[3] OMONDI, A., PREMKUMAR, B. *Residue Number Systems, Theory and Implementation*. Imperial College Press, 2007.

[4] ANANDA MOHAN, P.V. *Residue Number Systems, Algorithms and Architectures*. Kluwer Academic Publishers, 2002.

[5] JENKINS, W. K., KROGMEIER, J. V. The design of dual-mode complex signal processors based on quadratic modular number codes. *IEEE Transactions on Circuits and Systems*, 1987, vol. 34, no. 4, p. 354–364. DOI: 10.1109/TCS.1987.1086154

[6] KRISHAN, R., JULLIEN, G. A., MILLER, W. C. The modified quadratic residue number system (MQRNS) for complex high-speed signal processing. *IEEE Transactions on Circuits and Systems*, 1986, vol. 33, no. 3, p. 325–327. DOI: 10.1109/TCS.1986.1085897

[7] ULMAN, Z., CZYZAK, M. Highly parallel, fast scaling of numbers in nonredundant residue arithmetic. *IEEE Transactions on Signal Processing*, 1998, vol. 46, no. 2, p. 487–496. DOI: 10.1109/78.655432

[8] FRERKING, W. I., PARHI, K. K. Low-power FIR digital filters using residue arithmetic. In *Proceedings of the 31st Asilomar Conference on Signals, Systems and Computers ACSSC1997*. Pacific Grove (CA, USA), November 2 - 7, 1997, p. 739–743. DOI: 10.1109/ACSSC.1997.680542

[9] CARDARILLI, G. C., DEL RE, A., LOJACONO, R., NANNARELLI, A., RE, M. RNS implementation of high performance filter for satellite demultiplexing. In *Proceedings of the 2003 IEEE Aerospace Conference*. Big Sky (Montana, USA), March 8 - 15, 2003, vol. 3, p. 1365–1379. DOI: 10.1109/AERO.2003.1235253

[10] CARDARILLI, G. C., DEL RE, A. NANNARELLI, A., RE, M. Low-power implementation of polyphase filters in quadratic residue number system. In *Proceedings of the IEEE International Symposium on Circuits and Systems (ISCAS 2004)*. Vancouver (Canada), 2004, p. 725–728. DOI: 10.1109/ISCAS.2004.1329374

[11] CZYZAK, M., SMYK, R. FPGA implementation of the two-stage high-speed FIR filter in residue arithmetic. *Elektronika*, 2011, no. 12, p. 90–92.

[12] CZYZAK, M., SMYK, R. Radix-4 DFT butterfly realization with the use of the modified quadratic residue number system. *Poznan University of Technology Academic Journals. Electrical Engineering*, 2010, no. 63, p. 39–51.

[13] GARCIA, A., MEYER-BAESE, U., TAYLOR, F. J. Pipelined Hogenauer CIC filters using field-programmable logic and residue number system. In *Proceedings of the 1998 IEEE International Conference on Acoustics, Speech and Signal Processing.* Seattle (Washington, USA), May 12 - 15, 1998, p. 3085–3088. DOI: 10.1109/ICASSP.1998.678178

[14] BARRACLOUGH, S. R., SOTHERAN, M., BURGIN, K., WISE, A. P., et al. The design and implementation of the IMS A110 image and signal processor. In *Proceedings of the 1989 IEEE Custom Integrated Circuits Conference (ICICC 1989)*. San Diego (CA, USA), May 15 - 18, 1989, p. 24.5.1–24.5.4. DOI: 10.1109/CICC.1989.56826

[15] WEI, W., SWAMY, M.N.S., AHMAD, M.O. RNS application for digital image processing. In *Proceedings of the 4th IEEE International Workshop System-on-Chip for Real-Time Applications*. Banff (Alberta, Canada), July 19 - 21, 2004, p. 77 to 80. DOI: 10.1109/IWSOC.2004.1319854

[16] BANERJI, D. K. A novel implementation method for addition and subtraction in residue number systems. *IEEE Transactions on Computers*, 1974, vol. 23, no. 1, p. 106–109. DOI: 10.1109/T-C.1974.223790

[17] AGRAWAL, D. P., RAO, T. R. N. Modulo 2^n +1 arithmetic logic. *IEEE Journal on Electronic Circuits and Systems*, 1978, vol. 2, no. 6, p. 186–188. DOI: 10.1049/ij-ecs.1978.0037

[18] SODERSTRAND, M. A. A new hardware implementation of modulo adders for residue number systems. In *Proceedings of the 26th IEEE Midwest Symposium on Circuits and Systems.* Puebla (Mexico), August 15 - 16, 1983, p. 412–415.

[19] BAYOUMI, M., JULLIEN. G. A. VLSI implementation of residue adders. *IEEE Transactions Circuits and Systems*, 1987, vol. 34, no. 3, p. 284–288. DOI: 10.1109/TCS.1987.1086130

[20] DUGDALE, M. VLSI implementation of residue adders based on binary adders. *IEEE Transactions on Circuits and Systems II: Analog and Digital Signal Processing*, 1992, vol. 39, no. 5, p. 325–329. DOI: 10.1109/82.142036

[21] PIESTRAK, S. J. Design of high-speed residue-to-binary number system converter based on the chinese remainder theorem. In *Proceedings* of *the International Conference on Computer Design ICCD'94, VLSI in Computers and Processors.* Cambridge (MA, USA), 1994, p. 508–511. DOI: 10.1109/ICCD.1994.331962

[22] ZIMMERMANN, R. Efficient VLSI implementation of modulo (2^n ± 1) addition and multiplication. In *Proceedings of the 14th IEEE Symposium on Computer Arithmetic.* Adelaide (Australia), April, 1999, p. 158–167. DOI: 10.1109/ARITH.1999.762841

[23] HIASAT, A. A. High-speed and reduced-area modular adder structures for RNS. *IEEE Transactions on Computers*, 2002. vol. 51, no. 1, p. 84–89. DOI: DOI: 10.1109/12.980018

[24] PATEL, R. A., BENAISSA, M., POWELL, N., BOUSSAKTA, S. Novel power-delay-area-efficient approach to generic modular addition. *IEEE Transactions on Circuits and Systems–I Regular Papers*, June 2007, vol. 54, no. 6, p. 1279–1292. DOI: 10.1109/TCSI.2007.895369

[25] KELLIHER, T. P., OWENS, R.M., IRWIN, M. J., HWANG, T.-T. ELM - a fast addition algorithm discovered by a program. *IEEE Transactions on Computers*, 1992, vol. 41, no. 9, p. 1181–1184. DOI: 10.1109/12.165399

[26] KOGGE, P. M., STONE, H. S. A parallel algorithm for the efficient solution of a general class of recurrence equations. *IEEE Transactions on Computers*, 1973, vol. 22, no. 8, p. 786–793. DOI: 10.1109/TC.1973.5009159

[27] LADNER, R. E., FISCHER, M. J. Parallel-prefix computation. *Journal of Association of Computing Machinery*, 1980, vol. 27, no. 4, p. 831–838. DOI: 10.1145/322217.322232

[28] BRENT, R. P., KUNG, H. T. A regular layout for parallel adders. *IEEE Transactions on Computers*, 1982, vol. 31, no. 3, p. 260 to 264.

[29] *0.13 micron Standard Cell Logic Library STDH 150.* Samsung Inc., 2004.

[30] PARHAMI, B. *Computer Arithmetic: Algorithms and Hardware Designs.* New York: Oxford University Press, 2000.

About the Authors ...

Maciej CZYŻAK was born in 1950, Chorzow, Poland. He received his M. Sc. in Automatic Control and Computer Engineering from the Faculty of Electronics, Gdansk University of Technology (GUT) in 1973, and his Ph.D. in Automatic Control from the Faculty of Electrical Engineering, GUT in 1985. He authored and co-authored over 60 technical papers. His interests encompass digital signal processing with the use of residue arithmetic, VLSI and FPGA design.

Appendix A

Area [GE]	Delay [ns]
A_{AND2d1} = 1.67	t_{AND2d1}=0.054
A_{AND2d2} = 2.00	t_{AND2d2}=0.055
A_{AND3d1} =2.00	t_{AND3d1} = 0.066
A_{AND3d2} =2.67	t_{AND3d2} = 0.068
A_{AND4d1} =2.33	t_{AND4d1}=0.082
A_{AND4d2} =2.67	t_{AND4d2}=0.085
A_{FAd1}=8.00	t_{FAd1}=0.143
A_{FAd2}=9.00	t_{FAd2}=0.150
A_{HAd1}=4.67	t_{HAd1}=0.088
A_{HAd2}=5.67	t_{HAd2}=0.092
$A_{NAND2d1}$=1.00	$t_{NAND2d1}$=0.037
$A_{NAND2d2}$=2.00	$t_{NAND2d2}$=0.031
$A_{NAND3d1}$=1.67	$t_{NAND3d1}$=0.052
$A_{NAND3d2}$=3.00	$t_{NAND3d2}$=0.044
$A_{NAND4d1}$=2.00	$t_{NAND4d1}$=0.067
$A_{NAND4d2}$=3.67	$t_{NAND4d2}$=0.059
A_{NOR2d1} =1.33	t_{NOR2d1}= 0.050
A_{NOR2d2} =2.00	t_{NOR2d2}= 0.040
A_{OR2d1} =1.67	t_{OR2d1}=0.065
A_{OR2d2} = 2.00	t_{OR2d2}=0.069
A_{OR3d1} =2.00	t_{OR3d1}=0.090
A_{OR3d2} =2.67	t_{OR3d2}=0.090
A_{OR4d1} =3.00	t_{OR4d1}= 0.076
A_{OR4d2} =3.33	t_{OR4d2}= 0.082
A_{OR5d1} =3.33	t_{OR5d1}=0.094
A_{OR5d2} =3.67	t_{OR5d2}= 0.105
A_{XOR2d1}=3.00	t_{XOR2d1}=0.090
A_{IVd1} =1	t_{IVd1}=0.029
A_{NID6} =3.67	t_{NID6} =0.054
A_{MX2d1} =3.00	t_{MX2d1}=0.078
A_{MX2d2} =3.33	t_{MX2d2}=0.076
A_{MX2d4} = 4.33	t_{MX2d4} = 0.092
A_{MX4d1} =6.33	t_{MX4d1}=0.105
A_{FD1Q} = 5.67	$t_{FD1Q\ SU}$ = 0.094

Tab. 3. Hardware amount and time delays for STDH150 basic elements.

Half-adder (HA) delays [ns]	Full-adder (FA) delays [ns]
$t_{HAd1\ ACO}$=0.054	$t_{FAd1\ CICO}$=0.083
$t_{HAd1\ BCO}$=0.055	$t_{FAd1\ ACO}$=0.122
$t_{HAd1\ AS}$= 0.088	$t_{FAd1\ BCO}$=0.143
$t_{HAd1\ BS}$= 0.073	$t_{FAd1\ AS}$=0.121
$t_{HAd2\ ACO}$=0.057	$t_{FAd1\ BS}$=0.139
$t_{HAd2\ BCO}$=0.058	$t_{FAd2\ CICO}$=0.089
$t_{HAd2\ AS}$= 0.092	$t_{FAd2\ ACO}$=0.130
$t_{HAd2\ BS}$= 0.074	$t_{FAd2\ BCO}$=0.150
	$t_{FAd2\ AS}$=0.129
	$t_{FAd2\ BS}$=0.150

Tab. 4. Individual delays between input and output nodes for FAs and HAs (STDH150).

Jacek HORISZNY was born in 1962, Gdansk Poland. He received his M. Sc. in Electrical Engineering from the Faculty of Electrical Engineering, Gdansk University of Technology (GUT) in 1986, and his Ph.D. in Electrical Engineering from the Faculty of Electrical and Control Engineering, GUT in 1996. He is the author and co-author of over 40 technical papers. His current interests include the transformer design and applications of digital signal processing for transformer protection.

Robert SMYK was born in 1978, Malbork, Poland. He received his M. Sc. in Automatic Control from the Faculty of Electrical and Control Engineering, Gdansk University of Technology (GUT) in 2008, and his Ph.D. in Automatic Control from the Faculty of Electrical and Control Engineering, GUT in 2008. He co-authored and authored over 20 technical papers. His interests encompass digital signal processing, residue arithmetic, and FPGA design.

Exact Outage Performance Analysis of Multiuser Multi-relay Spectrum Sharing Cognitive Networks

Tao ZHANG, Weiwei YANG, Yueming CAI

College of Communications Engineering, PLA University of Science and Technology, Yudao Street, Nanjing City, Jiangsu Province, 210007, China.

ztcool@126.com, wwyang1981@163.com, caiym@vip.sina.com

Abstract. *In this paper, we investigate the outage performance of dual-hop multiuser multi-relay cognitive radio networks under spectrum sharing constraints. Using an efficient relay-destination selection scheme, the exact and asymptotic closed-form expressions for the outage probability are derived. From these expressions it is indicated that the achieved diversity order is only determined by the number of secondary user (SU) relays and destinations, and equals to $M+N$ (where M and N are the number of destination nodes and relay nodes, respectively). Further, we find that the coding gain of the SU network will be affected by the interference threshold $\bar{I}$ at the primary user (PU) receiver. Specifically, as the increases of the interference threshold, the coding gain of the considered network approaches to that of the multiuser multi-relay system in the non-cognitive network. Finally, our study is corroborated by representative numerical examples.*

Keywords

Spectrum sharing, cognitive relay, outage probability, diversity order, coding gain

1. Introduction

Recently, due to the ability to alleviate the spectrum shortage problem spectrum sharing cognitive radio has received much interests [1]. In spectrum sharing networks, the secondary users (SUs) are authorized to have a concurrent transmission with primary user (PU) as long as the generated interference is below an interference temperature tolerated by the primary system. In order to extend the coverage of secondary transmission and enhance system performance in spectrum sharing cognitive networks, cooperative relaying techniques can be further exploited. The performance of decode-and-forward (DF) and amplify-and-forward (AF) relaying in spectrum sharing network is widely investigated in the literature [2], [3]. However, all these prior works only consider a single SU user.

Moreover, multiuser diversity (MUD) has attracted significant attention in non-cognitive cooperative networks. In [4], Sun et al. proposed a joint source-relay selection scheme to select the best source-relay pair to access the channel. Furthermore, Ding et al. [5] proposed a source-relay selection scheme with lower system complexity compared to [4] and achieved the same diversity order. Recently, there were also several works to study the multiuser diversity in spectrum sharing cognitive relaying networks. In [6], the impact of multiuser diversity on the performance of SUs in DF spectrum sharing systems over Nakagami-m fading channels was investigated, while the system only consider a single SU relay and the multi-relay cooperative diversity could not be achieved. Combing multiuser diversity and multi-relay cooperative diversity, the authors of [7] analyzed the outage performance of the multiuser multi-relay networks using an efficient relay-destination selection scheme. However, the theoretical analysis in [7, Eq.(10)] assumes that the interference links from SU relays to primary destination are identical. As we known, there are multiple SU relays to primary destination and the interference links should not be identical due to the different locations of the secondary relays. However, the theoretical analysis in [7, Eq.(10)] assuming that the the interference links are identical, based on which the analysis of the system model is simplified. Such that, the derived closed-form expression in [7] is appropriate.

In this manuscript, we investigate the performance of multiuser multi-relay spectrum sharing cognitive networks with one SU source and *M* users and *N* relays in presence of primary receiver and considering both interference and peak power constraints on the SU networks. The exact and asymptotic outage performance are analyzed using the efficient relay-destination selection scheme. Specially, the contributions of this paper are summarized as follows:

1) We present a general analysis of the multiuser multi-relay spectrum sharing cognitive network using the efficient relay-destination selection scheme. Different to that in [7], the interference links from SU relays to primary destination are assumed to be not identical in our analysis. Moreover, we derive the generally exact closed-form expression for the outage probability compared to that in [7], which indicates that the result of [7, Eq.(10)] is the appropriate of our achieved exact closed-form expression for the outage probability.

2) Since the exact analysis is too complicated to ren-

der insight on the impact of the interference threshold and the number of the relays and users, the asymptotic analysis is investigated to indicate that the diversity order is $M+N$, which reveals that the diversity order is only affected by the number of SU relays and destinations. Moreover, we find that the interference threshold at PU receiver will affect the coding gain of the considered network. Specifically, as the increases of the interference threshold, the coding gain of the considered network will approach to that of the multiuser multi-relay system in the non-cognitive network.

3) In special cases, we further analyze the outage performance of the multi-user multi-relay networks based on the efficient relay-destination selection scheme without interference threshold. Moreover, we demonstrate the outage performance analysis in [5] for non-cognitive networks can be the special cases of our works without interference constraint. Finally, simulation results are presented to demonstrate the validity of our theoretical analysis.

2. System Model

We consider a spectrum sharing cognitive relay network, where the secondary network consists of one source $S-S$, N relays $S-R_n\,(n=1,\ldots,N)$, and M users $S-D_m\,(m=1,\ldots,M)$, whereas the primary network consists of a source $\mathrm{P-Tx}$ and a receiver $\mathrm{P-Rx}$ and all receivers are affected by additive white Gaussian noise (AWGN) which has zero mean and equal variance (N_0). The $\mathrm{P-Tx}$ transmitter is assumed to be far away from the SU nodes so that it does not interfere on the selection process of the relay and destination nodes. An example of such system is shown in Fig. 1. Each node is equipped with a single-antenna device and operates in a half-duplex mode. For the secondary network, the channels are mutually independent flat Rayleigh fading and we denote h_{KT} as the coefficients of the channels between the node K and the node T. And $|h_{KT}|^2$ is an exponentially distributed random with variance $\lambda_{KT} \propto d_{KT}^{-\rho}$, where d_{KT} is the distance between node K and node T, and ρ is the path loss factor.

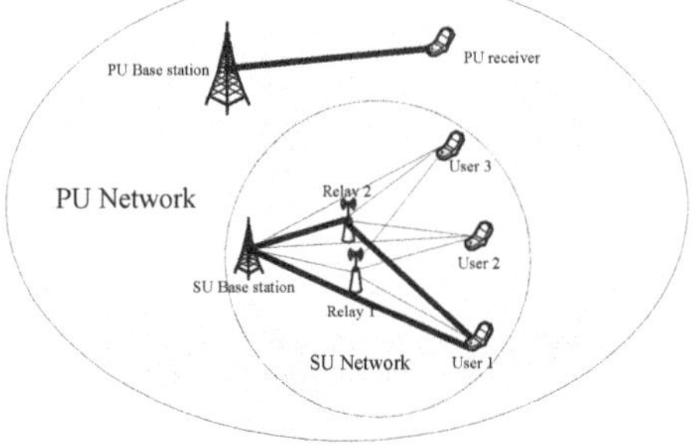

Fig. 1. An example of the considered system. The lines between any two nodes represent the communication links. In this example, the scheduler selects user 1 to access the channel with the help of relay 2 in SU network.

Under the underlay paradigm, $S-S$ and $S-R_n\,(n=1,\ldots,N)$ are allowed to use the same frequency as the primary system if the interference generated on $\mathrm{P-Rx}$ remains below the interference threshold $\bar{I}$, which is the maximum interference powers tolerable at $\mathrm{P-Rx}$. Thus, the transmit power of $S-S$ and $S-R_n\,(n=1,\ldots,N)$ must satisfy $P_S \le \min(\bar{I}/|h_{SP}|^2, P)$ and $P_{R_n} \le \min(\bar{I}/|h_{R_nP}|^2, P)$, respectively, where P is the maximum transmit power constraint of source and relays.

More specifically, the best SU destination D^* is first selected based on the direct links, i.e., $D^* = \arg\max_m [\gamma_{SD_m}]$, where $\gamma_{SD_m} = \min(\bar{I}/|h_{SP}|^2, P)|h_{SD_m}|^2/N_0$ is the instantaneous SNR between the $S-S$ and the mth user $S-D_m$. Secondly, the relay selection process is performed to chose the best relay $R^* = \arg\max_n [\min[\gamma_{SR_n}, \gamma_{R_nD^*}]]$, where $\gamma_{SR_n} = \min(\bar{I}/|h_{SP}|^2, P)|h_{SR_n}|^2/N_0$ is the instantaneous SNR between the $S-S$ and the nth relay $S-R_n$, $\gamma_{R_nD^*} = \min(\bar{I}/|h_{R_nP}|^2, P)|h_{R_nD^*}|^2/N_0$ is the instantaneous SNR between the nth relay $S-R_n$ and the best user $S-D^*$.

For DF protocol, using selection combining (SC) scheme, the achieved system SNR at the destination can be expressed as

$$\gamma_{end}^{DF} = \max\left[\max_m [\gamma_{SD_m}], \max_n [\min[\gamma_{SR_n}, \gamma_{R_nD^*}]]\right]. \qquad (1)$$

From (1), the max-min scheme chooses the relay node and the destination node for the multiuser multi-relay spectrum sharing cognitive network. As shown in $\gamma_{SR_n} = \min(\bar{I}/|h_{SP}|^2, P)|h_{SR_n}|^2/N_0$, because the transmission power from the source to every relay is limited according to the same interference threshold at the PU receiver, there exists a common term $|h_{SP}|^2$ in every γ_{SR_n}. This common term implies that the operation in (1) becomes correlated. Note that in [7, Eq(10)], the term $|h_{R_nP}|^2$ in every $\gamma_{R_nD^*} = \min(\bar{I}/|h_{R_nP}|^2, P)|h_{R_nD^*}|^2/N_0$ is assumed to be identical, based on which the analysis of the system model is simplified. However, this assumption is not correct because there are multiple secondary relays in the system, or there are multiple secondary relays to primary destination links. Due to the different locations of the secondary relays, the interference links from SU relays to the primary destination should not be identical.

3. Outage Probability Analysis

Outage event occurs when the achieved system SNR of the selected best channel is below a given threshold γ_{th}. Due to the common terms $|h_{SP}|^2$, the conditional outage probability can be formulated as

$$\Pr\left[\gamma_{end}^{DF} < \gamma_{th} \middle| |h_{SP}|^2\right] = \underbrace{\Pr\left[\max_m\{\gamma_{SD_m}\} < \gamma_{th} \middle| |h_{SP}|^2\right]}_{\Psi_1} \times \underbrace{\Pr\left[\max_{m,n}[\min[\gamma_{SR_n}, \gamma_{R_nD*}]] < \gamma_{th} \middle| |h_{SP}|^2\right]}_{\Psi_2} \qquad (2)$$

Firstly, since all the links from S to D_m are statistically independent, Ψ_1 in (2) can be rewritten as

$$\Psi_1 = \prod_{m=1}^{M} F^*_{\gamma_{SD_m}}(\gamma_{th}) \tag{3}$$

where $F^*_{\gamma_{SD_m}}(x)$ is the cumulative distribution function (CDF) of γ_{SD_m} conditioned on $|h_{SP}|^2 = z$, and we firstly derive that

$$\min(\bar{I}/z, P) = \begin{cases} P, \text{when} z \le \bar{I}/P \\ \bar{I}/z, \text{when} z > \bar{I}/P \end{cases} \tag{4}$$

Based on (4), we derive the CDF of $F^*_{\gamma_{SD_m}}(x)$ that

$$F^*_{\gamma_{SD_m}}(x) = \begin{cases} 1-\exp(-\lambda_{SD_m} x N_0/P), z \le \bar{I}/P \\ 1-\exp\left(-\lambda_{SD_m} x N_0/\bar{I}\cdot z\right), z > \bar{I}/P \end{cases} \tag{5}$$

Then, using the total probability theorem [8, Eq. (12)(13)], Ψ_2 in (2) can be rewritten as

$$\begin{aligned}\Psi_2 &= \Pr\left(\max_n \left[\min\left[\gamma_{SR_n}, \gamma_{R_n D^*}\right]\right] < \gamma_{th} \big| |h_{SP}|^2\right) \\ &= \sum_{m=1}^{M} \Pr(D^*{=}D_m) \prod_{n=1}^{N} \left[1-\left(1-F^*_{\gamma_{SR_n}}(\gamma_{th})\right)\left(1-F_{\gamma_{R_n D_m}}(\gamma_{th})\right)\right]\end{aligned} \tag{6}$$

where $\Pr(D^* = D_m)$ can be derived with the help of [8, Eq. (14)] as

$$\begin{aligned}&\Pr(D^* = D_m) \\ &= 1+\sum_{k=1}^{M-1} \sum_{\substack{A_k \subseteq \{1,\ldots,m-1,m+1,\ldots,M\} \\ |A_k|=k}} (-1)^k \frac{\lambda_{SD_m}}{\lambda_{SD_m}+\sum_{j\in A_k}\lambda_{SD_j}}\end{aligned} \tag{7}$$

Notice that $F^*_{\gamma_{SR_n}}(x)$ is the CDF of γ_{SR_n} conditioned on $|h_{SP}|^2 = z$. Based on (5), the conditioned CDF $F^*_{\gamma_{SR_n}}(x)$ and $F^*_{\gamma_{R_n D_m}}(x)$ can be written as

$$F^*_{\gamma_{SR_n}}(x) = \begin{cases} 1-\exp(-\lambda_{SR_n} x N_0/P), z \le \bar{I}/P \\ 1-\exp\left(-\lambda_{SR_n} x N_0/\bar{I}\cdot z\right), z > \bar{I}/P \end{cases} \tag{8}$$

$$F^*_{\gamma_{R_n D_m}}(x) = \begin{cases} 1-\exp(-\lambda_{R_n D_m} x N_0/P), z \le \bar{I}/P \\ 1-\exp\left(-\lambda_{R_n D_m} x N_0/\bar{I}\cdot z\right), z > \bar{I}/P \end{cases} \tag{9}$$

In addition, $F_{\gamma_{R_n D_m}}(x)$ in (6) is the CDF of $\gamma_{R_n D_m}$. Performing the integration in (9), $F_{\gamma_{R_n D_m}}(x)$ can be written as

$$\begin{aligned}F_{\gamma_{R_n D_m}}(x) &= \int_0^\infty F^*_{\gamma_{R_n D_m}}(x) f_{|h_{SP}|^2}(z)\,dz \\ &= \int_0^{\bar{I}/P} (1-\exp(-\lambda_{R_n D_m} x N_0/P))\lambda_{SP}\exp(-\lambda_{SP} z)\,dz \\ &+ \int_{\bar{I}/P}^\infty \left(1-\exp\left(-\lambda_{R_n D_m} x N_0/\bar{I}\cdot z\right)\right)\lambda_{SP}\exp(-\lambda_{SP} z)\,dz \\ &= 1-\exp(-\lambda_{R_n D_m} x N_0/P)\left(1-\frac{\exp(-\lambda_{R_n P}\bar{I}/P)}{\lambda_{R_n P}/\lambda_{R_n D_m}\bar{I}/N_0/x+1}\right)\end{aligned} \tag{10}$$

By substituting (3) and (6) into (2) and the general expression for the outage probability can be found by the following integral

$$\begin{aligned}P_{out}^{DL} &= \int_0^\infty \Pr\left(\gamma_{end}^{DF} < \gamma_{th} \big| |h_{SP}|^2 = z\right) f_{|h_{SP}|^2}(z)\,dz \\ &= \underbrace{\int_0^{\bar{I}/P} \prod_{m=1}^{M}\left[1-\exp(-\lambda_{SD_m}\gamma_{th}N_0/P)\right] \sum_{m=1}^{M}\Pr(D^*{=}D_m) \times \prod_{n=1}^{N}\left[1-\Phi_n\exp(-\lambda_{SR_n}\gamma_{th}N_0/P)\right]\lambda_{SP}\exp(-\lambda_{SP} z)\,dz}_{\xi_1} \\ &+ \underbrace{\int_{\bar{I}/P}^\infty \prod_{m=1}^{M}\left[1-\exp\left(-\lambda_{SD_m}\gamma_{th}N_0/\bar{I}\cdot z\right)\right] \sum_{m=1}^{M}\Pr(D^*{=}D_m) \times \prod_{n=1}^{N}\left[1-\Phi_n\exp\left(-\lambda_{SR_n}\gamma_{th}N_0/\bar{I}\cdot z\right)\right]\lambda_{SP}\exp(-\lambda_{SP} z)\,dz}_{\xi_2}\end{aligned} \tag{11}$$

where $\Phi_n = 1-F_{\gamma_{R_n D_m}}(\gamma_{th}) = \exp(-\lambda_{R_n D_m}\gamma_{th}N_0/P)\left(1-\frac{\exp(-\lambda_{R_n P}\bar{I}/P)}{\lambda_{R_n P}/\lambda_{R_n D_m}\bar{I}/N_0/\gamma_{th}+1}\right)$.

Thus, performing the appropriate substitutions in (11) we can derive that

$$\begin{aligned}\xi_1 &= \prod_{m=1}^{M}\left[1-\exp(-\lambda_{SD_m}\gamma_{th}N_0/P)\right] \\ &\times \sum_{m=1}^{M}\Pr(D^* = D_m)\prod_{n=1}^{N}\left[1-\Phi_n\exp(-\lambda_{SR_n}\gamma_{th}N_0/P)\right] \\ &\times \left(1-\exp\left(-\lambda_{SP}\bar{I}/P\right)\right)\end{aligned} \tag{12}$$

$$\xi_2 = \sum_{m=1}^{M}\Pr(D* = D_m)\left(\exp\left(-\lambda_{SP}\bar{I}/P\right)+\Delta_1+\Delta_2+\Delta_3\right) \tag{13}$$

where

$$\Delta_1 = \sum_{m=1}^{M}\frac{(-1)^m}{m!}\underbrace{\sum_{k_1=1}^{M}\cdots\sum_{k_m=1}^{M}}_{k_1\ne k_2\ne\cdots\ne k_m}\frac{\lambda_{SP}\exp\left(-\sum_{b=1}^{m}\lambda_{SD_{k_b}}\gamma_{th}N_0/P-\lambda_{SP}\bar{I}/P\right)}{\sum_{b=1}^{m}\lambda_{SD_{k_b}}\gamma_{th}N_0/\bar{I}+\lambda_{SP}},$$

$$\Delta_2 = \sum_{n=1}^{N}\frac{(-1)^n}{n!}\underbrace{\sum_{t_1=1}^{N}\cdots\sum_{t_n=1}^{N}}_{t_1\ne t_2\ne\cdots\ne t_n}\prod_{a=1}^{n}\Phi_{t_a}\cdot\frac{\lambda_{SP}\exp\left(-\sum_{a=1}^{n}\frac{\lambda_{SR_{t_a}}\gamma_{th}N_0}{P}-\frac{\lambda_{SP}\bar{I}}{P}\right)}{\sum_{a=1}^{n}\lambda_{SR_{t_a}}\gamma_{th}N_0/\bar{I}+\lambda_{SP}},$$

$$\begin{aligned}\Delta_3 &= \sum_{m=1}^{M}\sum_{n=1}^{N}\frac{(-1)^n}{n!}\frac{(-1)^m}{m!}\underbrace{\sum_{k_1=1}^{M}\cdots\sum_{k_m=1}^{M}}_{k_1\ne k_2\ne\cdots\ne k_m}\underbrace{\sum_{t_1=1}^{N}\cdots\sum_{t_n=1}^{N}}_{t_1\ne t_2\ne\cdots\ne t_n}\prod_{a=1}^{n}\Phi_{t_a} \\ &\times\frac{\lambda_{SP}\exp\left(-\sum_{b=1}^{m}\lambda_{SD_{k_b}}\gamma_{th}N_0/P-\sum_{a=1}^{n}\lambda_{SR_{t_a}}\gamma_{th}N_0/P-\lambda_{SP}\bar{I}/P\right)}{\sum_{b=1}^{m}\lambda_{SD_{k_b}}\gamma_{th}N_0/\bar{I}+\sum_{a=1}^{n}\lambda_{SR_{t_a}}\gamma_{th}N_0/\bar{I}+\lambda_{SP}}.\end{aligned}$$

The outage probability analysis of the multiuser multi-relay spectrum sharing cognitive networks can be used to assess the feasibility of the application of multi-relay transmission and the multiuser diversity techniques in the cognitive cellular networks [9]. In addition, it will guide the designment of the interference threshold of PU receiver and the number of the SU relays and users in practical system designing.

Special Case: *Multiuser Multi-relay Networks without Interference Threshold* $(\bar{I}\to\infty)$

In the case when the multiuser multi-relay networks without interference threshold, and the PU can tolerate an unlimited interference from the SU $(\bar{I}\to\infty)$, taking $(\bar{I}\to\infty)$

in $P_{out}^{DL} = \xi_1 + \xi_2$.

When $(\bar{I} \to \infty)$, $\exp\left(-\lambda_{SP}\bar{I}/P\right) \to 0$, ξ_1 in (12) can be written as

$$\xi_1 = \prod_{m=1}^{M} \left[1 - \exp\left(-\lambda_{SD_m}\gamma_{th}N_0/P\right)\right] \sum_{m=1}^{M} \Pr\left(D* = D_m\right) \times \prod_{n=1}^{N} \left[1 - \Phi_n \exp\left(-\lambda_{SR_n}\gamma_{th}N_0/P\right)\right] \tag{14}$$

where $\Phi_n = \exp\left(-\lambda_{R_nD_m}\gamma_{th}N_0/P\right)$.

When $(\bar{I} \to \infty)$, in order to obtain ξ_2 in (13), we firstly derive that

$\exp\left(-\lambda_{SP}\bar{I}/P\right) \to 0$ in (14)

$\exp\left(-\sum_{b=1}^{m} \lambda_{SD_{k_b}}\gamma_{th}N_0/P - \lambda_{SP}\bar{I}/P\right) \to 0$ in Δ_1

$\exp\left(-\sum_{a=1}^{n} \lambda_{SR_{t_a}}\gamma_{th}N_0/P - \lambda_{SP}\bar{I}/P\right) \to 0$ in Δ_2

$\exp\left(-\sum_{b=1}^{m} \lambda_{SD_{k_b}}\gamma_{th}N_0/P - \sum_{a=1}^{n} \lambda_{SR_{t_a}}\gamma_{th}N_0/P - \lambda_{SP}\bar{I}/P\right) \to 0$ in Δ_3

Therefore, $\xi_2 \to 0$ when $\bar{I} \to \infty$. Then, we can obtain the outage probability of the multiuser multi-relay network without interference threshold as

$$P_{out}^{DL} = \prod_{m=1}^{M} \left[1 - \exp\left(-\lambda_{SD_m}\gamma_{th}N_0/P\right)\right] \sum_{m=1}^{M} \Pr\left(D* = D_m\right) \times \prod_{n=1}^{N} \left[1 - \exp\left(-\lambda_{R_nD_m}\gamma_{th}N_0/P\right) \exp\left(-\lambda_{SR_n}\gamma_{th}N_0/P\right)\right] \tag{15}$$

The closed-form expression of the outage probability for this special case is obtained, which is equivalent to the result obtained in non-cognitive cooperative system [5, Eq. (18)].

4. Asymptotic Performance Analysis

Since the exact analysis is too complicated to render insight on the performance of the multiuser multi-relay spectrum sharing cognitive network, we turn our attention to the asymptotic outage probability at high SNR regime. Without loss of generality, let $\bar{\gamma} \triangleq P/N_0$ be the system SNR and assume that $P = \eta\bar{I}$. For given $|h_{R_nP}|^2$ and $|h_{SP}|^2$, the outage probability can be rewritten as

$$P_{out}^{DL}\left(\gamma_{th}|\alpha_n,\beta\right) = \prod_{m=1}^{M} \left(1 - \exp\left(-\frac{\lambda_{SD_m}N_0}{\beta}\gamma_{th}\right)\right) \times \Pr\left(D* = D_m\right) \prod_{n=1}^{N} \left\{1 - \exp\left(-\frac{\lambda_{SR_n}N_0\gamma_{th}}{\beta} - \frac{\lambda_{R_nD_m}N_0\gamma_{th}}{\alpha_n}\right)\right\} \tag{16}$$

where $\frac{1}{\alpha_n} = \frac{1}{\min\left[1/\left(\eta|h_{R_nP}|^2\right),1\right]}$, $\frac{1}{\beta} = \frac{1}{\min\left[1/\left(\eta|h_{SP}|^2\right),1\right]}$.

For high SNR regime, $\bar{\gamma} \to \infty$, with the help of the Taylor series $\lim_{x\to 0} e^x = 1 + x$, (16) can be approximated as

$$P_{out}^{DL,\infty}\left(\gamma_{th}|\alpha_n,\beta\right) \approx \frac{1}{\beta^M} \prod_{m=1}^{M} \left(\frac{\lambda_{SD_m}\gamma_{th}}{\bar{\gamma}}\right) \times \Pr\left(D* = D_m\right) \prod_{n=1}^{N} \left\{\frac{\lambda_{SR_n}\gamma_{th}}{\bar{\gamma}\beta} + \frac{\lambda_{R_nD_m}\gamma_{th}}{\bar{\gamma}\alpha_n}\right\} \tag{17}$$

Averaging over the random variables α_n and β, (17) can be approximated as

$$P_{out}^{DL,\infty}\left(\gamma_{th}\right) = \prod_{m=1}^{M} \left(\frac{\lambda_{SD_m}\gamma_{th}}{\bar{\gamma}}\right) \Pr\left(D* = D_m\right) \times \prod_{n=1}^{N} \left(\frac{\lambda_{R_nD_m}\gamma_{th}}{\bar{\gamma}}\right) E\left[\frac{1}{\beta^M} \prod_{n=1}^{N} \left\{\frac{1}{\mu_n\beta} + \frac{1}{\alpha_n}\right\}\right] \tag{18}$$

where $\mu_n = \lambda_{R_nD_m}/\lambda_{SR_n}$, in order to derive $E\left[\frac{1}{\beta^M} \prod_{n=1}^{N} \left\{\frac{1}{\mu_n\beta} + \frac{1}{\alpha_n}\right\}\right]$, using the binomial expansion, we firstly write $\frac{1}{\beta^M} \prod_{n=1}^{N} \left(\frac{1}{\alpha_n} + \frac{1}{\mu_n\beta}\right)$ as

$$\frac{1}{\beta^M} \prod_{n=1}^{N} \left(\frac{1}{\alpha_n} + \frac{1}{\mu_n\beta}\right) = \sum_{a=0}^{N} \sum_{\substack{A_a \subseteq \{1,\ldots,N\} \\ |A_a|=a, |\bar{A}_a|=N-a}} \prod_{i \in A_a} \frac{1}{\alpha_i} \prod_{j \in \bar{A}_a} \frac{1}{\mu_j} \frac{1}{\beta^{M+N-a}} \tag{19}$$

In order to derive $E\left[\frac{1}{\beta^M} \prod_{n=1}^{N} \left(\frac{1}{\mu_n\alpha_n} + \frac{1}{\beta}\right)\right]$, we perform the integration of α_i and β in (19) and obtain that

$$\begin{aligned} E\left[\frac{1}{\beta^M} \prod_{n=1}^{N} \left(\frac{1}{\alpha_n} + \frac{1}{\mu_n\beta}\right)\right] &= \sum_{a=0}^{N} \sum_{\substack{A_a \subseteq \{1,\ldots,N\} \\ |A_a|=a, |\bar{A}_a|=N-a}} \underbrace{\int_0^\infty \cdots \int_0^\infty}_{a+1} \prod_{i \in A_a} \frac{1}{\alpha_i} \frac{\prod_{j \in \bar{A}_a} \frac{1}{\mu_j}}{\beta^{M+N-a}} \lambda_{SP} \\ &\times \exp\left(-\lambda_{SP}x\right) \prod_{i \in A_a} \lambda_{R_iP} \exp\left(-\lambda_{R_iP}y_i\right) dx \prod_{i \in A_a} dy_i \\ &= \sum_{a=0}^{N} \sum_{\substack{A_a \subseteq \{1,\ldots,N\} \\ |A_a|=a, |\bar{A}_a|=N-a}} \underbrace{\prod_{i \in A_a} \int_0^\infty \frac{1}{\alpha_i} \lambda_{R_iP} \exp\left(-\lambda_{R_iP}y_i\right) dy_i}_{\Xi_1} \\ &\times \underbrace{\int_0^\infty \frac{\prod_{j \in \bar{A}_a} \frac{1}{\mu_j}}{\beta^{M+N-a}} \lambda_{SP} \exp\left(-\lambda_{SP}x\right) dx}_{\Xi_2} \end{aligned} \tag{20}$$

Along with some algebraic manipulations, Ξ_1 in (20) can be derived as

$$\begin{aligned} \Xi_1 &= \int_0^{\frac{1}{\eta}} \frac{1}{P} f_{|h_{R_iP}|^2}(y_i)\, dy_i + \int_{\frac{1}{\eta}}^{\infty} \eta y_i f_{|h_{R_iP}|^2}(y_i)\, dy_i \\ &= \left(1 - \exp(-\lambda_{R_iP}/\eta)\right) + \eta\lambda_{R_iP}^{-1}\Gamma\left(2, \lambda_{R_iP}/\eta\right) \end{aligned} \tag{21}$$

And Ξ_2 in (20) can be obtained by following similar lines as in (21), along with some simple algebraic manipulations, it can be written as

$$\begin{aligned} \Xi_2 &= \int_0^{\frac{1}{\eta}} \prod_{j \in \bar{A}_a} \frac{1}{\mu_j} f_{|h_{SP}|^2}(x)\, dx \\ &+ \int_{\frac{1}{\eta}}^{\infty} \prod_{j \in \bar{A}_a} \frac{1}{\mu_j} (\eta x)^{M+N-a} f_{|h_{SP}|^2}(x)\, dx \\ &= \left(1 - \exp(-\lambda_{SP}/\eta)\right) \prod_{j \in \bar{A}_a} \frac{1}{\mu_j} \\ &+ \eta^{M+N-a} \lambda_{SP}^{a-M-N} \Gamma\left(M+N-a+1, \lambda_{SP}/\eta\right) \prod_{j \in \bar{A}_a} \frac{1}{\mu_j} \end{aligned} \tag{22}$$

By substituting (20) into (18), the asymptotic outage probability can be finally obtained as

$$\begin{aligned} P_{out}^{DL,\infty}(\gamma_{th}) \approx & \sum_{a=0}^{N} \sum_{\substack{A_a \subseteq \{1,\ldots,N\} \\ |A_a|=a, |\overline{A_a}|=N-a}} \Bigg((1-\exp(-\lambda_{SP}/\eta)) \prod_{j\in \overline{A_a}} \frac{1}{\mu_j} \\ & + \eta^{M+N-a} \lambda_{SP}^{a-M-N} \Gamma(M+N-a+1, \lambda_{SP}/\eta) \prod_{j\in \overline{A_a}} \frac{1}{\mu_j} \Bigg) \\ & \times \prod_{i\in A_a} \left((1-\exp(-\lambda_{R_iP}/\eta)) + \eta \lambda_{R_iP}^{-1} \Gamma(2, \lambda_{R_iP}/\eta) \right) \\ & \times \prod_{m=1}^{M} \left(\frac{\lambda_{SD_m}\gamma_{th}}{\bar{\gamma}} \right) \sum_{m=1}^{M} \Pr(D* = D_m) \prod_{n=1}^{N} \left(\frac{\lambda_{R_nD_m}\gamma_{th}}{\bar{\gamma}} \right) \\ = & \, \varepsilon_A \left(\frac{\gamma_{th}}{\bar{\gamma}} \right)^{M+N} \end{aligned} \quad (23)$$

where

$$\begin{aligned} \varepsilon_A = & \sum_{a=0}^{N} \sum_{\substack{A_a \subseteq \{1,\ldots,N\} \\ |A_a|=a, |\overline{A_a}|=N-a}} \Bigg((1-\exp(-\lambda_{SP}/\eta)) \prod_{j\in \overline{A_a}} \frac{1}{\mu_j} \\ & + \eta^{M+N-a} \lambda_{SP}^{a-M-N} \Gamma(M+N-a+1, \lambda_{SP}/\eta) \prod_{j\in \overline{A_a}} \frac{1}{\mu_j} \Bigg) \\ & \times \prod_{i\in A_a} \left((1-\exp(-\lambda_{R_iP}/\eta)) + \eta \lambda_{R_iP}^{-1} \Gamma(2, \lambda_{R_iP}/\eta) \right) \\ & \times \prod_{m=1}^{M} (\lambda_{SD_m}) \sum_{m=1}^{M} \Pr(D* = D_m) \prod_{n=1}^{N} (\lambda_{R_nD_m}) \end{aligned}.$$

5. Remarks

As can be observed from (23), the diversity order of the SU network is only determined by the number of SU relays and destinations, as the diversity gain $\mathcal{G}_d \triangleq \lim_{\bar{\gamma}\to\infty} \frac{-\log P_{out}^{DL,\infty}}{\log \bar{\gamma}} = M+N$. It is noteworthy that the primary network affects the coding gain $\mathcal{G}_c = \frac{\varepsilon_A^{-\frac{1}{M+N}}}{\gamma_{th}}$ of the SU network. Moreover, we note that there exits a coding gain gap [10, Eq. (31)] between the multiuser multi-relay system in the cognitive network and the same multiuser multi-relay system in the non-cognitive system due to the interference threshold $\bar{I}$ at the PU receiver, which can be written as

$$G = 10\log_{10} \left(\frac{\varepsilon_A}{\varepsilon_B} \right)^{\frac{1}{M+N}} \quad (24)$$

where ε_B is the gain of the multiuser multi-relay system in the non-cognitive network, which can be derived with the help of (15) as

$$\begin{aligned} P_{out}^{DL,\infty} \approx & \prod_{m=1}^{M} [\lambda_{SD_m}\gamma_{th}/\bar{\gamma}] \sum_{m=1}^{M} \Pr(D* = D_m) \\ & \times \prod_{n=1}^{N} [\lambda_{R_nD_m}\gamma_{th}/\bar{\gamma} + \lambda_{SR_n}\gamma_{th}/\bar{\gamma}] \\ = & \prod_{m=1}^{M} [\lambda_{SD_m}] \sum_{m=1}^{M} \Pr(D* = D_m) \prod_{n=1}^{N} [\lambda_{R_nD_m} + \lambda_{SR_n}] \left(\frac{\gamma_{th}}{\bar{\gamma}} \right)^{M+N} \\ = & \, \varepsilon_B \left(\frac{\gamma_{th}}{\bar{\gamma}} \right)^{M+N} \end{aligned} \quad (25)$$

Thus, we can derive ε_B that

$$\varepsilon_B = \prod_{m=1}^{M} [\lambda_{SD_m}] \sum_{m=1}^{M} \Pr(D* = D_m) \prod_{n=1}^{N} [\lambda_{R_nD_m} + \lambda_{SR_n}] \quad (26)$$

This result indicates that for the same outage probability, the performance of multiuser multi-relay system in non-cognitive network outperforms the same system in the cognitive network by an gap of $10\log_{10}\left(\frac{\varepsilon_A}{\varepsilon_B}\right)^{\frac{1}{M+N}}$ dB, because of the multiuser multi-relay system in the cognitive network is affected by the interference threshold at the PU receiver.

6. Simulation Results and Analysis

In this section, we confirm our outage probability analysis through comparisons with simulation results. In all cases, we assume that the distance between the SU source and the first SU destination $S-D_1$ is $d_{SD_1} = 1$ without loss of generality. And those the links $S-D_2$, $S-R_1$, $S-R_2$, R_1-D_1, R_1-D_2, R_2-D_1 and R_2-D_2 are $d_{SD_2} = 1.1d_{SD_1}$, $d_{SR_1} = 0.5d_{SD_1}$, $d_{SR_2} = 0.6d_{SD_1}$, $d_{R_1D_1} = 0.5d_{SD_1}$, $d_{R_1D_2} = 0.55d_{SD_1}$, $d_{R_2D_1} = 0.5d_{SD_1}$, $d_{R_2D_2} = 0.55d_{SD_1}$, respectively. In addition, the distance between the SU source and the PU receiver $S-P$ is $d_{SP} = 0.8d_{SD_1}$, the distance between two SU relays and the PU receiver R_1-P and R_2-P are $d_{R_1P} = 0.5d_{SD_1}$ and $d_{R_2P} = 0.6d_{SD_1}$, respectively, so that the overall transmit power is governed by the interference at the PU receiver as well as by the maximum transmission power at the respective nodes. The variance of the Rayleigh channel fading between any two nodes is determined by the distance between them, and the path loss exponent $\rho = 3$. The threshold γ_{th} is set to 3 dB.

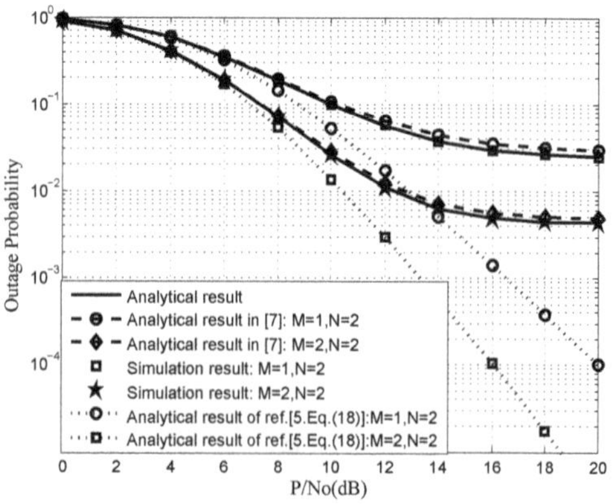

Fig. 2. Outage probability of the system for different numbers of M and N.

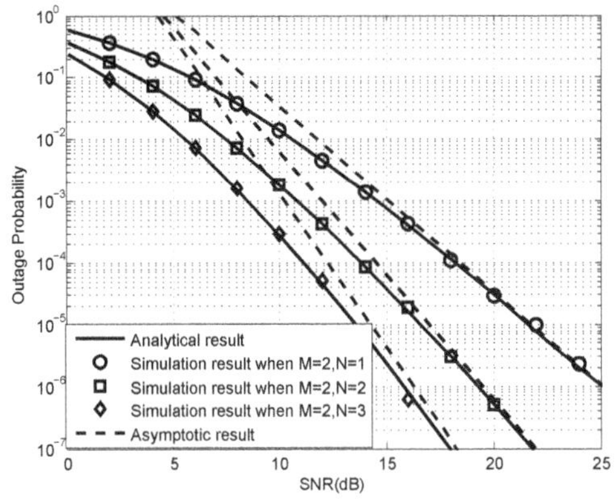

Fig. 3. Outage probability versus SNR $\bar{\gamma}$ with direct links assuming DF relays.

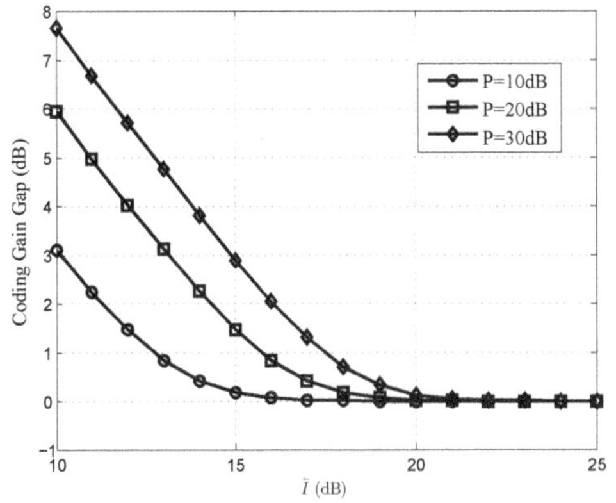

Fig. 4. Coding gain gap between the cognitive network and non-cognitive network versus $\bar{I}$ under different P.

Figure 2 shows that the outage probability when $(M=1, N=2)$ and $(M=2, N=2)$ and $\bar{I}$ =15 dB. We can see that the simulation results match exactly with our proposed analysis results and the outage performance is better than that in [7], which demonstrates that the result of [7, Eq.(10)] is the appropriate of our achieved exact closed-form expression for the outage probability. A floor in the outage performance curve is observed, which is due to the interference threshold constraint. Moreover, as interference threshold gets ∞, the analytical results of [5, Eq.(18)] are derived. It shows that the analysis of the multiuser multi-relay networks based on the efficient source-relay selection scheme in [5] can be the special cases of our works without interference constraint.

Figure 3 illustrates the outage probability and the asymptotic results based on the analysis in section IV. The following parameter values are used: $P = \eta\bar{I}$, $\eta = 1$ and $\bar{\gamma} \stackrel{\Delta}{=} P/N_0 \to \infty$. In addition, the asymptotic curves are shown to be very tight with the exact curves at high SNR regions, which confirm the correctness of our analysis. As can be observed from the figure that the diversity order of the considered system is only determined by the number of SU relays and destinations, and equals to $M+N$, which reveals that the diversity order of the considered system is not affected by the interference threshold at the PU receiver.

Figure 4 plots the coding gain gap between the multiuser multi-relay system in the cognitive network and the multiuser multi-relay system in the non-cognitive network. From the figure, we can see that the coding gain of the multiuser multi-relay system in the cognitive network is worse than that of the system in the non-cognitive system due to the impact of the interference threshold $\bar{I}$ at the PU receiver. However, the coding gain gap approaches to 0 dB for $\bar{I}$ beyond 20 dB. The main reason is that the performance of the system in the cognitive network approaches to that of the system in the non-cognitive network. This observation can also be analytically supported by (24). For given P, when $\bar{I} \to \infty$, $\eta = P/\bar{I} \to 0$. we have $\exp(-\lambda_{SP}/\eta) \to 0$, $\exp(-\lambda_{R_iP}/\eta) \to 0$, $\eta^{M+N-a}\lambda_{SP}^{a-M-N}\Gamma(M+N-a+1, \lambda_{SP}/\eta) \to 0$ and $\eta\lambda_{R_iP}^{-1}\Gamma(2, \lambda_{R_iP}/\eta) \to 0$ in ε_A, hence, we will derive that $\varepsilon_A = \varepsilon_B$.

7. Conclusions

In this paper, we derive the closed-form expressions for outage probability of the multiuser multi-relay spectrum sharing cognitive networks with both interference and maximum allowable transmit power constraint. We derive the exact closed-form expression for outage probability using the efficient relay-destination selection scheme compared to that in [7]. Furthermore, asymptotic analysis indicates that the achieved diversity order is $M+N$, which notes that the diversity order is same to that in non-cognitive cooperative system and the interference threshold at the PU receiver only affects the coding gain of the SU network. Our analysis can be used to guide the designment of the multiuser multi-relay spectrum sharing cognitive system in the cognitive cellular networks. Further study is to derive the optimum power allocation strategy and energy efficiency for the multiuser multi-relay spectrum sharing cognitive network.

Acknowledgements

This work is supported by the Project of Natural Science Foundations of China (No. 61301162 and 61301163) and the Jiangsu Provincial Natural Science Foundation of China (No. BK20130067).

References

[1] MITOLA, J. *Cognitive Radio: An Integrated Agent Architecture for Software Defined Radio.* Ph. D. dissertation, Royal Institute of Technology (KTH), Stockholm, Sweden, 2000.

[2] XU, W., ZHANG, J., ZHANG, P., TELLAMBURA, C. Outage probability of decode-and-forward cognitive relay in presence of primary user's interference, *IEEE Communication Letters,* vol. 16, no. 8, p. 1252–1255, 2012. DOI: 10.1109/LCOMM.2012.061912.120770

[3] XIA, M., AISSA, S. Cooperative AF relaying in spectrum-sharing systems: performance analysis under average interfer-

ence power constraints and Nakagami-m fading. *IEEE Transactions on Communications,* vol. 60, no. 6, p. 1523–1533, 2012. DOI: 10.1109/TCOMM.2012.042712.110410

[4] SUN, L., ZHANG, T., LU, L., NIU, H. On the combination of cooperative diversity and multiuser diversity in multi-source multi-relay wireless networks. *IEEE Signal Processing Letters*, vol. 17, no. 6, p. 535–538, 2010. DOI: 10.1109/LSP.2010.2046350

[5] DING, H., GE, J., DA COSTA, D. B., JIANG, Z. A new efficient low complexity scheme for multi-source multi-relay cooperative networks. *IEEE Transactions on Vehicular Technology*, vol. 60, no. 2, p. 716–722, 2011. DOI: 10.1109/TVT.2010.2100416

[6] HUANG, Y., AL-QAHTANI, F., WU, Q., ZHONG, C., WANG, J., ALNUWEIRI, H. Outage analysis of spectrum sharing relay systems with multi-secondary destinations under primary user's interference. *IEEE Transactions on Vehicular Technology*, vol. 63, no. 7, p. 3456–3463, 2014. DOI: 10.1109/TVT.2014.2297973

[7] GUIMARAES, F. R. V., DA COSTA, D. B., TSIFTSIS, T. A., CAVALCANTE, C. C., ET AL. Multi-user and multi-relay cognitive radio networks under spectrum sharing constraints. *IEEE Transactions on Vehicular Technology*, vol. 63, no. 1, p. 433–439, 2014. DOI: 10.1109/TVT.2013.2275201

[8] DE MELO, M. A. B., DA COSTA, D. B. An efficient relay-destination selection scheme for multiuser multirelay downlink cooperative networks. *IEEE Transactions on Vehicular Technology*, vol. 61, no. 5, p. 2354–2360, 2012. DOI: 10.1109/TVT.2012.2192488

[9] XU, T., GE, J., DING, H. Opportunistic scheduling for uplink cognitive cellular networks with outage protection of the primary user, *IEEE Communication Letters*, vol. 17, no. 1, p. 71–74, 2013. DOI: 10.1109/LCOMM.2012.111612.121872

[10] DUONG, T. Q., DA COSTA, D. B., ELKASHLAN, M., BAO, V. N. Q. Cognitive amplify-and-forward relay networks over Nakagami-m fading, *IEEE Transactions on Vehicular Technology*, vol. 61, no. 5, p. 2368–2374, 2012. DOI: 10.1109/TVT.2012.2192509

About the Authors...

Tao ZHANG received the B.S. degree in Communication Engineering from PLA University of Science and Technology, Nanjing, China in 2011. He is currently pursuing for the Ph.D. degree in Communications and Information Systems at the College of Communications Engineering, PLA University of Science and Technology, Nanjing, China. His current research interest includes cooperative communications, wireless sensor networks and physical layer security.

Weiwei YANG received the B.S. and M.S. degrees in College of Communications Engineering, PLA University of Science and Technology, Nanjing, China in 2003 and 2006, respectively. His research interests are mainly on OFDM systems, signal processing in communications, cooperative communications and wireless sensor networks and physical layer security.

Yueming CAI received the B.S. degree in Physics from Xiamen University, Xiamen, China in 1982, the M.S. degree in Micro-electronics Engineering and the Ph.D. degree in Communications and Information Systems both from Southeast University, Nanjing, China in 1988 and 1996 respectively. His current research interest includes MIMO systems, OFDM systems, signal processing in communications, cooperative communications, wireless sensor networks and physical layer security.

24

Optimal Power Allocation for Channel Estimation in MIMO-OFDM System with Per-Subcarrier Transmit Antenna Selection

Kaleeswaran RAJESWARI [1], *S. Jayaraman THIRUVENGADAM* [1,2]

[1] Dept. of Electronics and Communication Engg., Thiagarajar College of Engineering, 625 015 Madurai, TamilNadu, India
[2] TIFAC CORE in Wireless Technologies, Thiagarajar College of Engineering, 625 015 Madurai, TamilNadu, India

rajeswari@tce.edu, sjtece@tce.edu

Abstract. *A novel hybrid channel estimator is proposed for multiple-input multiple-output orthogonal frequency- division multiplexing (MIMO-OFDM) system with per-subcarrier transmit antenna selection having optimal power allocation among subcarriers. In practice, antenna selection information is transmitted through a binary symmetric control channel with a crossover probability. Linear minimum mean-square error (LMMSE) technique is optimal technique for channel estimation in MIMO-OFDM system. Though LMMSE estimator performs well at low signal to noise ratio (SNR), in the presence of antenna-to-subcarrier-assignment error (ATSA), it introduces irreducible error at high SNR. We have proved that relaxed MMSE (RMMSE) estimator overcomes the performance degradation at high SNR. The proposed hybrid estimator combines the benefits of LMMSE at low SNR and RMMSE estimator at high SNR. The vector mean square error (MSE) expression is modified as scalar expression so that an optimal power allocation can be performed. The convex optimization problem is formulated and solved to allocate optimal power to subcarriers minimizing the MSE, subject to transmit sum power constraint. Further, an analytical expression for SNR threshold at which the hybrid estimator is to be switched from LMMSE to RMMSE is derived. The simulation results show that the proposed hybrid estimator gives robust performance, irrespective of ATSA error.*

Keywords

MIMO-OFDM, per-subcarrier transmit antenna selection, antenna-to-subcarrier assignment, LMMSE, relaxed MMSE, hybrid estimator

1. Introduction

Orthogonal frequency division multiplexing (OFDM) is a popular method for high data rate wireless transmission [1]. Wireless standards such as digital audio broadcasting (DAB), digital video broadcasting-terrestrial (DVB-T), the IEEE 802.11a local area network (LAN) and the IEEE 802.16a metropolitan area network (MAN) have adopted OFDM technology. OFDM is also a potential candidate for fourth-generation (4G) mobile wireless systems. When OFDM is combined with multiple-input multiple-output (MIMO) system, it converts the frequency selective channel of the MIMO channel into a set of parallel frequency-flat channels. This decreases the MIMO receiver complexity [2]. Further, MIMO system provides spatial diversity by having spatially separated antennas [3]. Combination of OFDM and MIMO aims to increase the diversity gain and/or to enhance the system capacity [4]. Further, antenna selection in MIMO-OFDM systems provide considerable gains while only requiring a small amount of feedback to convey information to the receiver about the chosen transmit antennas [5–7]. The antenna to subcarrier assignment (ATSA) is signaled in a control channel from the transmitter to receiver.

In MIMO-OFDM systems, the antenna selection can be performed based on group-of-subcarriers or on a per-subcarrier. In bulk selection, one or more antennas are chosen among the available antennas, based on signal to noise ratio or based on performance metrics such as capacity or bit error rate (BER), for transmission on all frequencies. This reduces channel state information (CSI) feedback requirement and number of radio frequency (RF) chains. The per-tone selection is capable of achieving a much lower BER as it exploits an additional degree of freedom that allows the antenna selection to differ across the utilized bandwidth [8], [9].

In MIMO-OFDM system with per tone transmit antenna selection, channel estimation is the most essential task in compensating distortion from channels. LMMSE based channel estimation technique performs well in rapid dispersive fading channels and it is proved to be robust against channel power delay profile [10–12]. In [13], LMMSE channel estimator for a system employing per tone selection is derived. The performance of LMMSE based channel estimator in MIMO-OFDM system is improved by combining with optimal power allocation [15–18]. Extending this to per tone antenna selection scheme is not simple as the autocorrelation matrix is non-invertible due to fluctuations in its rank. This paper proposes a LMMSE based channel estimation along with optimal power allocation scheme for MIMO-OFDM with per tone antenna selection system. An eigenvalue decomposition (EVD) of the autocorrelation matrix is employed to obtain a scalar MSE expression such that optimal power allocation scheme can be applied. A convex optimization framework is formulated to minimize the channel estimation MSE subject to sum power constraint. The improvement in the performance of LMMSE estimate with optimal power allocation is validated through MSE and BER analysis.

When LMMSE channel estimator is used in the system employing per tone selection with ATSA error, it results in intolerable performance at high SNR. This is due to error introduced in channel frequency correlation [12]. To overcome this problem in the system with equal power allocation to subcarriers, a hybrid estimator is proposed in [13]. This paper extends the design of hybrid estimator with optimal power allocation to subcarriers. Hybrid estimator requires less knowledge about the channel correlation and robust to ATSA error. It improves the performance of LMMSE estimator by switching to relaxed MMSE which is less complex and easy to implement. A closed form expression for the SNR threshold at which the estimator switches from LMMSE to relaxed MMSE is derived analytically.

The paper is organized as follows. System model with per subcarrier transmit antenna selection and maximal ratio combining at the receiver is presented in Sec. 2. The effect of incorrect antenna selection in the channel estimation is analyzed in Sec. 3. The proposed method of estimating CSI is described in Sec. 4. The performance of the proposed technique is analyzed by simulations in Sec. 5. Section 6 concludes the paper. In this paper, boldface letters are used to denote matrices and vectors. Superscript H denotes Hermitian operations, $\otimes$ denotes the Kronecker product and $E[\ ]$ denotes the expectation operation. vec($\mathbf{C}$) transforms a matrix $\mathbf{C}=[\mathbf{c}_1,\mathbf{c}_2,\ldots,\mathbf{c}_n]$ into a column vector $[\mathbf{c}_1^T,\mathbf{c}_2^T,\ldots,\mathbf{c}_n^T]$, where $\mathbf{c}_i$ is the ith column vector of $\mathbf{C}$.

2. System Model

Consider a MIMO-OFDM system with n_t transmit antennas and n_r receive antennas. It is assumed that the OFDM symbol has N subcarriers. Let $\mathbf{g}(i,j)$ be a $L\times 1$ channel impulse response vector between the jth transmit antenna and the ith receive antenna. The corresponding channel frequency response (CFR) between the jth transmit antenna and the ith receive antenna at the kth subcarrier is given by

$$f_k(i,j)=\sum_{l=0}^{L-1} g_l(i,j)e^{-j2\pi lk/N},\quad k=0,1\ldots N-1. \quad (1)$$

At each subcarrier, one of the n_t transmit antennas is selected for transmission. The criterion for selecting the jth transmit antenna is given by

$$\hat{j}_k=\underset{j\in[1,2\ldots n_t]}{\arg\max}\left\{\sum_{i=1}^{n_r}\left|f_k(i,j)\right|^2\right\},\quad k=0,1\ldots N-1. \quad (2)$$

This ATSA information is signaled through a binary symmetric control channel to the receiver. The $N\times 1$ receive signal vector at the ith receive antenna is given by

$$\mathbf{y}^{(i)}=\mathbf{X}\mathbf{h}^{(i)}+\mathbf{w}^{(i)} \quad (3)$$

where $\mathbf{X}$ is a $N\times N$ diagonal matrix, defined as $\mathbf{X}=\text{diag}(x_0, x_1, \ldots, x_{N-1})$ and $x_k=\sqrt{p_k}b_k$. The data on the kth subcarrier b_k is assumed to be a random variable with zero mean and unit variance and p_k is power of the kth subcarrier. $\mathbf{h}^{(i)}$ is a $N\times 1$ selected CFR vector at the ith receive antenna, defined as

$$\mathbf{h}^{(i)}=\mathbf{S}_o\mathbf{f}^{(i)} \quad (4)$$

where

$$\mathbf{f}^{(i)}=\left[\mathbf{f}(i,1)^T\ \mathbf{f}(i,2)^T\ \ldots\ \mathbf{f}(i,n_t)^T\right] \quad (5)$$

$\mathbf{f}(i,j)$ is the $N\times 1$ CFR vector between the jth transmit antenna and the ith receive antenna. $\mathbf{S}_o$ is a $N\times Nn_t$ selection matrix given by

$$\mathbf{S}_o=\left[\ \text{diag}(\mathbf{a}_1)\ \ \text{diag}(\mathbf{a}_2)\ \ \ldots\ \ \text{diag}(\mathbf{a}_{n_t})\right] \quad (6)$$

where $\mathbf{a}_j$ is $N\times 1$ vector with its kth element being unity if the jth antenna is selected for the kth subcarrier transmission. $\mathbf{w}^{(i)}$ is the $N\times 1$ noise vector with mean zero and covariance matrix $\sigma_w^2\mathbf{I}_N$ of the ith receive antenna.

3. Optimal Power Allocation Algorithm

$\mathbf{R_h}^{(i)}$ is a Hermitian and positive definite matrix with the eigenvalue decomposition

$$\mathbf{R}_{\mathbf{h}}^{(i)}=\mathbf{U}^{(i)}\mathbf{\Lambda}^{(i)}\left(\mathbf{U}^{(i)}\right)^H \quad (7)$$

where $\mathbf{U}^{(i)}$ is a unitary matrix and $\mathbf{\Lambda}^{(i)}$ is $N\times N$ diagonal matrix with non zero diagonal elements $\left[\lambda_0^{(i)},\lambda_1^{(i)},\ldots,\lambda_{r_c-1}^{(i)}\right]$. The channel CFR vector $\mathbf{h}^{(i)}$ is assumed to be Gaussian with zero mean and auto correlation matrix $\mathbf{R_h}^{(i)}$ of size $N\times N$. The LMMSE estimation of selected CFR vector $\mathbf{h}^{(i)}$ is given by [19]

$$\hat{\mathbf{h}}^{(i)}=E\left[\mathbf{h}^{(i)}\left(\mathbf{y}^{(i)}\right)^H\right]E\left[\mathbf{y}^{(i)}\left(\mathbf{y}^{(i)}\right)^H\right]^{-1}\mathbf{y}^{(i)}. \quad (8)$$

Substituting (3), the statistical average of $E[\mathbf{h}^{(i)}(\mathbf{y}^{(i)})^H]$ and $E[\mathbf{y}^{(i)}(\mathbf{y}^{(i)})^H]$ are determined as

$$E\left[\mathbf{h}^{(i)}\left(\mathbf{y}^{(i)}\right)^H\right]=\mathbf{R}_{\mathbf{h}}^{(i)}\mathbf{X}^H$$

and
$$E\left[\mathbf{y}^{(i)}\left(\mathbf{y}^{(i)}\right)^H\right]=\mathbf{X}\mathbf{R}_{\mathbf{h}}^{(i)}\mathbf{X}^H+\sigma_w^2\mathbf{I}_N \quad (9)$$

where $\mathbf{R}_{\mathbf{h}}^{(i)}=E\left[\mathbf{h}^{(i)}\left(\mathbf{h}^{(i)}\right)^H\right]$. Using (4), $\mathbf{R}_{\mathbf{h}}^{(i)}$ is written as

$$\mathbf{R}_{\mathbf{h}}^{(i)}=\mathbf{S}_o E\left[\mathbf{f}^{(i)}\left(\mathbf{f}^{(i)}\right)^H\right]\mathbf{S}_o^H=\mathbf{S}_o\mathbf{R}_{\mathbf{f}}^{(i)}\mathbf{S}_o^H . \quad (10)$$

Substituting (9) and (10) in (8), the estimation of CFR vector $\mathbf{h}^{(i)}$ is given by

$$\hat{\mathbf{h}}^{(i)}=\mathbf{R}_{\mathbf{h}}^{(i)}\mathbf{X}^H\left(\mathbf{X}\mathbf{R}_{\mathbf{h}}^{(i)}\mathbf{X}^H+\sigma_w^2\mathbf{I}_N\right)^{-1}\mathbf{y}^{(i)} . \quad (11)$$

The MSE in estimating the CFR vector $\mathbf{h}^{(i)}$ is defined as

$$h_{\text{MSE}}^{(i)}=Tr\left(E\left[\left\|\mathbf{h}^{(i)}-\hat{\mathbf{h}}^{(i)}\right\|^2\right]\right). \quad (12)$$

Substituting (11) in (12), MSE is derived as

$$h_{\text{MSE}}^{(i)}=Tr\left(\mathbf{R}_{\mathbf{h}}^{(i)}-\mathbf{R}_{\mathbf{h}}^{(i)}\mathbf{X}^H\left(\mathbf{X}\mathbf{R}_{\mathbf{h}}^{(i)}\mathbf{X}^H+\sigma_w^2\mathbf{I}_N\right)^{-1}\mathbf{X}\mathbf{R}_{\mathbf{h}^{(i)}}\right). \quad (13)$$

Using matrix inversion lemma, $\mathbf{A}^{-1}-\mathbf{A}^{-1}\mathbf{B}\left(\mathbf{D}\mathbf{A}^{-1}\mathbf{B}+\mathbf{C}^{-1}\right)^{-1}\mathbf{D}\mathbf{A}^{-1}=\left(\mathbf{A}+\mathbf{B}\mathbf{C}\mathbf{D}\right)^{-1}$, and assuming $\mathbf{A}=\left(\mathbf{R}_{\mathbf{h}}^{(i)}\right)^{-1}$, $\mathbf{B}=\mathbf{X}^H$, $\mathbf{C}=\sigma_w^{-2}\mathbf{I}_N$ and $\mathbf{D}=\mathbf{X}$, (13) is simplified as

$$h_{\text{MSE}}^{(i)}=Tr\left(\left(\sigma_w^{-2}\mathbf{X}^H\mathbf{X}+\left(\mathbf{R}_{\mathbf{h}}^{(i)}\right)^{-1}\right)^{-1}\right) \quad (14)$$

On rearranging, (14) is rewritten as

$$h_{\text{MSE}}^{(i)}=Tr\left(\mathbf{R}_{\mathbf{h}}^{(i)}\left(\sigma_w^{-2}\mathbf{R}_{\mathbf{h}}^{(i)}\mathbf{X}^H\mathbf{X}+\mathbf{I}_N\right)^{-1}\right). \quad (15)$$

As $\mathbf{R}_{\mathbf{h}}^{(i)}$ is a Hermitian and positive definite matrix and $\mathbf{X}^H\mathbf{X}$ in (15) is a multiple of the identity matrix, $\left(\sigma_w^{-2}\mathbf{R}_{\mathbf{h}}^{(i)}\mathbf{X}^H\mathbf{X}+\mathbf{I}_N\right)$ is also Hermitian and positive definite matrix. It has the same basis of eigen decomposition as that of $\mathbf{R}_{\mathbf{h}}^{(i)}$ [21]. The eigen decomposition of $\left(\sigma_w^{-2}\mathbf{R}_{\mathbf{h}}^{(i)}\mathbf{X}^H\mathbf{X}+\mathbf{I}_N\right)$ is written as

$$\left(\sigma_w^{-2}\mathbf{R}_{\mathbf{h}}^{(i)}\mathbf{X}^H\mathbf{X}+\mathbf{I}_N\right)=\mathbf{U}^{(i)}\overline{\Lambda}^{(i)}\left(\mathbf{U}^{(i)}\right)^H \quad (16)$$

where $\overline{\Lambda}$ is $N\times N$ diagonal matrix with the r_c number of non zero diagonal elements $\left[\overline{\lambda}_0^{(i)},\overline{\lambda}_1^{(i)},\ldots,\overline{\lambda}_{r_c-1}^{(i)}\right]$. The value of $\overline{\lambda}_k^{(i)}$ can be written in terms of $\lambda_k^{(i)}$ as

$$\overline{\lambda}_k^{(i)}=\sigma_w^{-2}x_k^*x_k\lambda_k^{(i)}+1 , \ k=0,1,\ldots,r_c-1 . \quad (17)$$

Let $p_k=x_k^*x_k$. The vector expression for MSE in (15) is reduced to scalar expression as

$$h_{\text{MSE}}^{(i)}=\sum_{k=0}^{r_c-1}\frac{\lambda_k^{(i)}}{\sigma_w^{-2}p_k\lambda_k^{(i)}+1} . \quad (18)$$

Mean square error in (18) can be minimized by allocating optimal power to r_c subcarriers, with the sum power constraint $\sum_{k=0}^{r_c-1}p_k=P$ where P is the total power. Mathematically, the optimization problem is formulated as

$$\underset{p_k}{\text{Minimize}}\ \text{MSE}=\sum_{k=0}^{r_c-1}\frac{\lambda_k^{(i)}}{\sigma_w^{-2}p_k\lambda_k^{(i)}+1},\quad \text{s.t}\sum_{k=0}^{r_c-1}p_k=P . \quad (19)$$

The equation in (19) is a constrained optimization problem and can be solved using Lagrangian multiplier method. The Lagrangian associated with the minimization problem in (19) is given by

$$L\left(\mu,p_k\right)=\sum_{k=0}^{r_c-1}\frac{\lambda_k^{(i)}}{\sigma_w^{-2}p_k\lambda_k^{(i)}+1}+\mu\left(\sum_{k=0}^{r_c-1}p_k-P\right),\ k=0,1,\ldots,r_c-1 \quad (20)$$

where μ is Lagrangian multiplier.

The optimal power for the kth subcarrier p_k is computed by differentiating (20) with respect to p_k and equating to zero. It is determined as

$$p_k^{opt}=\left(\sqrt{\frac{\sigma_w^2}{\mu}}-\frac{\sigma_w^2}{\lambda_k^{(i)}}\right)^+,\quad k=0,1,\ldots,r_c-1 \quad (21)$$

where $(y)^+=\max(y,0)$. Using (21) and the sum power constraint in (19), the expression for Lagrangian multiplier μ is derived as

$$\mu=\sigma_w^2r_c^2\left(P+\sigma_w^2\sum_{k=0}^{r_c-1}\frac{1}{\lambda_k^{(i)}}\right)^{-2} . \quad (22)$$

Substituting the value of μ from (22) in (21), the expression for optimal power for the kth subcarrier that minimizes the mean square error of the LMMSE channel estimator is expressed as

$$p_k^{opt}=\left(\frac{P}{r_c}+\frac{\sigma_w^2}{r_c}\sum_{k=0}^{r_c-1}\frac{1}{\lambda_k^{(i)}}-\frac{\sigma_w^2}{\lambda_k^{(i)}}\right)^+,\ k=0,1,\ldots,r_c-1 . \quad (23)$$

4. Performance Analysis with ATSA Error

This paper deals with MIMO-OFDM system in Time Division Duplexing (TDD) environment. At the transmitter, antenna is selected for each subcarrier based on channel coefficients between each transmit antenna and receive antenna. The antenna selection information is transmitted to the receiver through a binary symmetric control channel. However, when the bandwidth of binary symmetric control channel is limited, it introduces ATSA error. Conventionally, LMMSE estimator is employed at the receiver to estimate the channel coefficients at the receiver. Although it performs well at low SNR, it introduces irreducible error at high SNR in the presence of ATSA error. This section characterizes the MSE due to ATSA error. Let $\mathbf{S}_{o,k}$ be $N \times Nn_t$ selection matrix with error in the kth subcarrier. The resultant CFR vector is $\mathbf{h}_k^{(i)} = \mathbf{S}_{o,k}\mathbf{f}^{(i)}$. The correlation matrix of $\mathbf{h}_k^{(i)}$ is $\mathbf{R}_{\mathbf{h},k}^{(i)} = \mathbf{S}_{o,k} E\left[\mathbf{f}^{(i)}\left(\mathbf{f}^{(i)}\right)^H\right]\mathbf{S}_{o,k}^H = \mathbf{S}_{o,k}\mathbf{R}_{\mathbf{f}}^{(i)}\mathbf{S}_{o,k}^H$. Then, LMMSE channel estimate in (11) is modified as

$$\hat{\mathbf{h}}_k^{(i)} = \mathbf{R}_{\mathbf{h},k}^{(i)}\mathbf{X}^H\left(\mathbf{X}\mathbf{R}_{\mathbf{h},k}^{(i)}\mathbf{X}^H + \sigma_w^2\mathbf{I}_N\right)^{-1}\mathbf{y}^{(i)}. \tag{24}$$

Using simple algebraic metric, the LMMSE estimate in (24) is modified as [20]

$$\hat{\mathbf{h}}_k^{(i)} = \mathbf{R}_{\mathbf{h},k}^{(i)}\mathbf{X}^H\mathbf{X}^{-1}\left(\mathbf{R}_{\mathbf{h},k}^{(i)} + \sigma_w^2\left(\mathbf{X}\mathbf{X}^H\right)^{-1}\right)^{-1}\mathbf{X}^{-H}\mathbf{y}^{(i)}. \tag{25}$$

With real data as pilots, (25) is rewritten as,

$$\hat{\mathbf{h}}_k^{(i)} = \mathbf{R}_{\mathbf{h},k}^{(i)}\left(\mathbf{R}_{\mathbf{h},k}^{(i)} + \sigma_w^2\left(\mathbf{X}\mathbf{X}^H\right)^{-1}\right)^{-1}\mathbf{X}^{-1}\mathbf{y}^{(i)}. \tag{26}$$

The mean square error of LMMSE estimator with ATSA error in the kth subcarrier is written as

$$\begin{aligned} h_{\text{MSE}(k,ATSA)}^{(i)} &= E\left[\left\|\mathbf{h}^{(i)} - \hat{\mathbf{h}}_k^{(i)}\right\|^2\right] \\ &= E\left[\left(\mathbf{h}^{(i)} - \hat{\mathbf{h}}_k^{(i)}\right)\left(\mathbf{h}^{(i)}\right)^H - \left(\mathbf{h}^{(i)} - \hat{\mathbf{h}}_k^{(i)}\right)\left(\hat{\mathbf{h}}_k^{(i)}\right)^H\right]. \end{aligned} \tag{27}$$

According to orthogonality principle, the second term in (27) is zero [19]. Then, $h_{\text{MSE}(k,ATSA)}^{(i)}$ is given by

$$h_{\text{MSE}(k,ATSA)}^{(i)} = E\left[\left(\mathbf{h}^{(i)} - \hat{\mathbf{h}}_k^{(i)}\right)\left(\mathbf{h}^{(i)}\right)^H\right]. \tag{28}$$

Substituting (26) in (28), $h_{\text{MSE}(k,ATSA)}^{(i)}$ is determined as

$$\begin{aligned} h_{\text{MSE}(k,ATSA)}^{(i)} &= \\ &= Tr\left(\mathbf{R}_{\mathbf{h}}^{(i)} - \sigma_w^{-2}\mathbf{R}_{\mathbf{h}}^{(i)}\mathbf{R}_{\mathbf{h},k}^{(i)}\mathbf{X}\mathbf{X}^H\left(\sigma_w^{-2}\mathbf{R}_{\mathbf{h},k}^{(i)}\left(\mathbf{X}\mathbf{X}^H\right) + \mathbf{I}_N\right)^{-1}\right) \end{aligned} \tag{29}$$

The expression for $h_{\text{MSE}(k,ATSA)}^{(i)}$ in (29) can be simplified further using eigenvalue decomposition of the terms $\mathbf{R}_{\mathbf{h},k}^{(i)}$ and $\sigma_w^{-2}\mathbf{R}_{\mathbf{h},k}^{(i)}\left(\mathbf{X}\mathbf{X}^H\right) + \mathbf{I}_N$. The EVD of $\mathbf{R}_{\mathbf{h},k}^{(i)}$ is $\mathbf{U}_k^{(i)}\mathbf{\Lambda}_k^{(i)}\left(\mathbf{U}_k^{(i)}\right)^H$ where $\mathbf{U}_k^{(i)}$ is a unitary matrix and $\mathbf{\Lambda}_k^{(i)}$ is $N \times N$ diagonal matrix with its eigenvalues as diagonal elements. $\mathbf{X}\mathbf{X}^H$ in (29) is a multiple of the identity matrix. Then, $\left(\sigma_w^{-2}\mathbf{R}_{\mathbf{h},k}^{(i)}\mathbf{X}\mathbf{X}^H + \mathbf{I}_N\right)$ is also Hermitian and positive definite matrix and it has the same basis of eigen decomposition as that of $\mathbf{R}_{\mathbf{h},k}^{(i)}$ [21]. Hence, the eigen decomposition of $\left(\sigma_w^{-2}\mathbf{R}_{\mathbf{h},k}^{(i)}\mathbf{X}\mathbf{X}^H + \mathbf{I}_N\right)$ can be written as

$$\left(\sigma_w^{-2}\mathbf{R}_{\mathbf{h},k}^{(i)}\mathbf{X}\mathbf{X}^H + \mathbf{I}_N\right) = \mathbf{U}_k^{(i)}\overline{\Lambda}_k^{(i)}\left(\mathbf{U}_k^{(i)}\right)^H \tag{30}$$

where $\overline{\Lambda}_k^{(i)}$ is the diagonal matrix with its jth diagonal element as

$$\overline{\lambda}_{k,j}^{(i)} = \sigma_w^{-2}x_j^*x_j\lambda_{k,j}^{(i)} + 1, \quad j = 0,1,\ldots,r_k - 1 \tag{31}$$

where r_k is the rank of $\mathbf{R}_{\mathbf{h},k}^{(i)}$. Substituting (30) in (29), $h_{\text{MSE}(k,ATSA)}^{(i)}$ is written as

$$\begin{aligned} h_{\text{MSE}(k,ATSA)}^{(i)} &= \sum_{j=0}^{r_c-1}\lambda_j^{(i)} - \\ &\sigma_w^{-2}Tr\left(\left(\mathbf{U}_k^{(i)}\right)^H\mathbf{R}_{\mathbf{h}}^{(i)}\mathbf{R}_{\mathbf{h},k}^{(i)}\mathbf{X}\mathbf{X}^H\mathbf{U}_k^{(i)}\left(\overline{\Lambda}_k^{(i)}\right)^{-1}\right) \end{aligned} \tag{32}$$

where r_c is the rank of $\mathbf{R}_{\mathbf{h}}^{(i)}$. Let

$$\left(\mathbf{U}_k^{(i)}\right)^H\mathbf{R}_{\mathbf{h}}^{(i)}\mathbf{R}_{\mathbf{h},k}^{(i)}\mathbf{X}\mathbf{X}^H\mathbf{U}_k^{(i)} = \mathbf{W}_k^{(i)}. \tag{33}$$

Using (31) and (33), (32) is simplified as

$$h_{\text{MSE}(k,ATSA)}^{(i)} = \sum_{j=0}^{r_c-1}\lambda_j^{(i)} - \sum_{j=0}^{r_k-1}\left(\frac{\left(\mathbf{W}_k^{(i)}\right)_{jj}}{\lambda_{k,j}^{(i)}x_jx_j^* + \sigma_w^2}\right). \tag{34}$$

With perfect ATSA, replacing $\mathbf{R}_{\mathbf{h},k}^{(i)}$ with $\mathbf{R}_{\mathbf{h}}^{(i)}$ in (29) and simplifying gives an expression for mean square error with perfect ATSA in the same format as (34). It is derived as,

$$h_{\text{MSE}(c)}^{(i)} = \sum_{j=0}^{r_c-1}\lambda_j^{(i)} - \sum_{j=0}^{r_c-1}\left(\frac{\left(\mathbf{W}_c^{(i)}\right)_{jj}}{\lambda_j^{(i)}x_jx_j^* + \sigma_w^2}\right) \tag{35}$$

where $\mathbf{W}_c^{(i)} = \left(\Lambda^{(i)}\right)^2\left(\mathbf{U}^{(i)}\right)^H\mathbf{X}\mathbf{X}^H\mathbf{U}^{(i)}$. When a binary symmetric channel with the crossover probability of q is utilized to send the selection information in $\mathbf{S}_o$, the probability of receiving a symbol with no ATSA error is $\frac{q_1}{q_1 + q_2r_c}$ [13], where $q_1 = (1-q)^{r_c}$ and $q_2 = q(1-q)^{r_c-1}$. The probability of receiving a symbol with error at the kth subcarrier is $\frac{q_2}{q_1 + q_2r_c}$. Then, the average mean square error of LMMSE estimator with crossover probability q of binary symmetric channel is given by

$$h_{\text{MSE}(q)}^{(i)} = \frac{q_1}{q_1 + q_2 r_c} h_{\text{MSE}(c)}^{(i)} + \frac{q_2}{q_1 + q_2 r_c} \sum_{k=1}^{r_c} h_{\text{MSE}(k,ATSA)}^{(i)} . \quad (36)$$

Substituting (34) and (35) in (36) and replacing $x_j x_j^*$ by the optimal power p_j derived in Sec. 3, $h_{\text{MSE}(q)}^{(i)}$ is given by

$$h_{\text{MSE}(q)}^{(i)} = \bar{q}_1 \left(\sum_{j=0}^{r_c-1} \lambda_j^{(i)} - \sum_{j=0}^{r_c-1} \left(\frac{\left(\mathbf{W}_c^{(i)}\right)_{jj}}{\lambda_j^{(i)} p_j + \sigma_w^2} \right) \right) + \bar{q}_2 \sum_{k=1}^{r_c} \left(\sum_{j=0}^{r_c-1} \lambda_j^{(i)} - \sum_{j=0}^{r_k-1} \left(\frac{\left(\mathbf{W}_k^{(i)}\right)_{jj}}{\lambda_{k,j}^{(i)} p_j + \sigma_w^2} \right) \right) \quad (37)$$

where $\bar{q}_1 = \frac{q_1}{q_1 + q_2 r_c}$ and $\bar{q}_2 = \frac{q_2}{q_1 + q_2 r_c}$.

Multiplying each element within summation in second term and fourth term of (37) by $\left(\lambda_j^{(i)} p_j - \sigma_w^2\right)$ and $\left(\lambda_{k,j}^{(i)} p_j - \sigma_w^2\right)$ respectively and ignoring the term σ_w^4 to get high SNR approximation results in

$$h_{\text{MSE}(q)}^{(i)} = \bar{q}_1 \sum_{j=0}^{r_c-1} \lambda_j^{(i)} - \bar{q}_1 \sum_{j=0}^{r_c-1} \frac{\left(\mathbf{W}_c^{(i)}\right)_{jj}}{\lambda_j^{(i)} p_j} + \bar{q}_1 \sigma_w^2 \sum_{j=0}^{r_c-1} \frac{\left(\mathbf{W}_c^{(i)}\right)_{jj}}{\left(\lambda_j^{(i)} p_j\right)^2} + \bar{q}_2 \sum_{k=1}^{r_c} \sum_{j=0}^{r_c-1} \lambda_j^{(i)} - \bar{q}_2 \sum_{k=1}^{r_c} \sum_{j=0}^{r_k-1} \frac{\left(\mathbf{W}_k^{(i)}\right)_{jj}}{\lambda_{k,j}^{(i)} p_j} + \bar{q}_2 \sigma_w^2 \sum_{k=1}^{r_c} \sum_{j=0}^{r_k-1} \frac{\left(\mathbf{W}_k^{(i)}\right)_{jj}}{\left(\lambda_{k,j}^{(i)} p_j\right)^2} . \quad (38)$$

5. Proposed Hybrid Estimator with Optimal Power Allocation

The performance degradation of LMMSE estimator due to the error in selection information can be overcome only if the correlation matrix $\mathbf{R_h}^{(i)}$ is independent of selection matrix. If suppose, the LMMSE estimator is relaxed to use the identity matrix instead of correlation matrix $\mathbf{R_h}^{(i)}$, then the performance degradation can be minimized significantly [22]. In (11), substituting $\mathbf{R_h}^{(i)} = \alpha\mathbf{I}$, the relaxed MMSE estimate is derived as

$$\hat{\mathbf{h}}_R^{(i)} = \mathbf{X}^H \left(\mathbf{X}\mathbf{X}^H + \sigma_w^2 \mathbf{I}_N\right)^{-1} \mathbf{y}^{(i)} . \quad (39)$$

By minimizing $E\left(\left\|\mathbf{h}^{(i)} - \hat{\mathbf{h}}_R^{(i)}\right\|^2\right)$, mean square error of the relaxed estimator is derived as

$$h_{\text{MSE}(R)}^{(i)} = \sum_{j=0}^{r_c-1} \lambda_j^{(i)} - \sum_{j=0}^{r_c-1} \frac{x_j x_j^* \lambda_j^{(i)}}{x_j x_j^* + \sigma_w^2} . \quad (40)$$

Replacing $x_j x_j^*$ by optimal power p_j derived in Sec. 3, (40) is rewritten as

$$h_{\text{MSE}(R)}^{(i)} = \sum_{j=0}^{r_c-1} \lambda_j^{(i)} - \sum_{j=0}^{r_c-1} \frac{p_j \lambda_j^{(i)}}{p_j + \sigma_w^2} . \quad (41)$$

Multiplying and dividing the term $\left(p_j - \sigma_w^2\right)$ inside the summation of the second term, and ignoring the term $\left(\sigma_w^2\right)^2$, (41) becomes

$$h_{\text{MSE}(R)}^{(i)} = \sigma_w^2 \sum_{j=0}^{r_c-1} \frac{\lambda_j^{(i)}}{p_j} . \quad (42)$$

It is observed that $h_{\text{MSE}(R)}^{(i)}$ is larger compared to $h_{\text{MSE}(q)}^{(i)}$ of LMMSE estimator at low SNR, as it depends only on diagonal elements of the correlation matrix. Its value decreases with increasing SNR, where LMMSE estimator gives irreducible error. Combining the merits of both LMMSE and RMMSE estimators, a hybrid estimator is proposed. It performs as LMMSE estimator till the noise dominates the effect of ATSA error. The proposed estimator switches to RMMSE estimator when $h_{\text{MSE}(q)}^{(i)}$ of LMMSE estimator becomes $\sigma_w^2 \sum_{j=0}^{r_c-1} \frac{\lambda_j^{(i)}}{p_j}$. The receive SNR per-sub-carrier at which hybrid estimator switches from LMMSE to RMMSE is determined from

$$h_{\text{MSE}(q)}^{(i)} = h_{\text{MSE}(R)}^{(i)} . \quad (43)$$

Substituting (38) and (42) in (43), it is rewritten as

$$\bar{q}_1 \sum_{j=0}^{r_c-1} \lambda_j^{(i)} - \bar{q}_1 \sum_{j=0}^{r_c-1} \frac{\left(\mathbf{W}_c^{(i)}\right)_{jj}}{\lambda_j^{(i)} p_j} + \bar{q}_1 \sigma_w^2 \sum_{j=0}^{r_c-1} \frac{\left(\mathbf{W}_c^{(i)}\right)_{jj}}{\left(\lambda_j^{(i)} p_j\right)^2} + \bar{q}_2 \sum_{k=1}^{r_c} \sum_{j=0}^{r_c-1} \lambda_j^{(i)} - \bar{q}_2 \sum_{k=1}^{r_c} \sum_{j=0}^{r_k-1} \frac{\left(\mathbf{W}_k^{(i)}\right)_{jj}}{\lambda_{k,j}^{(i)} p_j} + \bar{q}_2 \sigma_w^2 \sum_{k=1}^{r_c} \sum_{j=0}^{r_k-1} \frac{\left(\mathbf{W}_k^{(i)}\right)_{jj}}{\left(\lambda_{k,j}^{(i)} p_j\right)^2} = \sigma_w^2 \sum_{j=0}^{r_c-1} \frac{\lambda_j^{(i)}}{p_j} . \quad (44)$$

Replacing σ_w^2 by $1/\beta$ and solving for β results in

$$\text{SNR}_{TH} = \beta = \frac{c_3 - \bar{q}_1 c_4 - \bar{q}_2 c_5}{\bar{q}_1 c_1 + \bar{q}_2 c_2} \quad (45)$$

where

$$c_1 = \sum_{j=0}^{r_c-1} \lambda_j^{(i)} - \sum_{j=0}^{r_c-1} \frac{\left(\mathbf{W}_c^{(i)}\right)_{jj}}{\lambda_j^{(i)} p_j}, \quad c_2 = r_c \sum_{j=0}^{r_c-1} \lambda_j^{(i)} - \sum_{k=1}^{r_c} \sum_{j=0}^{r_k-1} \frac{\left(\mathbf{W}_k^{(i)}\right)_{jj}}{\lambda_{k,j}^{(i)} p_j},$$

$$c_3 = \sum_{j=0}^{r_c-1} \frac{\lambda_j^{(i)}}{p_j}, \quad c_4 = \sum_{j=0}^{r_c-1} \frac{\left(\mathbf{W}_c^{(i)}\right)_{jj}}{\left(\lambda_j^{(i)} p_j\right)^2}, \quad c_5 = \sum_{k=1}^{r_c} \sum_{j=0}^{r_k-1} \frac{\left(\mathbf{W}_k^{(i)}\right)_{jj}}{\left(\lambda_{k,j}^{(i)} p_j\right)^2} .$$

6. Results and Discussion

In this section, simulations are carried out to analyze the MSE performance of the proposed hybrid channel estimator with optimal power allocation in the presence of ATSA error in MIMO-OFDM system with per-subcarrier antenna selection. The simulation parameters are listed in Tab. 1.

Sl. No	Parameters	Values
	No. of transmit antennas (n_t)	2
	No. of receive antennas (n_r)	2
	Sampling frequency	30.72 MHz
	Channel power delay profile	Uniform, Extended Pedestrian-A, Extended Vehicular-A
	Length of channel (L)	4 (Uniform) 14 (Extended Pedestrian-A) [22] 78 (Extended Vehicular- A)
	No. of subcarriers (N)	16 64 (Extended Pedestrian-A) 128 (Extended Vehicular- A)

Tab. 1. Simulation parameters for MIMO-OFDM system with per-subcarrier antenna selection with optimal power allocation.

The effect of ATSA error in average MSE is studied by plotting the normalized error, which is defined as $\eta = \left| h^{(i)}_{\mathrm{MSE}(c)} - h^{(i)}_{\mathrm{MSE}(q)} \right| / h^{(i)}_{\mathrm{MSE}(c)}$. Figure 1 shows the effect of ATSA error in LMMSE estimator for the control channel crossover probabilities of 0.1 and 0.01, in a 2x2 MIMO-OFDM system with per-subcarrier transmit antenna selection. The number of subcarriers in the OFDM symbol is 16. With the crossover probability of $q = 0.1$, η is 0.2592 at SNR of 10 dB and it increases to 7.417 at SNR of 30 dB. Similarly, when $q = 0.01$, η increases from 0.02985 at 10 dB to 0.8685 at 30 dB. As the normalized error increases with increase in SNR, performance of LMMSE estimator decreases as SNR increases.

Figure 2 shows the MSE performance of LMMSE estimator with perfect ATSA, when the proposed optimal power allocation is applied at subcarrier level in 2x2 MIMO-OFDM system with per-subcarrier transmit antenna selection. It is assumed that the number of subcarriers in the OFDM signal is 16. The length of the channel is $L = 4$. The performance of the estimator is compared with MSE performance of the LMMSE estimator with equal power allocation. It is observed that the LMMSE estimator with optimal power allocation requires 1.2 dB less SNR compared to LMMSE estimator with equal power allocation at the MSE of -30 dB.

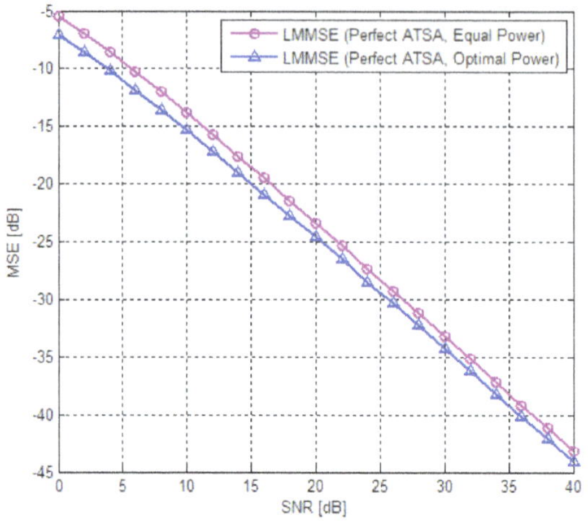

Fig. 2. MSE performance of LMMSE estimator with optimal and equal power allocation.

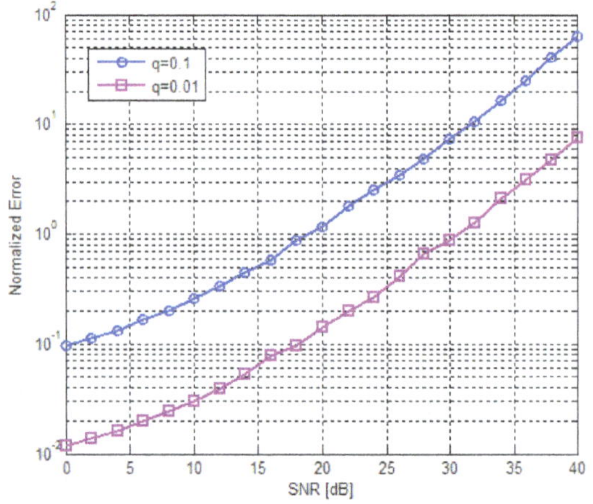

Fig. 1. Effect of ATSA error in the performance of LMMSE estimation with optimal power allocation.

Figure 3 shows the MSE performance of the proposed hybrid estimator with optimal power allocation. The MSE performance of the proposed hybrid estimator is compared with MSE performances of LMMSE estimator with perfect ATSA, LMMSE estimator with ATSA error and RMMSE estimator with optimal power allocation. It is assumed that the crossover probability of binary symmetric channel is, $q = 0.1$ which introduces ATSA error. The number of subcarriers used in OFDM signal is $N = 16$ and the channel length is taken as $L = 4$. The performance of RMMSE estimator is better than that of LMMSE with ATSA error, at high SNR. As the proposed hybrid estimator is designed such that it combines the merits of both LMMSE and

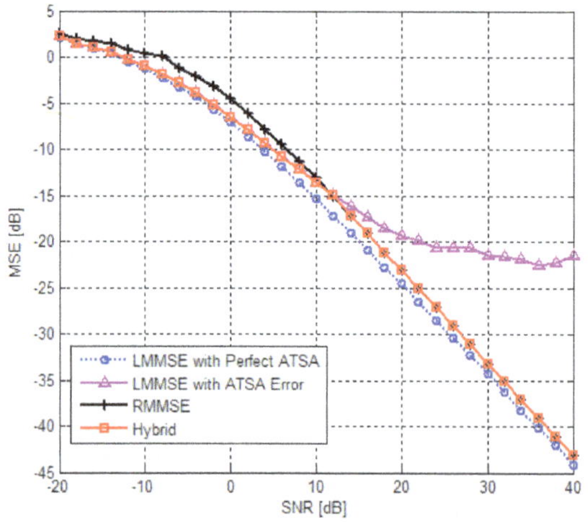

Fig. 3. MSE performance of estimators for $q = 0.1$ and $N = 16$ with optimal power allocation.

RMMSE estimators, the MSE performance of the proposed hybrid estimator is same as LMMSE estimator up to 2 dB SNR. At 2 dB, the performance of LMMSE estimator starts degrading. Further, MSE of LMMSE estimator becomes irreducible and almost constant from the SNR of 12 dB. The proposed hybrid estimator switches to RMMSE estimator at 12 dB. The threshold SNR, SNR_{TH} is calculated using (45) for various crossover probabilities of binary symmetric channel assuming that the number of subcarriers is $N = 16$ in a 2x2 MIMO-OFDM system with per-subcarrier antenna selection and with optimal power allocation. They are summarized in Tab. 2. At the threshold SNR, the proposed hybrid estimator switches from LMMSE estimator to RMMSE estimator to improve the MSE performance.

Crossover probability, q	0.1	0.09	0.07	0.05	0.03	0.01	0.001
With optimal power allocation SNR threshold SNR_{TH}(dB)	12	12.5	13	14	16	18	26
With equal power allocation SNR threshold SNR_{TH}(dB)	16	16.3	16.7	17	18	23	31

Tab. 2. Threshold SNR for different crossover probabilities q with optimal and equal power allocation.

Figure 4 shows the MSE performance of the proposed hybrid estimator for $q = 0.01$ for $N = 16$ and 128 with optimal power allocation among subcarriers. The hybrid estimator switches to RMMSE estimator at SNR of 18 dB and 22 dB with 16 and 128 subcarriers per OFDM symbol respectively.

Figure 5 shows the MSE performance of LMMSE, LMMSE with ATSA error, RMMSE and hybrid estimators for extended pedestrian-A power delay profile. The crossover probability of binary symmetric channel is assumed as $q = 0.01$. With use of 30.72 MHz sampling frequency, the length of extended pedestrian-A channel is 14. Number of subcarriers is to be higher than the length of the channel. Hence 64 subcarriers are used per OFDM symbol. The hybrid estimator switches to RMMSE estimator at SNR of 30 dB.

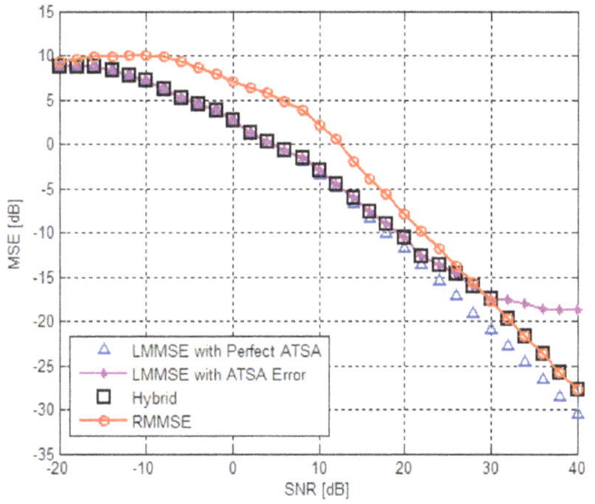

Fig. 5. MSE performance for extended pedestrian-A for $q = 0.01$ and $N = 64$ with optimal power allocation.

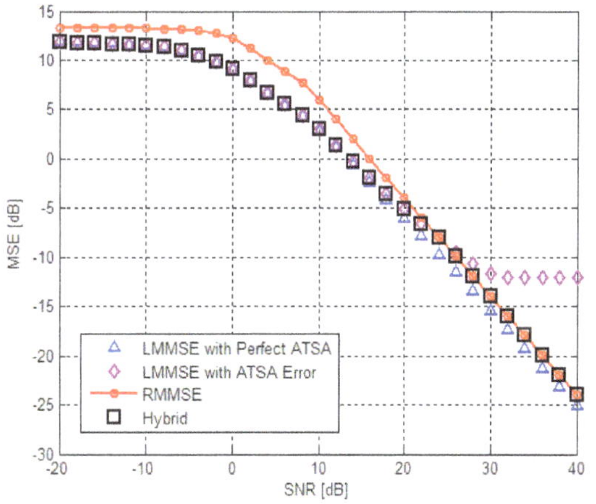

Fig. 6. MSE performance for extended vehicular-A for $q = 0.01$ and $N = 128$ with optimal power allocation.

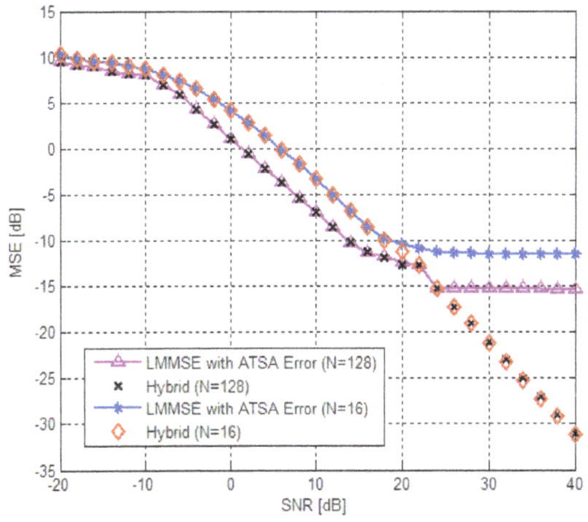

Fig. 4. MSE performance of estimators for $q = 0.01$ with optimal ower allocation for different number of subcarriers N.

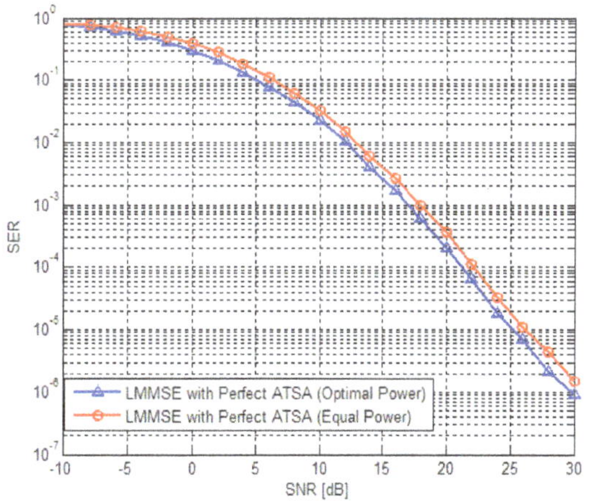

Fig. 7. SER performance of LMMSE estimator for $q = 0.1$ and $N = 16$ with optimal and equal power allocation.

Figure 6 shows the MSE performance of LMMSE, LMMSE with ATSA error, RMMSE and hybrid estimators for extended vehicular-A power delay profile. To simulate the MSE performance, an OFDM symbol with 128 subcarriers is considered. The crossover probability of binary symmetric channel is assumed as $q = 0.01$.The hybrid estimator switches to RMMSE estimator at SNR of 24 dB.

Figure 7 shows the symbol error rate (SER) performance of LMMSE estimator with perfect ATSA with optimal and equal power allocation in 2x2 MIMO-OFDM system with per-subcarrier antenna selection. The number of subcarriers in the OFDM signal is considered as 16. The length of the channel is $L = 4$. It is assumed that 16-QAM modulation is used. It is observed that the LMMSE estimator with optimal power allocation requires 1 dB less SNR compared to LMMSE estimator with equal power allocation at the SER of 10^{-3}.

Figure 8 shows the SER performance of LMMSE estimator with perfect ATSA, LMMSE estimator with ATSA error, RMMSE estimator and proposed hybrid estimator when the crossover probability $q = 0.1$, $N = 16$ and $L = 4$ with optimal power allocation. LMMSE estimator gives 1 dB SNR improvement over RMMSE estimator at SER of 10^{-1}. The performance of RMMSE estimator is better than that of LMMSE with ATSA error, at high SNR. As the proposed hybrid estimator combines the merits of both LMMSE and RMMSE estimators, the SER performance is same as LMMSE estimator for low SNR. SER of LMMSE estimator becomes irreducible after 12 dB. The proposed hybrid estimator switches to RMMSE estimator at 12 dB exactly.

Figure 9 shows the SER performance of the proposed hybrid estimator for crossover probability $q = 0.01$ with number of subcarriers $N = 16$ and 128. 16-QAM modulation is used. The hybrid estimator switches to RMMSE estimator at SNR of 18 dB with 16 subcarriers per OFDM symbol. The hybrid estimator switches to RMMSE estimator at SNR of 22 dB with 128 subcarriers per OFDM symbol.

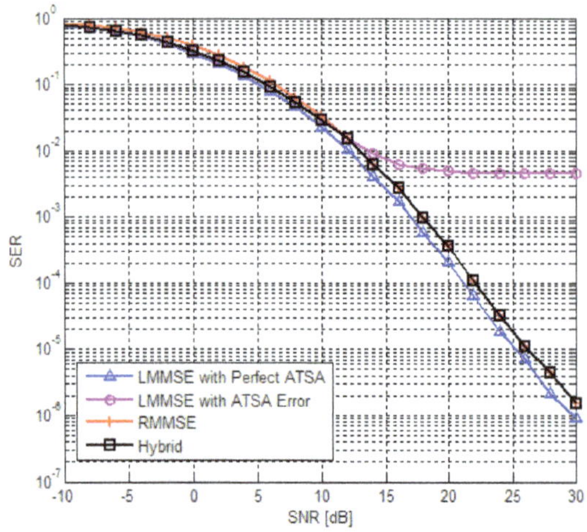

Fig. 8. SER performance of estimators for $q = 0.1$ and $N = 16$ with optimal power allocation.

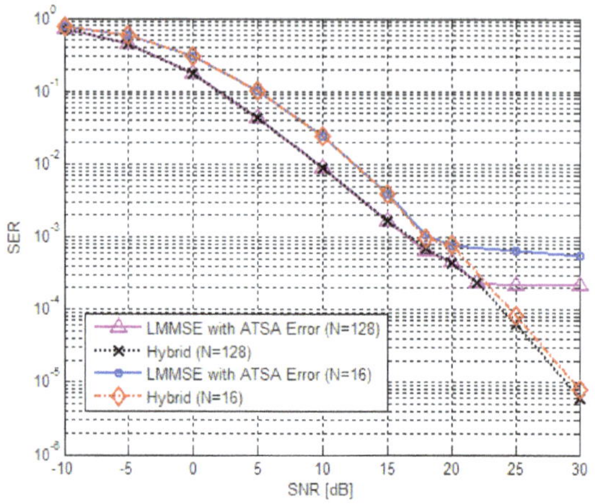

Fig. 9. SER performance of estimators for $q = 0.01$.

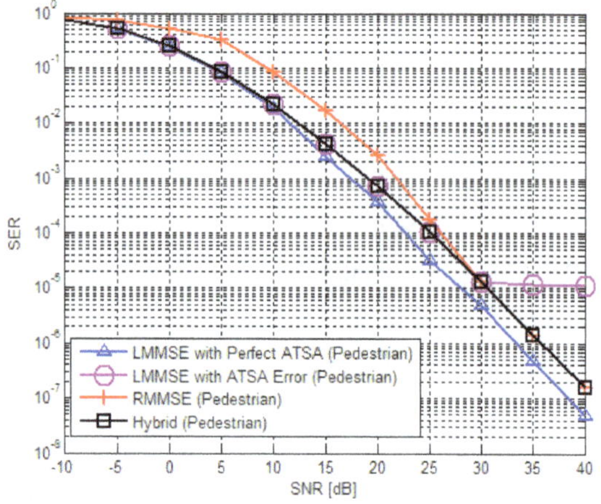

Fig. 10. SER performance for extended pedestrian-A PDP for $q = 0.01$ and $N = 64$.

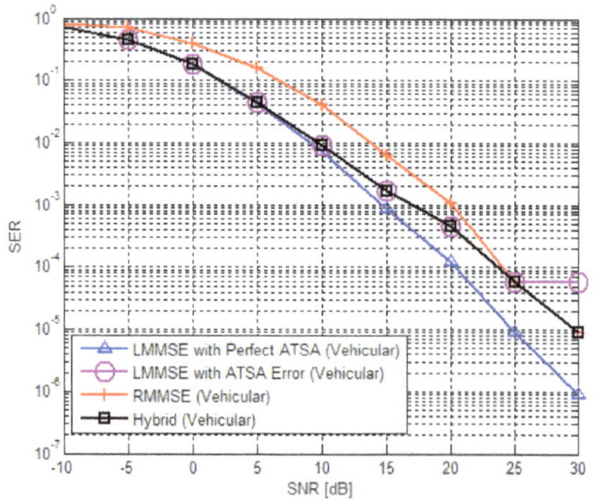

Fig. 11. SER performance for extended vehicular-A PDP for $q = 0.01$ and $N = 128$.

Figure 10 shows the SER performance of the proposed hybrid estimator with optimal power allocation for extended pedestrian-A power delay profile. It is assumed that the number of subcarriers in the system is $N = 64$ and the crossover probability of binary symmetric channel is $q = 0.01$. The SER performance with optimal power allocation is compared with SER performances of LMMSE estimator, LMMSE estimator with ATSA error and

RMMSE estimator. The hybrid channel estimator switches to RMMSE estimator at SNR of 30 dB.

Figure 11 shows the SER performance of LMMSE, LMMSE with ATSA error, RMMSE and hybrid estimators for extended vehicular-A power delay profile. To simulate the SER performance, an OFDM symbol with 128 subcarriers is considered and the crossover probability q is $q = 0.01$. The hybrid estimator switches to RMMSE estimator at SNR of 25 dB.

7. Conclusion

In this paper, a hybrid channel estimator along with optimal power allocation for per-subcarrier transmit antenna selection in MIMO-OFDM system in the presence of ATSA error is proposed. A scalar MSE expression for the channel estimation is derived to overcome the non invertibility of the autocorrelation matrix of CFR. A convex optimization problem is formulated with an objective to minimize the channel estimation MSE with a sum power constraint. The proposed hybrid channel estimator switches from LMMSE estimator to relaxed MMSE estimator exactly at the SNR where ATSA error dominates. An analytical expression for SNR threshold with optimal power allocation is derived. The SER performance of the proposed hybrid channel estimator is also found to be improved.

References

[1] STUBER, G. L., BARRY, J. R., MCLAUGHLIN, S. W., YE LI, G., INGRAM, M. A., PRATT, T. G. Broadband MIMO-OFDM wireless communications. In *Proceedings of the IEEE*, 2004, vol. 92, no. 2, p. 271–294.

[2] YENG, H. A road to future broadband wireless access: MIMO-OFDM based air interface. *IEEE Communication Magazine,* 2005, vol. 43, no. 1, p. 53–60.

[3] ALAMOUTI, S. A simple transmit diversity technique for wireless communications. *IEEE Journal on Selected Areas in Communications*, 1998, vol. 16, p. 1451–1458.

[4] BOLCSKEI, H., GESBERT, D., PAULRAJ, A. On the capacity of OFDM based spatial multiplexing systems. *IEEE Transactions on Communications*, 2002, vol. 50, no. 2, p. 225–234. DOI: 10.1109/26.983319

[5] ZHANG, H., NABAR, R. U. Transmit antenna selection in MIMO-OFDM systems: bulk versus per-tone selection. In *Proceedings of the IEEE ICC*, 2008, p. 4371–4375. DOI: 10.1109/ICC.2008.820

[6] XUN SHAO, JINHONG YUAN, RAPAJIC, P. Antenna selection for MIMO-OFDM spatial multiplexing system. In *Proceedings of the International Symposium on Information Theory ISIT 2003*. Yokohama (Japan), 2003, p. 90. DOI: 10.1109/ISIT.2003.1228104

[7] VITHANAGE, C. M., COON, J. P., PARKER, S. C. J. On capacity-optimal precoding for multiple antenna systems subject to EIRP restrictions. *IEEE Transactions on Wireless Communications*, 2008, vol. 7, no. 12, p. 5182–5187. DOI: 10.1109/T-WC.2008.070879

[8] COON, J. P., SANDELL, M. Combined bulk and per-tone transmit antenna selection in OFDM systems. *IEEE Communications Letters*, 2010, vol. 14, no. 5, p. 426–428. DOI: 10.1109/LCOMM.2010.05.100055

[9] COON, J. P., SANDELL, M. Performance of transmit antenna selection in space–time-coded OFDM systems. *IEEE Transactions on Vehicular Technology*, 2011, vol. 60, no. 6, p. 2824–2828.

[10] VAN DE BEEK, J. J., EDFORS, O., SANDELL, M., WILSON, S. K., BORJESSON, P. O. On channel estimation in OFDM systems. In *Proceedings of the 45th IEEE Vehicular Technology Conference*. Chicago (IL, USA), 1995, p. 815–819.

[11] LI, Y. G., CIMINI, L. J., SOLLENBERGER, N. R. Robust channel estimation for OFDM systems with rapid dispersive fading channels. *IEEE Transactions on Communication,* 1998, vol. 46, no. 7, p. 902–915.

[12] SRIVASTAVA, V., CHIN KEONG HO, HO WANG FUNG, P., SUMEI SUN. Robust MMSE channel estimation in OFDM systems with practical timing synchronization. In *IEEE Conf. on Wireless Communication and Networking WCNC 2004*. Atlanta (USA), 2004, p. 711 - 716. DOI: 10.1109/WCNC.2004.1311273

[13] VITHANAGE, C. M., DENIC, S., SANDELL, M. Robust linear channel estimation methods for per-subcarrier transmit antenna selection. *IEEE Transaction on Communications,* 2011, vol. 59, no. 7, p. 2018–2028. DOI: 10.1109/TCOMM.2011.051711.100420

[14] RAJESWARI, K., SASHIGANTH, M., THIRUVENGADAM, S.J. A hybrid channel estimator for MIMO-OFDM system with per-subcarrier transmit antenna selection. *Telecommunication Systems*, 2015, vol. 58, no. 1, p. 81–89. DOI: 10.1007/s11235-014-9894-3

[15] TUAN, H. D., KHA, H. H., NGUYEN, H. H., LUONG, V.J. Optimized training sequences for spatially correlated MIMO-OFDM. *IEEE Trans. on Wireless Communication*, 2010, vol. 9, no. 9, p. 2768–2778. DOI: 10.1109/TWC.2010.070710.081601

[16] LUONG, V. D., TRAN, N. N., TUAN, H. D. Optimal training sequence design for MIMO-OFDM in spatially correlated fading. In *Proceedings of the 2nd International Conference on Signal Processing and Communication Systems*. Australia (Gold Coast), Dec. 2008, p. 1–6. DOI: 10.1109/ICSPCS.2008.4813697

[17] GOLOVINS, E., VENTURA, N. Optimal training for the SM-MIMO-OFDM systems with MMSE channel estimation. In *Proceedings of the 6th Annual Communication Networks and Services Research Conf.* Halifax (Canada), 2008, p. 470–477. DOI:10.1109/CNSR.2008.57

[18] XIN GENG, HANYING HU, WEIJIA CUR, YANAN DUN. Optimal pilot design for MIMO OFDM channel estimation. In *Proceedings of the International Conference on Signal Processing Systems*, 2010, vol. 2, p. V2-404-V2-408.

[19] KAY, S. M. *Fundamentals of Statistical Signal Processing: Estimation Theory*. Prentice Hall, 1993.

[20] YUSHI SHEN, MARTINEZ, E. Channel estimation in OFDM systems. *Freescale Semiconductor Application Note*, 2006, Rev 0, 1/2006, p. 1–15.

[21] SAVAUX, V, LOUËT, Y, DJOKO-KOUAM, M., SKRZYP-CZAK, A. Minimum mean-square-error expression of LMMSE channel estimation in SISO OFDM systems. *IET Electronics Letters*, 2013, vol. 49, no. 18, p. 1152–1154. DOI: 10.1049/el.2013.1993

[22] BIGUESH, M., GERSHMAN, A. B. Training based MIMO channel estimation: A study of estimator tradeoffs and optimal training signals. *IEEE Transactions on Signal Processing*, 2006, vol. 54, no. 3, p. 884–893. DOI: 10.1109/TSP.2005.863008

[23] *3GPP TS 36.101. User Equipment (UE) Radio Transmission and Reception.* 3rd Generation Partnership Project; Technical Specification Group Radio Access Network; Evolved Universal Terrestrial Radio Access (E-UTRA).

About the Authors ...

Kaleeswaran RAJESWARI received A.M.I.E degree in Electronics and Communication Engineering from the Institute of Engineers (India) in 1996 and M.E from Madurai Kamaraj University, Madurai, India in 2003. She is currently doing PhD in Anna University, Chennai. She has a working experience as a lecturer for 12 years and as an assistant professor for 5 years. She is currently an Assistant Professor in the Electronics and Communication Engineering Department, Thiagarajar College of Engineering, Madurai. Her research interests include digital signal processing, statistical signal processing and MIMO-OFDM wireless communication.

S. Jayaraman THIRUVENGADAM received the B.E. degree in Electronics and Communication Engineering from Thiagarajar College of Engineering, Madurai, India in 1991, the M.E degree in Applied Electronics from Anna University, Chennai, India, in 1994, and the PhD degree from Madurai Kamaraj University, Madurai, India in 2005. From Jan. to Dec. 2008, he was a Visiting Associate Professor with the Dept. of Electrical Engineering, Stanford University, CA under a Postdoctoral Fellowship, by the Dept. of Science and Technology, Government of India. He is currently a Professor with the Dept. of Electronics and Communication Engineering, Thiagarajar College of Engineering. His areas of research interest include statistical signal processing and MIMO wireless communication.

Direction-of-Arrival Estimation Based on Sparse Recovery with Second-order Statistics

Hui CHEN, Qun WAN, Rong FAN, Fei WEN

School of Communication & Information Engineering, University of Electronic Science and Technology of China, Chengdu 611731, China

huichen0929@uestc.edu.cn, wanqun@uestc.edu.cn, uestc_fanrong@foxmail.com; wenfee@126.com

Abstract. *Traditional direction-of-arrival (DOA) estimation techniques perform Nyquist-rate sampling of the received signals and as a result they require high storage. To reduce sampling ratio, we introduce level-crossing (LC) sampling which captures samples whenever the signal crosses predetermined reference levels, and the LC-based analog-to-digital converter (LC ADC) has been shown to efficiently sample certain classes of signals. In this paper, we focus on the DOA estimation problem by using second-order statistics based on the LC samplings recording on one sensor, along with the synchronous samplings of the another sensors, a sparse angle space scenario can be found by solving an ℓ_1 minimization problem, giving the number of sources and their DOA's. The experimental results show that our proposed method, when compared with some existing norm-based constrained optimization compressive sensing (CS) algorithms, as well as subspace method, improves the DOA estimation performance, while using less samples when compared with Nyquist-rate sampling and reducing sensor activity especially for long time silence signal.*

Keywords

Direction-of-arrival estimation, level crossing, compressive sensing, dantzing selector, convex optimization

1. Introduction

Direction-of-arrival (DOA) estimation of propagating plane waves is an extensively studied problem in the field of array signal processing, sensor networks, remote sensing, etc. To determine signal source DOAs using multiple measurements vectors, minimum variance distortionless response (MVDR), and multiple signal classification (MUSIC) algorithms are commonly used [1]. By construction, all these traditional DOA estimation methods require Nyquist-rate sampling of the received signals, which may result in high storage and bandwidth requirements in many sensing systems.

Compressive Sensing (CS) [2] offers a framework for simultaneous sensing and compression of finite-dimensional vectors enabling a potentially significant reduction in the sampling and computation cost at a limited capability sensing system, which depends on linear dimensionality reduction. And the basis pursuit strategy has been used for formulating the DOA estimation problem as a dictionary selection problem where the dictionary entries are produced by discretizing the angle space and then synthesizing the sensor signals for each discrete angle. Sparseness in angle space implies that only a few of the dictionary entries will be required to match the measurements.

The compressive beamforming (CBF) approach based on ℓ_1 minimization [3, 4] can only take random projections of the received signals at the sensors and has a model for these as delayed and weighted combinations of multiple signal sources coming for different angles, which is substantially different from MVDR, MUSIC and some other convex optimization approaches based on regularization [5, 6] which require Nyquist-rate sampling. Although the CBF method does not require the Nyquist-rate sampling at all sensors, it still needs Nyquist-rate sampling at one sensor, which is called reference sensor (RS). Besides, the CBF method just considers the standard CS discrete framework without referring the hardware realization of compressed sampling for practical application. And some existing universal CS measurement instruments cannot be usually used in practice (at least cannot used for the real-time purpose) because of its time-consuming data collection and the difficulty of physical realization. The random convolution based sampling strategy has been investigated by [7], [8] and [9] to solve this drawback. To avoid hardware realization challenges, we acquire less samples from analog signal directly by using Level-crossing (LC) sampling [10, 11, 12, 13, 14, 15, 16, 17], in which an analog input signal is compared with a set of quantization levels (can also be called reference levels) and a sample is produced only when the input analog signal changes enough to cross a level, thus it results in nonuniform sampling and saves dynamic power in the analog-to-digital conversion (ADC) and following DSP for those powered by a very small battery and that involve "bursty" signals with varying activity over time, possibly including long periods of silence. Sampling by LC can mimic the behavior of such input signals. As such, the data collection rate is dictated by the signal itself, rather than Nyquist sampling frequency. One direct benefit of such sampling is that it can reduce number of samples without sampling in the non-bursty intervals. Higher instantaneous bandwidth can be offered when LC sampling is performed, and resolution is improved without overall increase in bit rate or power consumption because the significant information in the bursty intervals is sampled. The data transmission rate can be reduced by using LC sampling in communication systems [10, 11].

In order to obtain samples efficiently for certain types of input, the levels within the amplitude range of the input need to be appropriately assigned in the ADC. In this work, we assume the input dynamic range is known, and we implement a fixed scheme that uniformly assigns levels in the ADC. Usually, we are interested in reconstructing a single signal from LC samplings. In this paper, we consider the case for DOA estimation using a sensor array, where we try to reconstruct a vector of sources' positions using second-order statistics based on LC samplings of the RS and synchronous samplings of another sensors.

The paper is organized as follows, In Section 2, we provide the LC sampling scheme and LC ADC architecture. We then introduce the DOA estimation algorithm using second-order statistics based on multiple measurement vectors and a sparse vector finding in Sec. 3, where we also provide complete algorithmic descriptions and corresponding guaranteed performance analysis results. In Sec. 4, a number of simulation results of our proposed approach is compared with that of the state-of-art CS recovery methods and conventional method. Finally, we conclude with lots of simulation results described on speed signals which are collected using LC sampling of the RS and non-uniform samples of another sensors by keeping synchronous sampling of the RS in Section 5.

2. Nonuniform Signal Processing Tool

2.1 Level Crossing Sampling Scheme

The Level Crossing sampling scheme (LCSS) is one of the signal-dependent sampling schemes, and it is a better choice for sampling the time-varying signals. For LCSS, a sample on the RS is captured only when the input analog signal $x(t)$ crosses one of the quantization levels which are uniformly spaced by a quantum q, and the samples depend on $x(t)$ variations. The (x_n, t_n) is a sample pair with an amplitude x_n and a time t_n, x_n is exactly equal to one of the quantization levels, and current sampling instant t_n can be computed by adding the time elapsed dt_n between the t_n and t_{n-1} from the previous instant t_{n-1},

$$t_n = t_{n-1} + dt_n. \tag{1}$$

2.2 LC-based Analog-to-Digital Converter

Usually, the sampling instants are exactly known for conventional sampling, and the ADC number of bits determines the ADC resolution as sample amplitudes are quantization. In this paper, we consider a B-bit (2^B reference levels) LC ADC which equipped with an array of 2^B analog comparators, and the comparators compare with input with corresponding reference levels. Without loss of generality, we assume an amplitude-bound signal $x(t) \in$ [-A/2, +A/2] that is T second long, and the LC ADC has 2^B levels with uniform spacing $q = A/2^B$. Let $\{\ell_1, \ell_2, ..., \ell_{2^B}\}$ represent all the reference levels used by comparators. The ADC compares the input $x(t)$ with all the reference levels and it will record a level crossing with one of ℓ if the following comparison holds for a ℓ_i:

$$(x((n-1)\tau) - \ell_i)(x(n\tau) - \ell_i) < 0,\ i = 1, 2, ..., 2^B. \tag{2}$$

That is the LC ADC only records the quantization value $Q(s_i)$ of the true signal s_i in the interval $[(n-1)\tau, n\tau]$. In order to minimize the consuming power, we can also randomly choose the P of the 2^B comparators are on at any time according to [14], that is, a new set of P reference levels is picked and updated every v seconds (v is a constant), which can be accomplished by a digital circuit that periodically updates the set of on comparators by controlling power supply. Note that, in this paper, we need to add an external circuit to make the LC ADC sampling instants of the RS trigger the sampling instants of another sensors, which ensure the sampling instants on the RS and another sensors are synchronous.

3. DOA Estimation Method

Assume that we consider a far-field consisting of K sources and a sensor array of $(M+1)$ sensors with an arbitrary geometry. And we also assume the sensor positions are known and are given by $\eta_i = [x_i, y_i, z_i]^T (i = 0, 1, ..., M)$. Our goal is to determine the DOAs of the signal sources by using the received signals. We assume the RS receives a superposition of the time-domain source signals, $x_0(t) = \sum_{k=1}^{K} s_k(t) + n_0(t)$ and the $n_0(t)$ is the noise at the RS, then the sensor i observes the time-delayed superimposed source signal,

$$x_i(t) = \sum_{k=1}^{K} s_k(t + \Delta_i(\pi_k)) + n_i(t), \tag{3}$$

where $\pi_k = (\theta_k, \phi_k)$ is the angle pair consisting of the unknown azimuth and elevation of the source, and $\Delta_i(\pi_k)$ is the relative time delay at the i-th sensor for a source with DOA π_k with respect to the RS. Here finding the DOA is equivalent to finding the relative time delay, and the time delay Δ_i in (3) can be determined from geometry:

$$\Delta_i(\pi_k) = \frac{1}{c}\eta_i^T \begin{bmatrix} \cos\theta_k \sin\phi_k \\ \sin\theta_k \sin\phi_k \\ \cos\phi_k \end{bmatrix}, \quad (4)$$

where c is the speed of the propagating wave in the medium.

The source angle pair π_k lies in the product of space $[0,2\pi)_\theta \times [0,\pi)_\phi$, which must be discretized to form an angle dictionary (sparsity basis). Here, we enumerate a finite set of angles for both azimuth and elevation to generate a set of angle pairs $B = \{\pi_1, \pi_2, ..., \pi_{N_t}\}$, where N_t determines the angular resolution. Let $\mathbf{b}$ denote the sparsity pattern which selects members of the discretized angle-pair set B. A nonzero positive value at index j of $\mathbf{b}$ indicates the presence of a target at the angle pair π_j. In particular, assume the angle space is discretized in N_t points in all, the sparsity transform matrices $\{\Psi_i\}_{i=1}^M$ will be of dimension $N_s \times N_t (N_s \gg N_t)$, where N_s is the sample number. The Ψ_i for sensor i can be constructed using proper time shifts of $x_0(t)$ for each π_j in B, and the time shift for sensor i with respect to the RS using (4).

Assume the RS records the signal source using LC sampling as $\mathbf{x}_0 = [x_0(t_{lc1}), x_0(t_{lc2}), ..., x_0(t_{lcT_N})]$, then the sampling data on the sensor i can be described as

$$\mathbf{h}_i = [x_i(t_{lc1}), x_i(t_{lc2}), ..., x_i(t_{lcT_N})]^T, \quad (5)$$

where $t_{lci} (i = 1, ..., T_N)$ are the LC sample instants, $x_i(t_{lcj})$ denotes a sample at sample instant t_{lcj} for sensor i, and T_N is the number of LC samples. Besides, the j-th column of Ψ_i corresponding to the time shift of the sampled signal $\mathbf{x_0}$ corresponding to the j-th index of the sparsity pattern vector $\mathbf{b}$, which indicates the proper time shift corresponding to the angle pair π_j :

$$[\Psi_i]_j = [x_0(t_{lc1} + \Delta_i(\pi_j)), x_0(t_{lc2} + \Delta_i(\pi_j)), ..., x_0(t_{lcT_N} + \Delta_i(\pi_j))]. \quad (6)$$

The matrix Ψ_i is the sparsity basis corresponding to all discretized angle pairs B at the i-th sensor. Considering the effect of additive sensor noises, the sparsity pattern vector can be recovered using the Dantzing selector [18] convex optimization problem:

$$\hat{\mathbf{b}} = \arg\min \|\mathbf{b}\|_1 \text{ s. t. } \left\|\Psi^T(\mathbf{H} - \Psi\mathbf{b})\right\|_\infty < \varepsilon, \quad (7)$$

where $\mathbf{H} = [\mathbf{h}_1^T, ..., \mathbf{h}_M^T]^T$, and $\Psi = [\Psi_1^T, ..., \Psi_M^T]^T$. ε is a relaxation variable which makes the true $\mathbf{b}$ feasible with high probability, since the formulated problem in (7) is a convex optimization problem [19], so we solve it numerically using existing solver [20], and we can obtain a global optimum for the problem (7).

The terms $\Psi^T\mathbf{H}$ and $\Psi^T\Psi$ in the (7) constraint are actually auto- and cross-correlations, respectively. Take two signal sources $s_1(t)$ and $s_2(t)$ for example, the recorded LC sampling signal at the RS is

$$x(t_{lc}) = s_1(t_{lc}) + s_2(t_{lc}). \quad (8)$$

Assume the signal amplitudes are equal, and the shifted RS LC sample signal at the i-th sensor is

$$x(t_{lc} + \Delta_i(\pi_n)) = s_1(t_{lc} + \Delta_i(\pi_n)) + s_2(t_{lc} + \Delta_i(\pi_n)) \quad (9)$$

when the assumed DOA is π_n, and this time shift of the RS LC sample signal is used to populate the n-th column of the Ψ matrix. While the true received sample signal on the i-th sensor is

$$x_i(t_{lc}) = s_1(t_{lc} + \Delta_i(\pi_1)) + s_2(t_{lc} + \Delta_i(\pi_2)), \quad (10)$$

where there are different time shifts for the two signals. For $\Psi^T\mathbf{H}$ we get a column vector whose n-th element is

$$\sum_{i=1}^{M}[R_{11}(\Delta_i(\pi_n), \Delta_i(\pi_1)) + R_{12}(\Delta_i(\pi_n), \Delta_i(\pi_2)) + R_{12}(\Delta_i(\pi_n), \Delta_i(\pi_1)) + R_{22}(\Delta_i(\pi_n), \Delta_i(\pi_2))], \quad (11)$$

where R_{11} is the autocorrelation of signal $s_1(t_{lc})$, R_{22} is the autocorrelation of signal $s_2(t_{lc})$, R_{12} is the cross-correlation between signal $s_1(t_{lc})$ and $s_2(t_{lc})$. For the matrix $\Psi^T\Psi$, the element in the n-th row and r-th column is

$$\sum_{i=1}^{M}[R_{11}(\Delta_i(\pi_n), \Delta_i(\pi_r)) + R_{12}(\Delta_i(\pi_n), \Delta_i(\pi_r)) + R_{12}(\Delta_i(\pi_n), \Delta_i(\pi_r)) + R_{22}(\Delta_i(\pi_n), \Delta_i(\pi_r))]. \quad (12)$$

According to the two same assumptions as [3, 4]: small cross correlation for signals incoherent assumption, and small autocorrelations for signals decorrelate at small lags assumption. In order to make $\Psi^T\mathbf{H} - \Psi^T\Psi\mathbf{b}$ small, we should make sure that the large elements in the vector $\Psi^T\mathbf{H}$ are canceled by the large terms in $\Psi^T\Psi\mathbf{b}$. According to the assumptions, the two largest elements in $\Psi^T\mathbf{H}$ occur when $\pi_n = \pi_1$ and $\pi_n = \pi_2$, because these are two peaks in the autocorrelations $R_{11}(\Delta_i(\pi_1), \Delta_i(\pi_1))$ and $R_{22}(\Delta_i(\pi_2), \Delta_i(\pi_2))$. When we cancel the element $R_{11}(\Delta_i(\pi_1), \Delta_i(\pi_1))$ using the row of $\Psi^T\Psi$ corresponding to $\pi_n = \pi_1$, then the vector $\mathbf{b}$ must select the column where $\pi_r = \pi_1$. Likewise, to cancel the element $R_{22}(\Delta_i(\pi_2), \Delta_i(\pi_2))$, we use the $\pi_n = \pi_2$ row and the $\pi_r = \pi_2$ column. And all the other elements will be relatively small according to the above assumptions.

About the constraint parameter ε, it will allow the matching of the two signals at their true DOAs. Then the ℓ_1 minimization of the selector vector $\mathbf{b}$ will tend to pick the signals whose autocorrelation is large or the larger of the two for different signal amplitudes. Besides, the method can also be extended to the case with K unknown sources at DOAs $(\theta_1, \phi_1), (\theta_2, \phi_2), ..., (\theta_K, \phi_K)$ impinging on the array.

4. Simulation Results

The performance of our proposed approach is evaluated in this section using a linear array of 11 sensors uniformly placed on the x-axis, and the first sensor is selected as RS which is placed to be at the origin. For all the addressed scenarios, a DOA space is discretized into a 1° angular grid.

To demonstrate our proposed method, we will compare the DOA estimation performance with ℓ_1-SVD [21], CBF algorithm [3] and MUSIC method in respective of DOA estimation performance and sample number. Note that, in the CBF method, assume each sensor takes 20 compressive measurements along with a RS with high-rate sampling frequency (much higher than Nyquist-rate sampling) for constructing the dictionary $\boldsymbol{\Psi}$, and another algorithms use Nyquist sampling of each sensor in our simulation. Besides, to show the advantages of our proposed method, the following evaluation measures are employed to evaluate the signal reconstruction quality of the RS receiving data at LC sampling algorithm: the correlation coefficient (CC), the compressive ratio (CR), and the DOA estimation performance: probability of resolved. CC is used to evaluate the similarity between the original signal and its reconstruction by using the data samples:

$$CC = \left(\sum_{i=1}^{n}(x_i-\bar{x})(y_i-\bar{y})\right) \Big/ (n-1)s_x s_y, \tag{13}$$

where x_i are re-sampled values of the original signal $\mathbf{X} = [\mathbf{x_1}, \mathbf{x_2}, ..., \mathbf{x_n}]$, and y_i are re-sampled values of the reconstruction signal $\mathbf{Y} = [\mathbf{y_1}, \mathbf{y_2}, ..., \mathbf{y_n}]$, $\bar{x}$ and $\bar{y}$ are sample means of $\mathbf{X}$ and $\mathbf{Y}$, s_x and s_y are the sample standard deviations of $\mathbf{X}$ and $\mathbf{Y}$, respectively. The CR between the Nyquist sampling signal and LC sampling signal can be described as

$$CR = \frac{S_{orig}}{S_{lcs}}, \tag{14}$$

where S_{orig} and S_{lcs} represent Nyquist sampling number and LC sampling number respectively. And the probability of resolved is defined as follows, it is said to be resolved, if for any signal with DOA θ_k, its estimate $\hat{\theta}_k$ is such that $|\theta_k - \hat{\theta}_k| \leq 1°$.

4.1 Angular Resolution Analysis

First, we will analysis the angular resolution of our proposed method, and the achievable resolution of our approach is evaluated and compared with other tested methods in this subsection, two synthetic speech sources are taken and placed in the far-field of the array, and we set 60 reference levels for the RS ADC, that is we only make 60 comparators work for a 6-bit LC ADC. The two speed signal sources used in our simulation are shown in Fig. 1. The RS signal is the sum of the two source signals, and Fig. 2 gives the part of signal we used for DOA estimation and reconstructed signal using Akima interpolation [22] for the LC samplings.

Considering an SNR of 10 dB, subplots in Fig. 3 show two scenarios with angular separation of $\Delta\theta = [6°, 18°]$ between two uncorrelated sources, respectively. From Fig. 3, we can see that our approach, MUSIC method outperforms CBF approach and ℓ_1-SVD method in scenarios with closely spaced sources ($\Delta\theta = 6°$), and for a large separation angles ($\Delta\theta = 18°$) situation, all the methods can estimate the signal DOAs correctly, and thus the simulation curves are overlapped. But, for our sample signal, we just use 896 samples, that is, compared with Nyquist samples (1167 samples using $f_s = 7350$ Hz), the CR is 1.3025 (if we consider the whole signal, the CR is 4.7445). Meanwhile the CC is 0.9893, and the CC using Nyquist sampling is 0.9901,

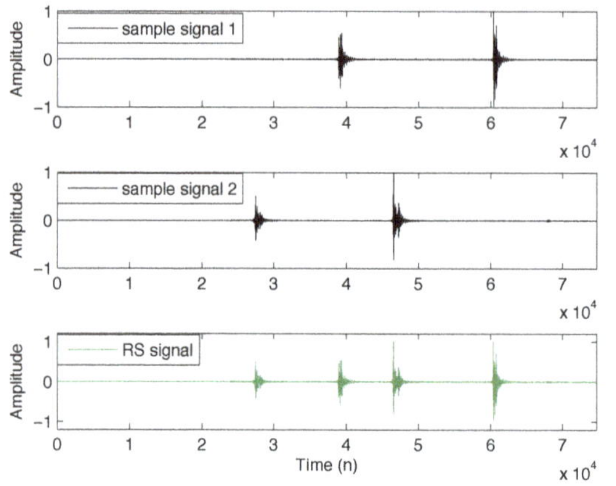

Fig. 1. Two input speech signal sources and reference signal.

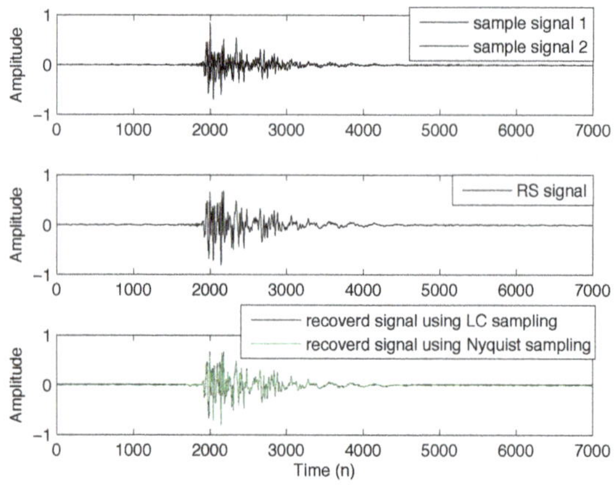

Fig. 2. Sample signals and reconstructed signals.

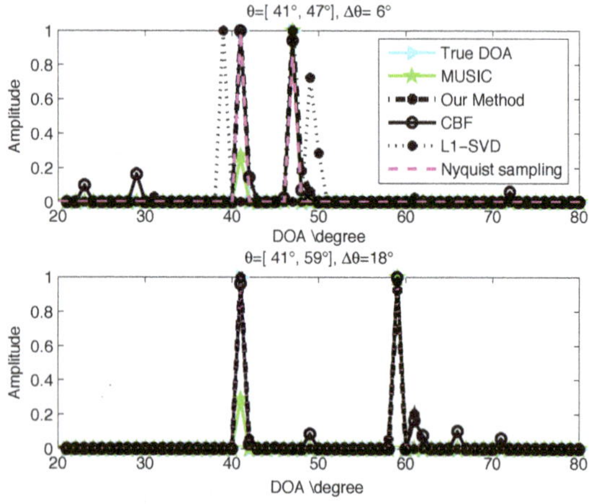

Fig. 3. Angular resolution of estimated spatial spectrum in two uncorrelated sources scenario with SNR=10 dB, and angular separation between source DOAs: $\Delta\theta = [6°, 18°]$.

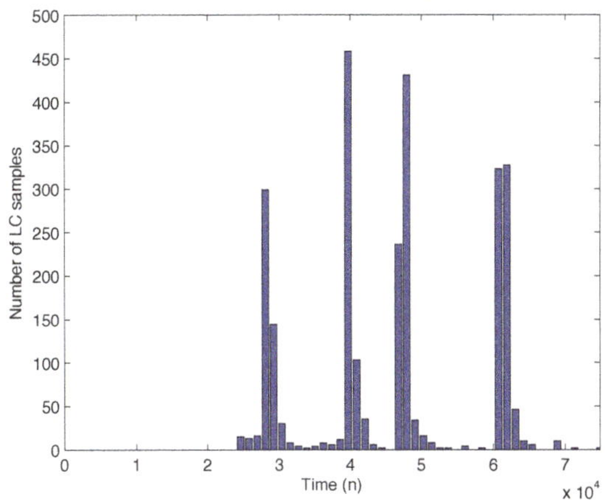

Fig. 4. The number of LC samples obtained using (2).

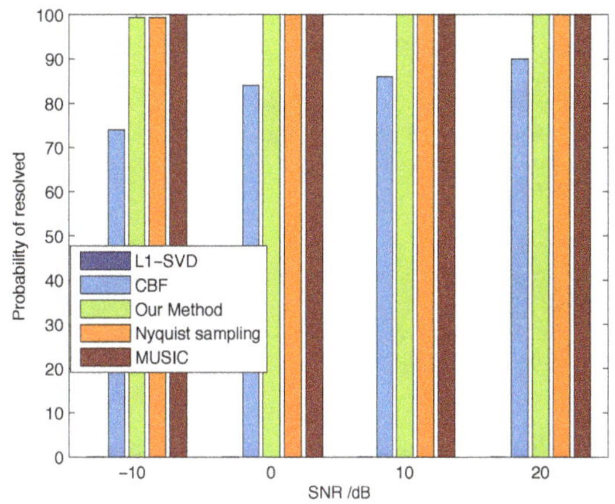

Fig. 5. Probability of resolved vs. SNR for three uncorrelated signal sources.

which demonstrate that the LC sampling method can obtain comparable signal reconstruction performance as that of the Nyquist sampling. Besides, our method will show more advantages in respect of using less samples and reducing the sensor activity especially for long time silence signal. Note that, we add the simulation result of our DOA estimation method using Nyquist samplings, which demonstrates our method using LC samplings can reduce sensor activity to save the sensor energy.

Furthermore, we can see in the Fig. 4 that the number of LC samples varies with input, which can be explained when we look at the sample signal in the third subpgraph of Fig. 1. More samples only when the utterance occurs. The LC's adaptive nature prevents it from registering many more samples during quiescent interval where there is no information, and enhances its efficiency.

4.2 DOA Estimation Performance

In this subsection, the influence of SNR on the spatial spectrum estimation performance of all tested methods was analyzed in this scenario via probability of resolved criterion in over 500 trials. Fig. 5 shows the probability of resolved of the spatial spectrum estimation at various levels of SNR for three uncorrelated signals $\theta = [-20°, 41°, 47°]$, and the main challenge in this scenario stems from the small separation between two sources $\theta = [41°, 47°]$. From Fig. 5, we can see that the MUSIC and our proposed method outperform much better than that of ℓ_1-SVD and CBF approaches, and the performance of our method coincides with that of MUSIC at all SNR levels, while the ℓ_1-SVD was unsuccessful in estimating all DOAs correctly for all values of the SNR, and thus ℓ_1-SVD is not visible in Fig. 5.

5. Conclusion

In this work, we demonstrate the feasibility of our proposed DOA estimation method by using sparse recovery algorithm with second-order statistics. In our solution, we first obtain the LC samplings on the RS and non-uniform samples of another sensors by keeping synchronous sampling of the RS; then we exploit second-order statistics and the sparsity of the sources in the angle domain, and the obtained sparse pattern by solving an ℓ_1 minimization problem determines the number of targets and their corresponding DOAs. We accomplish this by demonstrating that our wireless array scheme is robust against noise in the LC sampling samples, which also can be used to recover the data of the RS. And the fact that all array sensors uses even-based measurements will reduce the amount of data that must be communicated of sensors, which has potential in wireless sensor networks where arrays would be formed from distributed sensors.

Acknowledgements

This work was supported in part by the National Natural Science Foundation of China under grant 60772146, 61471103, the Applied Basic Research Program of Sichuan Province under grant 14JC0616 as well as the Program for New Century Excellent Talents in University under grant NCET-12-0095.

References

[1] JOHNSON, D. H., DUDGEON, D. E. *Array Signal Processing:Concepts and Techniques*. Prentice Hall, 1993.

[2] DONOHO, D. L. Compressed sensing. *IEEE Transactions on Information Theory*, 2006, vol. 52, no. 4, p. 1289–1306. DOI: 10.1109/TIT.2006.871582

[3] GURBUZ, A., MCCLELLAN, J. H., CEVHER, V. A compressive beamforming method. In *Proceedings of 2008 IEEE International Conference on Acoustics, Speech, and Signal Processing (ICASSP)*. Las Vegas (NV, USA), 2008, p. 2617–2620. DOI: 10.1109/ICASSP.2008.4518185

[4] GURBUZ, A. C., CEVHER, V., MCCLELLAN, J. H. Bearing estimation via spatial sparsity using compressive sensing. *IEEE Trans-*

actions on Aerospace and Electronic Systems*, 2012, vol. 48, no. 2, p. 1358–1369. DOI: 10.1109/TAES.2012.6178067

[5] FUCHS, J. J. On the application of the global matched filter to DOA estimation with uniform circular arrays. *IEEE Transactions on Signal Processing*, 2001, vol. 49, no. 4, p. 702–709. DOI: 10.1109/78.912914

[6] FUCHS, J. J. Linear programming in spectral estimation. Application to array processing. In *Proceedings of 1996 IEEE International Conference on Acoustics, Speech, and Signal Processing (ICASSP)*. Atlanta (GA, USA), 1996, vol. 6, p. 3161–3164. DOI: 10.1109/ICASSP.1996.550547

[7] TROPP, J. A., WAKIN, M. B., DUARTE, M. F., BARON, D., BARANIUK, R. G. Random filters for compressive sampling and reconstruction. In *Proceedings of 2006 IEEE International Conference on Acoustics, Speech, and Signal Processing (ICASSP)*. Toulouse (France), 2006. DOI: 10.1109/ICASSP.2006.1660793

[8] ROMBERG, J. Compressive sensing by random convolution. *SIAM Journal on Imaging Sciences*, 2008, vol. 4, no. 4, p. 1098–1128. DOI: 10.1137/08072975X

[9] LI, L. L., LI, F. Compressive sensing based robust signal sampling. *Applied Physics Research*, 2012, vol. 4, no. 1, p. 30–41. DOI: 10.5539/apr.v4n1p30

[10] GUAN, K. M., SINGER, A. C. A level-crossing sampling scheme for non-bandlimited signals. In *Proceedings of 2006 IEEE International Conference on Acoustics, Speech, and Signal Processing (ICASSP)*. Toulouse (France), 2006, vol. 3. DOI: 10.1109/ICASSP.2006.1660670

[11] GUAN, K. M., SINGER, A. C. Opportunistic sampling of bursty signals by level-crossing – an information theoretical approach. In *41st Annual Conference on Information Sciences and Systems (CISS)*. Baltimore (USA), 2007, p. 701–707. DOI: 10.1109/CISS.2007.4298396

[12] TSIVIDIS, Y. Digital signal processing in continuous time: a possibility for avoiding aliasing and reducing quantization error. In *Proceedings of 2004 IEEE International Conference on Acoustics, Speech, and Signal Processing (ICASSP)*. Montreal (Canada), 2004, vol. 2, p. 589–592. DOI: 10.1109/ICASSP.2004.1326326

[13] GUAN, K. M. *Opportunistic Sampling by Level-Crossing*, Ph.D. thesis. Urbana (USA): University of Illinois at Urbana-Champaign, 2008.

[14] GUAN, K. M., KOZAT, S. S., SINGER, A. C. Adaptive reference levels in a level-crossing analog-to-digital converter. *EURASIP Journal on Advances in Signal Processing*, 2008, vol. 2008, no. 183, Article ID 513706, 11 pages. DOI: 10.1155/2008/513706

[15] SENAY, S., OH, J. S., CHAPARRO, L. F. Regularized signal reconstruction for level-crossing sampling using Slepian functions. *Signal Processing*, 2012, vol. 92, no. 4, p. 1157–1165. DOI: 10.1016/j.sigpro.2011.11.017

[16] GRUNDE, U. Non-stationary signal reconstruction from level-crossing samples using Akima Spline. *Electronics and Electrical Engineering*, 2012, vol. 117, no. 1, p. 9–12. DOI: 10.5755/j01.eee.117.1.1044

[17] AKOPYAN, F., MANOHAR, R., APSEL, A. B. A level-crossing flash asynchronous analog-to-digital converter. In *Proceedings of 2006 IEEE International Symposium on Asynchronous Circuits and Systems*. Grenoble (France), 2006, p. 12–22. DOI: 10.1109/ASYNC.2006.5

[18] CANDES, E., TAO, T. The Dantzig Selector: Statistical estimation when p is much larger than n. *Annals of Statistics*, 2007, vol. 35, no. 6, p. 2313–2351. DOI: 10.1214/009053606000001523

[19] BOYD, S., VANDENBERGHE, L. *Convex Optimzation*. Cambridge (UK): Cambrige University Press, 2003.

[20] TOH, K. C., TODD, M. J., TUTUNCU, R. H. SDPT3-a Matlab software package for semidefinite programming. *Optimization Methods and Software*, 1999, p. 545–581.

[21] MALIOUTOV, D., CETIN, M., WILLSKY, A. A sparse signal reconstruction perspective for source localization with sensor arrays. *IEEE Transactions on Signal Processing*, 2005, vol. 53, no. 8, p. 3010–3022. DOI: 10.1109/TSP.2005.850882

[22] AKIMA, H. A new method of interpolation and smooth curve fitting based on local procedures. *Journal of the Association for Computing Machinery*, 1970, vol. 17, no. 4, p. 589–602. DOI: 10.1145/321607.321609

About the Authors...

Hui CHEN was born in Henan, China. She received the B.S. degree in electronics information engineering from Southwest University for Nationalities (SWUN) in 2007, and the Ph.D degree from University of Electronic Science and Technology of China (UESTC) in 2013. Her research interests include array signal processing, compressive sensing and wireless communication.

Qun WAN was born in Nanjing, China. He received the B.S. degree from Nanjing University in 1993, the M.S. degree from UESTC in 1996, and the Ph.D. degree from UESTC in 2001. During 2001-2002, he was a post-doctor at Tsinghua University, where he participated in cellular localization program. In 2003, he was a Technical Staff at UTstarcom. Since 2004, he has been a Professor in the Department of Electronic Engineering at the University of Electronic Science and Technology of China (UESTC). His research interests include array signal processing and compressed sensing, mobile and indoor localization. He is a Senior Member of CIE.

Rong FAN was born in Sichuan, China. He received the B.E. degree from Chengdu University of Technology in 2007. From September 2007 to July 2010, he was with the University of Electronic Science and Technology of China (UESTC), where he received the M.E degree in 2010. He received the Ph.D degree from University of Electronic Science and Technology of China (UESTC) in 2014. His specific research areas of current interest include sparse and array signal processing, adaptive beamforming, and parameter estimation with applications to radar and communications.

Fei WEN received the B.S. degree in electronic engineering from University of Electronic Science and Technology of China (UESTC) in 2006. He received the Ph.D degree in communications and information engineering at UESTC in 2013. He is part of the Teaching Staff at the Air Force Engineering University. His main research interests are statistical signal processing, communications, and estimation theory.

26

A Fast DOA Estimation Algorithm Based on Polarization MUSIC

Ran GUO[1], *Xing-Peng MAO*[1], *Shao-Bin LI*[1], *Yi-Ming WANG*[2], *Xiu-Hong WANG*[1]

[1]School of Electronic and Information Engineering, Harbin Institute of Technology, Harbin, Heilongjiang Province, China
[2]The First Institute of Oceanography, SOA, Qingdao, Shandong Province, China

mxp@hit.edu.cn

Abstract. *A fast DOA estimation algorithm developed from MUSIC, which also benefits from the processing of the signals' polarization information, is presented. Besides performance enhancement in precision and resolution, the proposed algorithm can be exerted on various forms of polarization sensitive arrays, without specific requirement on the array's pattern. Depending on the continuity property of the space spectrum, a huge amount of computation incurred in the calculation of 4-D space spectrum is averted. Performance and computation complexity analysis of the proposed algorithm is discussed and the simulation results are presented. Compared with conventional MUSIC, the proposed algorithm has considerable advantage in aspects of precision and resolution, with a low computation complexity proportional to a conventional 2-D MUSIC.*

Keywords

Polarization sensitive array, DOA, MUSIC, fast algorithm.

1. Introduction

The direction of arriving signals (DOA) estimation problem has raised great interest in various fields, including radar, mobile communications, and microphone array systems. In the past decades, numerous algorithms and techniques have been proposed. Some important classes of algorithms include the Maximum Likelihood Estimation (ML) [1, 2], the signal subspace based algorithms represented by Multiple Signal Classification (MUSIC) [3], and the computationally efficient Estimation of Signal Parameters via Rotational Invariance Techniques (ESPRIT) [4]. These algorithms help to enhance the performance of the DOA estimation substantially, and a great number of relevant algorithms are derived from them, for instance, Stochastic Maximum Likelihood Estimation (SML) [5], Root-MUSIC [6], Total Least Squares version of ESPRIT (TLS-ESPRIT) [7], etc.. In recent years, sparse signal presentation has been introduced into DOA estimation [8, 9, 10], making the DOA estimation with small number of snapshots possible.

Of all these algorithms, ML and its derivate algorithms are of the best performance in both accuracy and resolution, whereas the computation complexity is extremely high, preventing their practical applications. ESPRIT and MUSIC algorithms, proposed later, with precision error approaches to the Cramer-Rao bound (CRB) as snapshot number of array outputs accumulates to a certain threshold, and the calculation incurred in these 2 algorithms are acceptable.

Numerous factors should be considered to select a proper algorithm. First, the DOA algorithm should adapt to the form of the array. Generally, the signals' azimuths need to be estimated, and in certain circumstance, their elevation estimation are also demanded. Well-designed planar arrays, such as uniform circular array (UCA), can meet these requirements. However, ESPRIT and Root-MUSIC can only work on linear array, and algorithms derived from them for other array forms [11, 12, 13] can only work on particular array patterns, while only 1 direction parameter can be estimated. The MUSIC algorithm, which can be transplanted to more general forms of planar arrays and provides both azimuth and elevation estimation, seems to be more suitable. Current calculation capacity of DSP chips makes it practical as well.

For algorithm selection, another crucial factor is the performance, especially in aspects of precision and resolution. The results presented in [14, 15] suggested that the accuracy and resolution in MUSIC are both determined by three factors: SNR, the number of array's snapshots, and the array's aperture. In practical circumstances, the SNR is generally uncontrollable, the aperture is often restricted, and the numbers of samples is limited. From this perspective, the accuracy and resolution of MUSIC algorithm is confined, and further improvement in performance can only be sought within other methods.

The polarization phenomenon of electromagnetic wave has been noticed and studied by physicists for more than a century. However, it was only in 1980s that this phenomenon began to attract the researchers' attention and proved to be of great advantage in signal processing. The polarization sensitive array is introduced into DOA estimation by Nehorai and Li [16, 17]. In [17, 18], Jian Li integrated the polarization into ESPRIT successfully. The pencil-MUSIC algorithm is devised for DOA and polarization estimation in [19, 20, 21], which extend the ESPRIT-like algorithm to partially polarized sources. [22, 23] provide an algorithm based on sparse ULA, consisting of sensors which can measure all six electrical and magnetic components. [24] provide several algorithms to solve the correlation problem with the array of vector sensors. In [25, 26], the polynomial rooting algorithm was adopted in polarization sensitive arrays. Covariance tensor [27, 28] is introduced to define the array manifold in a new way and reduce the amount of calculation. These papers exploited the advantage of polarization of signals. However, to reduce the calculation complexity, all these algorithms assumed specific patterns of array, and most of them adopted the sensors with the capacity to detect all 6 electrical and magnetic components, which is not very applicable.

The algorithms presented in [29-33] require no assistance of magnetic components. But the algorithms mentioned in [29, 30] assumed that 1 polarization parameter and 1 direction parameter are predetermined, which cannot be satisfied generally. In [31, 32, 33], the amount of calculation is reduced by dividing the process of estimation into two steps: one concerns with the polarization, and the other with the direction. The drawback of this method is that the second step in the algorithm depends on the success of previous step. When the DOAs of multiple signals are very close, the second step inclines to fail, even when the signals are distinct in polarization domain.

To retain the advantage in resolving close signals offered combination of spatial and polarization domains, a new algorithm, which reveals the space-polarization spectrum of 4-D space in a new way, is proposed in this paper. Unlike the algorithms mentioned above, the proposed algorithm can be used on various forms of polarization sensitive arrays. Besides, in case the multiple signals are close in space domain, they can be resolved as they are distinctive in polarization domain. By analyzing the property of the space spectrum, the algorithm is designed to reduce the complexity of the 4-D space spectrum calculation to be proportional to a 2-D spectrum calculation [34]. The proposed algorithm makes the transplantation of MUSIC from arrays of ideal omnidirectional antennas to polarization sensitive arrays more practical.

This paper is organized as follows. In Sec. 2, the fundamental principle of MUSIC is introduced, and the transplantation of MUSIC into polarization sensitive array is discussed. Although the latter one is highly inapplicable because of the immense amount of calculation, it is essential in developing the faster new algorithm. In Sec. 3, the definition of zero spectrum and another form of space spectrum are given, and the continuous and differential properties of them are studied. Based on these properties, calculation complexity in the 2-D peaks searching of polarization domain can be reduced dramatically. In Sec. 4, the analysis on the accuracy and resolution performance is presented, and the analysis of algorithm complexity is discussed as well. In Sec. 5, simulation results are provided to demonstrate the continuous property mentioned in Sec. 3, and the accuracy and resolution performance.

2. MUSIC's Transplantation into Polarization Sensitive Array

2.1 Principle of Conventional MUSIC

Consider an arbitrary array (not a polarization sensitive array) which consists of M elements, and N narrow-band zero-mean signals of the same central frequency, arriving at the array from different directions. It is assumed that these signals are uncorrelated with each other, and $M > N$. The output of the ith element can be written as:

$$x_i(t) = \sum_{k=1}^{N} s_k(t)\exp(-j\omega_0\tau_{ki}) + n_i(t) \qquad (1)$$

where $s_k(t)$ is the kth signal, $n_i(t)$ is the Additive White Gaussian Noise (AWGN), and ω_0 is the central frequency of all the N signals. The parameter τ_{ki} is the delayed time of the kth signal at the ith element, whose form is determined by the direction of impinging signals and the pattern of the array. For a ULA, τ_{ki} is $(i-1)d\sin\varphi_k/c$, where φ_k represents the azimuth of the kth signal, d is the element spacing and c is the propagation velocity of signals; for a UCA with M elements, τ_{ki} becomes $r\sin\theta_k\cos(\varphi_k - 2\pi i/M)/c$, where θ_k and φ_k are the elevation and azimuth of the kth signal respectively, and r represents the radius of the UCA. For general planar arrays, the time elpase τ_{ki} becomes $(x_i\sin\theta_k\cos\varphi_k + y_i\sin\theta_k\sin\varphi_k)/c$, where x_i and y_i are used to indicate the location of the ith element. Using vector forms, (1) can be written as:

$$\mathbf{X}(t) = \mathbf{AS}(t) + \mathbf{N}(t) \qquad (2)$$

where the output vector is $\mathbf{X}(t) = [x_1(t),\dots,x_M(t)]^T$, the signal vector is $\mathbf{S}(t) = [s_1(t),\dots,s_N(t)]^T$, and the noise vector is $\mathbf{N}(t) = [n_1(t),\dots,n_M(t)]^T$; $\mathbf{A}$ is the matrix of array manifold, and the element on its kth row and ith column is $[\mathbf{A}]_{ki} = \exp(-j\omega_0\tau_{ki})$. The operator $(*)^T$ indicates the transposed matrix.

Then the covariance matrix of $\mathbf{X}(t)$ is:

$$\begin{aligned}\mathbf{R_x} &= E[\mathbf{X}(t)\mathbf{X}^H(t)]\\ &= \mathbf{AR_S A}^H + \sigma_N^2\mathbf{I}\\ &= \sum_{i=1}^{M}\lambda_i\mathbf{u}_i\mathbf{u}_i^H\end{aligned} \qquad (3)$$

where λ_i represents the eigenvalue of $\mathbf{R_X}$, arranged as $\lambda_1 \geqslant \lambda_2 \geqslant \cdots \geqslant \lambda_{N+1} = \cdots = \lambda_M = \sigma_N^2$. σ_N^2 is the power of white noise, and $\mathbf{u}_i$ denotes the eigenvector corresponding to λ_i. $\mathbf{I}$ is the identity matrix. From (3) we have:

$$\mathbf{AR_S A}^H = \sum_{i=1}^{N}(\lambda_i - \sigma_n^2)\mathbf{u}_i\mathbf{u}_i^H. \qquad (4)$$

Equation (4) indicates the fact that the column vectors of the matrix **A**, which are also referred as steering vectors, are within the N-dimension space spanned by the first N eigenvectors, perpendicular to the noise subspace spanned by the other $M-N$ eigenvectors.

In MUSIC algorithm [3], the form space spectrum is defined as:

$$P = \frac{1}{\sum_{i=N+1}^{M} \mathbf{a}^H \mathbf{u}_i} = \frac{1}{\|\mathbf{a}^H \mathbf{U}_N\|^2} \tag{5}$$

where the operator $\|*\|$ denotes the Euclidean norm, and $\mathbf{U}_N = [\mathbf{u}_{N+1}, \mathbf{u}_{N+2}, \ldots, \mathbf{u}_M]$, whose column vectors span the noise subspace. Corresponding to the signal subspace, $\mathbf{U}_S = [\mathbf{u}_1, \mathbf{u}_2, \ldots, \mathbf{u}_N]$. The form of vector **a** is similar to that of the column vectors of **A**. Every component of **a** is a function of the 2 direction parameters : the azimuth φ and elevation θ. The space spectrum P can be considered as a function of them as well. By calculating the spectrum in the space domain and searching the values of the parameters corresponding to the N maxima of P, the directions of the arriving signals can be estimated. The amount of calculation incurred in this process depends on the number of parameters determining the signal's direction. For ULA, all signals are assumed to be within a planar, and a 1-D search within the range of azimuth φ can estimate their directions; whereas for a UCA or a planar array, a 2-D search on the azimuth φ and elevation θ is required, and the amount of calculation increases considerably.

2.2 Polarization MUSIC

Consider a UCA of M polarized sensitive elements (as shown in Fig. 1, where $M = 8$). Assuming N signals arrive at the array, as mentioned in [17, 18], the ith column vector of **A** in (2) becomes:

$$\mathbf{a}(\theta_i, \varphi_i, \gamma_i, \eta_i) = \mathbf{a}_S(\theta_i, \varphi_i) \otimes \mathbf{a}_P(\theta_i, \varphi_i, \gamma_i, \eta_i) \tag{6}$$

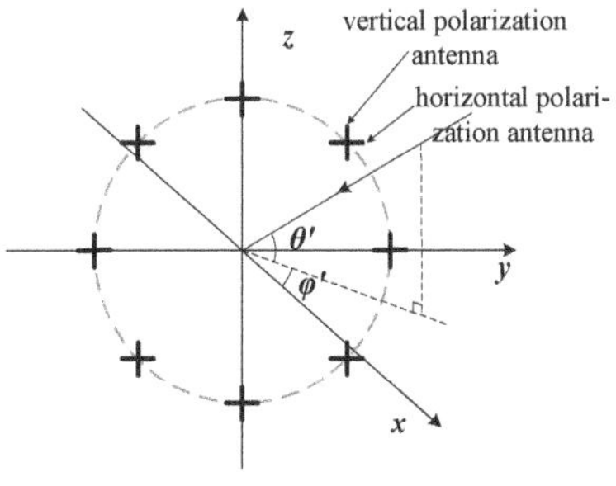

Fig. 1. An example of polarization sensitive array.

where $\otimes$ represents the Kronecker product, γ_i and η_i are the polarization parameters of ith signal, and $\mathbf{a}_S(\theta_i, \varphi_i)$ is the steering vector of the ith signal on behalf of its spatial direction, written as:

$$\mathbf{a}_S(\theta_i, \varphi_i) = \begin{bmatrix} q_0 & q_1 & q_2 \cdots & q_{M-1} \end{bmatrix}^T. \tag{7}$$

In a UCA the q_k is:

$$q_k = \exp\left(\frac{2\pi r}{\lambda} \sin\theta_i \cos\left(\varphi_i - \frac{2\pi k}{M}\right)\right). \tag{8}$$

The vector $\mathbf{a}_P(\theta_i, \varphi_i, \gamma_i, \eta_i)$ which represents the effect of polarization is:

$$\mathbf{a}_P(\theta_i, \varphi_i, \gamma_i, \eta_i) = \begin{bmatrix} \cos\theta_i \cos\varphi_i & -\sin\varphi_i \\ \cos\theta_i \sin\varphi_i & \cos\varphi_i \end{bmatrix} \begin{bmatrix} \sin\gamma_i e^{j\eta_i} \\ \cos\gamma_i \end{bmatrix}. \tag{9}$$

Substitute (6) into (2), following the similar program in [29] we can define a new joint space spectrum P_J:

$$P_J(\theta, \varphi, \gamma, \eta) = \frac{1}{\|\mathbf{a}^H(\theta, \varphi, \gamma, \eta)\mathbf{U}_N\|^2}. \tag{10}$$

Comparing (10) with (5), the only difference is that the space spectrum P_J becomes a function of 4 parameters: θ, φ, γ and η, specifying the elevation angle, azimuth angle, and the 2 polarization parameters respectively. It should be noticed that, even if polarization states of all arriving signals are the same, as the direction parameters θ_i and φ_i are involved in vector $\mathbf{a}_P(\theta_i, \varphi_i, \gamma_i, \eta_i)$, it still brings difference into the steering vector $\mathbf{a}(\theta_i, \varphi_i, \gamma_i, \eta_i)$ and helps to enhance the algorithm's resolution performance. However, when the combination of these 4 parameters makes $\mathbf{a}_P(\theta_i, \varphi_i, \gamma_i, \eta_i) = \mathbf{a}_P(\theta_j, \varphi_j, \gamma_j, \eta_j)$, $i \neq j$, the polarization states no longer provide assistance in resolving these 2 signals.

Theoretically, a peak search can be executed in a 4-D space spanned by these 4 parameters to obtain the estimation results. Benjamin Friedlander has proved that the performance of resolution will be improved considerably [35]. However, the calculation complexity of the 4-D space spectrum causes an unbearable computation complexity and makes the ordinary polarization MUSIC algorithm unpractical.

3. Fast Polarization MUSIC

In this part, properties of the space spectrum $P_J(\theta, \varphi, \gamma, \eta)$ of the polarization MUSIC is analyzed, and a more effective algorithm is developed to reduce the great amount of calculation in DOA estimation.

3.1 Property of the Joint Space Spectrum

As previously stated, the values of the 4 parameters corresponding to the N maximums of the space spectrum in (10) indicate the directions and polarization states of the N arriving signals. And a maximum in 4-D space spanned by the 4 parameters, whose position is $(\theta_0, \varphi_0, \gamma_0, \eta_0)$, remains to be a maximum in the 2-D subspace $(\theta_0, \varphi_0, \gamma, \eta)$, spanned by the two polarization parameters.

Suppose the values of the two direction parameters of P_J are constants, $\theta = \theta_0$ and $\varphi = \varphi_0$. Then the vector $\mathbf{a}(\theta_0, \varphi_0, \gamma, \eta)$ can be unfolded as:

$$\begin{aligned}
&\mathbf{a}(\theta_0,\varphi_0,\gamma,\eta)\\
&=\mathbf{a}_S(\theta_0,\varphi_0)\otimes\mathbf{a}_P(\theta_0,\varphi_0,\gamma,\eta)\\
&=\begin{bmatrix}q_0\\q_1\\\vdots\\q_{M-1}\end{bmatrix}\otimes\begin{bmatrix}\cos\theta_0\cos\varphi_0\sin\gamma e^{j\eta}-\sin\varphi_0\cos\gamma\\\cos\theta_0\sin\varphi_0\sin\gamma e^{j\eta}+\cos\varphi_0\cos\gamma\end{bmatrix}\\
&=\begin{bmatrix}u_1(\theta_0,\varphi_0)\sin\gamma e^{j\eta}+v_1(\theta_0,\varphi_0)\cos\gamma\\u_2(\theta_0,\varphi_0)\sin\gamma e^{j\eta}+v_2(\theta_0,\varphi_0)\cos\gamma\\\vdots\\u_{2M}(\theta_0,\varphi_0)\sin\gamma e^{j\eta}+v_{2M}(\theta_0,\varphi_0)\cos\gamma\end{bmatrix}
\end{aligned}\tag{11}$$

where:

$$u_{2k-1}(\theta_0,\varphi_0)=q_{k-1}\cos\theta_0\cos\varphi_0 \tag{12a}$$

$$v_{2k-1}(\theta_0,\varphi_0)=-q_{k-1}\sin\varphi_0 \tag{12b}$$

$$u_{2k}(\theta_0,\varphi_0)=q_{k-1}\cos\theta_0\sin\varphi_0 \tag{12c}$$

$$v_{2k}(\theta_0,\varphi_0)=q_{k-1}\cos\varphi_0 \tag{12d}$$

and $k=1,2,\ldots,M$.

As mentioned in [36], another form of the joint space spectrum in (10) can be derived:

$$\begin{aligned}P_J(\theta,\varphi,\gamma,\eta)&=Z^{-1}(\theta,\varphi,\gamma,\eta)\\&=[M-U(\theta,\varphi,\gamma,\eta)]^{-1}\end{aligned}\tag{13}$$

where:

$$\begin{aligned}
&Z(\theta,\varphi,\gamma,\eta)\\
&=\|\mathbf{a}^H(\theta,\varphi,\gamma,\eta)\mathbf{U}_N\|^2\\
&=\|\mathbf{a}(\theta,\varphi,\gamma,\eta)\|^2-\|\mathbf{a}^H(\theta,\varphi,\gamma,\eta)\mathbf{U}_S\|^2\\
&=M-M\sin^2\theta\sin^2\gamma-\|\mathbf{a}^H(\theta,\varphi,\gamma,\eta)\mathbf{U}_S\|^2
\end{aligned}\tag{14}$$

and:

$$U(\theta,\varphi,\gamma,\eta)=M\sin^2\theta\sin^2\gamma+\|\mathbf{a}^H(\theta,\varphi,\gamma,\eta)\mathbf{U}_S\|^2 \tag{15}$$

where $Z(\theta,\varphi,\gamma,\eta)$ is named as zero spectrum, as it approaches to 0 when values of the 4 parameters approach to the direction and polarization state of a signal. At the same time, $U(\theta,\varphi,\gamma,\eta)$ in (13) approaches to a local maximum, and a steep peak appears in the graph of $P_J(\theta,\varphi,\gamma,\eta)$. Although the maximums of $U(\theta,\varphi,\gamma,\eta)$ may not be so steep and distinct as those of $P_J(\theta,\varphi,\gamma,\eta)$, theoretically the property of the $U(\theta,\varphi,\gamma,\eta)$ can be used to reveal some characteristics of $P_J(\theta,\varphi,\gamma,\eta)$. So $U(\theta,\varphi,\gamma,\eta)$ is regarded as an alternative form of the space spectrum $P_J(\theta,\varphi,\gamma,\eta)$ in this subsection.

Suppose $\theta=\theta_0$ and $\varphi=\varphi_0$, and substitute (11) into (15), we have:

$$\begin{aligned}
&U(\theta_0,\varphi_0,\gamma,\eta)\\
&=M\sin^2\theta_0\sin^2\gamma+\|\mathbf{a}^H(\theta_0,\varphi_0,\gamma,\eta)\mathbf{U}_S\|^2\\
&=M\sin^2\theta_0\sin^2\gamma+\left\|\begin{bmatrix}u_1\sin\gamma e^{j\eta}+v_1\cos\gamma\\u_2\sin\gamma e^{j\eta}+v_2\cos\gamma\\\vdots\\u_{2M}\sin\gamma e^{j\eta}+v_{2M}\cos\gamma\end{bmatrix}^H\mathbf{U}_S\right\|^2\\
&=M\sin^2\theta_0\sin^2\gamma+\left\|\begin{bmatrix}x_1\sin\gamma e^{-j\eta}+y_1\cos\gamma\\x_2\sin\gamma e^{-j\eta}+y_2\cos\gamma\\\vdots\\x_N\sin\gamma e^{-j\eta}+y_N\cos\gamma\end{bmatrix}^T\right\|^2\\
&=M\sin^2\theta_0\sin^2\gamma+\sum_{r=1}^{N}\{|x_r|^2\sin^2\gamma+|y_r|^2\cos^2\gamma\\
&\quad+2\sin\gamma\cos\gamma[\mathrm{Re}(x_r^*y_r)\cos\eta+\mathrm{Im}(x_r^*y_r)\sin\eta]\}\\
&=a\sin^2\gamma+b\cos^2\gamma+\sin2\gamma(c\cos\eta+d\sin\eta)\\
&=\frac{a+b}{2}+\frac{a-b}{2}\cos2\gamma+\sin2\gamma\sqrt{c^2+d^2}\cos(\eta+\phi).
\end{aligned}\tag{16}$$

The operator $\mathrm{Re}(*)$ is the real part of a plural variable, whereas $\mathrm{Im}(*)$ is the imaginary part. u_i and v_i are determined by the form of the array, and the values of θ_0 and φ_0.

As the covariance matrix $\mathbf{R_X}$ is determined after the output of the array are sampled, $\mathbf{U_S}$ can be regarded as a constant matrix. Then the coefficients x_r, y_r, a, b, c, d and ϕ are all constants determined by θ_0 and φ_0, irrelevant with the value of γ and η. The detailed form of these coefficients can be deduced without much difficulties and omitted here.

As we know, the maximum of a continuous function with two variables has to meet the following expressions:

$$\frac{\partial U(\theta_0,\varphi_0,\gamma,\eta)}{\partial\eta}=0 \tag{17}$$

$$\frac{\partial U(\theta_0,\varphi_0,\gamma,\eta)}{\partial\gamma}=0 \tag{18}$$

$$\frac{\partial^2 U(\theta_0,\varphi_0,\gamma,\eta)}{\partial\eta^2}\leqslant0 \tag{19}$$

$$\frac{\partial^2 U(\theta_0,\varphi_0,\gamma,\eta)}{\partial\gamma^2}\leqslant0. \tag{20}$$

Substitute (16) into (17),

$$-\sqrt{c^2+d^2}\sin2\gamma\sin(\eta+\phi)=0. \tag{21}$$

Substitute (16) into (19),

$$-\sqrt{c^2+d^2}\sin2\gamma\cos(\eta+\phi)\leqslant0. \tag{22}$$

Notice that $\gamma\in[0,\pi/2]$, and $U(\theta_0,\varphi_0,\gamma,\eta)$ will be constant when $\gamma=0$, which rarely happens. Based on (21) and (22), when $\gamma\neq0$, the value of η corresponding to the maximum is:

$$\eta_m=2k\pi-\phi \tag{23}$$

where η_m indicates the value of η at the position of maximum. k should be 0 or 1, to ensure that $\eta_m \in [0, 2\pi]$. The value of η_m is not affected by γ.

Substitute (16) into (18):

$$\tan 2\gamma = \frac{2\sqrt{c^2+d^2}\cos(\eta+\phi)}{a-b}. \quad (24)$$

From (23), we have $\eta + \phi = 2k\pi$, then the value of γ of the maximum satisfies (25):

$$\tan 2\gamma_m = \frac{2\sqrt{c^2+d^2}}{a-b}. \quad (25)$$

Notice that (20) is true when (23) and (25) are satisfied, so there is one and only one set of η_m and γ_m which satisfy (17) to (20).

It should be noticed that, in the derivation of (23) and (25), no specific pattern of arrays are involved. This is the very reason that the algorithm proposed in this paper has no particular requirements in the form of arrays.

It is clear that when the direction parameters are constants, the function $U(\theta_0, \varphi_0, \gamma, \eta)$ and the joint space spectrum $P_J(\theta_0, \varphi_0, \gamma, \eta)$ has only one peak, respectively. So the peak search of $P_J(\theta, \varphi, \gamma, \eta)$ in the 4-D space can be executed in this way:

Step 1: Suppose $\theta = \theta_0$ and $\varphi = \varphi_0$, then run a 2-D search on γ and η to derive the single maximum value $P_D(\theta_0, \varphi_0)$, and the corresponding γ and η as $\Gamma(\theta_0, \varphi_0)$ and $H(\theta_0, \varphi_0)$, respectively:

$$H(\theta_0, \phi_0) = \arg\max_{\eta} P_J(\theta_0, \phi_0, \gamma, \eta), \quad (26)$$

$$\Gamma(\theta_0, \phi_0) = \arg\max_{\gamma} P_J(\theta_0, \phi_0, \gamma, \eta), \quad (27)$$

$$P_D(\theta_0, \phi_0) = \max_{(\gamma,\eta)} P_J(\theta_0, \phi_0, \gamma, \eta). \quad (28)$$

Step 2: Proceed this calculation throughout the ranges of θ and φ and complete the graph of $P_D(\theta, \varphi)$, $\Gamma(\theta, \varphi)$ and $H(\theta, \varphi)$.
Step 3: Take another 2-D peak search in $P_D(\theta, \phi)$. The direction parameters of the peaks are the estimation of the signals' direction, and the corresponding values of $\Gamma(\theta_0, \varphi_0)$ and $H(\theta_0, \varphi_0)$ provide the estimation of the signals' polarization states.

3.2 The Procedure of the Fast Algorithm

Obviously, the process mentioned in 3.1 cannot reduce the computation complexity. However, in the first 2-D search, based on the fact that only one peak exists in $P_J(\theta_0, \varphi_0, \gamma, \eta)$, we provide a straightforward method to reduce the amount of calculation.

According to (23), η_m is determined by ϕ at the location of the maximum of $U(\theta_0, \varphi_0, \gamma, \eta)$, which is independent of γ_m. So we can assign a constant to γ, and perform a 1-D search along the parameter η to get η_m; then set η as η_m, run another 1-D search to get γ_m. $P_D(\theta, \varphi)$, $\Gamma(\theta, \varphi)$ and $H(\theta, \varphi)$ can be obtained respectively.

Referring to (16), the a, b, c, d, and ϕ are all continuous functions of θ_0 and φ_0. From (23) and (25), it can be derived that γ and η, i.e., the $\Gamma(\theta, \varphi)$ and $H(\theta, \varphi)$, are also continuous functions of θ and φ under general conditions. In other words, the $\Gamma(\theta, \varphi)$ and $H(\theta, \varphi)$ will change smoothly if the parameters change gradually in the calculation process of $P_D(\theta, \varphi)$, and the location of the single peak in $P_J(\theta_0, \varphi_0, \gamma, \eta)$ does not change dramatically when the search step of θ and φ are limited to small values.

Therefore, in calculation of $P_D(\theta, \varphi)$, if $\Gamma(\theta, \varphi)$ and $H(\theta, \varphi)$ of a certain direction parameter are derived, $\Gamma(\theta, \varphi \pm \Delta\varphi)$ and $H(\theta, \varphi \pm \Delta\varphi)$, or $\Gamma(\theta \pm \Delta\theta, \varphi)$ and $H(\theta \pm \Delta\theta, \varphi)$, can be calculated very fast through a peak search within a considerably small area. Then the exhaustive computation employed in the 4-D peak search can be avoided.

In (16), It should be noticed that ϕ is derived from an inverse tangent function of d/c. This may cause a flip of π in the value of ϕ, and $H(\theta, \varphi)$ might be affected. However, (25) indicates that $\Gamma(\theta, \varphi)$ will remain unaffected. This flip does not happen frequently, and even if the flip occurs, as shown in (23), the values of $H(\theta, \varphi \pm \Delta\varphi)$ or $H(\theta \pm \Delta\theta, \varphi)$ corresponding to maximum will be near $H(\theta, \varphi) \pm \pi$, which is predictable.

The main steps of fast polarization MUSIC are summarized as follows:

Step 1: Calculate the covariance matrix $\mathbf{R_X}$ of the array's output, take an eigendecomposition to get the orthogonal basis of the noise subspace.
Step 2: Select a set of direction parameters θ and φ randomly, or according to the result of previous estimation. Use (10) to complete a peak search on the two polarization parameters. Record the maximum value and the corresponding values of $\Gamma(\theta, \varphi)$ and $H(\theta, \varphi)$.
Step 3: At directions nearby (θ, φ), e.x., $(\theta, \varphi \pm \Delta\varphi)$ or $(\theta \pm \Delta\theta, \varphi)$, search the maximum in a small adjacent area containing the $(\Gamma(\theta, \varphi), H(\theta, \varphi))$, which are obtained in step 2. If no obvious peak can be found, try another possible stationary point.
Step 4: Repeat step 3, until all $P_D(\theta, \varphi)$ are obtained.
Step 5: Exert a 2-D peak search in the $P_D(\theta, \varphi)$ to obtain all maximums. The directions of impinging signals can be estimated from direction parameters of these peaks.

4. Performance Analysis

4.1 Precision

The algorithm presented in 3.2 is an improved version of polarization MUSIC, so some analysis results for conventional MUSIC can be applied here. The analysis for the performance of MUSIC, given by Stoica and Nehorai in [15, 37], are adopted here.

The theorem 3.1 in [15] states that the MUSIC estimation errors $\hat{\omega}_i - \omega_i$ approach to jointly Gaussian distribution. Here ω_i indicates the parameter to be estimated in the MUSIC, which can be either θ_i or φ_i; and $\hat{\omega}_i$ is the corresponding estimation result. The theorem also states that the means of $\hat{\omega}_i - \omega_i$ are all 0. When only one signal arrives at the array, it becomes

$$E(\hat{\omega}_1 - \omega_1)^2 = \frac{\sigma_N^2}{2K}\left[\frac{\lambda_1 |\mathbf{a}^H(\omega_1)\mathbf{u}_1|^2}{(\sigma_N^2 - \lambda_1)^2}\right] \Big/ \left[\sum_{k=2}^{M} |\mathbf{d}^H(\omega_1)\mathbf{u}_k|\right]. \quad (29)$$

Under this condition, λ_1 is proportional the power of the signal, so it is clear that the variance is largely determined by the SNR and the number of snapshots. It is reasonable to conclude that the variance of results obtained by MUSIC and the proposed algorithm are of the same order if only one signal arrives. Computer simulations are presented to demonstrate the precision performance when multiple signals arrive.

4.2 Resolution

In this subsection, the resolution performance of the fast polarization MUSIC is discussed.

In the paper written by Benjamin Friedlander [35], a standard has been given to justify whether two close signals can be descried, when using a ULA. If there are more than 2 signals, the analysis will become extremely difficult, so we only consider a 2 signals' scenario, as the analysis presented in [35] did.

Assume that the direction and polarization parameters of the 2 signals are $(\theta_1, \varphi_1, \gamma_1, \eta_1)$ and $(\theta_2, \varphi_2, \gamma_2, \eta_2)$. Similar to [35], 2 signals can be distinguished when expression (30) is satisfied:

$$E\{\hat{Z}(\theta_0, \varphi_0, \gamma_0, \eta_0)\} > E\{\hat{Z}(\theta_i, \varphi_i, \gamma_i, \eta_i)\},\ i = 1, 2. \quad (30)$$

Here, $Z(\theta, \varphi, \gamma, \eta)$ is the zero spectrum, given in (14), $\hat{Z}(\theta, \varphi, \gamma, \eta)$ is the estimation of $Z(\theta, \varphi, \gamma, \eta)$, and $i = 1, 2$. Define:

$$(\theta_0, \varphi_0, \gamma_0, \eta_0) = \left(\frac{\theta_1 + \theta_2}{2}, \frac{\varphi_1 + \varphi_2}{2}, \frac{\gamma_1 + \gamma_2}{2}, \frac{\eta_1 + \eta_2}{2}\right). \quad (31)$$

After some deduction (see Appendix A), from expression (30) we have:

$$\|\mathbf{a}_0\|^2 - |\mathbf{a}_0^H\mathbf{u}_1|^2 - |\mathbf{a}_0^H\mathbf{u}_2|^2 > \frac{(2M-2)}{K} \cdot \frac{\lambda_2/\sigma_N^2}{(\lambda_2/\sigma_N^2 - 1)^2}\left[|\mathbf{a}_i^H\mathbf{u}_2|^2 - |\mathbf{a}_0^H\mathbf{u}_2|^2\right] \quad (32)$$

where $i = 1, 2$, $\mathbf{u}_1$ and $\mathbf{u}_2$ are the eigenvectors of the signal subspace, and λ_1 is the eigenvalue corresponding to $\mathbf{u}_1$. K is the number of snapshots, and σ_N^2 represents the power of noise. We adopt $\mathbf{a}_0 = \mathbf{a}(\theta_0, \varphi_0, \gamma_0, \eta_0)$, $\mathbf{a}_1 = \mathbf{a}(\theta_1, \varphi_1, \gamma_1, \eta_1)$ and $\mathbf{a}_2 = \mathbf{a}(\theta_2, \varphi_2, \gamma_2, \eta_2)$ to make the expression more clear and succinct.

With some conclusion from [38], we have (refer to Appendix B for more details):

$$K > \frac{2(M-1)}{\text{SNR}\left(\|\mathbf{a}_0\|^2 - |\mathbf{a}_0^H\mathbf{u}_1|^2 - |\mathbf{a}_0^H\mathbf{u}_2|^2\right)} \cdot \frac{|\mathbf{a}_k^H\mathbf{u}_2|^2 - |\mathbf{a}_0^H\mathbf{u}_2|^2}{\|\mathbf{a}_0\|^2\left[1 - |\mathbf{a}_1^H\mathbf{a}_2|/(\|\mathbf{a}_1\| \cdot \|\mathbf{a}_2\|)\right]}. \quad (33)$$

It can be seen that the value of the right side of expression (33) is determined by $\mathbf{a}_0$, $\mathbf{a}_1$ and $\mathbf{a}_2$ (for $\mathbf{u}_1$ and $\mathbf{u}_2$ are also determined by them). Then when the values of snapshot number K and SNR satisfies expression (33), the two close signals can be descried by the algorithm theoretically.

For the conventional MUSIC, some minor adjustment in deduction is required (see Appendix B), and the necessary snapshot number becomes

$$K > \frac{(M-2)\left[|\mathbf{a}_i^H\mathbf{u}_2|^2 - |\mathbf{a}_0^H\mathbf{u}_2|^2\right]}{\text{SNR}^2\|\mathbf{a}_0\|^4\left[1 - |\mathbf{a}_1^H\mathbf{a}_2|/(\|\mathbf{a}_1\|\|\mathbf{a}_2\|)\right]^2} \cdot \frac{1}{\|\mathbf{a}_0\|^2 - |\mathbf{a}_0^H\mathbf{u}_1|^2 - |\mathbf{a}_0^H\mathbf{u}_2|^2}. \quad (34)$$

Equations (33) and (34) provide a way to compare the numbers of snapshots needed in the polarization MUSIC with that of a conventional MUSIC to resolve two signals.

4.3 Complexity

In this subsection, the computation complexity of the proposed algorithm is discussed. In Subsection 3.2, the process of the algorithm are divided into 5 major steps. Step 1 and step 5 concern with the eigendecomposition and peak search in a 2-D matrix. Comparing with the amount of calculation incurred in step 2 to step 4, the computation in step 1 and step 5 is negligible. So we focus on the computation from step 2 to step 4.

The steering vector $\mathbf{a}(\theta, \varphi, \gamma, \eta)$ can be calculated offline, and the noise spectrum matrix $\mathbf{U}_N$ is calculated only once. Each value of space spectrum in (10) can be obtained by two matrix multiplications. Besides, the computation complexity for each point of spectrum is the same for both conventional MUSIC and polarization MUSIC, if the numbers of their array antennas are the same. So the overall computation complexity is determined by the numbers of spectrum values to be calculated. Elevation angle θ, azimuth angle φ and the polarization parameters γ and η are all continuous, whereas they need to be sampled in the process of space spectrum calculation. Assume that the sample numbers of 4 parameters are N_1, N_2, N_3 and N_4, respectively. To obtain the required space spectrum for a conventional 2-D MUSIC, the number of sampled space spectrum is

N_1N_2, whereas for the polarization MUSIC mentioned in 2.2, it becomes $N_1N_2N_3N_4$. For the fast polarization MUSIC algorithm, the required number of spectrum values is discussed as follows.

In subsection 3.2, it is stated that two 1-D peak searches along η and γ would be sufficient to obtain the maximum of $U(\theta_0,\varphi_0,\gamma_0,\eta_0)$. When the worst condition occurs, we need to go through both the ranges of η and γ to locate the peak, and the numbers of spectrum value required to be calculated in step 2 or step 3 would be no more than N_3+N_4.

In step 3, assume the positions of the two peaks of $U(\theta_0,\varphi_0,\gamma,\eta)$ and $U(\theta_0\pm\Delta\theta,\varphi,\gamma,\eta)$ or $U(\theta,\varphi\pm\Delta\varphi,\gamma,\eta)$ are close enough, and a peak search in a very small area will locate the latter one's peak position if the former's peak has been located. Consider the relationship between peaks of $U(\theta_0,\varphi_0,\gamma,\eta)$ and $U(\theta_0+\Delta\theta,\varphi,\gamma,\eta)$. The numbers of spectrum value required can be illustrated as follows:

Condition 1: The positions of the two peaks are the same, i.e.:

$$\Gamma(\theta_0+\Delta\theta,\varphi_0)=\Gamma(\theta_0,\varphi_0)=\gamma_0, \tag{35}$$

$$H(\theta_0+\Delta\theta,\varphi_0)=H(\theta_0,\varphi_0)=\eta_0. \tag{36}$$

Assume that $\Delta\gamma$ and $\Delta\eta$ are sampling intervals of γ and η, respectively. As $\Gamma(\theta_0,\varphi_0)$ and $H(\theta_0,\varphi_0)$ are already obtained, we need $U(\theta_0+\Delta\theta,\varphi_0,\gamma_0,\eta_0)$ and $U(\theta_0+\Delta\theta,\varphi_0,\gamma_0,\eta_0\pm\Delta\eta)$ to confirm the fact that:

$$U(\theta_0+\Delta\theta,\varphi_0,\gamma_0,\eta_0)\geqslant U(\theta_0+\Delta\theta,\varphi_0,\gamma_0,\eta_0\pm\Delta\eta). \tag{37}$$

$U(\theta_0+\Delta\theta,\varphi_0,\gamma_0\pm\Delta\gamma,\eta_0)$ are also needed to confirm that:

$$U(\theta_0+\Delta\theta,\varphi_0,\gamma_0,\eta_0)\geqslant U(\theta_0+\Delta\theta,\varphi_0,\gamma_0\pm\Delta\gamma,\eta_0). \tag{38}$$

Under such condition, 5 sample points of the space spectrum must be calculated to obtain the location of the peak in $U(\theta_0+\Delta\theta,\varphi_0,\gamma,\eta)$.

Condition 2: The positions of the two peaks are one step away from each other, for instance:

$$\Gamma(\theta_0+\Delta\theta,\varphi_0)=\Gamma(\theta_0,\varphi_0)+\Delta\gamma=\gamma_0+\Delta\gamma, \tag{39}$$

$$H(\theta_0+\Delta\theta,\varphi_0)=H(\theta_0,\varphi_0)+\Delta\eta=\eta_0+\Delta\eta. \tag{40}$$

Then we need to calculate $U(\theta_0+\Delta\theta,\varphi_0,\gamma_0,\eta_0)$, $U(\theta_0+\Delta\theta,\varphi_0,\gamma_0,\eta_0+\Delta\eta)$ and $U(\theta_0+\Delta\theta,\varphi_0,\gamma_0,\eta_0+2\Delta\eta)$ to confirm that $U(\theta_0+\Delta\theta,\varphi_0,\gamma_0,\eta_0+\Delta\eta)$ is the maximum value of them, i.e.:

$$U(\theta_0+\Delta\theta,\gamma_0,\gamma_0,\eta_0+\Delta\eta)\geqslant U(\theta_0+\Delta\theta,\varphi_0,\gamma_0,\eta_0), \tag{41}$$

$$U(\theta_0+\Delta\theta,\gamma_0,\gamma_0,\eta_0+\Delta\eta)\geqslant U(\theta_0+\Delta\theta,\varphi_0,\gamma_0,\eta_0+2\Delta\eta) \tag{42}$$

and calculate $U(\theta_0+\Delta,\varphi_0,\gamma_0+\Delta\gamma,\eta_0+\Delta\eta)$ and $U(\theta_0+\Delta,\varphi_0,\gamma_0+2\Delta\gamma,\eta_0+\Delta\eta)$ to confirm that:

$$\begin{aligned}&U(\theta_0+\Delta\theta,\varphi,\gamma_0+\Delta\gamma,\eta_0+\Delta\eta)\\&\geqslant U(\theta_0+\Delta\theta,\varphi,\gamma_0,\eta_0+\Delta\eta).\end{aligned} \tag{43}$$

$$\begin{aligned}&U(\theta_0+\Delta\theta,\varphi,\gamma_0+\Delta\gamma,\eta_0+\Delta\eta)\\&\geqslant U(\theta_0+\Delta\theta,\varphi,\gamma_0+2\Delta\gamma,\eta_0+\Delta\eta)\end{aligned} \tag{44}$$

It is noticed that sometimes $U(\theta_0+\Delta\theta,\varphi,\gamma_0,\eta_0-\Delta\eta)$ and $U(\theta_0+\Delta\theta,\varphi,\gamma_0-\Delta\gamma,\eta_0+\Delta\eta)$ also need to be calculated. So 5 to 7 points of the spectrum are required.

Other conditions are omitted here, since their deductions are similar.

Based on the discussion above, it can be concluded that to obtain the value of each $P_D(\theta,\varphi)$, at least 5 points of the spectrum are required. So the overall number of spectrum values calculated from step 2 to step 4 is between $5N_1N_2$ and $(N_3+N_4)N_1N_2$. In Section 5, simulation results show that the required number of spectrum is approximately $6N_1N_2$.

5. Simulation Results

5.1 Character of Polarization Spectrum

The fast polarization MUSIC algorithm can be used on various sorts of polarization sensitive arrays. In this section, a UCA of 8 polarization sensitive elements with a radius of 0.667 λ are adopted in the following simulation. The output of the array is contaminated by uncorrelated AWGN. Most of the following simulation results are based on these conditions. The directions of the arriving signals can be expressed in various forms, and the form adopted in this section is illustrated in Fig. 1, with a minor difference with those in (8) and (9). The azimuth and elevation here are written as φ' and θ', respectively, and the relationship between the two forms can be expressed as:

$$\tan\varphi'=\tan\theta\sin\varphi, \tag{45a}$$

$$\sin\theta'=\sin\theta\cos\varphi. \tag{45b}$$

In this section, the direction parameters in Fig. 1 are used, but polarization parameters remain the same.

Fig. 2 shows the contours of the polarization spectrum $P(\theta_0',\varphi_0',\gamma,\eta)$, to demonstrate the assertion in subsection 3.1, i.e., $U(\theta_0',\varphi_0',\gamma,\eta)$ or $P(\theta_0',\varphi_0',\gamma,\eta)$ has only one peak.

In this simulation, 3 signals impinge onto the array. Their directions written as (φ',θ') are $(-4,-3)$, $(-3,3)$ and $(3,-3)$, whereas their polarization parameters written as (γ,η) are $(12.5,30)$, $(22.5,50)$ and $(32.5,70)$, respectively. The number of snapshots is 500, and the SNR is 15 dB. $P(0,0,\gamma,\eta)$ and $P(2.5,2.5,\gamma,\eta)$ are plotted in Fig. 2(a) and Fig. 2(b), respectively. Fig. 2 indicates that regardless of the number of arriving signals and their polarization, in $P(\theta_0,\varphi_0,\gamma,\eta)$, one and only one maximum exists. From the

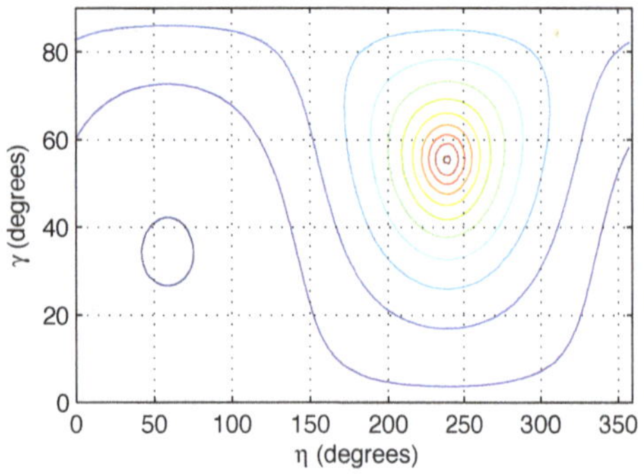

(a) contour of $P(0,0,\gamma,\eta)$

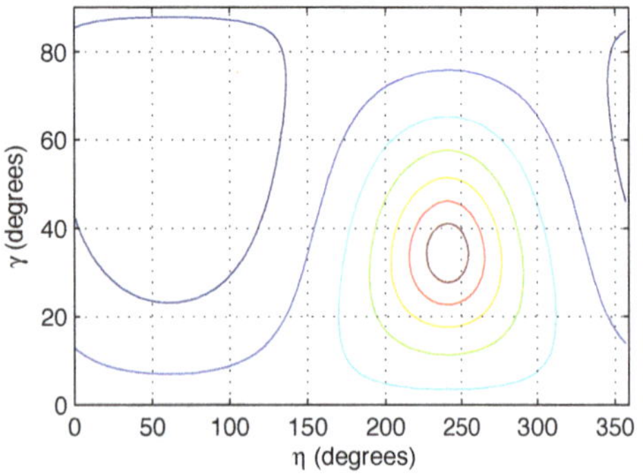

(b) contour of $P(2.5,2.5,\gamma,\eta)$

Fig. 2. Contour of spectrum $P(\theta_0,\varphi_0,\gamma,\eta)$.

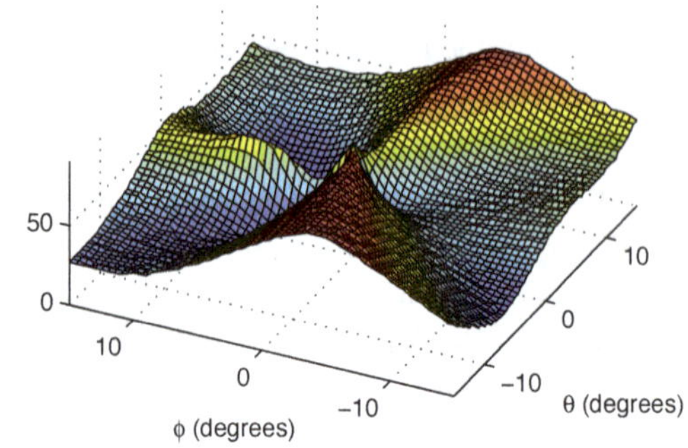

(a) $\Gamma(\theta,\varphi)$ with 3 signals

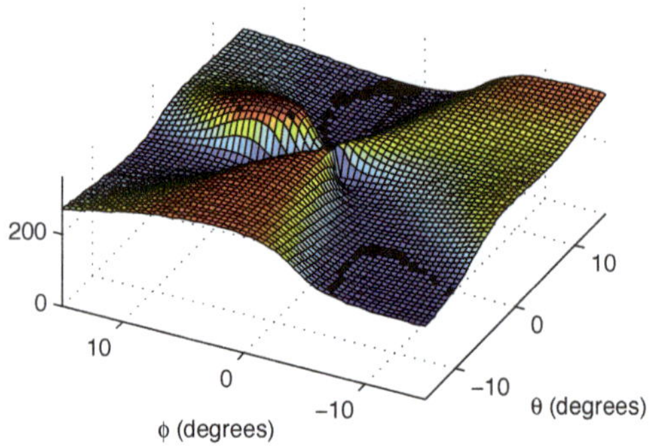

(b) $H(\theta,\varphi)$ with 3 signals

Fig. 3. Polarization at all directions.

contour graphs, the symmetrical character of parameter η can be observed more clearly. Based on this, the maximum can be located after two 1-D searches.

5.2 The Continuity in $\Gamma(\theta',\varphi')$ and $H(\theta',\varphi')$

Fig. 3 presents $\Gamma(\theta',\varphi')$ and $H(\theta',\varphi')$ with 3 arriving signals of various parameters, to illustrate the continuity property of them. Fig. 3(a) presents the graph of $\Gamma(\theta',\varphi')$, whereas Fig. 3(b) present $H(\theta',\varphi')$. The result of 3 signals, whose direction and polarization parameter written in the form of $(\theta',\varphi',\gamma,\eta)$ are $(-3,-4,12.5,30)$, $(3,-3,45.5,50)$ and $(-3,4,70.5,70)$, is drawn in Fig. 3(a) and 3(b). It can be observed that in the two figures, the $\Gamma(\theta',\varphi')$ and $H(\theta',\varphi')$ are continuous in most area, whereas the steep slopes occur near the directions of the impinging signals.

5.3 Precision and Resolution

Fig. 4, Fig. 5 and Fig. 6 provide stimulation results to compare the performances of the fast polarization MUSIC algorithm and other algorithms. The performance of the fast polarization MUSIC is compared with conventional MUSIC, and the rank reduction (RR) algorithm in [25, 26], which can be extended to arbitrary form of array as well. The performances of these algorithms are presented through Monte Carlo simulations.

Fig. 4 draws snapshot numbers required to descry 2 close signals with various angle differences, calculated by (33) and (34). Their azimuth angles are 0°, one signal's elevation angle is set to 0°, and the other is modified. The simulation is based on a UCA of 16 sensors, whose radius is 0.667 λ. The polarization parameters are $(30,15)$ and $(50,100)$. In Fig. 4, the calculated results of the fast polarization MUSIC and conventional 2-D MUSIC are given according to the SNR. It can be seen that less snapshots are required to descry 2 close signals in the proposed algorithm, particularly with relatively low SNR.

Fig. 5 compares different algorithms' precision performance with various number of signals. RMSE stands for root mean square error. In Fig. 5, the estimation's RMSE for the fast polarization MUSIC, conventional 2-D MUSIC and RR Algorithm are depicted. In 1-signal scenario, the signal's parameters are $(0.1,-3.1,12.5,30)$. In 2-signal scenario, the parameters are $(0,-3.1,12.5,30)$ and $(0,3.2,60,60)$, whereas in the 3-signal scenario, they are $(-3.1,-4.1,12.5,30)$, $(3.3,-3.1,45.5,50)$ and $(-3.2,3.1,70.5,70)$. SNR is 20 dB, and Monte Carlo experiment number is 200. In 1-signal scenario, the precision performance for all algorithms are relatively good. When the number of signals increases, the precision degrades, but the RMSE of fast polarization MUSIC remains lower than those of conventional MUSIC and RR Algorithm. RMSE of the fast polarization MUSIC is about 0.1 degree lower than that of RR algorithm in 3-signal scenario.

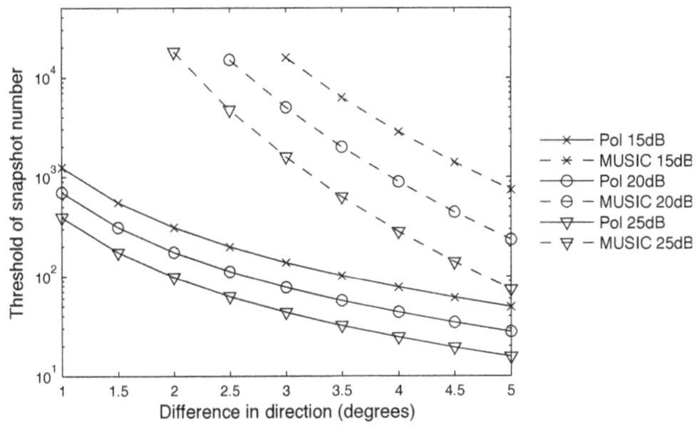

Fig. 4. Snapshot number required to resolve close signals.

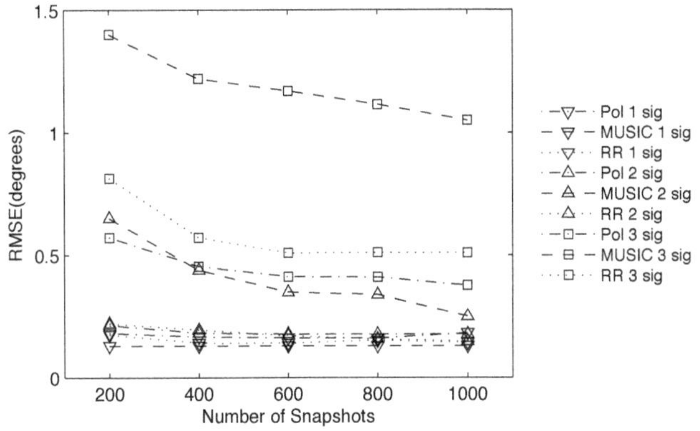

Fig. 5. Precision performance.

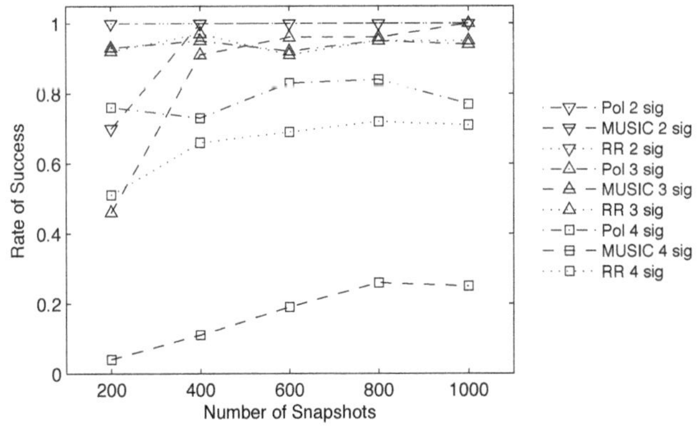

Fig. 6. Rate of success in resolution.

In Fig. 6, the success rates in multiple signals resolution are presented. The signals' parameters are $(-6,-4.2,12.5,30)$, $(3,-5.1,22.5,50)$, $(-3,3.3,41,70)$ and $(6,6.2,62,90)$ in the 4-signal scenario. In other scenarios, they are the same as Fig. 5. Fig. 6 shows that the success rates of the fast polarization MUSIC are generally higher than that of conventional 2-D MUSIC and RR algorithm, particularly with small snapshot numbers. Especially, the success rate of the proposed algorithm is approximately 80% in the 4-signal scenario, higher than that of RR algorithm and the conventional 2-D MUSIC.

Figure 7 gives the space spectrum $P(\theta',\varphi')$ of the conventional 2-D MUSIC, the RR algorithm and the graph of $P_D(\theta',\varphi')$ of the fast polarization MUSIC . The parameters of the 4 incoming signals are the same as the 4-signal scenario in Fig. 6, with 500 snapshots. Fig. 7(a) shows that the peaks of the conventional 2-D MUSIC are eclipsed severely and cannot be distinguished. For the RR algorithm, the peaks in Fig. 7(b) can be located, although not very distinct. But in the space spectrum of the fast polarization MUSIC, all 4 peaks are quite clear.

5.4 Computation Complexity

In simulation results presented in this article, the ranges of azimuth and elevation are sampled with an interval of 0.5°, and the ranges of azimuth and elevation angles are both limited from -15° to 15°, i.e., $\theta' \in [-15,15]$ and $\varphi' \in [-15,15]$. Then the 2-D space spanned by azimuth and elevation will be sampled with $61 \times 61 = 3721$ samples, i.e., if the conventional MUSIC is adopted, 3721 points of the space spectrum are required to be calculated to estimate the arriving signals' directions.

In the fast polarization MUSIC, suppose γ and η are sampled with intervals of 3 degrees and 6 degrees, respectively. Based on the analysis in Subsection 4.3, the lower limit of sample number needed will be $5N_1N_2 = 18605$. When the worst case happens, the number becomes $(N_3+N_4)N_1N_2 = 338611$, whereas the number of samples needed in the conventional 4-D polarization MUSIC is $N_1N_2N_3N_4 = 6921060$, approximately 20 times more than the worst case of the proposed algorithm.

Tab. 1 presents the average number of samples of space spectrum calculated in polarization MUSIC. The Monte Carlo experiments number is 100. The impinging signals' parameters and the configuration of the array are the same as Fig. 5 and Fig. 6. The data in Tab. 1 shows that the average sample numbers is approximately $6N_1N_2$, and is insensitive to the snapshot number and/or the signals' number. So the calculation amount in the fast polarization MUSIC is as 6 times as that of conventional 2-D MUSIC needs. The fast polarization MUSIC has the same performance in precision and resolution as the original 4-D search polarization MUSIC, but the low computation complexity property makes it more attractive in practical application.

The computation complexity of the RR algorithm, is approximately 4 times of the conventional 2-D MUSIC [26], whereas the amounts of processed points of them are same. It should be noticed that the computation complexity of fast polarization MUSIC is a bit higher than that of RR algorithm, but they are of the same order of magnitude, whereas the former has a better performance in precision and resolution.

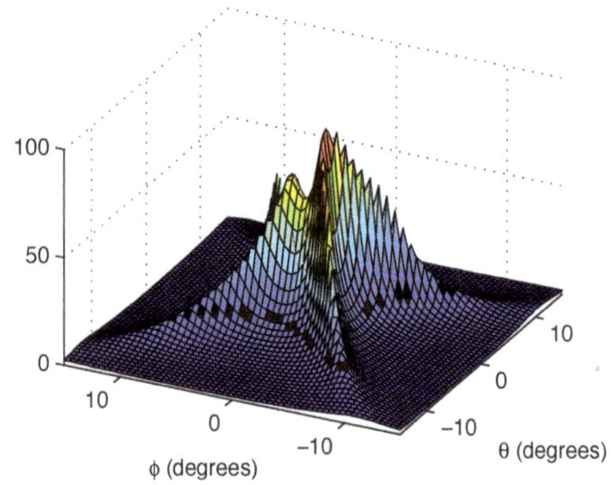

(a) conventional 2-D MUSIC

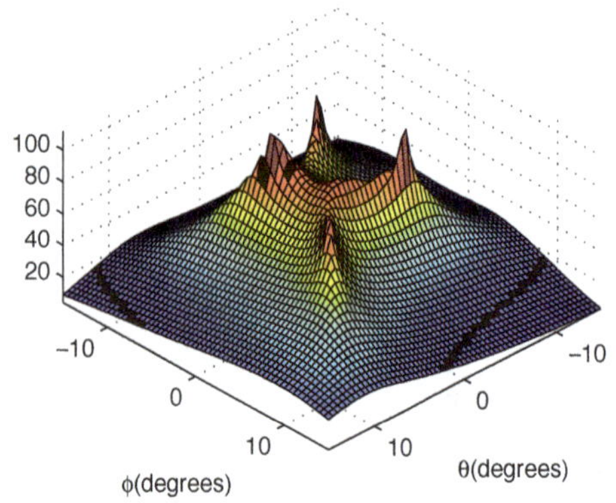

(b) rank reduction algorithm

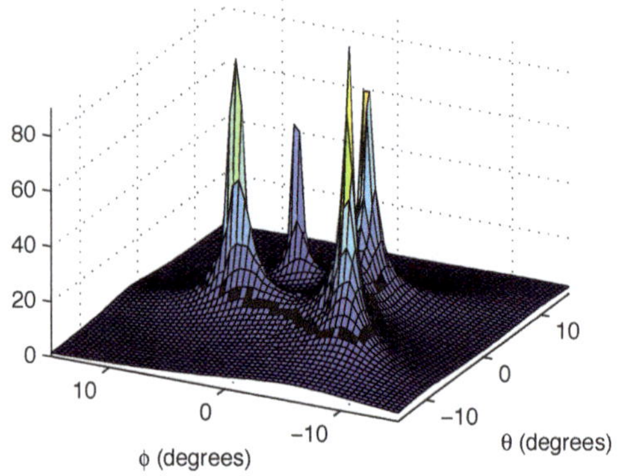

(c) proposed algorithm

Fig. 7. Space spectrum for conventional MUSIC, RR algorithm and the proposed algorithm.

Snapshots	200	400	600	800	1000
1 signal	23288	23432	23105	23203	23149
2 signals	22747	22843	22771	22866	22818
3 signals	22189	22236	22374	22241	22382
4 signals	21324	21302	21257	21149	21173

Tab. 1. Number of samples calculated in the proposed algorithm.

6. Conclusion

In this paper, a novel fast polarization MUSIC algorithm, which could reduce the considerable amount of computation, is proposed. Based on the fact that the difference of polarization between two close directions is minor or predictable, the daunting 4-D search is simplified by converting into two 2-D peak searches. The performance of fast polarization MUSIC is evaluated by theoretical analyses and simulations. It delivers improved precision and resolution with low computation complexity. The fact that this proposed algorithm is not confined to any specific pattern of polarization sensitive arrays expands its range of application.

Acknowledgements

The authors thank the support from National Natural Science Foundation of China (No. 61171180) for the research. We also gratefully acknowledge the assistance from Zhi-min Lin, Hong Hong, Zuo-liang Yin and Hui-jun Hou.

Appendix A: Proof of (32)

Define the noise space projection matrix as

$$\mathbf{P}_N = \mathbf{U}_N \mathbf{U}_N^H. \tag{46}$$

Using the expression given in [35], we have

$$E\{\hat{\mathbf{P}}_N - \mathbf{P}_N\} = -\left[\frac{\sigma_n^2}{K}\operatorname{tr}(\Lambda_{\mathbf{S}}\Gamma^{-2})\right]\mathbf{P}_N + (2M-2)\frac{\sigma_N^2}{K}\mathbf{U}_{\mathbf{S}}\Lambda_{\mathbf{S}}\Gamma^{-2}\mathbf{U}_{\mathbf{S}}^H \tag{47}$$

where K is the number of snapshots, $\hat{\mathbf{P}}_N$ the estimation of $\mathbf{P}_N$ and σ_N^2 is the power of white noise. Elaborate (47) further, we obtain:

$$E\{\hat{\mathbf{P}}_N - \mathbf{P}_N\} = -\frac{\sigma_N^2}{K}(\omega_1+\omega_2)\mathbf{P_N} + \frac{(2M-2)\sigma_N^2}{K}\cdot(\omega_1\mathbf{u}_1\mathbf{u}_1^H + \omega_2\mathbf{u}_2\mathbf{u}_2^H) \tag{48}$$

where $\omega_i = \lambda_i/(\lambda_i - \sigma_N^2)^2$, $i = 1,2$. According to the definition of zero spectrum $Z(\theta,\varphi,\gamma,\eta)$, we have

$$E\{\hat{Z}(\theta,\varphi,\gamma,\eta)\} = \mathbf{a}^H E\{\hat{\mathbf{P}}_\mathbf{N}\}\mathbf{a} = \left[1 - \frac{\sigma_N^2}{K}(\omega_1+\omega_2)\right]Z(\theta,\varphi,\gamma,\eta) + e_1 + e_2 \tag{49}$$

where $\mathbf{a} = \mathbf{a}(\theta,\varphi,\gamma,\eta)$, and

$$e_i = e_i(\theta,\varphi,\gamma,\eta) = \frac{(2M-2)\sigma_N^2\omega_i}{K}|\mathbf{a}^H\mathbf{u}_i|^2, \ i = 1,2. \tag{50}$$

The estimation of the zero spectrum of the first signal is

$$E\{\hat{Z}(\theta_1,\varphi_1,\gamma_1,\eta_1)\} = e_1(\theta_1,\varphi_1,\gamma_1,\eta_1) + e_2(\theta_1,\varphi_1,\gamma_1,\eta_1) \tag{51}$$

where we use $Z(\theta_1,\varphi_1,\gamma_1,\eta_1) = 0$. Similarly, the estimation of the zero spectrum of the second one is $E\{\hat{Z}(\theta_2,\varphi_2,\gamma_2,\eta_2)\} = e_1(\theta_2,\varphi_2,\gamma_2,\eta_2) + e_2(\theta_2,\varphi_2,\gamma_2,\eta_2)$.

At $(\theta_0,\varphi_0,\gamma_0,\eta_0)$, we have:

$$\begin{aligned} &E\{\hat{Z}(\theta_0,\varphi_0,\gamma_0,\eta_0)\} \\ &= \left[1 - \frac{\sigma_N^2}{K}(\omega_1+\omega_2)\right] Z(\theta_0,\varphi_0,\gamma_0,\eta_0) \\ &\quad + e_1(\theta_0,\varphi_0,\gamma_0,\eta_0) + e_2(\theta_0,\varphi_0,\gamma_0,\eta_0). \end{aligned} \tag{52}$$

Neglect the influence of $e_1(\theta,\varphi,\gamma,\eta)$, we get:

$$Z(\theta_0,\varphi_0,\gamma_0,\eta_0) > e_2(\theta_i,\varphi_i,\gamma_i,\eta_i) - e_2(\theta_0,\varphi_0,\gamma_0,\eta_0). \tag{53}$$

The value of the zero spectrum at $(\theta_0,\varphi_0,\gamma_0,\eta_0)$ is:

$$Z(\theta_0,\varphi_0,\gamma_0,\eta_0) = \|\mathbf{a_0}\|^2 - |\mathbf{a}_0^H\mathbf{u}_1|^2 - |\mathbf{a}_0^H\mathbf{u}_2|^2 \tag{54}$$

where $\mathbf{a}_i = \mathbf{a}(\theta_i,\varphi_i,\gamma_i,\eta_i)$, $i = 0,1,2$.

Substitute (50) and (54) into (53), we have:

$$\begin{aligned} &\|\mathbf{a}_0\|^2 - |\mathbf{a}_0^H\mathbf{u}_1|^2 - |\mathbf{a}_0^H\mathbf{u}_2|^2 \\ &> \frac{(2M-2)}{K} \cdot \frac{\lambda_2/\sigma_N^2}{(\lambda_2/\sigma_N^2-1)^2}\left[|\mathbf{a}_i^H\mathbf{u}_2|^2 - |\mathbf{a}_0^H\mathbf{u}_2|^2\right]. \end{aligned} \tag{55}$$

Appendix B: Proof of (33) and (34)

From some results concerning the eigenvalue calculation in [38]:

$$\begin{aligned} &\frac{\lambda_2/\sigma_N^2}{(\lambda_2/\sigma_N^2-1)^2} > \frac{1}{\lambda_2/\sigma_N^2-1} \\ &\approx \frac{1}{\|\mathbf{a}_0\|^2\,\mathrm{SNR}\left[1 - |\mathbf{a}_1^H\mathbf{a}_2|/(\|\mathbf{a}_1\|\cdot\|\mathbf{a}\|)\right]} \end{aligned} \tag{56}$$

where SNR is the signal-noise ratio of each signal. Substitute (56) into (32):

$$\begin{aligned} &\|\mathbf{a}_0\|^2 - |\mathbf{a}_0^H\mathbf{u}_1|^2 - |\mathbf{a}_0^H\mathbf{u}_2|^2 \\ &> \frac{2(M-1)\left[|\mathbf{a}_k^H\mathbf{u}_2|^2 - |\mathbf{a}_0^H\mathbf{u}_2|^2\right]}{K\cdot\mathrm{SNR}\left[1 - |\mathbf{a}_1^H\mathbf{a}_2|/(\|\mathbf{a}_1\|\cdot\|\mathbf{a}_2\|)\right]}. \end{aligned} \tag{57}$$

Then:

$$\begin{aligned} K &> \frac{2(M-1)}{\mathrm{SNR}\left(\|\mathbf{a}_0\|^2 - |\mathbf{a}_0^H\mathbf{u}_1|^2 - |\mathbf{a}_0^H\mathbf{u}_2|^2\right)} \\ &\cdot \frac{|\mathbf{a}_k^H\mathbf{u}_2|^2 - |\mathbf{a}_0^H\mathbf{u}_2|^2}{\|\mathbf{a}_0\|^2\left[1 - |\mathbf{a}_1^H\mathbf{a}_2|/(\|\mathbf{a}_1\|\cdot\|\mathbf{a}_2\|)\right]}. \end{aligned} \tag{58}$$

This is expression (33).

For the conventional MUSIC, the value of $1 - |\mathbf{a}_1^H\mathbf{a}_2|/(\|\mathbf{a}_1\|\|\mathbf{a}_2\|)$ is very close to zero, so the approximation in (56) should be rewritten as:

$$\begin{aligned} &\frac{\lambda_2/\sigma_N^2}{(\lambda_2/\sigma_N^2-1)^2} \\ &\approx \frac{1}{\mathrm{SNR}^2\|\mathbf{a}_0\|^4\left[1 - |\mathbf{a}_1^H\mathbf{a}_2|/(\|\mathbf{a}_1\|\|\mathbf{a}_2\|)\right]^2}. \end{aligned} \tag{59}$$

The expression (57) becomes

$$\begin{aligned} &\|\mathbf{a}_0\|^2 - |\mathbf{a}_0^H\mathbf{u}_1|^2 - |\mathbf{a}_0^H\mathbf{u}_2|^2 \\ &> \frac{(M-2)\left[|\mathbf{a}_i^H\mathbf{u}_2|^2 - |\mathbf{a}_0^H\mathbf{u}_2|^2\right]}{K\cdot\mathrm{SNR}^2\|\mathbf{a}_0\|^2\left[1 - |\mathbf{a}_1^H\mathbf{a}_2|/(\|\mathbf{a}\|\|\mathbf{a}_2\|)\right]^2}. \end{aligned} \tag{60}$$

Then in MUSIC, the required number of snapshots is:

$$\begin{aligned} K &> \frac{(M-2)\left[|\mathbf{a}_i^H\mathbf{u}_2|^2 - |\mathbf{a}_0^H\mathbf{u}_2|^2\right]}{\mathrm{SNR}^2\|\mathbf{a}_0\|^4\left[1 - |\mathbf{a}_1^H\mathbf{a}_2|/(\|\mathbf{a}_1\|\|\mathbf{a}_2\|)\right]^2} \\ &\cdot \frac{1}{\|\mathbf{a}_0\|^2 - |\mathbf{a}_0^H\mathbf{u}_1|^2 - |\mathbf{a}_0^H\mathbf{u}_2|^2}. \end{aligned} \tag{61}$$

References

[1] BOHME, J. F. Estimation of source parameters by maximum likelihood and nonlinear regression. In *Acoustics, Speech, and Signal Processing, IEEE International Conference on ICASSP'84*. 1984, vol. 9, p. 271–274. DOI: 10.1109/ICASSP.1984.1172397

[2] BOHME, J. F. Estimation of spectral parameters of correlated signals in wavefields. *Signal Processing*, 1986, vol. 11, no. 4, p. 329–337. DOI: 10.1016/0165-1684(86)90075-7

[3] SCHMIDT, R. Multiple emitter location and signal parameter estimation. *IEEE Transactions on Antennas and Propagation*, 1986, vol. 34, no. 3, p. 278–280. DOI: 10.1109/TAP.1986.1143830

[4] ROY, R., PAULRAJ, A., KAILATH, T. ESPRIT-A subspace rotation approach to estimation of parameters of cisoids in noise. *IEEE Transactions on Acoustics, Speech and Signal Processing*, 1986, vol. 34, no. 5, p. 1340–1342. DOI: 10.1109/TASSP.1986.1164935

[5] OTTERSTEN, B., VIBERG, M., STOICA, P., NEHORAI, A. *Exact and Large Sample ML Techniques for Parameter Estimation and Detection in Array Processing*. New York: Springer-Verlag, 1993.

[6] RAO, B. D., HARI, K. V. S. Performance analysis of root-MUSIC. *IEEE Transactions on Acoustics, Speech and Signal Processing*, 1989, vol. 37, no. 12, p. 1939–1949. DOI: 10.1109/29.45540

[7] ROY, R., KAILATH, T. ESPRIT-estimation of signal parameters via rotational invariance techniques. *IEEE Transactions on Acoustics, Speech and Signal Processing*, 1989, vol. 37, no. 7, p. 984–995. DOI: 10.1109/29.32276

[8] YARDIBI, T., LI, J., STOICA, P., CATTAFESTA III, L. N. Sparsity constrained deconvolution approaches for acoustic source mapping. *The Journal of the Acoustical Society of America*, 2008, vol. 123, p. 2631. DOI: 10.1121/1.2896754

[9] YARDIBI, T., LI, J., STOICA, P., XUE, M., BAGGEROER, A. B. Source localization and sensing: A nonparametric iterative adaptive approach based on weighted least squares. *IEEE Transactions on Aerospace and Electronic Systems*, 2010, vol. 46, no. 1, p. 425–443. DOI: 10.1109/TAES.2010.5417172

[10] MISHALI, M., ELDAR, Y.C., ELRON, A. J. Xampling: Signal acquisition and processing in union of subspaces. *IEEE Transactions on Signal Processing*, 2011, vol. 59, no. 10, p. 4719–4734. DOI: 10.1109/TSP.2011.2161472

[11] GAO, F., GERSHMAN, A. B. A generalized ESPRIT approach to direction-of-arrival estimation. *IEEE Signal Processing Letters*, 2005, vol. 12, no. 3, p. 254–257. DOI: 10.1109/LSP.2004.842276

[12] ELKADER, S. A., GERSHMAN, A. B., WONG, K. M. Rank reduction direction-of-arrival estimators with an improved robustness against subarray orientation errors. *IEEE Transactions on Signal Processing*, 2006, vol. 54, no. 5, p. 1951–1955. DOI: 10.1109/TSP.2006.872321

[13] RUBSAMEN, M., GERSHMAN, A. B. Direction-of-arrival estimation for nonuniform sensor arrays: from manifold separation to Fourier domain MUSIC methods. *IEEE Transactions on Signal Processing*, 2009, vol. 57, no. 2, p. 588–599. 10.1109/TSP.2008.2008560

[14] KAVEH, M., BARABELL, A. The statistical performance of the MUSIC and the minimum-norm algorithms in resolving plane waves in noise. *IEEE Transactions on Acoustics, Speech and Signal Processing*, 1986, vol. 34, no. 2, p. 331–341. DOI: 10.1109/TASSP.1986.1164815

[15] STOICA, P., ARYE, N. MUSIC, maximum likelihood, and Cramer-Rao bound. *IEEE Transactions on Acoustics, Speech and Signal Processing*, 1989, vol. 37, no. 5, p. 720–741. DOI: 10.1109/29.17564

[16] NEHORAI, A., PALDI, E. Vector-sensor array processing for electromagnetic source localization. *IEEE Transactions on Signal Processing*, 1994, vol. 42, no. 2, p. 376–398. DOI: 10.1109/78.275610

[17] LI, J., COMPTON, J. Angle and polarization estimation using ESPRIT with a polarization sensitive array. *IEEE Transactions on Antennas and Propagation*, 1991, vol. 39, no. 9, p. 1376–1383. DOI: 10.1109/8.99047

[18] LI, J., COMPTON, J. Two-dimensional angle and polarization estimation using the ESPRIT algorithm. *IEEE Transactions on Antennas and Propagation*, 1992, vol. 40, no. 5, p. 550–555. DOI: 10.1109/8.142630

[19] HUA, Y. A pencil-MUSIC algorithm for finding two-dimensional angles and polarizations using crossed dipoles. *IEEE Transactions on Antennas and Propagation*, 1993, vol. 41, no. 3, p. 370–376. DOI: 10.1109/8.233122

[20] HO, K.-C, TAN, K.-C., TAN, B. T. G. Efficient method for estimating directions-of-arrival of partially polarized signals with electromagnetic vector sensors. *IEEE Transactions on Signal Processing*, 1997, vol. 45, no. 10, p. 2485–2498. DOI: 10.1109/78.640714

[21] HO, K.-C., TAN, K.-C., NEHORAI, A. Estimating directions of arrival of completely and incompletely polarized signals with electromagnetic vector sensors. *IEEE Transactions on Signal Processing*, 1999, vol. 47, no. 10, p. 2845–2852. DOI: 10.1109/78.790664

[22] ZOLTOWSKI, M., WONG, K. ESPRIT-based 2-D direction finding with a sparse uniform array of electromagnetic vector sensors. *IEEE Transactions on Signal Processing*, 2000, vol. 48, no. 8, p. 2195–2204. DOI: 10.1109/78.852000

[23] HE, J., LIU, Z. Computationally efficient 2D direction finding and polarization estimation with arbitrarily spaced electromagnetic vector sensors at unknown locations using the propagator method. *Digital Signal Processing*, 2009, vol. 19, no. 3, p. 491–503. DOI: 10.1016/j.dsp.2008.01.002

[24] RAHAMIM, D., TABRIKIAN, J., SHAVIT, R. Source localization using vector sensor array in a multipath environment. *IEEE Transactions on Signal Processing*, 2004, vol. 52, no. 11, p. 3096–3103. DOI: 10.1109/TSP.2004.836456

[25] WEISS, A. J., FRIEDLANDER, B. Direction finding for diversely polarized signals using polynomial rooting. *IEEE Transactions on Signal Processing*, 1993, vol. 41, no. 5, p. 1893–1905. DOI: 10.1109/78.215307

[26] WONG, K., LI, L., ZOLTOWSKI, M. D. Root-MUSIC-based direction-finding and polarization estimation using diversely polarized possibly collocated antennas. *IEEE Antennas and Wireless Propagation Letters*, 2004, vol. 3, no. 1, p. 129–132. DOI: 10.1109/LAWP.2004.831083

[27] MIRON, S., GUO, X., BRIE, D. DOA estimation for polarized sources on a vector-sensor array by PARAFAC decomposition of the fourth-order covariance tensor. In *Proceedings of the EUSIPCO*. 2008, p. 25–29.

[28] GONG, X., LIU, Z., XU, Y., ISHTIAQ, A. Direction-of-arrival estimation via twofold mode-projection. *Signal Processing*, 2009, vol. 89, no. 5, p. 831–842. DOI: 10.1016/j.sigpro.2008.10.034

[29] YUAN, Q., CHEN, Q., SAWAYA, K. MUSIC based DOA finding and polarization estimation using USV with polarization sensitive array antenna. In *IEEE Radio and Wireless Symposium*. 2006, p. 339–342. DOI: 10.1109/RWS.2006.1615164

[30] LIU, S., JIN, M., QIAO, X. Joint polarization-DOA estimation using circle array. In *IET International Radar Conference*. 2009, p. 1–5.

[31] COSTA, M., KOIVUNEN, V., RICHTER, A. Azimuth, elevation, and polarization estimation for arbitrary polarimetric array configurations. In *Statistical Signal Processing*. Cardiff (UK), 2009, p. 261–264. DOI: 10.1109/SSP.2009.5278590

[32] COSTA, M., RICHTER, A., KOIVUNEN, V. DoA and polarization estimation for arbitrary array configurations. *IEEE Transactions on Signal Processing*, 2012, vol. 60, no. 5, p. 2330–2343. DOI: 10.1109/TSP.2012.2187519

[33] YANG, P., YANG, F., NIE, Z., ZHOU, H., LI, B., TANG, X. Fast 2-d DOA and polarization estimation using arbitrary conformal antenna array. *Progress In Electromagnetics Research C*, 2012, vol. 25, p. 119–132. DOI: 10.2528/PIERC11070706

[34] GUO, R., MAO, X., LI, S., LIN, Z. Fast four-dimensional joint spectral estimation with array composed of diversely polarized elements. In *IEEE Radar Conference (RADAR)*. 2012 , p. 919–923. DOI: 10.1109/RADAR.2012.6212268

[35] FRIEDLANDER, B., WEISS, A. J. The resolution threshold of a direction-finding algorithm for diversely polarized arrays. *IEEE Transactions on Signal Processing*, 1994, vol. 42, no. 7, p. 1719–1727. DOI: 10.1109/78.298279

[36] ZHOU, C., HABLER, F., JAGGARD, D. L. A resolution measure for the MUSIC algorithm and its application to plane wave arrivals contaminated by coherent interference. *IEEE Transactions on Signal Processing*, 1991, vol. 39, no. 2, p. 454–463. DOI: 10.1109/78.80829

[37] STOICA, P., NEHORAI, A. MUSIC, maximum likelihood, and Cramer-Rao bound: further results and comparisons. *IEEE Transactions on Acoustics, Speech and Signal Processing*, 1990, vol. 38, no. 12, p. 2140–2150. DOI: 10.1109/29.61541

[38] HUDSON, J. E. *Adaptive Array Principles*. Peter Peregrinus Press, 1981.

About the Authors . . .

Ran GUO was born in Dalian, Liaoning Province, China. He received his M.Sc. from HIT (Harbin Institute of Technology) in 2009. His research interests include array signal processing, polarization sensitive array and super-resolution DOA estimation. Email: guoran1112@gmail.com.

Xingpeng MAO was born in Panjin, Liaoning Province, China. He received his doctor's degree from HIT in 2004. He is a doctoral advisor and professor in School of Electronic and Information Engineering, HIT, at present. His research interests include radar signal processing, polarization signal processing, modern signal processing and embedded system application. Email: mxp@hit.edu.cn.

First-Order Statistics Prediction for a Propagation Channel of Arbitrary Non-Geostationary Satellite Orbits

Milan KVICERA, Pavel PECHAC

Faculty of Electrical Engineering, Czech Technical University in Prague, Technicka 2, 166 27 Prague, Czech Republic

kvicemil@fel.cvut.cz, pechac@fel.cvut.cz

Abstract. *A method enabling the prediction of the first-order statistics of received signal levels for arbitrary non-geostationary satellite orbits based on a reference dataset for a wide range of elevation angles is introduced for azimuth-independent scenarios and high elevation angles. The method is further validated by experimental data obtained during measurements with a remote-controlled airship utilized as a pseudo-satellite. These experimental trials were performed at a frequency of 2.0 GHz at two scenarios at Stromovka Park in Prague, Czech Republic, in August 2013 and March 2014. An excellent match between the predicted and actual cumulative distribution functions of received signal levels was identified for both scenarios.*

Keywords

Satellite-to-Earth propagation, channel measurements, modeling, vegetation

1. Introduction

Central to the design of a satellite system are the constellation and selected orbit which determine the elevation and azimuth angles at which a user shall receive a direct signal. The elevation and azimuth angle toward a particular satellite can remain the same for a fixed receiver, as in case of a geostationary Earth orbit, or change dramatically when considering for example a highly elliptical orbit. This has a strong influence on the corresponding satellite-to-Earth propagation channel as, based on user surroundings, the signal may be shadowed differently determined by the direction of the incoming signal. To predict such behavior, a number of various propagation channel models for different scenarios can be found in the literature, see for example [1–8]. Generally, such models need to be based on available experimental data; however, obtaining suitable experimental data is demanding. It is common to utilize a so-called pseudo-satellite which may be in a form of a transmitter (Tx) placed on a helicopter [9–11] or a remote-controlled airship [7], [12], [13], at a crane or the upper-most point in the surroundings for fixed-elevation measurements [14–16], or even collect data from an existing satellite [17–19].

However, it is not feasible to obtain experimental data for all the combinations of azimuth and elevation angles observed by a user on Earth when considering a particular non-geostationary satellite system. Instead, respecting high costs of experimental campaigns when using a pseudo-satellite, pre-defined flight paths, such as a star-pattern [7], circle [11], or hemisphere [10], are chosen for selected scenarios.

Considering azimuth-independent scenarios identified in the text as regular, such as a densely vegetated area, only one set of reference experimental data in a vast range of elevation angles at an arbitrary azimuth should be sufficient to predict received signal characteristics for arbitrary non-geostationary satellite orbits leading to less demanding experimental campaigns. Such a novel approach would follow [20], where a probability density function (PDF) of elevation angles between a user and a low Earth orbit satellite is utilized to obtain resulting rain attenuation time series. However, such an experiment has not been documented in the literature and needs to be validated. Thus, a series of measurements at 2.0 GHz at Stromovka Park in Prague, Czech Republic, were carried out in 2013 and 2014. Throughout these trials, a remote-controlled airship was utilized as a pseudo-satellite following pre-defined flight paths according to sub-satellite points of selected Galileo and Iridium satellites. Both the left- (LHCP) and right-handed (RHCP) circularly polarized signals were transmitted from the airship towards a receiver located at two different scenarios. Unlike [21], where four different distributions were fitted to the first-order statistics of experimental data obtained previously at Stromovka Park with a low sampling rate of 100 Hz, this paper presents a method how to obtain a cumulative distribution function (CDF) of received signal levels for various non-geostationary satellite orbits based on a reference dataset for the case of azimuth-independent scenarios. Details of the experimental campaign are provided in Sec. 2, the data processing method is described in Sec. 3, while the results and discussion are given in Sec. 4.

2. Measurement Setup and Trials

The measurement setup used during the trials was as follows. A remote-controlled airship carried a Tx, the same type as in [7], connected to an LHCP and an RHCP planar wideband antenna attached to a positioner enabling an instant pointing towards the receiver (Rx) location based on the airship GPS coordinates. Unmodulated continuous wave signals with a fixed output power of 27 dBm were transmitted at frequencies of 2.00106 GHz and 2.00086 GHz by the LHCP and RHCP antenna, respectively. Unlike [7] and [12], to obtain the received signal levels of both the co-polarized and cross-polarized components of the transmitted signals, a dual-polarized rectangular patch antenna was connected by an H-hybrid and two power splitters to a sensitive, custom-made, four-channel receiver with a low noise floor of -126 dBm for a measurement bandwidth of 12.5 kHz. The receiver provided a 10-kHz sampling rate and its first two channels were tuned to 2.00106 GHz and the remaining two were tuned to 2.00086 GHz. The height of the upwards-pointing receiving antenna was 1.5 meters and the altitude of the airship was kept approximately 200 meters above ground level at a near-constant speed of 8 m/s. Similar to [7], recorded signal levels were recalculated to a uniform distance of 20 km to eliminate the influence of free space loss for different distances between the Tx and Rx. Further, data obtained during periods of airship pitch and roll of more than 15 degrees were removed as they represent gusty conditions during which the Tx antenna positioner did not perfectly keep the direction towards Rx.

The airship followed pre-defined flight paths over the vegetated area of Stromovka Park according to the selected typical sub-satellite points of the Galileo PFM and Iridium 98 satellite for the location of Prague (50.08° N, 14.43° E), see Fig. 1. Two scenarios, marked as A and B in Fig. 1 and shown in more detail in Fig. 2, were selected to represent regular scenarios independent of azimuth: a receiver located inside coniferous trees and within a group of tall deciduous trees, respectively. It should be noted that all the flight paths in Fig. 1 refer to scenario A and were thus slightly shifted for the case of scenario B. By performing the measurements in July 2013 and March 2014, representative experimental data were obtained for the actual satellite passes during both the summer and winter season. As scenario A is evergreen, it demonstrates the repeatability of the measurements while defoliation of scenario B represents different propagation conditions. Thus only these two scenarios can be utilized to obtain a statistically significant amount of experimental data and validate the method presented below in Sec. 3.

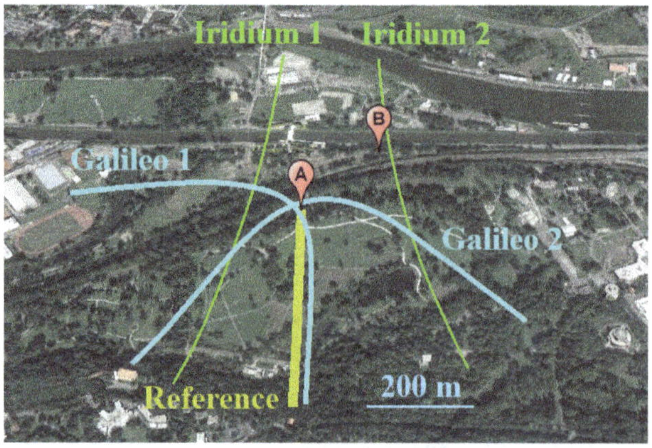

Fig. 1. Regular scenarios A and B at Stromovka Park together with the pre-defined airship flyovers simulating the Iridium 98 satellite (two almost north-south thin lines), Galileo PFM satellite (two wider curved lines) and the reference north-south flyover (widest line). (Image from Google Earth).

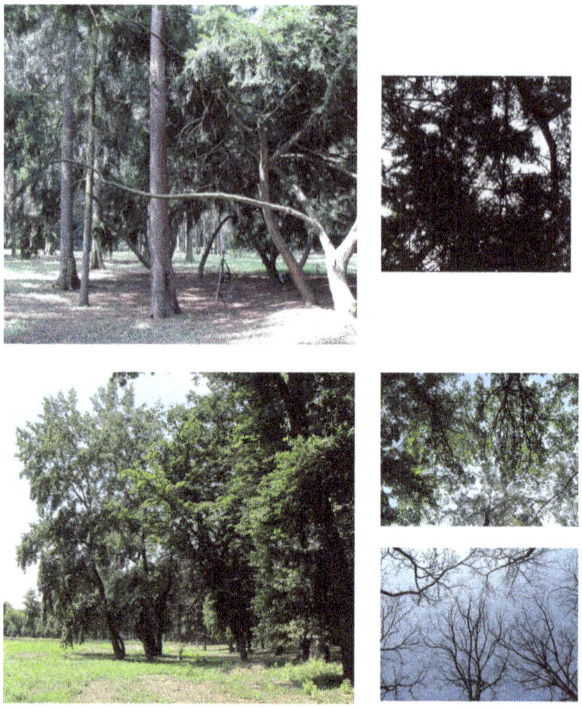

Fig. 2. Scenarios A (top left) and B (bottom left) in detail during the summer season together with the upwards views from the Rx during the summer and winter season (top right and bottom right). As scenario A is evergreen, only one upwards view is shown for both seasons.

It should be noted that although the utilized frequencies are at the upper part of L-band, higher than the actual Galileo or Iridium frequencies, the method introduced in the next section is limited by the azimuthal symmetry of a particular scenario rather than by a selected frequency. In addition, even though the airship simulated actual non-geostationary satellite systems, arbitrary flyovers could have been selected.

3. Data Processing

The experimental data were processed in the following way. The measurements were aimed at high elevation angles and, thus, the lowest elevation considered was 30 degrees from the Rx point of view. On the other hand, to avoid an insufficient amount of data for high elevation angles, the maximum considered elevation was 80 degrees. For the analysis described below, only the co-polarized components of the received LHCP and RHCP signals were considered.

As a subsequent step, the entire range of elevation angles achieved during the north-south reference flyover (shown by the thick line in Fig. 2) was divided into five-degree intervals and a PDF $pdf_{reference,i}(r)$ of received signal levels r in dB was calculated by means of histogram for each elevation interval i. In contrast to [20], where segments of constant time are utilized, our approach requires constant elevation intervals as the simulated flyovers do not follow the actual time duration of the particular satellite pass. The choice of the interval size respected a statistically significant number of data samples (about 80000 and 40000 for the lowest and highest elevation interval, respectively) and, in addition, propagation conditions and antenna patterns are considered as not significantly varying within such an interval, similar to [7].

Consequently, for the Galileo or Iridium flyovers shown in Fig. 1, a PDF of the elevation intervals $pdf_{\Theta}(\theta)$ achieved during a particular flyover was calculated and thus the probability $p_{\Theta,i}(\theta)$ of each elevation interval i was obtained. Then, following (1), a PDF of signal levels received during the reference flyover within an elevation interval i, $pdf_{reference,i}(r)$, was multiplied by $p_{\Theta,i}(\theta)$ so that a PDF of received signal level $pdf_{predicted,i}(r)$ was predicted for this elevation interval. After that, a PDF of received signal levels $pdf_{predicted}(r)$ for the whole range of elevation angles from 30 degrees up to 80 degrees was obtained according to (2), simply as a sum of the PDFs for every elevation interval.

$$pdf_{predicted,i}(r) = p_{\Theta,i}(\theta) \cdot pdf_{reference,i}(r) \text{ for } i = 1 \ldots n, \quad (1)$$

$$pdf_{predicted}(r) = \sum_{i=1}^{n} pdf_{predicted,i}(r) \quad (2)$$

where n is the total number of elevation intervals. Obtaining a corresponding CDF is then straightforward by using (3)

$$\Pr\{r \le r_{th}\} = \int_{-\infty}^{r_{th}} pdf_{predicted}(r)dr \,. \quad (3)$$

4. Results and Discussion

To analyze and quantify the quality of the overall match of the predicted and actual received signal level CDF for a particular satellite flyover, we calculated the mean and standard deviation of their difference for the whole range of percentages with a step of 0.01. Corresponding results are then given in Tab. 1 with the exception of Iridium flyover number 2 for scenario B during the summer season as the experimental data are not available.

Based on this table it is evident that overall similar results were obtained for scenarios A and B during the summer and winter season. Absolute values of maximum and minimum mean of differences are below 5 dB and about 0 dB, respectively. Together, with maximum standard deviations below 2.5 dB, these values indicate an excellent match of the predicted and actual CDFs. Such results also reflect the fact that even scenario B, consisting of only five deciduous high-rose trees, with non-uniformly distributed sparse trunks and branches without leaves in the winter season, can still be considered as regular, although it could be assumed to be more dependent on azimuth than scenario A.

Scenario		Mean ± standard deviation (dB)			
		Galileo 1	Galileo 2	Iridium 1	Iridium 2
A, sum.	LHCP	-1.4 ± 1.6	0.8 ± 1.5	1.1 ± 2.4	-2.5 ± 0.8
	RHCP	-1.8 ± 1.3	1.2 ± 0.8	-0.2 ± 2.1	-2.1 ± 0.7
A, winter	LHCP	3.6 ± 1.2	1.9 ± 1.4	3.0 ± 1.7	0.0 ± 0.8
	RCHP	2.4 ± 0.9	1.2 ± 1.4	0.9 ± 2.2	1.2 ± 1.1
B, sum.	LHCP	-2.9 ± 0.8	-0.6 ± 1.8	0.8 ± 2.1	Not avail.
	RHCP	-4.9 ± 0.9	-3.1 ± 2.0	-3.8 ± 1.3	Not avail.
B, winter	LHCP	-2.3 ± 1.0	-2.9 ± 2.7	-1.4 ± 1.5	-2.3 ± 2.2
	RHCP	-3.6 ± 1.0	-1.7 ± 2.1	-2.7 ± 1.3	0.8 ± 1.7

Tab. 1. Mean and standard deviation of difference between predicted and measured CDF.

It should be noted that a certain level of inaccuracy is introduced into the predicted results due to the Rx antenna pattern not being perfectly azimuth independent. However, respecting the Rx antenna design, the maximum difference of gain for both the LHCP and RHCP polarization can be identified for the lowest elevations as only about 2 dB for an elevation of 30 degrees.

To illustrate results from Tab. 1, Figs. 3 and 4 represent selected actual, predicted and reference signal level CDFs for the Galileo satellite flyover number 2, RHCP signal at scenario A and B during the summer season, respectively. For these cases, the mean value of the differences is 1.2 dB and -3.1 dB, whereas the standard deviation of the differences is 0.8 dB and 2.0 dB, respectively. Based on Figs. 3 and 4, such values lead to a very good visual match of the predicted and actual signal level CDF, even for the worse case.

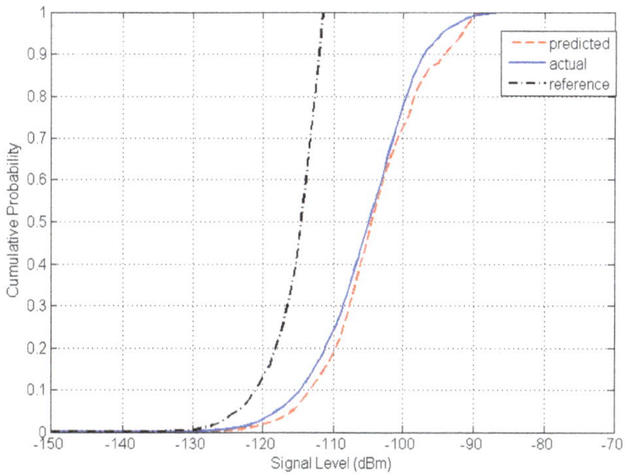

Fig. 3. An example of a match between the predicted and actual signal level CDF for the case of the RHCP signal during the Galileo satellite flyover number 2 at scenario A within the summer season. It corresponds to values of (1.2±0.8) dB taken from Tab. 1.

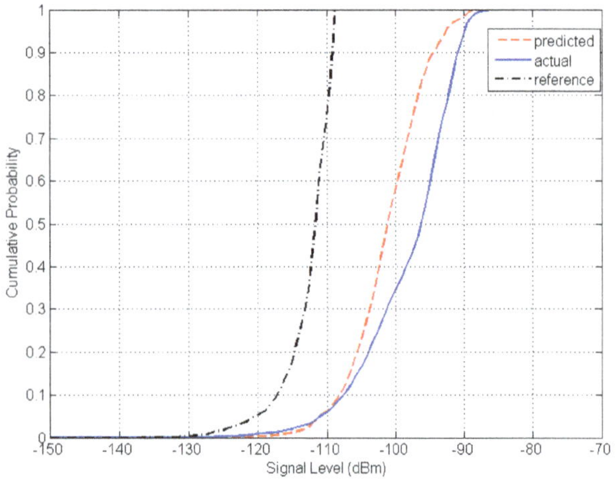

Fig. 4. An example of a match between the predicted and actual signal level CDF for the case of the RHCP signal during the Galileo satellite flyover number 2 at scenario B within the summer season. It corresponds to values of (-3.1±2.0) dB taken from Tab. 1.

To summarize, based on Tab. 1, it can be stated that an overall very good match of the prediction was achieved for both scenarios A and B during summer and winter. Based on the large experimental dataset which has been processed, the presented prediction method can be considered as successfully validated for the case of high elevation angles and regular scenarios.

5. Conclusion

We have presented a method enabling the prediction of the first-order statistics of signal levels received at high elevation angles considering arbitrary satellite orbits and azimuth-independent scenarios. This method is based on utilizing a reference dataset for a wide range of elevation angles at one azimuth only. The validity of this method was demonstrated at a frequency of 2.0 GHz on a significant number of experimental datasets obtained by using a remote-controlled airship simulating selected Galileo and Iridium satellite passes over two vegetated scenarios during both the summer and winter season. It was also shown that the proposed method can be utilized for common, not perfectly azimuth-independent scenarios, which would result in a significant decrease of complexity of corresponding experimental campaigns. Furthermore, as a consequence of the proposed method, if data obtained for the case of an existing satellite system are available, they may be utilized as a reference dataset to predict the first-order statistics of received signal levels at a similar frequency for a different constellation of the satellite system, at least as an initial estimation, without the considerable expense of a measurement campaign.

Acknowledgements

This publication was supported by the European social fund within the framework of realizing the project „Support of inter-sectoral mobility and quality enhancement of research teams at the Czech Technical University in Prague“, CZ.1.07/2.3.00/30.0034.

References

[1] JOST, T., CARRIE, G., PEREZ-FONTAN, F., WANG, W. FIEBIG, U.-C. A Deterministic satellite-to-indoor entry loss model. *IEEE Transactions on Antennas and Propagation*, 2013, vol. 61, no. 4, p. 2223–2230. DOI: 10.1109/TAP.2012.2232898

[2] CHEFFENA, M., PEREZ-FONTAN, F. Land mobile satellite channel simulator along roadside trees. *IEEE Antennas and Wireless Propagation Letters*, 2010, vol. 9, p. 748–751. DOI: 10.1109/LAWP.2010.2060465

[3] AIT-IGHIL, M., LEMORTON, J., PEREZ-FONTAN, F., LACOSTE, F., THEVENON, P., BOURGA, C., BOUSQUET, M. SCHUN - a hybrid land mobile satellite channel simulator enhanced for multipath modelling applied to satellite navigation systems. In *Proceedings of the 7th European Conference on Antennas and Propagation*. Gothenburg (Sweden), 2013, p. 692–696.

[4] CARRIE, G., PEREZ-FONTAN, F., LACOSTE, F., LEMORTON, J. A generative MIMO channel model: encompassing single satellite and satellite diversity cases. In *Proceedings of the 6th European Conference on Antennas and Propagation*. Prague (Czech Rep.), 2012, p. 2454–2458. DOI: 10.1109/EuCAP.2012.6206627

[5] JEANNIN, N., PEREZ-FONTAN, F., MAMETSA, H.-J., CASTANET, L. Physical-statistical model for the LMS channel at Ku/Ka Band. In *Proceedings of the 5th European Conference on Antennas and Propagation*. Rome (Italy), 2011, p. 3571–3575.

[6] ABELE, A., PEREZ-FONTAN, F., BOUSQUET, M., VALTR, P., LEMORTON, J., LACOSTE, F., CORBEL, E. A new physical-statistical model of the land mobile satellite propagation channel. In *Proceedings of the 4th European Conference on Antennas and Propagation*. Barcelona (Spain), 2010, p. 1-5.

[7] KVICERA, M., PECHAC, P. Building penetration loss for satellite services at L-, S- and C-band: measurement and modeling. *IEEE Transactions on Antennas and Propagation*, 2011, vol. 59, no. 84, p. 3013–3021. DOI: 10.1109/TAP.2011.2158963

[8] ARNDT, D., HEYN, T., HEUBERGER, A., PRIETO-CERDEIRA, R., EBERLEIN, E. State modeling of the land mobile satellite channel with angle diversity. In *Proceedings of the 6th European Conference on Antennas and Propagation*. Prague (Czech Rep.) 2012, p. 3130–3144. DOI: 10.1109/EuCAP.2012.6206141

[9] MOLNAR, B., FRIGYES, I., BODNAR, Z., HERCZKU, Z., KORMANYOS, Z., BERCES, J., PAPP, I., JUHASZ, L. Characterisation of the satellite-to-indoor channel based on narrow-band scalar measurements. In *Proceedings of the 8th IEEE International Symposium on Personal, Indoor and Mobile Radio Communications*. Helsinki (Finland), 1997, p. 1014–1018.

[10] PEREZ-FONTAN, F., HOVINEN, V., SCHONHUBER, M., PRIETO-CERDEIRA, R., DELGADO-PENIN, J. A., TESCHL, F., KYROLAINEN, J., VALTR, P. Building entry loss and delay spread measurements on a simulated HAP-to-indoor link at S-band. *EURASIP Journal on Wireless Communications and Networking*, 2008, vol. 2008, no. 5, 6 p. DOI:10.1155/2008/427352

[11] TESCHL, F., PEREZ-FONTAN, F., SCHONHUBER, M., PRIETO-CERDEIRA, R., TESCHL, R. Attenuation of spruce, pine, and deciduous woodland at C-band. *IEEE Antennas and Wireless Propagation Letters*, 2012, vol. 11, p. 109–112. DOI: 10.1109/LAWP.2012.2184253

[12] HORAK, P., PECHAC, P. Excess loss for high elevation angle links shadowed by a single tree: measurements and modeling.

IEEE Transactions on Antennas and Propagation, 2012, vol. 60, no.7, p. 3541–3545. DOI: 10.1109/TAP.2012.2196944

[13] STEINGASS, A., LEHNER, A. Measuring the navigation multipath channel – a statistical analysis. In *Proceedings of the 17th International Technical Meeting of the Satellite Division of he Institute of Navigation (ION GNSS 2004)*. Long Beach (CA, USA), 2004, p. 1157–1164.

[14] LACOSTE, F., LEMORTON, J., CASADEBAIG, L., ROUSSEAU, F. Measurements of the land mobile and nomadic satellite channels at 2.2 GHz and 3.8 GHz. In *Proceedings of the 6th European Conference on Antennas and Propagation*. Prague (Czech Rep.), 2012, p. 2422–2426. DOI: 10.1109/EuCAP.2012.6206356

[15] JOST, T., WANG, W., FIEBIG, U.-C., PEREZ-FONTAN, F. Comparison of L- and C-Band satellite-to-indoor broadband wave propagation for navigation applications. *IEEE Transactions on Antennas and Propagation*, 2011, vol. 59, no. 10, p. 3899–3909. DOI: 10.1109/TAP.2011.2163753

[16] KING, P. R., STAVROU, S. Low elevation wideband land mobile satellite MIMO channel characteristics. *IEEE Transactions on Wireless Communications*, 2007, vol. 6, no. 70, p. 2712–2720. DOI: 10.1109/TWC.2007.051018

[17] TESCHL, F., HOVINEN, V., PEREZ-FONTAN, F., SCHONHUBER, M., PRIETO-CERDEIRA, R. Narrow- and wideband land mobile satellite channel statistics for various environments at Ku-band. In *Proceedings of the 6th European Conference on Antennas and Propagation*. Prague (Czech Rep.), 2012, p. 2464–2468. DOI: 10.1109/EuCAP.2012.6206670

[18] JOST, T., WANG, W., SCHUBERT, F., ANTREICH F., FIEBIG, U.-C. Channel sounding using GNSS signals. In *Proceedings of the 5th European Conference on Antennas and Propagation*. Rome (Italy), 2011, p. 3724–3728.

[19] VELTSISTAS, P., KALABOUKAS, G., KONITOPOULOS, G., DRES, D., KATIMERTZOGLOU, E., CONSTANTINOU, P. Satellite-to-indoor building penetration loss for office environment at 11 GHz. *IEEE Antennas and Wireless Propagation Letters*, 2007, vol. 6, p. 96–99. DOI: 10.1109/LAWP.2007.893070

[20] ARAPOGLOU, P.-D. M., PANAGOPOULOS, A. D. A tool for synthesizing rain attenuation time series in LEO Earth observation satellite downlinks at Ka band. In *Proceedings of the 5th European Conference on Antennas and Propagation*. Rome (Italy), 2011, p. 1467–1470.

[21] KOUROGIORGAS, C. I., KVICERA, M., SKRAPARLIS, D., KORINEK, T., SAKARELLOS, V. K., PANAGOPOULOS, A. D., PECHAC, P. Modeling of first-order statistics of the MIMO dual polarized channel at 2 GHz for land mobile satellite systems under tree shadowing. *IEEE Transactions on Antennas and Propagation*, 2014, vol. 62, no. 10, p. 5410–5415. DOI: 10.1109/TAP.2014.2346186

About the Authors ...

Milan KVICERA was born in 1983. He received the M.Sc. degree and the Ph.D. degree in Radio Electronics from the Czech Technical University in Prague, Czech Republic, in 2008 and 2012, respectively. He is currently working as a postdoc researcher in the Department of Electromagnetic Field, Czech Technical University in Prague. His research interests are focused on radiowave propagation and satellite communication.

Pavel PECHAC received the M.Sc. degree and the Ph.D. degree in Radio Electronics from the Czech Technical University in Prague, Czech Republic, in 1993 and 1999 respectively. He is currently a Professor in the Department of Electromagnetic Field, Czech Technical University in Prague. His research interests are in the field of radiowave propagation and wireless systems.

Statistical Multirate High-Resolution Signal Reconstruction using the EMD-IT Based Denoising Approach

Aydin KIZILKAYA, Adem UKTE, Mehmet Dogan ELBI

Dept. of Electrical and Electronics Engineering, University of Pamukkale, 20070 Denizli, Turkey

akizilkaya@pau.edu.tr, ademukte@pau.edu.tr, melbi@pau.edu.tr

Abstract. *The reconstruction problem of a high-resolution (HR) signal from a set of its noise-corrupted low-resolution (LR) versions is considered. As a part of this problem, a hybrid method that consists of four operation units is proposed. The first unit applies noise reduction based on the empirical mode decomposition interval-thresholding to the noisy LR observations. In the second unit, estimates of zero-interpolated HR signals are obtained by performing up-sampling and then time shifting on each noise reduced LR signal. The third unit combines the zero-interpolated HR signals for attaining one HR signal. To eliminate the ripple effect, finally, median filtering is applied to the resulting reconstructed signal. As compared to the work that employs linear periodically time-varying Wiener filters, the proposed method does not require any correlation information about desired signal and LR observations. The validity of the proposed method is demonstrated by several simulation examples.*

Keywords

Empirical mode decomposition (EMD), statistical multirate signal reconstruction, noise reduction, high-resolution, median filtering

1. Introduction

Multirate statistical signal processing have found great interest over the last two decades for coping with the problems such as spectrum estimation, prediction, sensor fusion, time-delay estimation, and reconstruction or estimation of a stationary random process from multiple observations measured at different sampling rates [1]. For the solution of these problems, different methods are suggested by several researchers. Because the focus of our study is concerned with high-resolution (HR) signal reconstruction, in this paper, we shall concentrate on the methods related to this topic.

Using the principle of maximum entropy (ME), in [2], an optimal least-mean-squares estimator is developed for estimating the samples of a wide-sense stationary (WSS) random signal from multiple observations at low sampling rates. Within the scope of the reconstruction or estimation of a WSS random signal through its noisy LR measurements observed at different sampling rates, studies that employ the optimal linear filtering referred as Wiener filtering are carried out under the guidance of Therrien [3–9]. The problem of estimating a WSS random signal from two sequences observed separately at full-rate and half-rate, and both of which are affected by measurement noise is investigated in [3]. Using the least-squares (LS) approach, in [4], three optimal filter structures are proposed for the estimation of the HR random process from two observations, one of which is measured at full-rate with a low signal-to-noise ratio (SNR) and the other is measured at arbitrary low-rate with a high SNR. A generalized version of [3] is considered in [5] that is a preliminary work of dissertation presented in [6] where optimal multirate filtering procedure is formulated taking the second-order statistics into consideration for the goal of estimating an HR signal from LR measurements observed at different sampling rates. In [7], derivations in [5] and [6] are adapted to reconstruct an HR signal from a set of its noisy and distorted LR versions. Eventually, on the basis of the LS approach, formulations of multirate optimal filtering derived in [5–7] are extended to the two-dimensional case [8] (refer to [9] for detailed information). It is remarked that both ME and Wiener filter based methods require the knowledge of second-order statistics related to the desired HR signal and the noisy LR observations.

In addition to these mentioned two main groups of methods, recently, adaptive filtering and compressive sensing based approaches dealing with the statistical multirate signal estimation are suggested [10–12].

This paper aims at providing the reconstruction of an HR signal from a set of its noisy LR observations, without requiring the knowledge of any second-order statistics. In this context, a hybrid method consisting of four operation units is proposed. The paper is organized as follows. The statement of the problem and its relation to prior work are addressed in Sec. 2. Section 3 is devoted to the presentation of the proposed HR signal reconstruction method. To demonstrate the validity of the proposed method, two simulation examples are considered in Sec. 4. Paper ends up with concluding remarks.

2. Problem Statement and Relation to Prior Work

The problem studied here is concerned with the reconstruction of an HR unobservable random signal $d[n]$ from a set of its LR measurements $\{x_i[m_i] | i = 0,1,\ldots,M-1\}$ corrupted by additive white Gaussian noise (AWGN). In other words, these measurements are noisy, time shifted and down-sampled versions of $d[n]$ and are generated by the model shown in Fig. 1, where AWGN, time shifting and down-sampling factor are indexed to the particular signal of interest [9]. Thus, the ith LR observation signal $x_i[m_i]$ is defined by $x_i[m_i] = s_i[n_i]$ with $n = L_i m_i$. On the other hand, throughout this paper, LR signals $\{x_i[m_i]\}$ are assumed as maximally decimated versions of $d[n]$, i.e. $L_i = L$, $m_i = m$, and $M = L$.

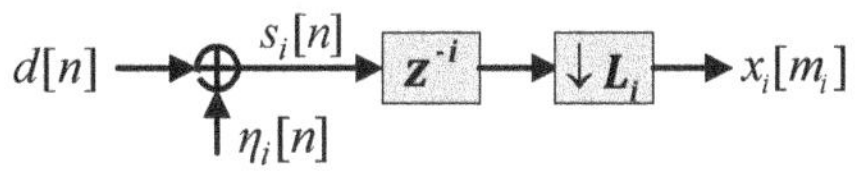

Fig. 1. Model used to generate the ith LR signal [9].

One of the existing methods to solve the problem emphasized above employs linear periodically time-varying (LPTV) Wiener filters that are optimal in mean-squares sense [7]. As shown in Fig. 2a, this method requires knowing down-sampling rate L, LPTV Wiener filter order P, cross-correlations between desired HR signal and LR signals, and cross-correlations among LR signals. However, the proposed method does not require any information related to the second order statistics of $d[n]$ and LR signals $\{x_i[m]\}$. To know down-sampling rate L alone is enough for performing signal reconstruction task (see Fig. 2b).

3. Proposed Method

Under the assumption of down-sampling rate L is known, the method proposed for the aim of reconstructing a desired HR signal from a set of its noisy LR observations consists of four signal processing units.

The first signal processing unit serves the purpose of reducing the noise effect on noisy LR observations. For this aim, empirical mode decomposition (EMD) based denoising approach is used. As a part of Hilbert–Huang transform, EMD is a method of decomposing any complicated signal into a monotonic residual component and a series of intrinsic mode functions (IMFs) with degree of frequencies in descending order [13], [14]. Note that the first IMF is related to the highest-frequency sequence of the ith noisy LR signal $x_i[m]$, [14]. Taking account of the fact that the frequency components of AWGN spread out

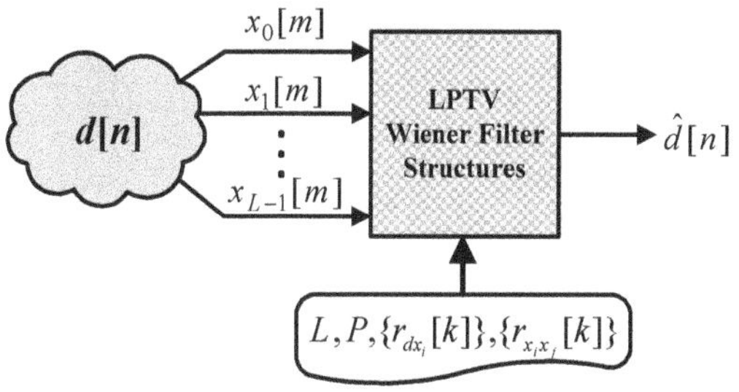

(a) HR signal reconstruction based on the method of [7].

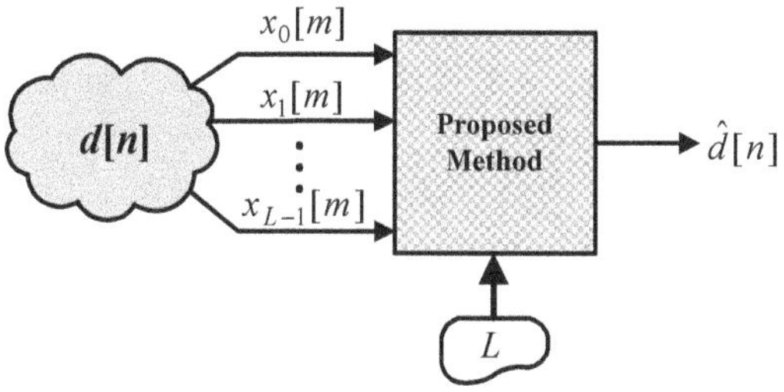

(b) HR signal reconstruction based on the proposed method.

Fig. 2. Schematic representations of statistical multirate HR signal reconstruction procedures.

over the entire frequency range, the first IMF of each noisy LR observation signal can be considered as it is mainly associated with the AWGN [15]. Thus, a noise reduced version of the signal can be obtained by extracting its first IMF from the noisy signal. Such an attempt has recently been made in [16] for the purpose of reducing the AWGN effect on each of the noisy LR observation signals before using them with the LPTV Wiener filter structures for carrying out the HR signal reconstruction task.

Nevertheless, there are still effects of noise on other IMFs. That is, separating the first IMF from noisy signal alone may not usually be sufficient for filtering noise in an effective manner. Taking this fact into consideration, in the first subunit of the proposed method, the noise reduced versions of the noisy LR observations are obtained on the basis of the EMD interval thresholding (EMD-IT) algorithm [17]. In the framework of this algorithm, each noisy LR signal is decomposed into a monotonic residual sequence $r^{(K_i)}[m]$ and K_i empirical modes termed IMFs $\{c_i^{(k)}[m] | k = 1,2,\ldots,K_i\}$ by applying the EMD technique to it. The ith LR observation signal can then be expressed as

$$x_i[m] = \sum_{k=1}^{K_i} c_i^{(k)}[m] + r^{(K_i)}[m]. \qquad (1)$$

Let $z_j^{(k)}$ be the instant of the jth zero-crossing in the kth IMF. For isolated IMF samples that are located in the closed interval $\mathbf{z}_j^{(k)} = [z_j^{(k)}, z_{j+1}^{(k)}]$ of the kth IMF, $c_i^{(k)}[m]$, it is possible to decide whether this interval is noise- or signal-dominant by considering the single extrema $c_i^{(k)}[p_j^{(k)}]$ in this interval. Note that $p_j^{(k)}$ denotes the instant of the extrema for the kth IMF. If the signal is weak as compared with the noise, the absolute value of this extrema will lie below the predefined threshold. Otherwise, in the presence of strong signal, it can be expected that the absolute value of this extrema will exceed the threshold [17].

Let $T_i^{(k)}$ be the threshold value used for denoising the kth IMF of the ith LR signal. Application of the interval thresholding to the kth IMF of the ith LR signal is performed as follows [17]:

$$\tilde{c}_i^{(k)}[\mathbf{z}_j^{(k)}] = \begin{cases} c_i^{(k)}[\mathbf{z}_j^{(k)}], & \left|c_i^{(k)}[p_j^{(k)}]\right| > T_i^{(k)} \\ 0, & \left|c_i^{(k)}[p_j^{(k)}]\right| \le T_i^{(k)} \end{cases} \quad (2)$$

for $j = 1,2,\ldots,N^{(k)}$, where $N^{(k)}$ indicates the number of zero-crossings in the kth IMF. In (2), $c_i^{(k)}[\mathbf{z}_j^{(k)}]$ denotes the samples of the kth IMF belonging to the interval $\mathbf{z}_j^{(k)}$. The threshold value required in (2) is selected as [17]

$$T_i^{(k)} = C\sqrt{E_i^{(k)} 2\ln M^{(i)}} \quad (3)$$

where C is a multiplication constant and $M^{(i)}$ indicates the sample number of the ith LR observation signal. The IMF energies denoted by $\{E_i^{(k)}\}$ in (3) can be computed as in the following [17]:

$$E_i^{(k)} = \frac{\rho^{-k} E_i^{(1)}}{\beta}, \quad k = 2,3,\ldots \quad (4)$$

where

$$E_i^{(1)} = \left[\frac{median(c_i^{(1)}[m])}{0.6745}\right]^2 \quad (5)$$

corresponds to the energy of the first IMF and can be defined as the variance estimate of the AWGN, as well. The parameters β and ρ in (4) depend on the number of sifting iterations performed as a part of the EMD algorithm. Thus, at the end of the EMD-IT based denoising procedure, each noise reduced LR signal can be obtained by

$$\tilde{x}_i[m] = \sum_{k=M_1}^{M_2} \tilde{c}_i^{(k)}[m] + \sum_{k=M_2+1}^{K_i} c_i^{(k)}[m] + r^{(K_i)}[m]. \quad (6)$$

It appears from (6) that, after the EMD-IT algorithm [17], the noise reduced version of the ith LR signal is acquired by extracting the first $M_1 - 1$ low-order noisy IMFs from (1) and by replacing the high-order less noisy IMFs between M_1 and M_2 in (1) with their thresholded counterparts.

In the second signal processing unit of the proposed method, each denoised LR signal produced by (6) is firstly upsampled by $L_i = L$ and then each resulting signal is time shifted with z^i in order to get estimates of zero interpolated HR signals $\tilde{d}_i[n]$ for $i = 0,1,\ldots, L-1$.

In the third signal processing unit, the nonzero valued samples of the zero interpolated HR signals are combined in terms of the following rule:

$$\tilde{d}[n] = \tilde{d}_i[n]\big|_{i=\langle n\rangle_L} \quad (7)$$

where $\langle n\rangle_L$ stands for the common residue of n modulo L.

Finally, in the last stage of the proposed method, the HR signal obtained by (7) is applied to a median filter for eliminating the ripples appearing on it. Eventually, an estimate of the desired HR signal, $\hat{d}[n]$, is acquired.

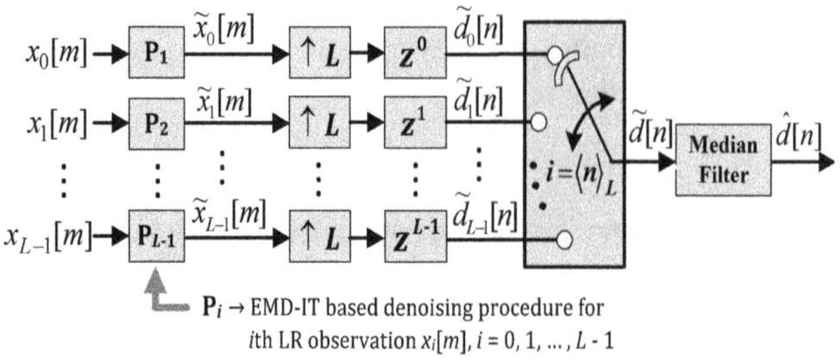

Fig. 3. Block diagram representation of the proposed multirate HR signal reconstruction method.

A combination of all signal processing units constituting the proposed method for achieving an estimate of the desired HR signal $d[n]$ from its measured L noisy LR signals is shown by the block diagram in Fig. 3.

4. Simulation Results

In order to inspect the validity and the performance of the proposed method, we try to reconstruct two different signals from its three LR versions that are corrupted by AWGN. For a reliable evaluation, the results obtained by the proposed method are also compared to those of method [7]. Since the LR signals are maximally decimated (downsampled) forms of the considered HR signals, the decimation factor is $L_i = L = 3$ and the number of Wiener filter coefficients required in [7] is taken as $P = 8$. In the context of the proposed method, the IMFs of each LR signals are obtained by eight sifting iterations and the values of β and ρ constants are respectively fixed to 0.719 and 2.01 during the computation of IMF energies [17]. Considering the minimum normalized mean square error (NMSE) related to the original and estimated HR signals, the multiplication constant C in (3) is determined via grid searching in the closed interval [0.1, 1.4] with 0.1 increments such that it meets the minimum NMSE. In order to avoid the ripple effect on the samples of signals reconstructed by the proposed method, a median filter with a window size of 3 is used. Furthermore, the simulation results provided by the method of [7] and the proposed method are compared according to the squared errors $e^2[n]$ and the NMSE criteria defined by

$$NMSE = \sum_{n=PL-1}^{N-1} |e[n]|^2 \Big/ \sum_{n=PL-1}^{N-1} |d[n]|^2$$

where $e[n]$ expresses the estimation error equaling to the difference between desired $\{d[n]\}$ and reconstructed HR signals $\{\hat{d}[n]\}$. Note that, since the first $PL - 1$ samples of the desired signal cannot be reconstructed with the method of [7], for making one-by-one comparison between the proposed method and the method of [7], the NMSE values are calculated over the last $N - PL - 1$ samples. On the other hand, all samples of the desired signal can be estimated by the proposed signal reconstruction method.

Example 1: In this example, we try to reconstruct the HR triangular waveform signal $d[n]$ with $N = 4000$ samples from its three noisy LR versions obtained by the model shown in Fig. 1, where zero-mean WGN is added on $d[n]$ to make SNR = –5 dB. The following Matlab code with signal frequency $f_0 = 4$ Hz and sampling frequency $f_s = 4000$ Hz is used to generate signal $d[n]$:

```
n = 0:1:fs;
d = sawtooth(2*pi*f0*(n/fs+0.625), 0.5);
```

The desired signal $d[n]$ and its three noisy LR forms are plotted in Fig. 4.

As a part of the EMD-IT based denoising procedure [17] used in the proposed method, the multiplication constant C in (3) is determined as 0.3 by grid searching and the values of M_1 and M_2 in (6) are taken as 6 and $K_i - 2$, respectively. The signal reconstruction results and the squared errors provided by the compared two methods are depicted in Fig. 5. It is observed from these figures that the proposed method exhibits better performance than that of method [7]. The NMSE values calculated for the proposed method and the method of [7] are 0.031 and 0.109, respectively, support this thought as well. On the other hand, it is worthwhile to note that performances of the EMD-based methods are influenced by the frequency of signal, f_0, and the sampling frequency, f_s (the number of signal samples). Further details related to this fact can be found in [16].

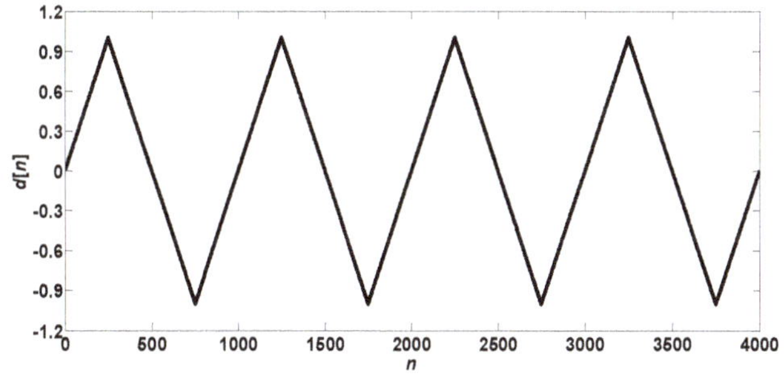

(a) HR signal $d[n]$.

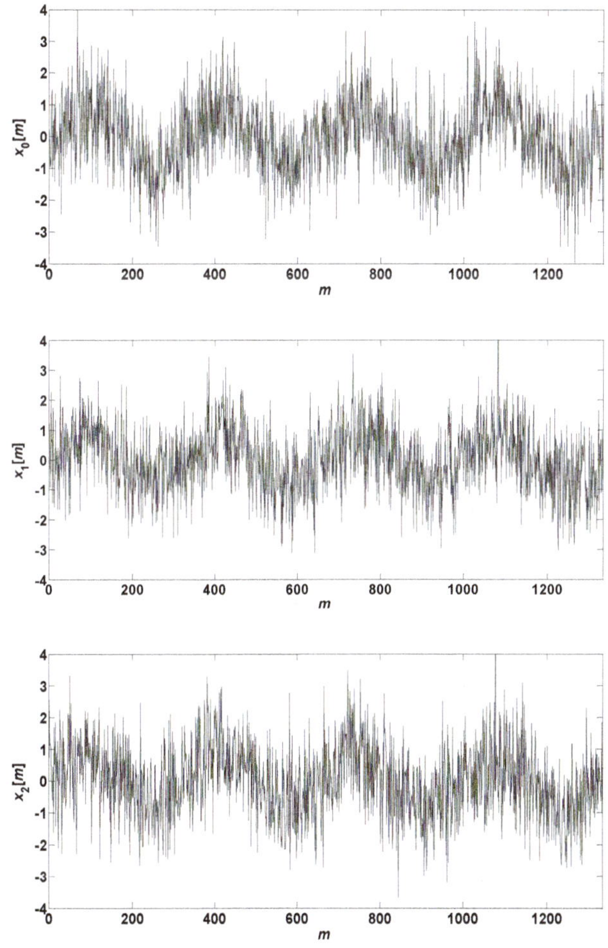

(b) Three noisy LR versions of $d[n]$.

Fig. 4. *Example 1:* Triangular waveform signal and its three noisy LR forms.

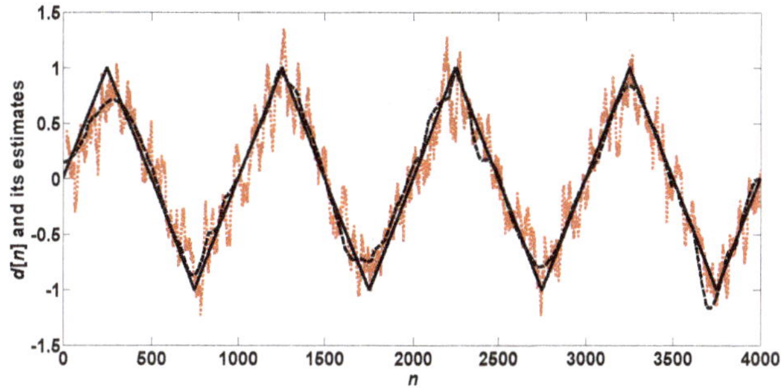

(a) Reconstruction results reached by the method of [7] (red dotted line) and the proposed method (blue dashed line).

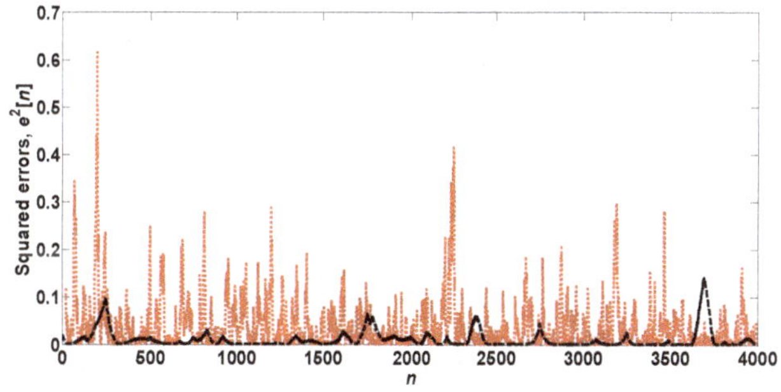

(b) Squared errors for the method of [7] (red dotted line) and the proposed method (blue dashed line).

Fig. 5. *Example 1:* Reconstruction of the triangular waveform signal shown in Fig. 4a from its three noisy LR versions shown in Fig. 4b.

Example 2: This example considers the reconstruction of an audio signal recording that consists of five spoken words 'hello' at certain intervals. This signal is 6.9 seconds long and sampled at 44.1 kHz. In line with the reconstruction purpose, three noisy LR versions of this audio signal with SNR = 0 dB are synthetically produced by the model given in Fig. 1. The original audio signal and its three LR (down-sampled) waveforms are shown in Fig. 6. For this example, on the point of using in the EMD-IT based denoising procedure, the multiplication constant C required in (3) is determined as 0.4 by grid searching and the values of M_1 and M_2 in (5) are taken as 2 and $K_i - 2$, respectively.

Using the LR signals shown in Fig. 6b along with the proposed method and the method of [7] lead to the recon-

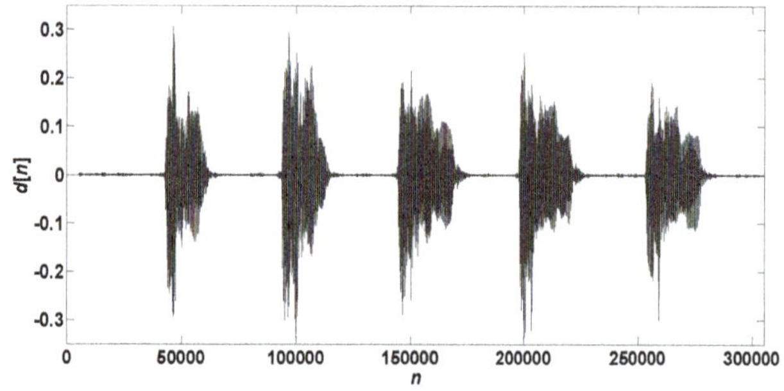

(a) HR audio signal $d[n]$.

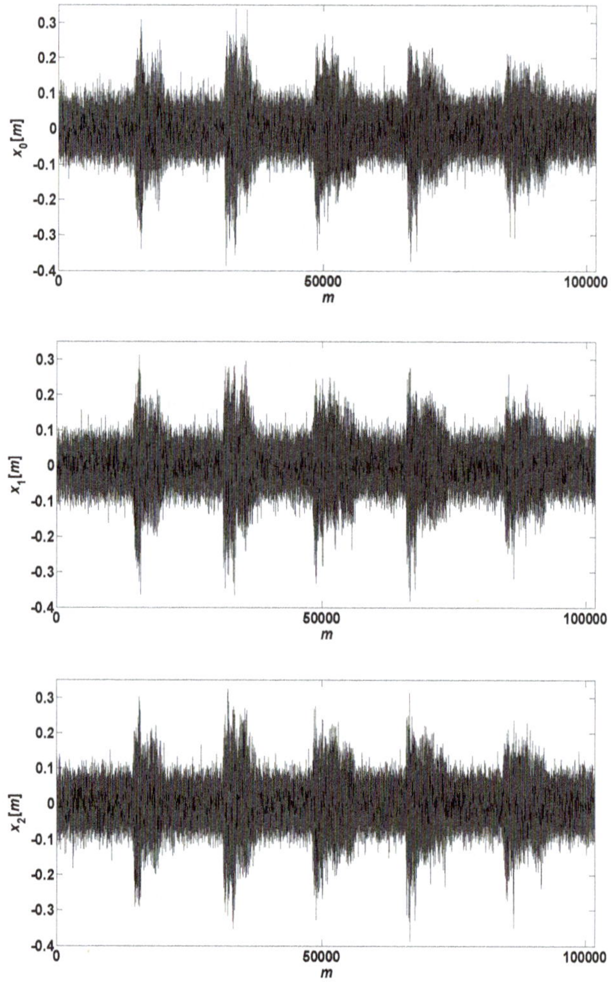

(b) Three noisy LR versions of $d[n]$ in (a).

Fig. 6. *Example 2:* The waveform of the audio signal and its three noisy LR forms.

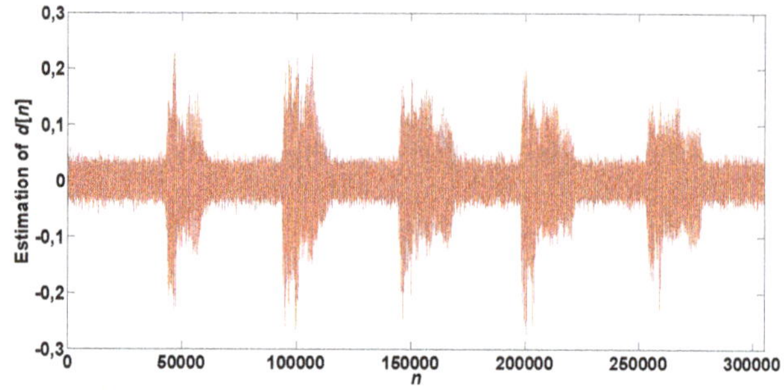

(a) Reconstruction result obtained by the method of [7].

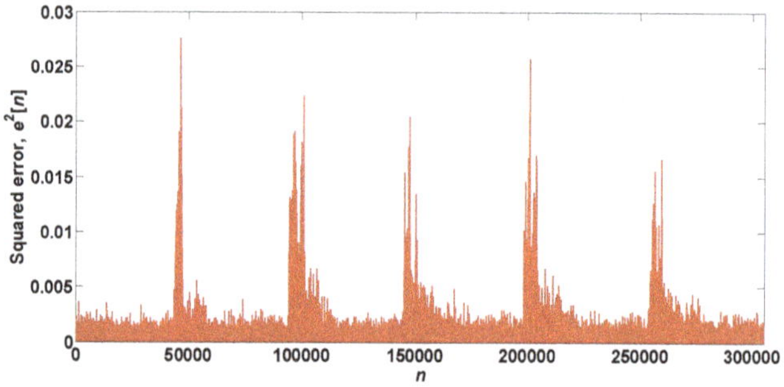

(b) Squared error resulted from using the method of [7].

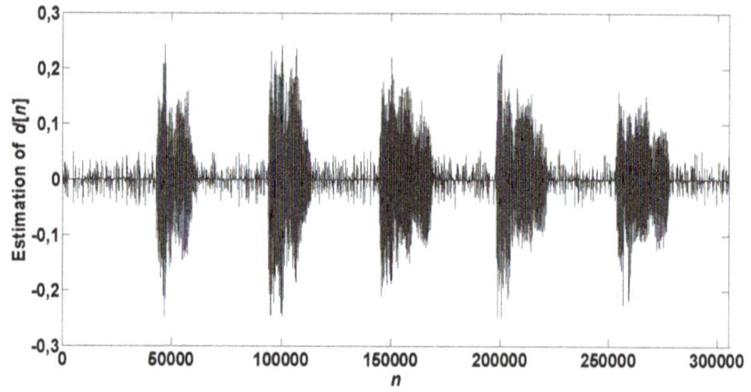

(c) Reconstruction result obtained by the proposed method.

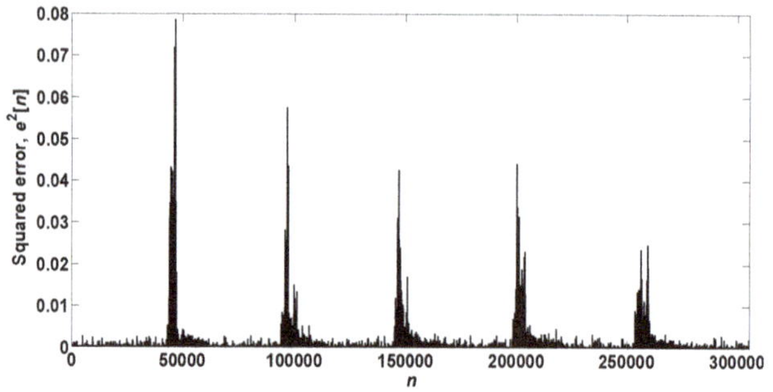

(d) Squared error resulted from using the proposed method.

Fig. 7. *Example 2:* Reconstruction of the HR audio signal shown in Fig. 6a from its three noisy LR versions shown in Fig. 6b.

struction results and the squared errors depicted in Fig. 7. It appears from this figure that, all in all, the proposed method introduces better performance than that of method [7]. The respective NMSE values calculated for the proposed method and the method of [7] are 0.156 and 0.223, confirm this claim as well. In the NMSE sense, the reason of providing the proposed method better result than that of the method of [7] can be explained in the following manner.

When we compare Fig. 7a to Fig. 7c, it can clearly be seen that, at the out of the time intervals corresponding to 'hello' sayings (idle channel case), the proposed method reconstructs the audio signal with less noise. The NMSE values presented in Tab. 1 reflect this case. Note that the numerical values tabulated in Tab. 1 are obtained by detecting the time intervals related to in and out of 'hello' sayings (speeches) in the original and reconstructed audio signals. In order to fulfill this detection, a rectangular window having a size of 100 samples is moved on the original signal. After each window movement, variance of 100 samples is calculated. The samples whose variance values are greater than 10^{-4} are considered as speeches.

	in	out
Method of [7]	0.140	34.607
Proposed method	0.144	5.021

Tab. 1. Computed NMSE values that correspond to the time intervals in and out of "hello" sayings.

5. Conclusions

A new method for the reconstruction of an HR signal from a set of its available noisy LR observations is proposed. Unlike popular methods that are based on the principle of ME or Wiener filter theory; the proposed method does not need of knowing any second-order statistics related to the desired HR signal and its LR versions. To know the value of down-sampling rate alone is enough for fulfilling the signal reconstruction by using the proposed method. Furthermore, it is demonstrated by several simulations that satisfactory performance has been provided by the proposed method. Thus, the method introduced in this paper can be an option to available methods for HR signal reconstruction. Applying this method to image reconstruction will be a subject of our further direction.

Acknowledgements

This work is based on the project supported by Pamukkale University Scientific Research Projects Coordination Unit, under Grant number of 2011FBE053

References

[1] JAHROMI, O. S. *Multirate Statistical Signal Processing.* Dordrecht: Springer-Verlag, 2007.

[2] JAHROMI, O. S., FRANCIS, B. A., KWONG, R. H. Multirate signal estimation. In *Proc. IEEE Canadian Conf. Electrical Computer Eng.* Toronto (ON, Canada), 2001, p. 147–152.

[3] CRISTI, R., KOUPATSIARIS, D. A., THERRIEN, C. W. Multirate filtering and estimation: the multirate Wiener filter. In *Proc. IEEE Thirty-Fourth Asilomar Conf. Signals, Systems and Computers.* Pacific Grove, CA (USA), 2000, p. 450–454. DOI: 10.1109/ACSSC.2000.910995

[4] THERRIEN, C. W., HAWES, A. H. Least squares optimal filtering with multirate observations. In *Proc. IEEE Thirty-Sixth Asilomar Conf. Signals, Systems and Computers.* Pacific Grove, CA (USA), 2002, p. 1782–1786. DOI: 10.1109/ACSSC.2002.1197081

[5] KUCHLER, R. J., THERRIEN, C. W. Optimal filtering with multirate observations. In *Proc. IEEE Thirty-Seventh Asilomar Conf. Signals, Systems and Computers.* Pacific Grove, CA (USA), 2003, p. 1208–1212. DOI: 10.1109/ACSSC.2003.1292116

[6] KUCHLER, R. J. Theory of multirate statistical signal processing and applications. *Ph.D. thesis,* Naval Postgraduate School, Monterey, CA (USA), 2005.

[7] SCROFANI, J. W., THERRIEN, C. W. A stochastic multirate signal processing approach to high-resolution signal reconstruction. In *Proc. 30th IEEE Int. Conf. Acoustics, Speech, and Signal Process.* Philadelphia, PA (USA), 2005, p. 561–564. DOI: 10.1109/ICASSP.2005.1416070

[8] SCROFANI, J. W., THERRIEN, C. W. A multirate approach to high-resolution image reconstruction. In *Proc. 7th IASTED Int. Conf. Signal and Image Process.* Honolulu, (HI, USA), 2005, p. 538–542.

[9] SCROFANI, J. W. *Theory of Multirate Signal Processing with Application to Signal and Image Reconstruction.* Ph.D. thesis, Naval Postgraduate School, Monterey, CA (USA), 2005.

[10] HAWES, A. H., THERRIEN, C. W. LMS adaptive filtering with multirate observations. In *Proc. IEEE Thirty-Seventh Asilomar Conf. Signals, Systems and Computers.* Pacific Grove, CA (USA), 2003, p. 567–570.

[11] TANC, A. K., EKSIOGLU, E. M., KAYRAN, A. H. Adaptive multirate signal estimation with lattice orthogonalization. In *Proc. 16th IEEE International Conf. Electronics, Circuits and Systems.* Yasmine Hammamet (Tunesia), 2009, p. 117–119. DOI: 10.1109/ICECS.2009.5410937

[12] EKSIOGLU, E. M., TANC, A. K., KAYRAN, A. H. A compressive sensing framework for multirate signal estimation. In *Proc. 10th IEEE Int. Conf. Information Sciences Signal Processing and their Applications.* Kuala Lumpur (Malaysia), 2010, p. 716– 719. DOI: 10.1109/ISSPA.2010.5605572

[13] HUANG, N. E., LONG, S. R., SHEN, Z. The mechanism for frequency downshift in nonlinear wave evolution. *Advances in Applied Mechanics,* 1996, vol. 32, p. 59–117C. DOI: 10.1016/S0065-2156(08)70076-0

[14] HUANG, N. E., SHEN, Z., LONG, S. R., WU, M. L. C., SHIH, H. H., ZHENG, Q. N., YEN, N. C., TUNG, C. C., LIU, H. H. The empirical mode decomposition and the Hilbert spectrum for nonlinear and non-stationary time series analysis. In *Proc. R. Soc. London Ser. A-Math. Phys. Eng. Sci.* vol. 454, no. 1971, 1998, p. 903–995. DOI: 10.1098/rspa.1998.0193

[15] WU, Z. H., HUANG, N. E. A study of the characteristics of white noise using the empirical mode decomposition method. In *Proc. R. Soc. London Ser. A-Math. Phys. Eng. Sci.* vol. 460, no. 2046, 2004, p. 1597–1611. DOI: 10.1098/rspa.2003.1221

[16] UKTE, A., KIZILKAYA, A., ELBI, M. D. Two empirical methods for improving the performance of statistical multirate high-resolution signal reconstruction. *Digital Signal Processing,* 2014, vol. 26, p. 36–49. DOI: 10.1016/j.dsp.2013.11.014

[17] KOPSINIS, Y., MCLAUGHLIN, S. Development of EMD-based denoising methods inspired by wavelet thresholding. *IEEE Transactions on Signal Processing,* 2009, vol. 57, no. 4, p. 1351 to 1362. DOI: 10.1109/TSP.2009.2013885

About the Authors ...

Aydin KIZILKAYA was born in Augsburg, Germany, in 1972. He received the B. Sc. degree from the Karadeniz Technical University, Trabzon, Turkey, in 1994, the M. S. degree from the Pamukkale University, Denizli, Turkey, in 1997, all in Electrical and Electronics Engineering, and the Ph. D. degree from the Technical University of Istanbul, Istanbul, Turkey, in 2006, in Electronics and Communication Engineering. In 1995, he joined the Pamukkale University, where he is now an Associate Professor of Communication Engineering, and has been a head of communication engineering subdivision since 2009. He was a deputy head of the Electrical and Electronics Engineering Department from 2009 to 2012. His research interests include theory and applications of statistical digital signal processing and multidimensional signal processing.

Adem UKTE was born in Istanbul, Turkey, in 1977. He received the B.Sc. and M.Sc. degrees, both in Electrical and Electronics Engineering, from the Pamukkale University, Denizli, Turkey, in 1999 and 2002, respectively. Since 1999, he is a Research Assistant at the Pamukkale University. Since 2008, he has been a Ph.D. student at the Pamukkale University, Denizli, Turkey. His research interests include statistical digital signal processing, and multirate signal and image processing.

Mehmet Dogan ELBI was born in Denizli, Turkey, in 1988. He received the B.Sc. degree in Electrical and Electronics Engineering, from the Pamukkale University, Denizli, Turkey, in 2010. Since 2012, he is a Research Assistant at the Pamukkale University. Since 2011, he has been a M.Sc. student at the Pamukkale University. His research interests include digital signal and image processing, FPGA architecture, VHDL parallel programming, nonlinear modeling, and machine learning.

Multiproduct Uniform Polar Quantizer

Milan R. DINCIC, Zoran H. PERIC

Faculty of Electronic Engineering, University of Nis, Aleksandra Medvedeva 14, 18000 Nis, Serbia

mdincha@hotmail.com, zoran.peric@elfak.ni.ac.rs

Abstract. *The aim of this paper is to reduce the complexity of the unrestricted uniform polar quantizer (UUPQ), keeping its high performances. To achieve this, in this paper we propose the multiproduct uniform polar quantizer (MUPQ), where several consecutive magnitude levels are joined in segments and within each segment the uniform product quantization is performed (i.e. all levels within one segments have the same number of phase levels). MUPQ is much simpler for realization than UUPQ, but it achieves similar performances as UUPQ. Since MUPQ has low complexity and achieves much better performances than the scalar uniform quantizer, it can be widely used instead of scalar uniform quantizers to improve performances, for any signal with the Gaussian distribution.*

Keywords

Unrestricted uniform polar quantization, multiproduct uniform polar quantization, polar coordinates, Gaussian distribution

1. Introduction

Digital transmission and processing are dominant nowadays, while the most signals are analog. Therefore, an inevitable part of almost all modern telecommunication systems is A/D (analog-to-digital) converter. The main part of A/D converters is a quantizer. The quality of the digitized signal mostly depends on the quantizer. Also, with the appropriate design of the quantizer, the compression of signals can be achieved, which is very important since resources for the transmission and storage of signals (channel bandwidth and memory space) are limited. Therefore, the designing of quantizers is very important and topical subject.

There are two main types of quantizers [1], [2]: scalar quantizers (where each sample of the signal is separately quantized) and vector quantizers (where several samples are jointly quantized; for example, if n samples are jointly quantized, we can say that they represent an n-dimensional vector in n-dimensional vector space, therefore we have an n-dimensional vector quantizer). Vector quantizers have much better performances (i.e. they can achieve much higher SQNR (signal-to-quantization noise ratio) for the same bit-rate) than scalar quantizers [2]. On the other hand, vector quantizers have one drawback: high complexity, which exponentially increases with the increasing of the dimension n. Therefore, the most used vector quantizers are two-dimensional quantizers, which are the simplest of all vector quantizers, but they still have much better performances than scalar quantizers. Almost all signals of interest are random and can be described with some probability density function (pdf). A lot of signals can be modeled with the Gaussian pdf. For signals with the Gaussian pdf, it is easier to design two-dimensional vector quantizer in the polar coordinates (magnitude r and phase ϕ) than in the Cartesian coordinates. Such quantizers, designed in the polar coordinates, are called polar quantizers.

There are two main types of polar quantizers: restricted (also called product) and unrestricted. In product polar quantizers, the number of phase levels is the same for all magnitude levels, while in unrestricted polar quantizers the numbers of phase levels are different for different magnitude levels, i.e. the number of phase levels are optimized for each magnitude level. Unrestricted polar quantizers have better performances than product polar quantizers, but they are more complex. In polar quantizers, the magnitude r can be quantized in different ways (using uniform or nonuniform quantization), while the phase ϕ is always quantized using the uniform quantization since it has the uniform distribution.

Polar quantization has been considered in many papers. In [3], [4] unrestricted polar quantizers were analyzed, using the optimal companding function for the quantization of the magnitude r. In [5], [6] the product uniform polar quantization was considered. The product polar quantizer with the companding function optimal for scalar but not for polar quantization, was considered in [7]. Embedded product and unrestricted polar quantizers were considered in [8]. Product polar quantizers with A-law companding function were analyzed in [9] with the application for audio signals. In [10], [11] product polar quantizers with μ-law companding function were considered. The solution in [11] is compatible with the ITU-T G.711 standard. Unrestricted polar quantizers with square cells were analyzed in [12], applying μ-law companding function for the quantization of the magnitude r.

The unrestricted uniform polar quantizer (UUPQ) is the polar quantizer with uniformly quantized magnitude r, where the optimization of numbers of phase levels is done for each magnitude levels. It was analyzed in [13], [14]. It is two-dimensional polar counterpart of the scalar uniform quantizer. UUPQ can achieve much better performances (i.e. much higher SQNR for the same bit-rate) than the scalar uniform quantizer. UUPQ can be used instead of the scalar uniform quantizer (which is widely used), to improve performances. However, UUPQ has high complexity, since numbers of phase levels for all magnitude levels should be calculated and stored in memory.

The aim of this paper is to reduce the complexity of UUPQ, keeping its good performances. Therefore, in this paper we propose the multiproduct uniform polar quantizer (MUPQ), where several magnitude levels are joined into segments. Within each segment the product polar quantization is done, i.e. all magnitude levels in one segment have the same number of phase levels. Therefore, for MUPQ we have to calculate numbers of phase levels for each segment, not for each magnitude level (which is the case in UUPQ). Since the number of segments is much smaller than the number of magnitude levels, it follows that for MUPQ we have to calculate much smaller number of parameters than for UUPQ, thus MUPQ is much simpler for realization than UUPQ. Furthermore, we will show that performances of MUPQ are very close to performances of UUPQ. Therefore, the main contribution of this paper is the designing of MUPQ, which is much simpler for realization than UUPQ, but which can achieve excellent performances, very close to performances of UUPQ.

The comparison between MUPQ and PUPQ (product uniform polar quantizer) [5], [6] is also presented. It is shown that MUPQ is better solution since MUPQ is slightly more complex than PUPQ but MUPQ achieves higher SQNR compared to PUPQ. Slight increase of complexity provides appreciable increasing of SQNR.

The asymptotic analysis is usually used for the designing of polar quantizers [3-14]. Hence, the asymptotic analysis will be applied in this paper. We will consider medium and high bit-rates (roughly speaking, higher than 4 bps (bits-per-sample)), since the asymptotic analysis is valid for these bit-rates.

Simulations are done in MATLAB both for UUPQ and MUPQ. Simulation and theoretical results are matched very well, which proves the correctness of the developed theory.

Theory is also proven by the experiment performed on the speech signal, for MUPQ, PUPQ and UUPQ. We use the speech signal since it can be modeled with the Gaussian distribution very well [1].

This paper is organized in the following way. Polar quantizers are defined in Section 2. In Section 3, the analysis of performances of UUPQ is presented shortly. The designing of MUPQ, which is the main contribution of this paper, is presented in Section 4. Numerical results and discussion are given in Section 5. Section 6 concludes the paper.

2. The Definition of Polar Quantizer

Let's consider a signal with the Gaussian distribution, which is defined with the probability density function (pdf) $f(x)=\frac{1}{\sqrt{2\pi}\sigma}\exp(-x^2/(2\sigma^2))$, where σ^2 denotes the power (variance) of the signal. The design of quantizers is usually done for the unit variance $\sigma^2=1$, hence we will use this approach in this paper. For the designing of two-dimensional quantizers, the joint probability density function of two consecutive samples x_1 and x_2 should be used. In Cartesian coordinates, the joint pdf is defined as $f(x_1,x_2)=\exp\left(-(x_1^2+x_2^2)/2\right)/(2\pi)$. Let's define polar coordinates: magnitude $r=\sqrt{x_1^2+x_2^2}$ and phase $\phi=\arctan(x_2/x_1)$. The joint pdf in polar coordinates becomes $f(r,\phi)=r\exp\left(-r^2/2\right)/(2\pi)$ [12]. We can see that $f(r,\phi)$ depends only on r but not on ϕ. The pdf of the magnitude r is:

$$f_r(r)=\int_0^{2\pi} f(r,\phi)d\phi = re^{-r^2/2} \tag{1}$$

which is the pdf of the Rayleigh distribution. The pdf of the phase ϕ is $f_\phi(\phi)=\int_0^{+\infty} f(r,\phi)dr=\frac{1}{2\pi}$. We can see that the phase ϕ has the uniform distribution, which means that the two-dimensional source of information is circularly-symmetric. One often used approach of the designing of vector quantizers is the geometric principle [15]. According to this principle, quantization cells are deployed on contours where the joint pdf is constant. Since the joint pdf of the Gaussian source in polar coordinates $f(r,\phi)$ depends only on r, it follows that $f(r,\phi)$ is constant where $r=const.$, i.e. contours where $f(r,\phi)$ is constant are concentric circles. Applying the geometric principle, the two-dimensional quantizer for the Gaussian source is much easier to design in polar than in Cartesian coordinates, by deploying cells on concentric circles. Quantizers designed in polar coordinates are called polar quantizers.

Let's define some parameters of polar quantizers: N denotes the total number of quantization cells, $r_{\max}$ denotes the maximal magnitude, L denotes the number of magnitude levels, r_i, $i=0,\ldots,L$ denote thresholds and m_i, $i=1,\ldots,L$ denote representation levels for the quantization of the magnitude r. On each magnitude level the uniform quantization of the phase ϕ is done, since the phase ϕ has the uniform distribution. Let P_i denote the number of phase levels on the i-th magnitude level. It holds that $\sum_{i=1}^{L} P_i = N$. Let $\phi_{i,j}=j2\pi/P_i$, $j=0,\ldots,P_i$ denote thresholds and $\psi_{i,j}=(j-1/2)2\pi/P_i$, $j=1,\ldots,P_i$ denote representation levels for the quantization of the phase ϕ for the i-th magnitude level.

Within each cell there is one representation point. Let's consider an arbitrary quantization cell $S_{i,j} = \{(r,\phi) \mid r_{i-1} \le r < r_i \wedge \phi_{i,j-1} \le \phi < \phi_{i,j}\}$, $i = 1,\dots,L$, $j = 1,\dots,P_i$. Within this cell there is the representation point $(m_i, \psi_{i,j})$. All points (r,ϕ) from the cell $S_{i,j}$ are mapped to the representation point $(m_i, \psi_{i,j})$.

There are two basic types of polar quantizers: unrestricted (where numbers of phase levels are different for different magnitude levels, i.e. the optimization of the number of phase levels is done for each magnitude level) and restricted (also called product, where the numbers of phase levels are the same for all magnitude levels, i.e. $P_1 = P_2 = \dots = P_L$).

During the quantization process, an irreversible error is made, which is expressed by the distortion. The total distortion D is equal to the sum of the granular distortion D_g (which is made in the area $0 \le r \le r_{max}$) and the overload distortion D_{ov} (which is made in the area $r > r_{max}$), i.e. $D = D_g + D_{ov}$. Distortions in this paper will be defined per one dimension. It is usual that performances of vector quantizers are defined per one dimension, to simplify comparison of performances of vector quantizers with different dimensions. The quality of the quantized signal is defined with SQNR (signal-to-quantization noise ratio), which is defined as:

$$\mathrm{SQNR[dB]} = 10\log_{10}(1/D). \tag{2}$$

The bit-rate (the average number of bits required for the coding of one sample) is defined as:

$$R = \frac{1}{2}\log_2 N \text{ [bps]}. \tag{3}$$

3. Unrestricted Uniform Polar Quantizer (UUPQ)

This is the unrestricted polar quantizer where the uniform quantization of the magnitude r is performed, i.e. the magnitude range $[0, r_{max}]$ is uniformly divided into L intervals. Thresholds for the uniform quantization of the magnitude r are defined as $r_i = i r_{max}/L$, $i = 0,\dots,L$, while representation levels are defined as $m_i = (i-1/2) r_{max}/L$, $i = 1,\dots,L$. Let $\Delta_r = r_i - r_{i-1} = r_{max}/L$ denotes the stepsize for the uniform quantization of the magnitude r.

UUPQ was already analyzed in [13], [14]. We will recall some results from [13], [14]. The expression for the number of magnitude levels was derived in [14]:

$$L = r_{max}\sqrt{\frac{N}{6\pi\sqrt{3}}}\cdot\left(1 - e^{-r_{max}^2/6}\right)^{-3/4}. \tag{4}$$

In [13], the following expression for the number of phase levels on the i-th magnitude level was presented: $P_i = N r_{max}\sqrt[3]{m_i^2 f_r(m_i)}\left(L\int_0^{r_{max}}\sqrt[3]{r^2 f_r(r)}dr\right)^{-1}$. For $f_r(r)$ defined with (1), it follows that:

$$P_i = \sqrt{\frac{2\pi N}{\sqrt{3}}}\frac{m_i\exp(-m_i^2/6)}{\sqrt[4]{(1 - e^{-r_{max}^2/6})}}, i = 1,\dots,L. \tag{5}$$

The expression for the granular distortion was derived in [14]:

$$D_g = \frac{\sqrt{3}\pi}{2N}(1 - e^{-r_{max}^2/6})^{3/2}. \tag{6}$$

However, the analysis of UUPQ presented in [13] and [14] was incomplete and not quite correct from the following reasons: in [13] the overload distortion was neglected and the optimization of r_{max} was not done (instead of that, the value of r_{max} for the scalar uniform quantizer was used); in [14] the expression for the overload distortion was not given and the method for calculation of r_{max} was not explained. Due to these reasons, performances of UUPQ calculated using the analysis from [13], [14] is not accurate enough. Since we need accurate performances of UUPQ, we will present some new results for UUPQ (calculation of the overload distortion and the optimization of r_{max}) which are missing in [13] and [14].

Using the similar procedure as for the granular distortion in [13], we can define the overload distortion as: $D_{ov} = \frac{1}{2}\int_{r_{max}}^{+\infty}\left[(r - m_L)^2 + r m_L \pi^2/(3P_L^2)\right] f_r(r)dr$. The article 1/2 on the beginning of this expression denotes the fact that D_{ov} is defined per one dimension. Using approximation $m_L \approx r_{max}$, which is valid for the asymptotic analysis, it is obtained that:

$$D_{ov} = \frac{1}{2}\int_{r_{max}}^{+\infty}\left[(r - r_{max})^2 + \frac{r r_{max}\pi^2}{3P_L^2}\right] f_r(r)dr. \tag{7}$$

Using (5) for P_L, the following final expression for the overload distortion is obtained:

$$D_{ov} = X + \frac{\sqrt{3}\pi e^{r_{max}^2/3}}{12 N r_{max}}\sqrt{(1 - e^{-r_{max}^2/2})(1 - e^{-r_{max}^2/6})}\cdot Y \tag{8}$$

where

$$X = \frac{1}{2}\left(2e^{-r_{max}^2/2} - \sqrt{2\pi}r_{max}\,\mathrm{erfc}(r_{max}/\sqrt{2})\right), \quad Y = r_{max}e^{-r_{max}^2/2} + \sqrt{\pi/2}\,\mathrm{erfc}(r_{max}/\sqrt{2}). \tag{9}$$

The total distortion D is obtained by summing expressions (6) and (8). SQNR and the bit-rate R are calculated using (2) and (3).

Now, we will shortly summarize the design process. The parameter N is defined in advance, i.e. N is known on the beginning of the design process. Optimal values of other parameters (r_{max}, L and P_i) should be found during the design process, in the following way. The parameter r_{max} should be found firstly. Based on (6) and (8), we can see that the total distortion D depends only on one unknown parameter r_{max}. Optimal value of r_{max} is obtained by the minimization of the D, i.e. by numerical solving of the equation $(dD/dr_{max})=0$. When r_{max} is found, we can calculate all other unknown parameters: L is calculated using (4); after that thresholds and representation levels are calculated as $r_i = ir_{max}/L$ and $m_i=(i-1/2)r_{max}/L$; finally, values for P_i, $i=1,...,L$ are calculated using (5).

Numerical values for UUPQ are given in Tab. 1. Simulation of UUPQ is done in MATLAB. In Tab. 1 we present values of SQNR obtained by the theory ($SQNR_{th}$) and by the simulation ($SQNR_{sim}$). We can see that values of SQNR obtained by the theory and by the simulation are matched very well, which proves the previously developed theory.

R [bps]	5	6	7	8
r_{max}	3.55	4.00	4.42	4.82
L	22	47	102	219
SQNRth [dB]	26.28	32.06	37.95	43.90
SQNRsim [dB]	26.25	32.05	37.92	43.89

Tab. 1. Numerical results for UUPQ.

R [bps]	5	6	7	8
SQNR [dB]	24.57	29.83	35.13	40.34

Tab. 2. SQNR for the scalar uniform quantizer [1].

For the purpose of comparison, values of SQNR for the scalar uniform quantizer for different values of R are presented in Tab. 2. These values are taken from [1]. We can see that UUPQ has much better performances (i.e. much higher SQNR) than the corresponding uniform scalar quantizer, for the same bit-rate R. Due to its very good performances, UUPQ could be very important in many applications. However, UUPQ has one drawback: high complexity. Namely, due to the fact that the numbers of phase levels P_i are different for different magnitude levels, we have to calculate and store parameters P_i for all magnitude levels, which increases complexity and requires large memory space, both in transmitter and receiver.

In the aim to decrease the complexity, but also to keep good performances in the same time, we will present the multiproduct uniform polar quantizer (MUPQ) in the next section.

4. Multiproduct Uniform Polar Quantizer

The meaning of parameters N, r_{max}, L, r_i, m_i, and Δ_r is the same as in UUPQ. In MUPQ, the uniform quantization of the magnitude r is performed (i.e. the magnitude range [0, r_{max}] is uniformly divided into L magnitude intervals) and the uniform quantization of the phase ϕ is done for each magnitude interval, as in the previously described UUPQ. But, in MUPQ, L_0 consecutive magnitude intervals are joined into segments. There are $K = L/L_0$ segments. Within one segment, the product polar quantization is performed, which means that all magnitude intervals within one segment have the same number of phase levels. Since this polar quantizer consists of K product uniform polar quantizers, it is called 'multiproduct'. Let M_j, $j=1,...,K$, denote the number of phase levels on magnitude intervals within the j-th segment. It holds that $\sum_{j=1}^{K} M_j = N/L_0 = NK/L$. Let $t_j = jr_{max}/K$, $j=1,...,K$, denote borders between segments. Let $r_{j,l}=t_{j-1}+l\Delta_r$, $j=1,...,K$, $l=0,...,L_0$ denote thresholds and $m_{j,l}=t_{j-1}+(l-1/2)\Delta_r$, $j=1,...,K$, $l=1,...,L_0$ denote representation levels for the quantization of the magnitude r in the j-th segment. Parameters N and K are given in advance, i.e. they are known at the beginning of the designing process. Optimal values of other parameters should be found through the designing process.

The total distortion D is equal to the sum of the granular D_g and overload D_{ov} distortions. The granular distortion D_g can be written as $D_g = D_{g1}+D_{g2}$. It was shown in [13] that $D_{g1} = \frac{1}{2}\sum_{i=1}^{L}\int_{r_{i-1}}^{r_i}(r-m_i)^2 f_r(r)dr$. For the asymptotic analysis it holds that $D_{g1} = \frac{1}{2}\sum_{i=1}^{L} f_r(m_i)\int_{r_{i-1}}^{r_i}(r-m_i)^2 dr$. Since $r_{i-1}=m_i-\Delta_r/2$ and $r_i = m_i+\Delta_r/2$, it follows that $D_{g1} = (1/2)\cdot\sum_{i=1}^{L} f_r(m_i)\Delta_r^3/12 = (\Delta_r^2/24)\cdot\sum_{i=1}^{L} f_r(m_i)\Delta_r$. Changing summation with integration, it is obtained that $D_{g1}=(\Delta_r^2/24)\int_0^{r_{max}} f_r(r)dr$. Solving this integral for $f_r(r)$ defined with (1), it is obtained that:

$$D_{g1} = \frac{r_{max}^2}{24L^2}\left(1-\exp(-r_{max}^2/2)\right). \quad (10)$$

Based on [13], D_{g2} can be expressed as: $D_{g2} = (\pi^2/6)\sum_{j=1}^{K}\sum_{l=1}^{L_0}\int_{r_{j,l-1}}^{r_{j,l}}(rm_{j,l}/M_j^2)f_r(r)dr$. Applying the asymptotic analysis, it follows that $D_{g2}=(\pi^2/6)\cdot\sum_{j=1}^{K}(1/M_j^2)\sum_{l=1}^{L_0} m_{j,l}f_r(m_{j,l})\int_{r_{j,l-1}}^{r_{j,l}} rdr$. Since $\int_{r_{j,l-1}}^{r_{j,l}} rdr = (r_{j,l}+r_{j,l-1})(r_{j,l}-r_{j,l-1})/2 = m_{j,l}\Delta_r$, we obtain that:

$$D_{g2} = \frac{\pi^2}{6}\sum_{j=1}^{K}\frac{1}{M_j^2}\sum_{l=1}^{L_0} m_{j,l}^2 f_r(m_{j,l})\Delta_r . \quad (11)$$

Changing summation with integration, it follows that:

$$D_{g2} = \frac{\pi^2}{6}\sum_{j=1}^{K}\frac{1}{M_j^2}\int_{t_{j-1}}^{t_j} r^2 f_r(r)dr = \frac{\pi^2}{6}\sum_{j=1}^{K}\frac{I_j}{M_j^2} \quad (12)$$

where

$$I_j = \int_{t_{j-1}}^{t_j} r^2 f_r(r)dr = -\left(2+\frac{j^2 r_{\max}^2}{K^2}\right)\exp\left(-\frac{j^2 r_{\max}^2}{2K^2}\right) + \left(2+\frac{(j-1)^2 r_{\max}^2}{K^2}\right)\exp\left(-\frac{(j-1)^2 r_{\max}^2}{2K^2}\right). \quad (13)$$

The expression for the granular distortion becomes $D_g = r_{\max}^2/(24L^2)\cdot(1-\exp(-r_{\max}^2/2))+(\pi^2/6)\sum_{j=1}^{K} I_j/M_j^2$. To find optimal values of M_j, minimization of D_g will be done with the constraint that $\sum_{j=1}^{K} M_j = NK/L$. Function $G = D_g + \lambda\left(\sum_{j=1}^{K} M_j - NK/L\right)$ is formed, where λ is the Lagrange multiplier. Solving the equation $(\partial G/\partial M_j)=0$, it is obtained that $M_j = \sqrt[3]{\pi^2/(3\lambda)}\cdot\sqrt[3]{I_j}$. Since $\sum_{q=1}^{K} M_q \equiv \sqrt[3]{\pi^2/(3\lambda)}\cdot\sum_{q=1}^{K}\sqrt[3]{I_q} = NK/L$, it follows that $\sqrt[3]{\pi^2/(3\lambda)} = NK/\left(L\cdot\sum_{q=1}^{K}\sqrt[3]{I_q}\right)$. Therefore, the following expression is obtained:

$$M_j = \frac{N\cdot K}{L}\cdot\frac{\sqrt[3]{I_j}}{\sum_{q=1}^{K}\sqrt[3]{I_q}}, \quad j=1,\dots,K. \quad (14)$$

Putting (14) into (12), it follows that:

$$D_{g2} = \frac{\pi^2}{6}\frac{L^2}{N^2K^2}\left(\sum_{q=1}^{K}\sqrt[3]{I_q}\right)^3. \quad (15)$$

The granular distortion D_g can be written as $D_g = C_1/L^2 + C_2L^2$, where $C_1 = (r_{\max}^2/24)\left(1-\exp\left(-r_{\max}^2/2\right)\right)$ and $C_2 = \pi^2/(6N^2K^2)\cdot\left(\sum_{q=1}^{K}\sqrt[3]{I_q}\right)^3$. Solving the equation $(\partial D_g/\partial L)=0$, it is obtained that:

$$L = \sqrt[4]{\frac{C_1}{C_2}} = \sqrt{\frac{r_{\max}NK}{2\pi}}\sqrt[4]{\frac{1-e^{-r_{\max}^2/2}}{\left(\sum_{q=1}^{K}\sqrt[3]{I_q}\right)^3}}. \quad (16)$$

Putting (16) into (14), it follows that:

$$M_j = \frac{\sqrt{2\pi NK/r_{\max}}\cdot\sqrt[3]{I_j}}{\sqrt[4]{(1-e^{-r_{\max}^2/2})\left(\sum_{q=1}^{K}\sqrt[3]{I_q}\right)}}, \quad j=1,\dots,K. \quad (17)$$

Putting $L=\sqrt[4]{C_1/C_2}$ into $D_g = C_1/L^2 + C_2L^2$, the following final expression for the granular distortion is obtained:

$$D_g = 2\sqrt{C_1C_2} = \frac{r_{\max}\pi}{6NK}\sqrt{1-e^{-r_{\max}^2/2}}\left(\sum_{q=1}^{K}\sqrt[3]{I_q}\right)^{3/2}. \quad (18)$$

Based on (7), the overload distortion is defined as $D_{ov} = \frac{1}{2}\int_{r_{\max}}^{+\infty}\left[(r-r_{\max})^2 + rr_{\max}\pi^2/(3M_K^2)\right]f_r(r)dr$. Using (17) for M_K, the following final expression for the overload distortion is obtained:

$$D_{ov} = X + \frac{r_{\max}^2\pi^2\sqrt{1-e^{-r_{\max}^2/2}}}{12NK(I_K)^{2/3}}\sqrt{\sum_{q=1}^{K}\sqrt[3]{I_q}}\cdot Y \quad (19)$$

where X and Y are defined with (9).

The total distortion D is obtained by summation of (18) and (19). D depends on only one unknown parameter $r_{\max}$. Optimal value of $r_{\max}$ is obtained by the numerical minimization of D, i.e. by solving the equation $(dD/dr_{\max})=0$. The designing process of MUPQ will be summarized in the following algorithm (recall that values of N and K are given in advance):

1. Optimal value of $r_{\max}$ is calculated by the numerical solving of the equation $(dD/dr_{\max})=0$.
2. Based on (16), the optimal value of L is calculated. For the practical realization, L is rounded to the nearest integer divisible with K. This rounding has negligible effect on performances, since K takes small values (usually, $4 \le K \le 8$).
3. Thresholds $r_i = ir_{\max}/L$ and representation levels $m_i = (i-1/2)r_{\max}/L$, $i=1,\dots,L$, as well as borders between segments $t_j = jr_{\max}/K$, $j=1,\dots,K$ are calculated.
4. Using (17), the optimal number of phase levels M_j, $j=1,\dots,K$ for each segment is calculated. For the practical realization, values of M_j should be rounded to the nearest integers. This rounding has negligible effect on performances.
5. Thresholds $\phi_{j,l} = l2\pi/M_j$ and representation levels $\psi_{j,l} = (l-1/2)2\pi/M_j$, $j=1,\dots,K$, $l=1,\dots,M_j$, are calculated, for the quantization of the phase ϕ.

5. Numerical Results and Discussion

In Tab. 3, numerical results for MUPQ are presented, for different values of N and K. Simulation of this quantizer is done in MATLAB. Let $SQNR_{th}$ denote value of SQNR obtained by the theory and $SQNR_{sim}$ denote value of SQNR obtained by the simulation. We can see that

R [bps]	K	$r_{\max}$	L	$SQNR_{th}$ [dB]	$SQNR_{sim}$ [dB]
6	3	3.87	45	31.65	31.65
	4	3.90	44	31.79	31.77
	5	3.92	45	31.87	31.85
	6	3.94	48	31.89	31.90
	7	3.95	49	31.90	31.87
	8	3.96	48	31.95	31.93
7	3	4.24	93	37.43	37.41
	4	4.28	96	37.60	37.58
	5	4.31	95	37.7	37.72
	6	4.33	96	37.76	37.74
	7	4.35	98	37.80	37.77
	8	4.36	96	37.83	37.79
8	3	4.59	195	43.28	43.25
	4	4.64	200	43.48	43.48
	5	4.67	205	43.60	43.62
	6	4.69	210	43.67	43.67
	7	4.71	210	43.72	43.71
	8	4.73	208	43.75	43.77

Tab. 3. Numerical results for MUPQ.

values of SQNR obtained by the theory and by the simulation are matched very well, which proves the previously developed theory.

Comparison between MUPQ and UUPQ is presented in Tab. 4 (values of $SQNR_{th}$ for UUPQ are taken from Tab. 1). From Tab. 4 we can see that SQNR increases with the increasing of K, becoming closer to SQNR of UUPQ. On the other hand, the increasing of K leads to the increasing of complexity. Therefore, optimal choice of K should be made, taking into account both SQNR and complexity. We propose values K = 5, 6 and 7 as very good solutions.

In the aim of completeness, the comparison between MUPQ and PUPQ (product uniform polar quantizer) is also presented in Tab. 4. PUPQ is well known in literature [5], [6]; this is the polar quantizer where the uniform quantization of magnitude is performed (as well as in MUPQ and UUPQ), but in PUPQ the number of phase levels is the same for all magnitude levels, i.e. $P_1 = P_2 = \ldots = P_L = P$. PUPQ is the simplest of those three quantizers (PUPQ, UUPQ, MUPQ) since only one parameter P has to be stored and calculated, but it has the smallest SQNR. MUPQ is slightly more complex than PUPQ (since we have to calculate and store K parameters, but K is very small number, less than 8), but MUPQ achieves higher SQNR compared to PUPQ. Slight increase of complexity provides appreciable increasing of SQNR.

Let's consider three quantizers: UUPQ, PUPQ and MUPQ with K = 7 segments, for the same bit-rate of R = 8 bps. UUPQ achieves SQNR of 43.90 dB and it requires calculation and storage of L = 219 different values of P_i (see Tab. 1). PUPQ achieves SQNR of 42.80 dB and it requires calculation and storage of one value P. MUPQ achieves SQNR of 43.72 dB and it requires calculation and storage of only K = 7 different values of M_j. We can see that MUPQ achieves SQNR which is very close to SQNR of UUPQ (decreasing of SQNR is only 0.18 dB), while

R [bps]		MUPQ		UUPQ		PUPQ	
	K	$SQNR_{th}$ [dB]	$SQNR_e$ [dB]	$SQNR_{th}$ [dB]	$SQNR_e$ [dB]	$SQNR_{th}$ [dB]	$SQNR_e$ [dB]
6	3	31.65	31.75	32.06	32.21	31.37	31.27
	4	31.79	31.86				
	5	31.87	31.96				
	6	31.89	32.04				
	7	31.90	32.13				
	8	31.95	32.10				
7	3	37.43	37.40	37.95	38.10	37.06	36.83
	4	37.60	37.69				
	5	37.70	37.73				
	6	37.76	37.83				
	7	37.80	37.90				
	8	37.83	37.88				
8	3	43.28	43.28	43.90	44.21	42.80	42.58
	4	43.48	43.65				
	5	43.60	43.80				
	6	43.67	43.90				
	7	43.72	43.94				
	8	43.75	43.93				

Tab. 4. Comparison of MUPQ with UUPQ and PUPQ and experimental results.

MUPQ is much simpler for realization since it requires calculation and storage of drastically smaller number of parameters (7 instead of 219). Therefore, MUPQ is much better solution than UUPQ. On the other hand, MUPQ is slightly more complex than PUPQ (7 parameters are calculated and stored instead of 1) but SQNR of MUPQ is appreciable higher (for 0.92 dB) than SQNR of PUPQ. Therefore, MUPQ is better solution than PUPQ.

One application scenario is also considered, i.e. an experiment is performed applying developed theory on the speech signal. We choose the speech signal since it can be modeled very well with the Gaussian distribution [1]. Experimentally obtained values of SQNR are presented in Tab. 4 ($SQNR_e$). We can see two things. Firstly, experimental results are matched well with the theoretical results, which proves the developed theory. Secondly, experimental results confirm our previous conclusion: SQNR of MUPQ is very close to SQNR of UUPQ; on the other hand, SQNR of MUPQ is appreciable higher than SQNR of PUPQ.

MUPQ can be considered as a generalized uniform polar quantizer, whose special cases are UUPQ (for $K = L$) and PUPQ (for K = 1). The aim of MUPQ is to achieve the best ratio between SQNR and complexity, i.e. to achieve SQNR near to SQNR of UUPQ and to achieve complexity near to complexity of PUPQ.

6. Conclusion

The main goal of this paper is the design of the multi-product uniform polar quantizer (MUPQ), using the asymptotic analysis. It has been known from the literature that the unrestricted uniform polar quantizer (UUPQ) could achieve very good performances, much better than the scalar uniform quantizer, but it is very complex for realization since large number of parameters should be calculated and stored in memory. Therefore, in this paper we proposed the multiproduct uniform polar quantizer which can achieve performances very close to performances of UUPQ, while it is much simpler for realization than UUPQ. It was shown that MUPQ is also better solution than PUPQ. The aim of MUPQ is to achieve the best ratio between SQNR and complexity. MUPQ can be used for any signal with the Gaussian distribution (a lot of real signals belong to this category).

Acknowledgements

This paper is supported by the Serbian Ministry of Science, projects TR-32045 and TR-32035.

References

[1] JAYANT, N. S., NOLL, P. *Digital Coding of Waveforms.* New Jersey: Prentice-Hall, 1984.

[2] GERSHO, A., GRAY, R. M. *Vector Quantization and Signal Compression.* Massachusetts: Kluwer Academic Publishers, 1992.

[3] SWASZEK, P., KU, T. Asymptotic performances of unrestricted polar quantizer. *IEEE Transactions on Information Theory*, 1986, vol. 32, no. 2, p. 330–333. DOI: 10.1109/TIT.1986.1057141

[4] NEUHOFF, D. Polar quantization revisited. In *Proceedings of IEEE International Symposium on Information Theory.* Ulm (Germany), 1997, p. 60. DOI: 10.1109/ISIT.1997.612975

[5] SWASZEK, P. Uniform spherical coordinate quantization of spherically symmetric sources. *IEEE Transactions on Communications*, 1985, vol. 33, no. 6, p. 518–521. DOI: 10.1109/TCOM.1985.1096333

[6] MOO, P. W., NEUHOFF, D. Uniform polar quantization revisited. In *Proceedings of IEEE International Symposium on Information Theory ISIT'98.* Cambridge (USA), 1998. DOI: 10.1109/ISIT.1998.708687

[7] HAMKINS, J., ZEGER, K. Gaussian source coding with spherical codes. *IEEE Transactions on Information Theory*, 2002, vol. 48, no. 11, p. 2980–2989. DOI: 10.1109/TIT.2002.804056

[8] RAVELLI, E., DAUDET, L. Embedded polar quantization. *IEEE Signal Processing Letters*, 2007, vol. 14, no. 10, p. 657–660. DOI: 10.1109/LSP.2007.896379

[9] MATSCHKAL, B., HUBER, J. B. Spherical logarithmic quantization. *IEEE Transactions on Audio, Speech and Language Processing*, 2010, vol. 18, no. 1, p. 126–140. DOI: 10.1109/TASL.2009.2024383

[10] DINCIC, M., PERIC, Z., PETKOVIC, M., DENIC, D. Design of product polar quantizers for A/D conversion of measurement signals with Gaussian distribution. *Measurement*, 2013, vol. 46, no. 8, p. 2441–2446.

[11] DINCIC, M., PERIC, Z. Log-polar quantizer with embedded the ITU-T G.711 codec. *Radioengineering*, 2010, vol. 19, no. 4, p. 712–717.

[12] PERIC, Z., DINCIC, M., PETKOVIC, M. The general design of asymptotic unrestricted polar quantizers with square cells. *Digital Signal Processing*, 2013, vol. 23, no. 5, p. 1731–1737. DOI: 10.1016/j.dsp.2013.06.001

[13] PERIC, Z., STEFANOVIC, M. Asymptotic analysis of optimal uniform polar quantization. *International Journal of Electronics and Communications AEU*, 2002, vol. 56, no. 5, p. 345–347. DOI:10.1078/1434-8411-54100111

[14] STOJANOVIC, D., ALEKSIC, D., PERIC, Z., JOVANOVIC, A. Analysis of uniform polar quantization over Bennett's integral. *Elektronika ir elektrotechnika*, 2005, vol. 8(64), p. 5–9.

[15] JOVANOVIC, A., PERIC, Z. Geometric piecewise uniform lattice vector quantization of the memoryless Gaussian source. *Information Sciences*, 2011, vol. 181, no. 14, p. 3043–3053. DOI: 10.1016/j.ins.2011.03.012

About the Authors ...

Milan DINCIC was born in Nis, Serbia, in 1983. He received the B. S. degree from the Faculty of Electronic Engineering, Nis, Serbia, in 2007, and Ph.D. degree from the University of Nis, in 2012. His current research interests include the information theory, signal processing and compression, scalar and vector quantization. He is an author of about 20 papers in digital communications.

Zoran PERIC was born in Nis, Serbia, in 1964. He received the B. Sc., M. Sc. and Ph. D. from the Faculty of Electronic Engineering, University of Nis, Serbia, in 1989, 1994 and 1999, respectively. He is currently a Full Professor at the Department of Telecommunications and vicedean of the Faculty of Electronic Engineering Nis. His current research interests include the information theory, source and channel coding and signal processing. He is particularly working on scalar and vector quantization techniques in speech and image coding. He is an author and coauthor of over 170 papers in digital communications. Dr Peric has been a Reviewer for IEEE Transactions on Information Theory, IEEE Transactions on Signal Processing, Informatica, Information Technology and Control, Electronics and Electrical Engineering and The International Journal for Computation and Mathematics in Electrical Engineering (COMPEL).

30

Copyright Protection of Color Imaging Using Robust-Encoded Watermarking

Manuel CEDILLO-HERNANDEZ [1], Antonio CEDILLO-HERNANDEZ [1], Francisco GARCIA-UGALDE[1], Mariko NAKANO-MIYATAKE[2], Hector PEREZ-MEANA[2]

[1]Electric Engineering Division, Engineering Faculty, National Autonomous University of Mexico, Circuito Exterior, Ciudad Universitaria, Coyoacan 04510, Mexico City, Mexico
[2] Postgraduate Section, Mechanical Electrical Engineering School, National Polytechnic Institute of Mexico, 1000 Santa Ana Avenue, San Francisco Culhuacan, Coyoacan 04430, Mexico City, Mexico

mcedillohdz@hotmail.com, antoniochz@hotmail.com, fgarciau@unam.mx, mariko@infinitum.com.mx , hmpm@prodigy.net.mx

Abstract. *In this paper we present a robust-encoded watermarking method applied to color images for copyright protection, which presents robustness against several geometric and signal processing distortions. Trade-off between payload, robustness and imperceptibility is a very important aspect which has to be considered when a watermark algorithm is designed. In our proposed scheme, previously to be embedded into the image, the watermark signal is encoded using a convolutional encoder, which can perform forward error correction achieving better robustness performance. Then, the embedding process is carried out through the discrete cosine transform domain (DCT) of an image using the image normalization technique to accomplish robustness against geometric and signal processing distortions. The embedded watermark coded bits are extracted and decoded using the Viterbi algorithm. In order to determine the presence or absence of the watermark into the image we compute the bit error rate (BER) between the recovered and the original watermark data sequence. The quality of the watermarked image is measured using the well-known indices: Peak Signal to Noise Ratio (PSNR), Visual Information Fidelity (VIF) and Structural Similarity Index (SSIM). The color difference between the watermarked and original images is obtained by using the Normalized Color Difference (NCD) measure. The experimental results show that the proposed method provides good performance in terms of imperceptibility and robustness. The comparison among the proposed and previously reported methods based on different techniques is also provided.*

Keywords

Digital watermarking, image normalization, geometric and signal processing operations convolutional encoder, Viterbi decoder, discrete cosine transform

1. Introduction

During the last decades, demand for multimedia data such as digital image, video and audio technologies, has increased dramatically due to its wide use within home computers, mobile devices and open networks. However the growth of digital multimedia has several advantages, it has created problems related to the preservation of copyright. Illegal media contents have the same visual quality as the original ones and they are very easy to reproduce and distribute. Digital watermarking is considered as a suitable solution for copyright protection of digital multimedia. In not visible digital watermarking, a short message called "watermark" is embedded into an image, audio or video signal without affecting the perceptive quality such that it can be detected using a detection algorithm. In public digital image watermarking methods, the synchronization loss, between the embedding and detection stages, causes watermark detection errors. Geometric distortions are frequently the principal factors of this problem.

Different techniques have been proposed in the literature to address the problem related to geometrically resilient image watermarking [1–7]. However, almost all the algorithms [1–7] are designed to grayscale images and their application to color images is often inadequate since they usually work with an individual color channel [8]. The use of color in image and video processing systems has become a key element in the recent years within security, steganography, and watermarking applications of multimedia data [9]. Unlike other applications, in the image watermarking field, the color cannot be considered as a simple RGB color model decomposition and the whole intrinsic information must be integrated into the embedding and detection processes. While several methods have been proposed to watermark grayscale images, only a few have been specifically designed for color images [8]. In the recent years, several color image watermarking methods have been proposed in the literature, these methods could be classified as those which are based on the frequency transform domain [10], pixel modification in the spatial domain [11], histogram modification [12] and image normalization [13]. Almost all the methods mentioned above have shown watermark robustness to geometrical distortion; however they cannot provide enough robustness against common signal processing, such as filtering, noise contamination and image compression neither some aggressive combinations of geometric attacks and common signal processing operations. In order to increase the watermark robustness, an effective technique consists in developing hybrid algorithms which combine the use of a frequency domain in conjunction with an image normalization process based on geometric moments, introduced by [14]. Image normalization is based on a concept that the normalized version obtained from the original image and their geometric distorted versions are exactly the same. Based on this concept, the watermark embedding and detection processes may be carried out in the normalized version of the image, and thereby; theoretically the problem of synchronization loss between both processes not occurs. However in practice, because the image normalization process contains interpolation; the visual quality distortion in the watermarked image is significant [15], [16].

To solve this inconvenience, authors in [5] proposed two image watermarking schemes for copyright protection purposes, the first one is based on image normalization and blind detection, meanwhile, the second one is based on an elastic graph and non-blind detection. Thus, in the first method proposed in [5], to avoid the visual quality distortion of the watermarked image, instead of the original image the watermark signal is normalized and is embedded in the spatial domain. This algorithm gets robustness against several geometric and signal processing distortions; however, obtained bit error rate (BER) is not quite low in the detection stage.

In order to obtain a lower BER value, watermark strength must be increased and, as consequence, quality distortion over the watermarked image cannot be avoided. On the other hand, together with the image normalization procedure, in order to obtain geometric invariance the authors in [13] proposed a robust color watermarking method based on support vector regression (SVR) and nonsubsampled contourlet transform (NSCT). Here, color image is decomposed in three RGB color model components and a significant region is obtained from the normalized components using the invariant centroid theory. Then, the NSCT is performed on the G channel of the significant region. Finally, the watermark is embedded into the original color image by modifying the low frequency NSCT coefficients, in which a human visual system (HVS) masking is used to control the watermark embedding strength. According to the high correlation among different channels of the color image, digital watermark can be recovered using the SVR technique. This algorithm presents robustness against several geometric and signal processing distortions, including cropping attacks; however, the method presents an important drawback: high computation time is needed for the SVR training, the computation of the NSCT as well as the image normalization process.

In this context, our proposal presents a robust-encoded watermarking method applied to color images for copyright protection. The scope is oriented in the same way as the related works presented in [5] and [13], where using the image normalization procedure together within a frequency domain. Nevertheless, our proposed algorithm presents significant improvements compared with them; with respect to [5]: a) the improvement of watermark imperceptibility in terms of PSNR metric, without affecting the robustness of the method, b) the design of a robust-encoded watermarking scheme based on the DCT domain; preserving the robustness against JPEG compression obtained in [5] and at the same time improving the robustness against aggressive geometric attacks such as cropping image, and c) the application of the proposed method to color images. On the other hand, with regard to [13]: a) the proposed method can save high computational time by providing a simple design which gets a better imperceptibility and robustness performance.

The rest of this paper is organized as follows. Sec. 2 shows the basic theory about the image normalization procedure. Sec. 3 describes the embedding and detection process of the proposed algorithm. Experimental results including comparison with the previously reported watermarking algorithms are presented in Sec. 4. Finally, Sec. 5 concludes this work.

2. Background

In the proposed algorithm, original color image and the watermark pattern are normalized in order to develop a watermarking scheme robust against geometric and signal processing distortions. In this section, a description about the image normalization technique is provided, which is based on the invariant moment theory proposed in [14]. Here we describe in detail the procedures of invariant moment calculi and image normalization procedure.

2.1 Invariant Moments

The geometric moments $m_{p,q}$ and the central moments $\mu_{p,q}$ of a given image $f(x,y)$ are defined by (1) and (2) respectively.

$$m_{pq} = \sum_{x=0}^{M-1} \sum_{y=0}^{N-1} x^p y^q f(x,y), \quad (1)$$

$$\mu_{pq} = \sum_{x=0}^{M-1} \sum_{y=0}^{N-1} (x-\bar{x})^p (y-\bar{y})^q f(x,y), \quad (2)$$

where p, q = 1,2,…., and $(\bar{x},\bar{y})$ is the centroid of the image, which is obtained as follows:

$$\bar{x} = \frac{m_{10}}{m_{00}}, \bar{y} = \frac{m_{01}}{m_{00}}. \quad (3)$$

2.2 Image Normalization Procedure

The image normalization procedure of a given image $f(x,y)$ consists of the following stages:

1) Translate the original image $f(x,y)$ with the values (d_x,d_y) in order to generate a new image $f_1(x,y) = f(x_a,y_a)$, in which the center is equal to the centroid of $f(x,y)$. This operation is given by (4) using an affine transformation.

$$\begin{pmatrix} x_a \\ y_a \end{pmatrix} = \mathbf{A} \cdot \begin{pmatrix} x \\ y \end{pmatrix} - \mathbf{d}, \quad (4)$$

where:

$$A = \begin{pmatrix} 1 & 0 \\ 0 & 1 \end{pmatrix}, \; d = \begin{pmatrix} d_x \\ d_y \end{pmatrix}, \quad (5)$$

whose elements are:

$$d_x = \frac{m_{10}}{m_{00}}, d_y = \frac{m_{01}}{m_{00}}. \quad (6)$$

This procedure normalizes an image with respect to the translation.

2) Apply a shearing transformation in x-direction to the image $f_1(x,y)$ using:

$$A_x = \begin{pmatrix} 1 & \beta \\ 0 & 1 \end{pmatrix}. \quad (7)$$

The resultant image is denoted as $f_2(x,y)$, which is normalized against the effect caused by the shearing in the x-direction.

3) Apply a shearing transformation in y-direction to the image $f_2(x,y)$ using:

$$A_y = \begin{pmatrix} 1 & 0 \\ \gamma & 1 \end{pmatrix}. \quad (8)$$

The resultant image is denoted as $f_3(x,y)$, which is normalized against the effect caused by the shearing in the y-direction.

4) Change the size of the image $f_3(x,y)$ in both directions, using:

$$A_s = \begin{pmatrix} \alpha & 0 \\ 0 & \delta \end{pmatrix}. \quad (9)$$

The resultant image is denoted as $f_4(x,y)$, which is normalized against the effect caused by the scaling in both directions.

5) Rotate the image $f_4(x,y)$ using:

$$A_r = \begin{pmatrix} \cos(\phi) & \sin(\phi) \\ -\sin(\phi) & \cos(\phi) \end{pmatrix}. \quad (10)$$

The resultant image is denoted as $f_5(x,y)$, which is normalized against the effect caused by the rotation. Finally, the image $f_5(x,y)$ is the normalized version of the image obtained from the original one $f(x,y)$. It is noteworthy that the above normalization procedures are totally reversible, applying the inverse matrices A_r, A_s, A_y, A_x and adding the vector d to the normalized image $f_5(x,y)$, getting the original image $f(x,y)$. However, for a discrete implementation of both processes, the normalization and the inverse procedures require of an interpolation, which causes visual quality degradation of the image. The determination of the parameters α, β, ϕ, γ and δ is given by the following theory [14].

If the image $g(x,y)$ is a transformed image of $f(x,y)$, applying a general affine transformation:

$$A = \begin{bmatrix} a_{11} & a_{12} \\ a_{21} & a_{22} \end{bmatrix}, \; d = 0 \quad (11)$$

$g(x,y)$ is given by $g(x,y) = A \cdot f(x,y)$. The invariant moments of image $g(x,y)$, $m'_{p,q}$ are given by:

$$m'_{pq} = \sum_{x=0}^{M-1}\sum_{y=0}^{N-1} x^p y^q g(x,y) = \sum_{x=0}^{M-1}\sum_{y=0}^{N-1} A \cdot x^p y^q f(x,y). \quad (12)$$

Applying the general affine transformation A in a coordinate (x,y), the transform coordinate (x',y') is obtained by:

$$\begin{aligned} x' &= a_{11}x + a_{12}y \\ y' &= a_{21}x + a_{22}y \end{aligned} \quad (13)$$

Replacing the coordinates (x', y') in (12), we obtain:

$$m'_{pq} = \sum_{x=0}^{M-1}\sum_{y=0}^{N-1} (a_{11}x + a_{12}y)^p (a_{21}x + a_{22}y)^q f(x,y). \quad (14)$$

Applying the binomial theorem in (14), we have:

$$\begin{aligned} (a_{11}x + a_{12}y)^p &= \sum_{i=0}^{p} \binom{p}{i} a_{11}^i x^i a_{12}^{p-i} y^{p-i} \\ &= \sum_{i=0}^{p} \binom{p}{i} a_{11}^i a_{12}^{p-1} x^i y^{p-i} \end{aligned} \quad (15)$$

$$\begin{aligned} (a_{12}x + a_{22}y)^q &= \sum_{j=0}^{q} \binom{q}{j} a_{21}^j x^j a_{22}^{q-j} y^{q-i} \\ &= \sum_{j=0}^{q} \binom{q}{j} a_{21}^j a_{22}^{q-1} x^j y^{q-j} \end{aligned} \quad (16)$$

Replacing (15) and (16) in (14), we obtain:

$$\begin{aligned} m'_{pq} &= \sum_{x=0}^{M-1}\sum_{y=0}^{N-1}\sum_{i=0}^{p} \binom{p}{i} a_{11}^i a_{12}^{p-1} x^i y^{p-i} \sum_{j=0}^{q} \binom{q}{j} a_{21}^j a_{22}^{q-1} x^j y^{q-j} f(x,y) \\ &= \sum_{x=0}^{M-1}\sum_{y=0}^{N-1}\sum_{i=0}^{p}\sum_{j=0}^{q} \binom{p}{i}\binom{q}{j} a_{11}^i a_{12}^{p-1} x^i y^{p-i} a_{21}^j a_{22}^{q-1} x^j y^{q-j} f(x,y) \\ &= \sum_{x=0}^{M-1}\sum_{y=0}^{N-1}\sum_{i=0}^{p}\sum_{j=0}^{q} \binom{p}{i}\binom{q}{j} a_{11}^i a_{12}^{p-1} a_{21}^j a_{22}^{q-1} x^{i+j} y^{p+q-i-j} f(x,y) \\ &= \sum_{i=0}^{p}\sum_{j=0}^{q} \binom{p}{i}\binom{q}{j} a_{11}^i a_{12}^{p-1} a_{21}^j a_{22}^{q-1} \sum_{x=0}^{M-1}\sum_{y=0}^{N-1} x^{i+j} y^{p+q-i-j} f(x,y) \end{aligned} \quad (17)$$

Using the definition in (1), we obtain:

$$m_{i+j,p+q-i-j} = \sum_{x=0}^{M-1}\sum_{y=0}^{N-1} x^{i+j} y^{p+q-i-j} f(x,y). \quad (18)$$

Finally, replacing (17) in (18), we obtain:

$$m'_{p,q} = \sum_{i=0}^{p}\sum_{j=0}^{q}\binom{p}{i}\binom{q}{j} a_{11}^{i} a_{12}^{p-i} a_{21}^{j} a_{22}^{q-j} m_{i+j,p+q-i-j} \quad (19)$$

In the same manner, we obtain $\mu'_{p,q}$ as follows:

$$\mu'_{p,q} = \sum_{i=0}^{p}\sum_{j=0}^{q}\binom{p}{i}\binom{q}{j} a_{11}^{i} a_{12}^{p-i} a_{21}^{j} a_{22}^{q-j} \mu_{i+j,p+q-i-j}. \quad (20)$$

Using this relationship between the invariant moments of the normalized and the original image, we estimate the parameters α, β, ϕ, γ, δ and the normalization matrices A_r, A_s, A_y, A_x. Interested readers can refer to [5] for more details. Fig. 1 shows an example of image normalization and the visual quality degradation of the reconstructed image produced by the interpolation used in the inverse procedure, where (c) is the normalized version of both images (a) original and (b) geometric distorted, (d) is the image reconstructed after the inverse procedure. For illustrative purposes, partial content obtained from the blue circular patch is zoomed and shown within a square at the left down corner of the images (a) original and (d) reconstructed, in order to perceive the visual degradation with more detail.

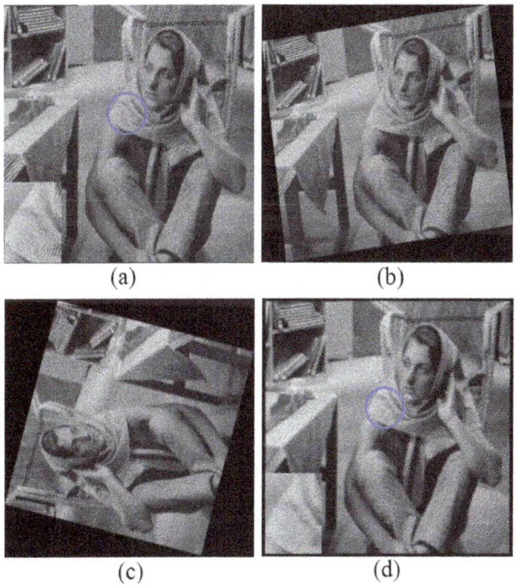

Fig. 1. (a) Original image. (b) Geometric distorted version of (a). (c) Normalized version of (a) and (b). (d) Reconstructed image after inverse normalization procedure.

3. Proposed Algorithm

The robust-encoded proposed watermarking method has been designed in order to get the following requirements: a) Blind detection, which means the original image is not needed during the detection process of the watermark algorithm; b) Robustness against common signal processing operations such as median and Gaussian filtering, as well as JPEG compression, impulsive and Gaussian noise contamination, among others; c) Robustness against geometric attacks such as rotation, scaling, aspect ratio, affine transformations, cropping, among others; d) Robustness against combined distortions composed by common signal processing operations together with geometric attacks; e) Improved watermark imperceptibility; and f) Oriented to color images applications.

3.1 Color Model Determination

One of the major issues in color image processing is to find the appropriated color model for the problem being addressed [17]. Whereas the application context often defines the original color model, the color model used for embedding a watermark has to be discussed according to the objective of the watermark scheme, i.e., fragile, semi-fragile or robust scheme. According to [8], the RGB color model has the most correlated components while the YCbCr color model components are less correlated. Also, the forward and backward transformations between RGB and YCbCr color models are linear. Considering this fact, if the correlated color model such as RGB is used, the modification of one component independently to the others is not the best choice, because the perceived colors are dependent of the three components together. This is the reason why the RGB model is called a correlated color model. On the other hand, YCbCr model allows to obtain non-correlated components and has the advantage of separating the luminance information from the chrominance information [8], [17]. According to this, YCbCr model is adopted as suitable color model in the watermarking method proposed.

3.2 Embedding Process

The embedding process is described as follows: **1)** Convert the RGB color model of the original image I to the YCbCr color model representation. Separate the luminance component (Y) from the YCbCr color model representation. **2)** Get a normalized version of the luminance component (Y) using the normalization procedure described in Sec. 2.2. **3)** Build a binary pseudo-random prototype pattern generated by any secret key, whose dimensions are the same as the original color image. **4)** The watermark is a zero mean 1-D binary pseudo-random pattern composed by {1, 0} values generated by a user's secret key k_1, $W = \{w_n| \ n=1, \ldots, L\}$, where L is the length of the watermark. **5)** Encode the watermark W using the convolutional code, proposed in [18] with the following characteristics: code rate = 1/3, constraint length $K = 5$, free distance $d_f = 12$ and the generator polynomials $g_1 = [11111]$, $g_2 = [11011]$ and $g_3 = [10101]$. The encoder output is then $E = \{e_1, e_2, \ldots, e_h\}$. In a digital communication system, the convolutional encoding adds redundancy to the information bits [18]. These additional redundant bits are used at the receiver to perform a forward error correction (FEC) with the purpose of achieving a lower bit error rate. In the case of the watermark system, the convolutional encoding increases the robustness of the watermark against attacks. Fig. 2 shows the convolutional encoder used in the proposed method.

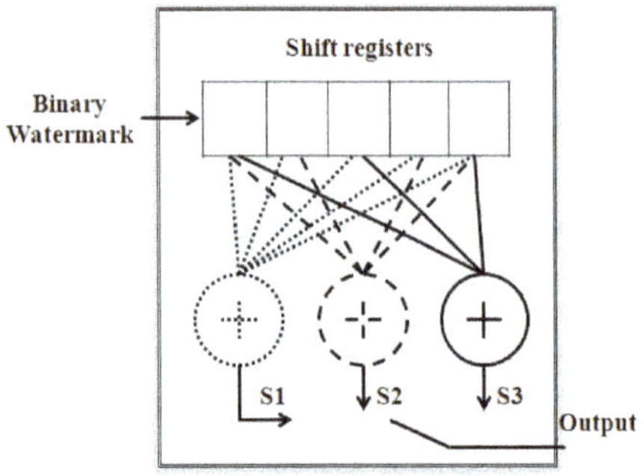

Fig. 2. Convolutional encoder.

6) Once the prototype pattern is generated, it is masked using the normalized version of the luminance component (Y) (obtained in the step 2 of this procedure) as a template. The template used in this masking operation is a binary pattern composed by {1, 0} values and their implementation considers to assign the value '1' in places that contain information of the luminance component (Y) normalized (i.e. white area of the "Template") and preserve the value '0' on the area outside the support of the normalized version (i.e. black area of the "Template"). This masking ensures that the watermark data bits are embedded only into information of the original image to be protected. On the other hand, the information of the area outside the support of the normalized version will be discarded by the inverse normalization procedure when is applied, as is illustrated in Fig. 1 of this paper. Later, it defines a region of interesting (ROI) using the centroid resulting from the normalization procedure. It divides this ROI in blocks of 8x8 pixels. The ROI's dimensions and the amount of blocks are in direct relation of the convolutional encoder output E. In this way, we determinate the watermark length experimentally as $L = 108$, in consequence, the amount of elements of E is $h = 399$ (including the zero padding needed to put the convolutional decoder at the final nul state), thereby, in the proposed method, 399 blocks of 8x8 pixels are needed for the watermark embedding. **7)** Once the prototype pattern is masked and the ROI together with 8x8 pixel blocks are defined, the watermark embedding process is performed as follows. In order to preserve the trade-off between imperceptibility and robustness, firstly, the encoder output E is multiplied by a first watermark strength factor α_1, $E = \{e_1 \cdot \alpha_1, e_2 \cdot \alpha_1, \ldots, e_h \cdot \alpha_1\}$. Later, each 8x8 block of the ROI is processed as shown in Fig. 3, this process is described as follows: a) replace the DC coefficient by its corresponding value of the encoder output E and then b) apply the inverse discrete cosine transform (IDCT).

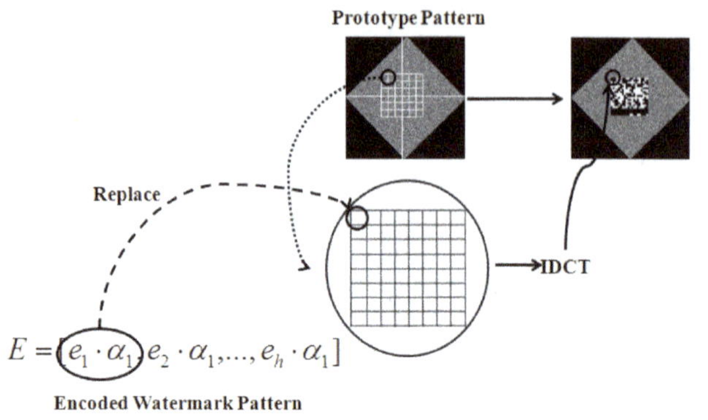

Fig. 3. Watermark embedding stage.

8) Apply the inverse image normalization procedure to the prototype pattern which contains the encoded watermark. The final prototype pattern is denoted as P. **9)** In order to preserve high watermark imperceptibility, the pattern P is multiplied by a second watermark strength factor α_2. Finally, the pattern P is embedded into the original image using an additive insertion in the spatial domain, given by (21):

$$Y_w = Y_o + P \cdot \alpha_2 , \tag{21}$$

where Y_w and Y_o are the watermarked and the original luminance components respectively. Restore the watermarked luminance component Y_w in conjunction with the chrominance components $CbCr$ to RGB color model representation and obtain the final watermarked color image I_w. The general diagram of the embedding process is shown in Fig. 4. The above procedure is equivalent to embedding the

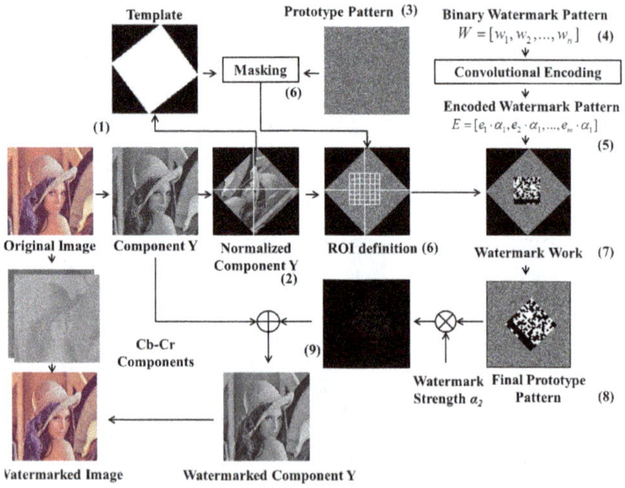

Fig. 4. General diagram of embedding process.

encoded watermark pattern into the DCT domain of the normalized image. It is important to note that within the proposed algorithm we have chosen to transform the prototype pattern which contains the watermark signal in order to fit the cover image instead of embedding directly the watermark into the normalized image. The above characteristic represents some advantages for the proposed algorithm. Firstly, it avoids any perceptual quality distortion which might otherwise have incurred to the color image.

Additionally, another advantage consists in discarding the information within the area outside the support of the prototype pattern obtained in the step 6 (i.e. black area). This can save computational time and provide an easier implementation. Because every DC coefficient replacement is carried out using only prototype pattern information, we can avoid any visual quality perceptual distortion into the color image controlling the watermark strength factors and at the same time improving the robustness against JPEG compression.

3.3 Detection Process

The detection process is described as follows:

1) Convert the watermarked color image I_w from RGB to YCbCr color model representation. Separate the watermarked luminance component (Y_w) from the YCbCr color model representation.

2) Get the normalized version from the Y_w component using the normalization procedure presented in Sec. 2.2. Later, reconstruct the region of interesting (ROI) using the centroid resulting from the normalization procedure and divide this in blocks of 8x8 pixels.

3) Apply the DCT transform to each block and extract the first coefficient (DC coefficient) from each DCT block. Concatenate these coefficients into a vector $\mathbf{V} = (v_1, v_2, \ldots, v_h)$.

4) The encoded watermarked coefficients $\hat{E}$ are recovered performing the following operation to the elements of the vector V:

$$\hat{E} = [\hat{e}_1, \hat{e}_2, \ldots, \hat{e}_{h-1}, \hat{e}_h] \tag{22}$$

where $\hat{e}_{k\text{-}th} = \text{sign}(v_{k\text{-}th})$, and sign is the sign function.

5) The decoding is performed using the Viterbi algorithm with hard decision [19], thus obtaining the retrieved watermark pattern W'.

6) Once that W' was recovered, compute the BER between the original W and the recovered W' watermarks. Assuming ergodicity the BER is defined as the ratio between the number of incorrectly decoded bits and the total number of embedded bits. A threshold value T_{BER} must be defined to determine if the watermark W is present or not into the image. In this concern, considering a binomial distribution with success probability equal to 0.5, the false alarm probability P_{fa} for L bits embedded watermark data is given by (23), and a threshold value T must be controlled in order to get a smaller value of P_{fa} than a predetermined one.

$$P_{fa} = \sum_{z=T}^{L} \left(\frac{1}{2}\right)^{L} \cdot \left(\frac{L!}{z!(L-z)!}\right) \tag{23}$$

where L is the total number of watermark data bits, whose value is empirically set to 108 and, based on the Bernoulli trials assumption, z is an independent random variable with binomial distribution [20]. The false alarm probability must be less than $P_{fa} = 4.9516 \times 10^{-11}$ which is set to satisfy the requirements of most watermarking applications for a reliable detection [20], and then an adequate threshold value T_{BER} ($=1-(T/L)=1-(87/108)$) is equal to 0.20, according to the fact that the bit error rate (BER) + the bit correct rate (BCR) must be equal to 1. If the BER value between W and W' is greater than 20% (more than 21 percent error bits), the watermark detection is failed, otherwise, the watermark detection is successful and the detection process is terminated. The general diagram of the detection procedure is shown Fig. 5.

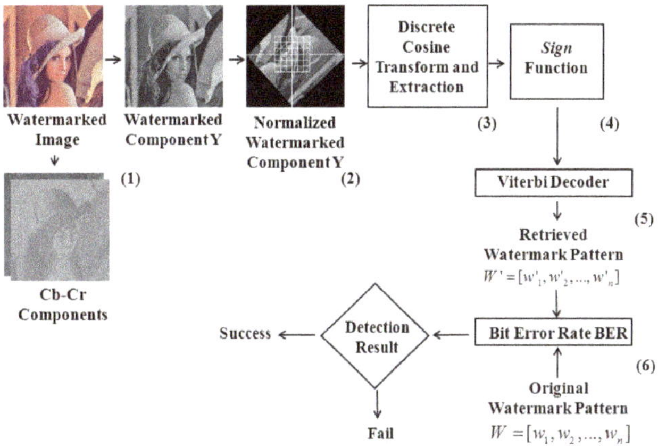

Fig. 5. General diagram of detection process.

4. Experimental Results

In this section, the performance of the proposed algorithm is evaluated considering watermark imperceptibility and robustness grades and using a variety of well-known digital images in the literature. We have used different images with a variety of texture content (e.g., Goldhill, Sailboat, Lena, Airplane, Baboon, Peppers, among others) of 512 × 512 pixels size and color resolution of 24 bits per pixel which can be found in the following academic database: http://sipi.usc.edu/database/ and some of them in the Matlab © Integrated Development Environment (IDE). Our experiments are carried out on a personal computer running Microsoft Windows 7 OS© with an Intel© Core i7 processor (@3.4 GHz) and 8 GB RAM memory while the embedding and extracting procedures were implemented on Matlab© version 7.10. In these conditions, the average computing time for the embedding procedure has been measured in 6.22 seconds while an average of 5.46 seconds was needed for the detection procedure. A 1D binary pseudorandom sequence of size $L = 108$ bits is used as the watermark pattern W. The false alarm probability is set to $P_{fa} = 4.9516 \times 10^{-11}$ when $T_{BER} = 0.20$. The convolutional code described in the embedding process has been used. The watermark strength parameters used in the watermark embedding process has been determined to be $\alpha_1 = 35$ and $\alpha_2 = 0.003$, whose setting procedure will be explained in the next section. The watermarked image visual quality distortion is measured using the following well known indices: Peak Signal to Noise Ratio (PSNR), Visual Information Fidelity (VIF) and Structural Similarity Index (SSIM). Finally, our experimental results are compared with some of the most important methods reported previously in the literature whose aim is to be robust against most common geometric and signal processing attacks.

4.1 Watermark Imperceptibility and Setting of Watermark Strengths α_1, and α_2

Using a watermark strength factor $\alpha_2 = 0.003$, and the parameters mentioned above, a watermark length $L = 108$, and a variable watermark strength value α_1 from 5 to 55; the watermark imperceptibility was evaluated in terms of the PSNR, VIF [21] and SSIM [22] visual image quality metrics defined by (24), (25) and (26) respectively.

$$PSNR(dB) = 10\log_{10}\left(\frac{MaxPixelValue^2}{(MSE_Y + MSE_{Cb} + MSE_{Cr})/3}\right). \quad (24)$$

$$VIF = \frac{\sum_{\omega \in channels} I(\vec{C}^{Z,\omega};\vec{G}^{Z,\omega} \mid s^{Z,\omega})}{\sum_{\omega \in channels} I(\vec{C}^{Z,\omega};\vec{E}^{Z,\omega} \mid s^{Z,\omega})}. \quad (25)$$

In (25) we sum over the channels of interest, where $\vec{C}^{Z,\omega}$ represent Z elements of the random field RF C_ω which describes the coefficients from channel ω, and so on [21]. E and G denote the visual signal at the output of the Human Visual System model (HVS) from the original and the watermarked images respectively, from which the human brain extracts cognitive information. The terms $I(\vec{C}^{Z,\omega};\vec{E}^{Z,\omega} \mid s^{Z,\omega})$ and $I(\vec{C}^{Z,\omega};\vec{G}^{Z,\omega} \mid s^{Z,\omega})$ represent the information that can ideally be extracted by the brain from a particular channel in the original and the watermarked images respectively [21].

$$SSIM(I_o, I_w) = \frac{(2\mu_{I_o}\mu_{I_w} + C_1)(2\sigma_{I_o I_w} + C_2)}{(\mu_{I_o}^2 + \mu_{I_w}^2 + C_1)(\sigma_{I_o}^2 + \sigma_{I_w}^2 + C_2)}. \quad (26)$$

In (26) I_o, and I_w are the original and watermarked images respectively and C_1, C_2 are small constant values [22]. As it is known in the literature the VIF value reflects perceptual distortions more precisely than PSNR. The range of VIF is [0, 1] and the closer value to 1 represents the best fidelity respect to the original image. It is also well known in the literature that the SSIM value reflects perceptual distortions more precisely than PSNR. The range of SSIM is [0, 1], and the closer value to 1 represents an identical quality respect to the original image, i.e. a value of 1 means that the original and the reference images are the same. In Figs. 6 and 7, the PSNR of ten test color images and average VIF-SSIM are plotted versus the variable watermark strength α_1 ranging from 5 to 55 respectively.

A greater value of α_1 could increase the robustness of the watermark, but as shown in Figs. 6 and 7, the watermark imperceptibility is diminished. Hence, there is a trade-off between robustness and imperceptibility. To preserve the trade-off between robustness and imperceptibility, based on our experiments, we considered a watermark strength of $\alpha_1 = 35$ as a suitable value, obtaining the following average values: PSNR = 44.19 dB, VIF = 0.9559 and SSIM = 0.9967.

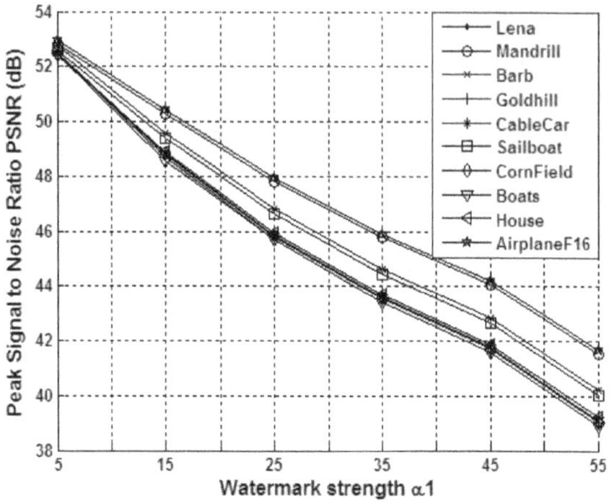

Fig. 6. PSNR obtained with variable watermark strength α_1.

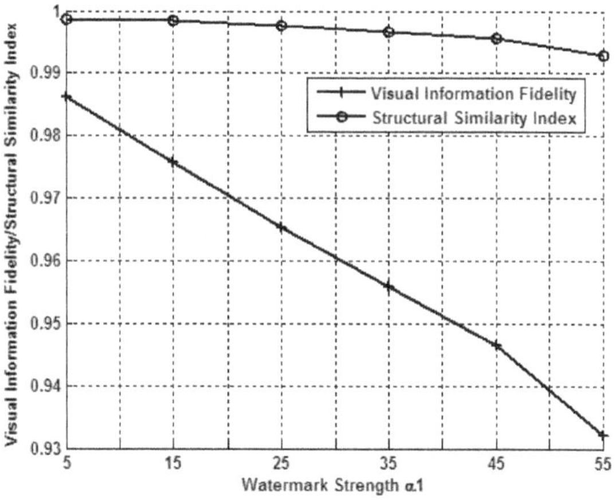

Fig. 7. Average VIF and SSIM values obtained with variable watermark strength α_1.

Using a watermark strength $\alpha_1 = 35$, the parameters mentioned above, a watermark length $L = 108$, and a variable watermark strength α_2 from 0.001 to 0.005; the watermark imperceptibility was evaluated in terms of the PSNR image quality metric and a subjective test based on the Mean Opinion Score (MOS) statistically supported in our experiments supported by the opinion of 100 persons of different academic levels. In Figs. 8 and 9, the PSNR and the average MOS of the ten test color images used are plotted versus the variable watermark strength α_2 respectively.

A bigger value of α_2 could increase the robustness of the watermark, but as shown in Figs. 8 and 9, the watermark imperceptibility is diminished. Hence ones again there is a trade-off between robustness and imperceptibility. To preserve the trade-off between robustness and imperceptibility, based on our experiments, we considered a watermark strength of $\alpha_2 = 0.003$ as a suitable value.

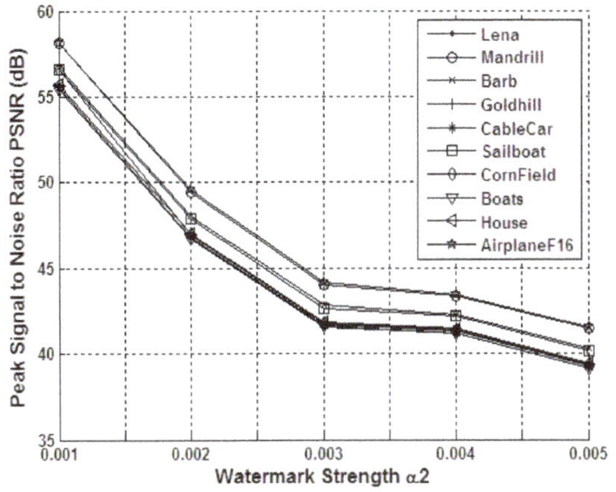

Fig. 8. PSNR obtained with variable watermark strength α_2.

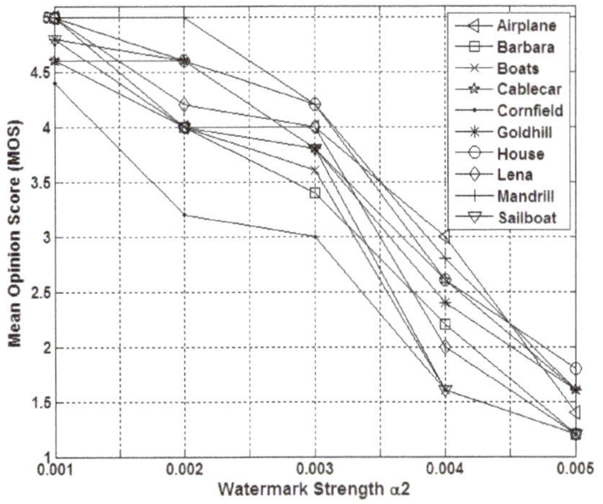

Fig. 9. Average MOS values obtained with variable watermark strength α_2.

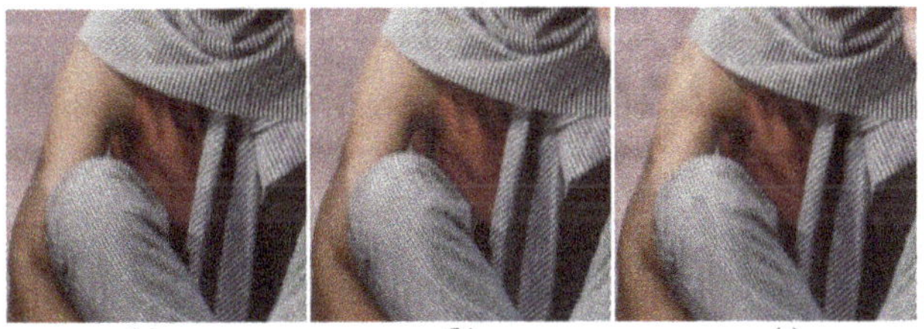

(a) (b) (c)

Fig. 10. a) Central content from Barb original color image, b) Watermarked version with α_1 = 35 and α_2 = 0.003, c) Watermarked version with α_1 = 35 and α_2 = 0.005.

An example of perceptual distortion is shown in Fig. 10, when the watermark strength is set to $\alpha_2 = 0.005$ and $\alpha_1 = 35$. To illustrative purposes, the central content of Barb image is zoomed. Fig. 10(c) shows the distortion into the watermarked color image when $\alpha_1 = 35$.

To conclude the setting of the watermark strengths values, a pair of experiments related with the robustness against signal processing distortions, are performed with the following characteristics: a) using a variable watermark strength α_1 ranging from 5 to 55 and α_2 = 0.003, b) using a variable watermark strength α_2 ranging from 0.001 to 0.005 and $\alpha_1 = 35$; both in conjunction with a watermark length $L = 108$, and a threshold value $T_{BER} = 0.20$. The robustness of the proposed method when ten watermarked

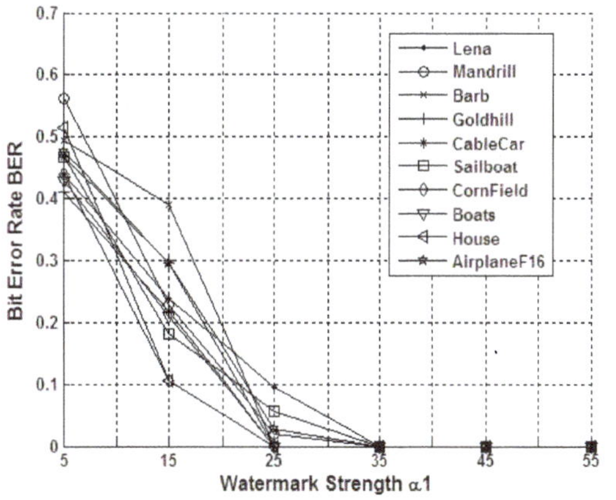

Fig. 11. BER values obtained with variable watermark strength α_1 after contamination by Gaussian noise with $\mu = 0$ and $\sigma^2 = 0.006$.

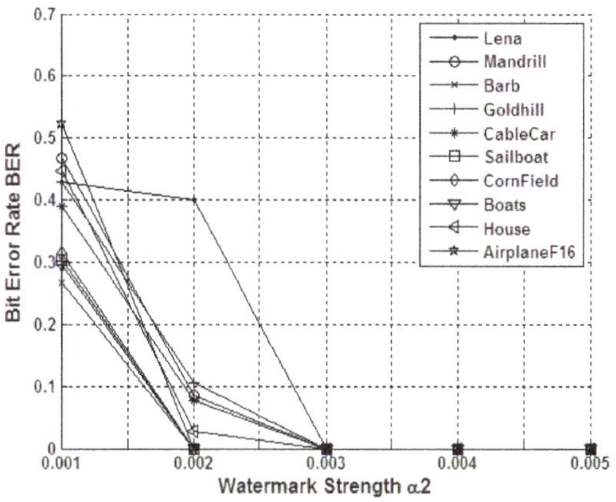

Fig. 12. BER values obtained with variable watermark strength α_2 after contamination by Gaussian noise with $\mu = 0$ and $\sigma^2 = 0.006$.

test color images are contaminated with Gaussian noise with $\mu = 0$ and $\sigma^2 = 0.006$ is shown in Figs. 11 and 12, respectively.

A larger value of α_1 and α_2 increases the robustness of the watermark, but as shown in Figs. 6-9, the watermark imperceptibility is decreased. To preserve the trade-off between robustness and imperceptibility, based on our experiments, we considered a watermark strengths of $\alpha_1 = 35$ and $\alpha_2 = 0.003$ as suitable values. Three test color images (a) with their watermarked version (b) are shown in Figs. 13, 14 and 15 respectively.

(a) (b)

Fig. 13. a) Lena image, b) Watermarked version.

Fig. 14. a) Gold hill image, b) Watermarked version.

Fig. 15. a) Airplane F-16 image, b) Watermarked version.

Algorithms [5] and [13] are two of the most robust watermarking techniques published with similar purposes as our proposed scheme. To get a proper comparison, we consider a homogeneous format of color images of 512×512×24 bits. The imperceptibility performance is compared in our simulations with those reported by algorithms in [5] and [13].

	Proposed method	Method in [5]	Method in [13]
PSNR (dB)	44.19	40.11	41.18

Tab. 1. Average comparative results of imperceptibility using the PSNR (dB) metric.

From Tab. 1 it follows that the proposed scheme provides a fairly good fidelity of the watermarked color image, achieving a PSNR greater than 44 dB, avoiding the visual quality distortions in the color images.

On the other hand, the normalized color difference (NCD) metric [23], [24] is based on the CIELAB color space and it is applied to measure the difference of color between two images. NCD is given by (27) [23], [24]:

$$NCD = \frac{\sum_{x=1}^{N_1}\sum_{y=1}^{N_2}\left(\sqrt{(L_o(x,y)-L_w(x,y))^2+(a_o(x,y)-a_w(x,y))^2+(b_o(x,y)-b_w(x,y))^2}\right)}{\sum_{x=1}^{N_1}\sum_{y=1}^{N_2}\left(\sqrt{(L_o(x,y))^2+(a_o(x,y))^2+(b_o(x,y))^2}\right)}, \quad (27)$$

where L_o, L_w represent lightness values and a_o, a_w, b_o, b_w the chrominance values corresponding to the original I_o and the watermarked I_w images expressed in the CIE Lab color space. Finally, in Tab. 2, we show NCD values calculated between the ten watermarked testing images used and its original versions.

Image	NCD
Lena	0.0053
Mandrill	0.0050
Barbara	0.0057
Gold hill	0.0076
Cable car	0.0091
Sailboat	0.0066
Cornfield	0.0083
Boats	0.0082
House	0.0058
Airplane F-16	0.0035

Tab. 2. NCD values obtained from ten test watermarked images.

From Tab. 2 and Figs. 13, 14 and 15, it can be seen that the proposed scheme provides good imperceptibility since the difference of colors between the watermarked image and the original one is insignificant [25].

4.2 Watermark Robustness

To evaluate the watermark robustness of the proposed algorithm we use the StirMark Benchmark [26], performing attacks of several geometrical distortions, signal processing and a combination of them. In order to carry out a representative comparison in terms of watermark robustness with respect to the related works reported in [5] and [13]; we have used one thousand color images with resolution of 24 bits per pixel of size 512×512.

Experimental results are classified in geometric, common signal processing, and combined distortions. For illustrative purposes, in Tab. 3 we show the average BER obtained in our proposed method and in those schemes reported in [5] and [13], after applying geometric distortions, and combined distortions composed by a JPEG compression with quality factor QF = 20 together with one or more geometric distortions. When the BER value is more than the predefined threshold value T_{BER} = 0.20 (more than 21 error bits), the watermark detection is reported with italics and corresponds to the case where the watermark detection has failed.

Analyzing the results in Tab. 3, we show that our proposed method presents good robustness against several geometric distortions, including scaling from 0.5 to 2, rotation by several angles, general affine transformation, shearing 5% in x and y directions, cropping attack with 50%, aspect ratio in x and y directions by 1.2 and 0.9, respectively, horizontal and vertical flipping and combined attacks all of them with JPEG 20 compression. In all cases we have obtained BER values less than the predefined threshold value T_{BER} = 0.20, except when the image is distorted by the Stirmark Random Bending Attack (RBA).

Geometric and combined distortions	Bit Error Rate (BER)		
	Proposed method	Method in [5]	Method in [13]
Without distortion	0	0	0
Rotation by 5°	0	0	0.03
Rotation by 45°	0.01	0	0.03
Scale 0.5	0.02	0	0.04
Scale 2	0.007	0.04	0.02
Cropping 50 %	0.03	*0.52*	0.12
Affine [0.9,-0.2,0;0.1,1.2,0;0,0,1]	0.01	0	0.02
Affine [-1.1,-0.2,0;-0.2,0.8,0;0,0,1]	0.02	0	0.02
Shearing (5%, 5%)	0.004	0	0.02
Aspect ratio (1.2,0.9)	0.03	0	0.03
Aspect ratio (1.2,1.0)	0	0	0.02
Horizontal flip	0.01	0	0.10
Vertical flip	0	0	0.15
JPEG 20 + Rotation 45 °	0.02	0.03	0.05
JPEG 20 + Scale 1.5	0.03	0.02	0.09
JPEG 20 + Affine [0.9,0.2,0;0.1,1.2,0;0,0,1]	0.02	0.01	0.05
JPEG 20 + Cropping 50%	0.03	*0.53*	0.15
JPEG 20 + Shearing (5%, 5%)	0.01	0.01	0.04
JPEG 20 + Aspect ratio (1.0,1.2)	0.02	0.01	0.03
JPEG 20 + Horizontal flip	0	0	0.12
Stirmark Random Bending Attack	*0.53*	*0.50*	*0.51*

Tab. 3. Average BER obtained from one thousand watermarked images after geometric and combined distortions.

On the other hand, the method proposed in [5] obtains a similar performance against the above mentioned geometric and combined distortions; however, the method is not robust against cropping attacks as well as RBA attack. In both cases, the obtaining BER values are greater than the threshold value $T_{BER} = 0.20$. The weakness against cropping attacks is due to the fact that the watermark is embedded into the mid-range DCT coefficients of the whole image, thus, when the image is cropped from the center or the edges, several DCT blocks are removed and the watermark cannot be recovered adequately. In the case of the method proposed in [13], it obtains a similar performance against the above mentioned geometric and combined distortions, except against RBA attack. Although the performance is similar, the method in [13] is outperformed in detection accuracy when the image is cropped or distorted by vertical and horizontal flipping, obtaining BER values greater than 0.1, whereas that, our proposed method obtains BER values closer up to 0. Additionally, Tab. 4 shows the average BER obtained in our proposed method and the schemes reported in [5] and [13] respectively, after applying common signal processing, and combined attacks all of them with JPEG 20 compression. Similarly, when the BER value is more than the predefined threshold value $T_{BER} = 0.20$ (more than 21 error bits) the watermark detection is reported in with italics and corresponds to the case where the watermark detection has failed.

Common signal processing and combined distortions	Bit Error Rate (BER)		
	Proposed method	Method in [5]	Method in [13]
JPEG compression QF = 50	0.001	0	0.03
JPEG compression QF = 20	0.003	0.004	0.04
Median filter 3x 3	0.001	*0.23*	0.03
Sharpening 3 x 3	0.005	0.064	0.02
Gaussian noise, $\mu = 0$, $\sigma^2 = 0.006$	0	*0.35*	0.02
Impulsive noise, density = 0.01	0.01	*0.21*	0.1
Gaussian filter 3x3	0.009	0	0.03
JPEG 20 + Median filter 3× 3	0	*0.25*	0.06
JPEG 20 + Sharpening 3 × 3	0.01	0.08	0.05
JPEG 20 + Gaussian noise, $\mu = 0$, $\sigma^2 = 0.006$	0.01	*0.37*	0.05
JPEG 20 + Impulsive noise, density = 0.01	0.004	*0.20*	0.15
JPEG 20 + Gaussian filter 3×3	0	0	0.06

Tab. 4. Average BER obtained from one thousand watermarked images after common signal processing, and combined distortions.

Analyzing the results in Tab. 4 we show that the algorithm [13] and ours present good robustness against several common signal processing operations, including JPEG compression with several quality factors ranging from 50 to 20. Also they present robustness against median, Gaussian and sharpen filters, all of them with window size of 3×3, impulsive noise with a density of 0.01 and Gaussian noise contamination with zero mean and variance 0.006. In all cases both algorithms have a BER value less than the predefined normalized threshold value $T_{BER} = 0.20$. Also, we have considered combined attacks composed by a JPEG 20 compression together with the common signal processing mentioned above. The robustness of both methods is not affected by this kind of combined attacks, obtaining BER values less than $T_{BER} = 0.20$. However, the method in [13] is outperformed in detection accuracy when the image is contaminated by impulsive noise with a density of 0.01, obtaining BER values equal to or greater than 0.1, whereas our proposed method obtains BER values closer up to 0. On the other hand, the method proposed in [5] obtains a similar performance against the above mentioned common signal processing and combined distortions; however, the method is not robust against median filtering, impulsive and Gaussian noise contamination, obtaining average BER values greater than $T_{BER} = 0.20$, i.e., 0.23, 0.21 and 0.35 respectively.

5. Conclusions

Using the combination of the image normalization process, the convolutional encoding and the robustness of the DCT domain, we have developed a robust encoded watermarking method for copyright protection applied to color images. We combined five key elements in our method: a) the image normalization procedure, b) DCT embedding domain, c) convolutional encoding, d) Viterbi decoder, e) the masking procedure. The watermark imperceptibility of the proposed algorithm was evaluated in terms of three well-known image quality distortion metrics (PSNR, VIF and SSIM), concluding that the visual distortion caused by the proposed watermarking algorithm is imperceptible, providing 44.19 dB, 0.9559 and 0.9967 values respectively. Watermark robustness of the proposed algorithm is evaluated using a wide range of attacks and a fairly good performance is obtained. From the evaluation results, we can conclude that our proposed algorithm outperforms the algorithms proposed in [5] and [13] that are two of the most efficient algorithms recently proposed for image watermarking, in terms of imperceptibility and detection accuracy. Experimental results show that our proposed method is robust against very aggressive attacks, such as scaling, rotation by several angles, general affine transformation, shearing in x and y directions, cropping attack, aspect ratio in x and y directions, horizontal and vertical flipping and combined attacks all of them with JPEG 20 compression. Obtained robustness against several geometric, signal processing and combined attacks, high imperceptibility and the compact design of our proposed method make it to be applied in a wide range of applications. As a future work, the proposed method will be extended to the authentication of digital video.

Acknowledgements

The authors thank the Post-Doctorate Scholarships Program and the PAPIIT IN-112513 project from DGAPA in the National Autonomous University of Mexico (UNAM), and the National Polytechnic Institute of Mexico (IPN) by the support provided during the realization of this research.

References

[1] AGOYI, M., ÇELEBI, E., ANBARJAFARI, G. A watermarking algorithm based on chirp z-transform, discrete wavelet transform, and singular value decomposition. *Signal, Image and Video Processing*, 2014, DOI: 10.1007/s11760-014-0624-9.

[2] TAO, H., CHONGMIN, L., ZAIN, J. M., ABDALLA, A. N. Robust image watermarking theories and techniques: A review. *Journal of Applied Research and Technology (JART)*, 2014, vol. 12, p. 122–138.

[3] ZHANG, Y., ZHOU, Z-H., ZHANG, C., LI, Y. An anti-geometric digital watermark algorithm based on histogram grouping and fault-tolerance channel. *Intelligent Science and Intelligent Data Engineering Lecture Notes in Computer Science*, 2012, vol. 7202, p. 753–760. DOI: 10.1007/978-3-642-31919-8_96

[4] YAN, L., JIYING, Z. A new video watermarking algorithm based on 1D DFT and Radon transform. *Signal Processing*, 2010, vol. 90, p. 626–639. DOI: 10.1016/j.sigpro.2009.08.001

[5] DONG, P., BRANKOV, J. B., GALATSANOS, N. P., YANG, Y., DAVOINE, F. Digital watermarking to geometric distortions. *IEEE Transactions on Image Processing*, 2005, vol. 14, no. 12, p. 2140–2150. DOI: 10.1109/TIP.2005.857263

[6] CEDILLO, M., NAKANO, M., PEREZ, H. A robust watermarking technique based on image normalization. *Rev. Fac. Ing. Univ. Antioquia / Journal of the Engineering Faculty University of Antioquia*, 2010, vol. 52, p. 147–160.

[7] MOUSAVI, S. M., NAGHSH, A., ABU-BAKAR, S. A. R. Watermarking techniques used in medical images: A survey. *Journal of Digital Imaging*, 2014, vol. 27, no. 6, p. 714–729. DOI: 10.1007/s10278-014-9700-5

[8] CHAREYRON, G., DA RUGNA, J., TRÉMEAU, A. Color in image watermarking. In *A. Al-Haj (Ed.), Advanced Techniques in Multimedia Watermarking: Image, Video and Audio Applications*, 2010, p. 36–56. DOI: 10.4018/978-1-61520-903-3.ch003

[9] TRÉMEAU, A., TOMINAGA, S., PLATANIOTIS, K. Color in image and video processing: Most recent trends and future research directions. *EURASIP Journal on Image and Video Processing*, 2008, p. 1–26. DOI:10.1155/2008/581371

[10] MAITY, S. P., KUNDU, M. K. DHT domain digital watermarking with low loss in image informations. *AEU - International Journal of Electronics and Communications*, 2009. DOI: 10.1016/j.aeue.2008.10.004

[11] ZHANG, X., WANG, S. Fragile watermarking scheme using a hierarchical mechanism. *Signal Processing*, 2009, vol. 89, no. 4, p. 675–679. DOI: 10.1016/j.sigpro.2008.10.001

[12] CEDILLO, M., GARCIA, F., NAKANO, M., PEREZ, H. Robust hybrid color image watermarking method based on DFT domain and 2D histogram modification. *Signal Image and Video Processing*, 2013, vol. 8, no. 1, p. 49–63. DOI: 10.1007/s11760-013-0459-9.

[13] PAN-PAN NIU, XIANG-YANG WANG, YI-PING YANG, MING-YU LU. A novel color image watermarking scheme in non sampled contourlet-domain. *Expert Systems with Applications*, 2011, vol. 38, no. 3, p. 2081–2098. DOI: 10.1016/j.eswa.2010.07.147

[14] HU, M.-K. Visual pattern recognition by moment invariants. *IRE Transactions on Information Theory*, 1962, vol. 8, no. 2, p. 179–187. DOI: 10.1109/TIT.1962.1057692

[15] ALGHONIEMY, M., TEWFIK, A. H. Geometric invariance in image watermarking. *IEEE Transactions on Image Processing*, 2004, vol. 13, no. 2, p. 145–153. DOI: 10.1109/TIP.2004.823831

[16] QI SONG, GUANG-XI ZHU, HANG-JIAN LUO. Geometrically robust image watermarking based on image normalization. In *Proceedings of International Symp. on Intelligent Signal Processing and Communication Systems ISPACS 2005*. Hong-Kong (China), 2005, p. 333–336. DOI: 10.1109/ISPACS.2005.1595414

[17] LUKAC, R., PLATANIOTIS, K. *Color Image Processing*. CRC Press, 2007.

[18] SKLAR, B. *Digital Communications: Fundamentals and Applications*. Second ed. System View, 2001.

[19] BATSON, B. H., MOOREHEAD, R. W. *Simulation Results for the Viterbi Decoding Algorithm*. NASA-TR-R-396, Technical report, 1972.

[20] TANG, C. W., HANG, H. M. A feature-based robust digital image watermarking scheme. *IEEE Transactions on Signal Processing*, 2003, vol. 51, no. 4, p. 950–959. DOI: 10.1109/TSP.2003.809367

[21] SHEIKH, H. R., BOVIK, A. C. Image information and visual quality. *IEEE Transactions on Image Processing*, 2006, vol. 15, no. 2, p. 430–444. DOI: 10.1109/TIP.2005.859378

[22] WANG ZHOU, BOVIK, A. C., SHEIKH, H. R., SIMONCELLI, E. P. Image quality assessment: From error measurement to structural similarity. *IEEE Transactions on Image Processing*,

2004, vol. 13, no. 4, p. 600–612. DOI: 10.1109/TIP.2003.819861

[23] CHANG, H. A., CHEN, H. H. Stochastic color interpolation for digital cameras. *IEEE Transactions on Circuits and Systems for Video Technology*, 2007, vol. 17, no. 8, p. 964–973. DOI: 10.1109/TCSVT.2007.897471

[24] PLATANIOTIS, K. N., VENETSANOPOULOS, A. N. *Color Image Processing and Applications*. Berlin: Springer Verlag, 2000.

[25] SAHOO, A., SINGH, M. P. Fuzzy weighted adaptive linear filter for color image restoration using morphological detectors. *International Journal on Computer Science and Engineering*, 2009, vol. 1, no. 3, p. 217–221.

[26] PETITCOLAS, F. A. P. Watermarking schemes evaluation. *IEEE Signal Processing*, 2000, vol. 17, no. 5, p. 58–64. DOI: 10.1109/79.879339

About the Authors ...

Manuel CEDILLO-HERNANDEZ was born in Mexico. He received the B.S. degree in Computer Engineering, M.S. degree in Microelectronics Engineering and his PhD in Communications and Electronic from the National Polytechnic Institute of Mexico in 2003, 2006 and 2011, respectively. From August 2005 to August 2011 he was in Federal Electoral Institute (IFE) and Government Secretary (SEGOB) of Mexico in several information technologies areas. Currently, he is a full-time researcher at the Electric Engineering Division, Engineering Faculty, National Autonomous University of Mexico. His principal research interests are image and video processing, watermarking, software development and related fields.

Antonio CEDILLO-HERNANDEZ was born in Mexico. He received the B.S. degree in Computer Engineering, M.S. degree in Microelectronics Engineering and his PhD in Communications and Electronic from the National Polytechnic Institute of Mexico in 2005, 2007 and 2013, respectively. He holds a certification as Project Management Professional (PMP®). He worked in Mexican Government Ministry (SEGOB) as a Contract and Project Management for five years. From September 2013 to December 2014 he was with the Polytechnic Metropolitan University of Hidalgo where he was a full-time researcher. Currently, he courses a postdoctoral residence at the Electric Engineering Division, Engineering Faculty, National Autonomous University of Mexico. His principal research interests are video processing, watermarking, software development and related fields.

Francisco GARCIA-UGALDE was born in Mexico. He obtained his bachelor in 1977 in Electronics and Electrical System Engineering from National Autonomous University of Mexico, his Diplôme d'Ingénieur in 1980 from SUPELEC France, and his PhD in 1982 in Information Processing from Université de Rennes I, France. Since 1983, he is a full-time professor at UNAM, Engineering Faculty. His current interest fields are: Digital filter design tools, analysis and design of digital filters, image and video coding, image analysis, watermarking, theory and applications of error control coding, turbo coding, applications of cryptography, parallel processing and data bases.

Mariko NAKANO-MIYATAKE was born in Japan. She received the M.E. degree in Electrical Engineering from the University of Electro-Communications, Tokyo, Japan in 1985, and her Ph. D in Electrical Engineering from the Universidad Autonoma Metropolitana (UAM), Mexico City, in 1998. From July 1992 to February 1997 she was with the Department of Electrical Engineering, UAM Mexico. In February 1997, she joined the Graduate Department of the Mechanical and Electrical Engineering School, National Polytechnic Institute of Mexico, where she is now a Professor. Her research interests are in information security, image processing, pattern recognition and related field.

Hector PEREZ-MEANA was born in Mexico. He received his M.S: Degree in Electrical Engineering from the Electro-Communications University of Tokyo, Japan in 1986 and his Ph. D. degree in Electrical Engineering from the Tokyo Institute of Technology, Tokyo, Japan, in 1989. From March 1989 to September 1991, he was a visiting researcher at Fujitsu Laboratories Ltd, Kawasaki, Japan. From September 1991 to February 1997, he was with the Electrical Engineering Department of the Metropolitan University of Mexico City where he was a Professor. In February 1997, he joined the Graduate Studies and Research Section of the Mechanical and Electrical Engineering School, National Polytechnic Institute of Mexico, where he is now a Professor. His principal research interests are adaptive systems, image processing, pattern recognition, watermarking and related fields.

Efficient Spectral Power Estimation on an Arbitrary Frequency Scale

Filip ZAPLATA, Miroslav KASAL

Dept. of Radio Electronics, Brno University of Technology, Technická 12, 612 00 Brno, Czech Republic

xzapla00@stud.feec.vutbr.cz, kasal@feec.vutbr.cz

Abstract. *The Fast Fourier Transform is a very efficient algorithm for the Fourier spectrum estimation, but has the limitation of a linear frequency scale spectrum, which may not be suitable for every system. For example, audio and speech analysis needs a logarithmic frequency scale due to the characteristic of a human's ear. The Fast Fourier Transform algorithms are not able to efficiently give the desired results and modified techniques have to be used in this case. In the following text a simple technique using the Goertzel algorithm allowing the evaluation of the power spectra on an arbitrary frequency scale will be introduced. Due to its simplicity the algorithm suffers from imperfections which will be discussed and partially solved in this paper. The implementation into real systems and the impact of quantization errors appeared to be critical and have to be dealt with in special cases. The simple method dealing with the quantization error will also be introduced. Finally, the proposed method will be compared to other methods based on its computational demands and its potential speed.*

Keywords

Goertzel algorithm, Mel frequency cepstral coefficients, MFCCs, Q-constant transform, spectral power estimation

1. Introduction

Many techniques for spectral analysis are based on the short time Fourier spectra and since the Fast Fourier Transform (FFT) algorithms have been developed, the estimation of the spectra is even more suitable. Spectral resolution of the algorithms depends on the number of time domain samples that are processed, and the number of complex samples directly equals the resulted number of complex spectral lines (bins). The bins are equidistantly spread along the linear frequency scale in the range constrained by the sampling theorem.

There are situations in which the linear frequency scale is not convenient, e.g. speech or music analysis. Such inconvenience originates from the natural character of a human's ear the frequency characteristic of which is far different from the linear spectral scale. The model of a human's ear [1] says that the spectral resolution should be logarithmically distributed over the audible spectrum. Several techniques are used for the logarithmically distributed frequency scale spectrum estimation; the Mel-Frequency Cepstrum Coefficients (MFCC) [2] and the constant Q-transform [3] are the main ones.

The MFFC algorithm uses the spectral estimation of a sufficient frequency resolution computed by the FFT. The spectrum is then successively processed by triangular windows in accordance to spectral requirements. Windowing may be provided by a predefined filter bank. Filter bank outputs are converted to a logarithmic amplitude scale and processed by the Discrete Cosine Transform (DCT). Each output of the DCT then corresponds to the spectral line of the desired frequency scale. Over the years many computational forms have been developed which are more or less effective as e.g. in [4] or [5].

The constant Q-transform originates from the basic definition of the Discrete Fourier Transform (DFT), where the computation is not provided for all bins. J. Brown in [3] defined the algorithm for computing DFT parameters to fulfill predefined requirements. The original calculation can also be efficiently modified as in [6] or for audio signals in [7].

M. Tröbs and G. Heinzel also published a different approach [8], where they applied an averaging over modified periodogram. Their improved method computes the Fourier transform optimally for the logarithmic frequency scale, but is more convenient for a high resolution analysis and therefore is very complex.

The first step of mentioned algorithms is to estimate the spectrum of a huge frequency resolution to catch the differences between spectral components at low frequencies. The proposed algorithm uses a separate setting for each desired spectral component as similar to a bank of filters and is also useful for estimating the spectrum of an arbitrary frequency scale distribution while maintaining a high computational efficiency. On the other hand the price paid for its simplicity is a spectral leakage when using a rectangular weighting window. The main contribution of the text is a simple way to partially eliminate the leakage at frequencies in between the Fourier spectrum bins described in Sec. 3. For better possibility of comparison, the logarithmic frequency scale distribution will be studied in the following text.

2. Efficient Approach to Spectral Power Estimation

In Fig. 1 the problem of using conventional Fourier transform algorithms for spectral estimation is depicted. Under the log-scale at low frequencies there is an obvious lack of information whereas at high frequencies there are many redundant bins that have to be averaged. Thus, even though the FFT is a very efficient algorithm for estimation of linearly distributed spectra, its computational demands grow rapidly if high resolution at low frequencies is required.

As obvious from the definition of the discrete Fourier transform

$$X(k)=\sum_{n=0}^{N-1}x(n)e^{-j2\pi n\frac{k}{N}} \tag{1}$$

individual bins can be calculated separately. The resolution of the spectrum depends on the number of processed samples N and k indexes the bins.

Obviously, this property helps to eliminate unnecessary bins and additionally set the resolution for each bin computation separately. The Fourier transform can be viewed as a bank of band-pass filters with an impulse response

$$h(n)=u_{\mathrm{N}}(n)e^{j2\pi n\frac{k}{N}} \tag{2}$$

where $u_{\mathrm{N}}(n)$ denotes the constraints of the sum (1), i.e. rectangular window. The power spectral density (PSD) characteristics of the filters can be derived by the method in [9] and an example is shown in Fig. 2.

From PSD it is clear that the length of the filter N directly influences its bandwidth, further description of this relation is in [10]; notice also the zero transfer at frequen-

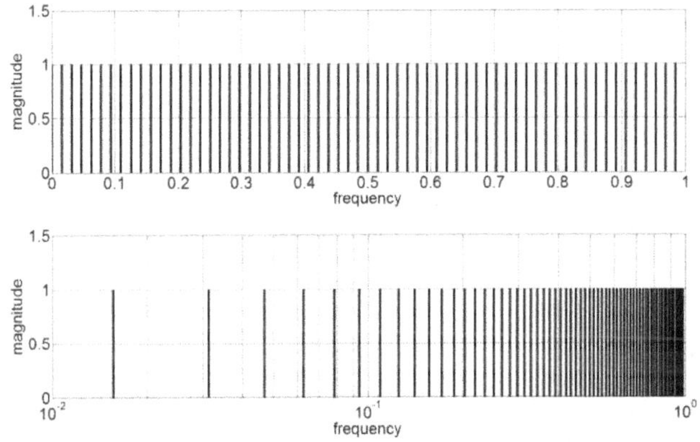

Fig. 1. The upper image shows the linear distribution of the Fourier transform and the bottom image shows the same distribution under the log-scale.

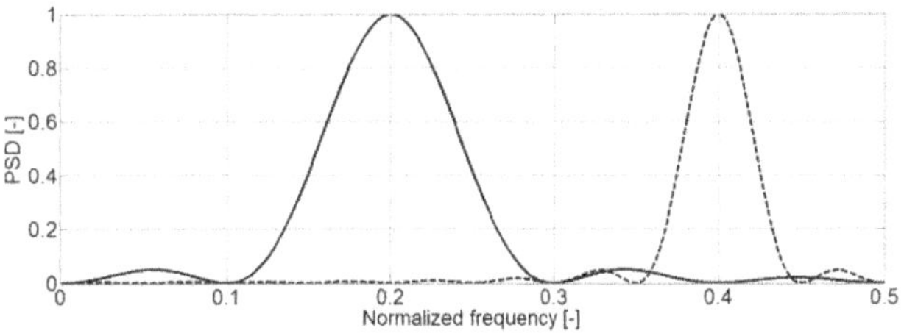

Fig. 2. PSD example for $N = 10$ and $k = 2$ (solid line) and $N = 20$ and $k = 8$ (dashed line).

cies of adjacent bins. The filters can then be set to analyze the signal under the desired scale; a nice example is a log-scale of base 2. The k coefficient is constant over all bins and length N is halved with each bin going to the highest frequencies and it causes the desired widening of its bandwidth as depicted in Fig. 3.

The direct evaluation of (1) can be replaced by the faster Goertzel algorithm. The algorithm is a convolutional form of (1) and the derivation can be found in [9]. It is used for fast calculation of selected bins and hence very suitable for our purpose. Its transfer function is shown in the following equation

$$H(z)=\frac{1-z^{-1}e^{-j2\pi\frac{k}{N}}}{1-2z^{-1}\cos\left(2\pi\frac{k}{N}\right)+z^{-2}} \tag{3}$$

where the output sample is valid for every N^{th} sample after which the filter has to be cleared. This condition provides the rectangular windowing of the impulse response.

The Goertzel algorithm is useful not only for discrete frequencies defined by N, i.e. $k\in\mathbb{Z}$, but as declared in [11] it can be used for frequencies in between the bins, i.e. k may be real. This statement is correct, but suffers from spectral leakage more than for integer k. The reason why the leakage is higher is that since the filter is complex, the main passing lobe of the PSD is present only on one side of the complex spectrum. Thus, shifting the frequency response of non-integer k causes the zero of the transfer function not to be at the position of the opposite frequency image, which then causes aliasing. The effect is clear from Fig. 4.

This effect is more apparent when the filter is tuned near the edges of a sampled region, where the images are very close, and when the spectral leakage is high, which is for a rectangular window. Another window with better

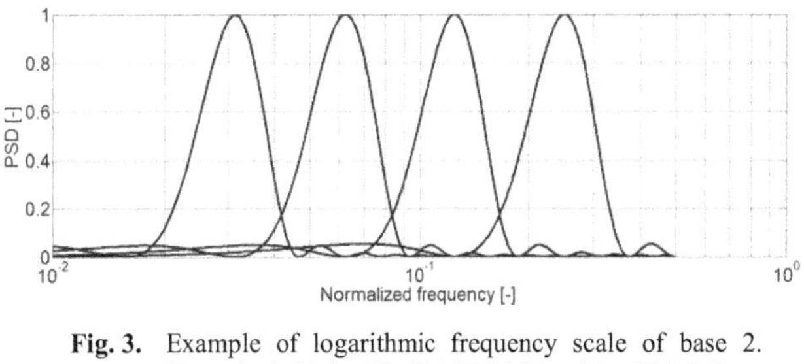

Fig. 3. Example of logarithmic frequency scale of base 2. From the left lobe $N = [64\ 32\ 16\ 8]$ and $k = [2\ 2\ 2\ 2]$.

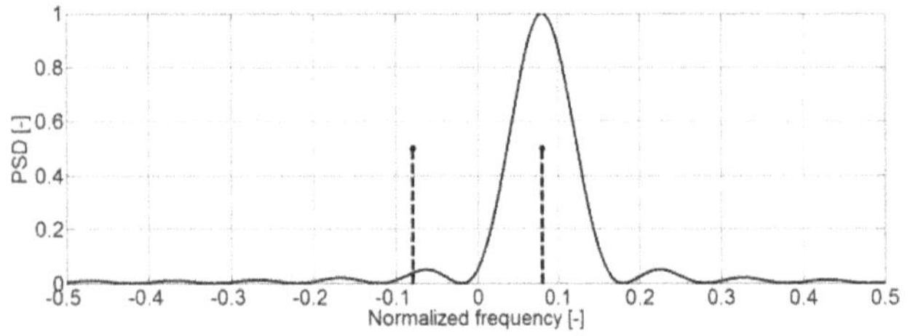

Fig. 4. Aliasing of mirror images. Filter of $N = 10$ is tuned to $k = 0.8$, corresponding harmonic images are shown by the dashed lines.

spectral leakage characteristics can be used by weighting the input signal because the impulse response is inaccessible within the effective realization of the Goertzel algorithm. The impact of weighting by Hanning window on the PSD has been presented in [10]. The additional windowing has to be applied before the filter and therefore it degrades useful power of the signal and significantly increases computational demands because each log-bin needs different N and consequently the window size is also different.

3. A Novel Technique for Elimination of Additional Spectral Leakage for Arbitrary Frequencies

The spectral leakage is usually treated by windowing. Even in [11], dealing with arbitrary frequencies defined by real k, some weighting window is supposed to be used. Authors in [8] also met this issue and due to its minimization by windowing considered it as negligible. The following text presents a novel and more efficient approach to eliminating leakage of the mirror image of the desired frequency and therefore providing better characteristic for real k and rectangular weighting window.

First, we have to find an analytic response of the bin calculation to a harmonic signal of the same frequency. The solution of the limited convolution is

$$\begin{aligned} Y_\mathrm{k}(n) &= \sum_{m=0}^{N-1} M \cdot \cos\left(2\pi\frac{k}{N}m + \varphi\right) e^{j2\pi\frac{k}{N}(n-m)} \\ &= \frac{M}{2}\left(Ne^{-j\varphi} + e^{j\varphi}\frac{1-e^{-j4\pi k}}{1-e^{-j4\pi\frac{k}{N}}}\right). \end{aligned} \tag{4}$$

The bin value is newly denoted as Y and is proposed to be dependent on time samples n instead of one time sample X in (1). The applied harmonic signal has magnitude M and phase φ. Note that for integer k the second term of the solution is zero and the result simplifies to a complex value expected for the Fourier bin power. Substituting the parasitic value

$$A = \frac{1-e^{-j4\pi k}}{1-e^{-j4\pi\frac{k}{N}}} \tag{5}$$

and the desired value

$$B = Me^{-j\varphi} \tag{6}$$

the response (4) can be rewritten as

$$Y_\mathrm{k}(n) = \frac{1}{2}\left(BN + B^*A\right). \tag{7}$$

Equation (7) can now be solved for unknown B as

$$B = \frac{2}{AA^* - N^2}\left(AY_\mathrm{k}^* - NY_\mathrm{k}\right) \tag{8}$$

where Y_k is the output of the uncompensated Goertzel filter.

The optimized way for signal power computation of the original Goertzel algorithm

$$\left|Y_\mathrm{k}(n)\right|^2 = d(n)^2 + d(n-1)^2 - Cd(n)d(n-1) \tag{9}$$

has been obtained by powering the complex Goertzel output X_k [12]. Variables $d(n)$ and $d(n\text{-}1)$ denote delayed samples within the recursive part of the algorithm [12]. C denotes the frequency coefficient (10), which can also be evaluated by corresponding absolute frequencies f_c (tuned frequency) and f_S (sampling frequency),

$$C = 2\cos\left(2\pi\frac{k}{N}\right) = 2\cos\left(2\pi\frac{f_\mathrm{c}}{f_\mathrm{S}}\right). \tag{10}$$

The solution of (8) can be put into (9) in the form of additional coefficients resulting in

$$\begin{aligned} \left|Y_\mathrm{k}(n)\right|^2 &= d(n)^2 Q_\mathrm{M1} + d(n-1)^2 Q_\mathrm{M2} \\ &-Cd(n)d(n-1)Q_\mathrm{M3}. \end{aligned} \tag{11}$$

The compensation coefficients are then

$$Q_\mathrm{M1} = \frac{4\left(A_\mathrm{R}^2 + A_\mathrm{I}^2 + N^2 - 2A_\mathrm{R}N\right)}{\left(A_\mathrm{R}^2 + A_\mathrm{I}^2 - N^2\right)^2}, \tag{12}$$

$$Q_\mathrm{M2} = \frac{4\left(A_\mathrm{R}^2 + A_\mathrm{I}^2 + N^2 - 2N\left(A_\mathrm{R}\cos\left(4\pi\frac{k}{N}\right) + A_\mathrm{I}\sin\left(4\pi\frac{k}{N}\right)\right)\right)}{\left(A_\mathrm{R}^2 + A_\mathrm{I}^2 - N^2\right)^2}, \tag{13}$$

$$Q_\mathrm{M3} = \frac{4\left(A_\mathrm{R}^2 + A_\mathrm{I}^2 + N^2 - 2N\left(A_\mathrm{R} + A_\mathrm{I}\frac{\sin\left(2\pi\frac{k}{N}\right)}{\cos\left(2\pi\frac{k}{N}\right)}\right)\right)}{\left(A_\mathrm{R}^2 + A_\mathrm{I}^2 - N^2\right)^2}. \tag{14}$$

For more convenient evaluation, the substituted complex constant A is split into real and imaginary parts A_R and A_I respectively. Notice again the state where the frequency coefficient k is integer and A is then zero. All coefficients simplify to only weighting constant $(2/N)^2$ that is common for the Fourier transform power computation.

The evolution of the coefficients is shown in Fig. 6.

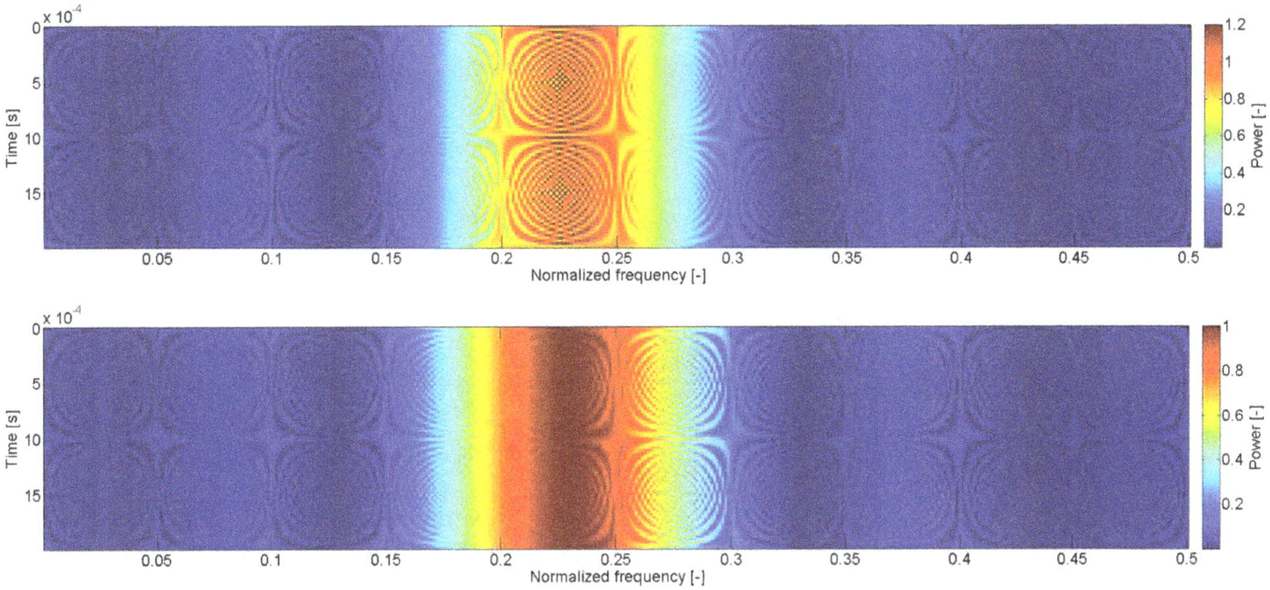

Fig. 5. Waterfall spectrogram of the original method and compensated method for $N = 10$ and $k = 2.3$.

It is obvious that the most critical are the frequencies near the edges of the sampled region where the coefficients grow to infinity. Otherwise the coefficients slightly fluctuate around the weighting value $(2/N)^2$, which equals 1/9 in this example. The compensation effect is clearly seen on the waterfall spectrogram in Fig. 5; the upper image is for the original uncompensated algorithm, where strong fluctuations at the tuned frequency ($k = 2.3$) are seen. Notice also the magnitude range, which goes even over the unity value. The bottom image shows the output signal after compensation, the fluctuations are effectively suppressed at the tuned frequency.

4. Quantization Error of the Coefficients

When implementing whatever signal processing algorithm, the issues about limited accuracy computing has to be taken into account. The general problem of overflowing and rounding error has been discussed e.g. in [12]. The solution of these issues mainly lies in properly set accuracy of the fixed point calculations and appropriate order of arithmetic operations. However, the quantization error of the coefficients is a more complicated issue which may cause total failure of the system and therefore it should be given attention.

In Fig. 7, there are two lines, the solid one shows the real quantization error of coefficient C when 4 bits of memory are allocated for its fractional part. The dashed curve shows the maximal error given by

$$E_{qC} = \frac{\cos^{-1}\left(\cos\left(2\pi\frac{k}{N}\right) - \frac{0.5}{R}\right)}{2\pi} \tag{15}$$

and is the envelope of the real error. The coefficient resolution is given by the quantum count R, which is $R = 2^4$ for 4-bit resolution. The equation was obtained by the reverse tuning coefficient computation, i.e. the absolute frequency error is the frequency calculated back from the ideal tuning coefficient distorted by a half quantum of fixed point num-

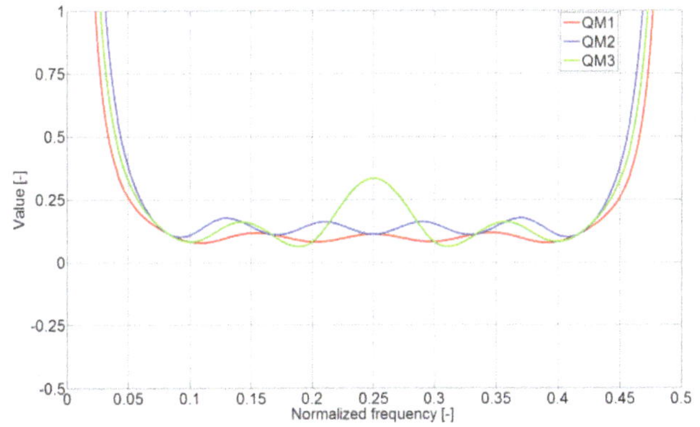

Fig. 6. Evolution of the magnitude compensation coefficients: example for $N = 6$.

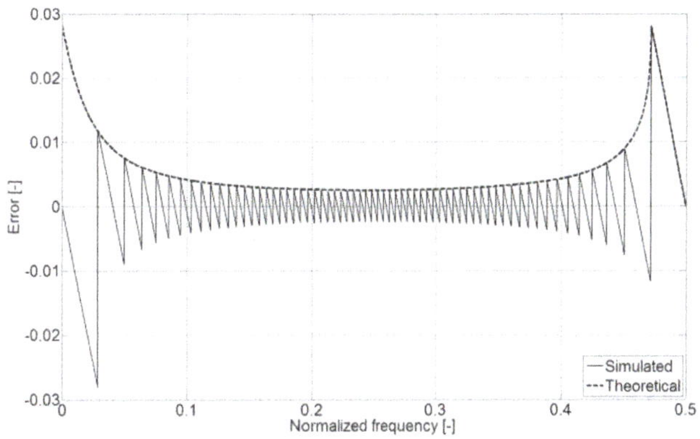

Fig. 7. Absolute quantization error of coefficient C for quantum count $R = 16$.

ber representation. It is now clear that the most critical are the edges of the frequency range, where the error grows rapidly. Note also, that equation (15) does not have a real solution for all frequencies, where the argument of the $\cos^{-1}(x)$ function can run out of the range $\langle -1; 1\rangle$. Such an occurrence has to be prevented during the design.

The quantization error of coefficient C is particularly critical for log-scale frequency spectral estimation, where the need of accuracy in frequency grows at low frequencies. A possible solution for this issue is to raise the quantization accuracy of the system, i.e. raise R, or to move the ratio k/N to a less critical region. The second way can be maintained by pre-filtering and decimation.

A simple low-pass filter is the moving average, which in fact is a limiting case of the Goertzel algorithm and therefore the equation

$$H_{\mathrm{G}}(f)=\frac{1}{N}\frac{\sin\left(\pi N\left(f-\frac{k}{N}\right)\right)}{\sin\left(\pi\left(f-\frac{k}{N}\right)\right)}e^{j2\pi\left(f-\frac{k}{N}\right)(1-N)} \tag{16}$$

[9] can be used for calculating the frequency response of the moving average filter

$$\left|H_{\mathrm{MA}}(f)\right|=H_{\mathrm{G}}(f)_{k=0}=\frac{\sin\left(\pi N_{\mathrm{MA}}f\right)}{N_{\mathrm{MA}}\sin\left(\pi f\right)}. \tag{17}$$

The impulse response is then

$$h_{\mathrm{MA}}(n)=\frac{1}{N_{\mathrm{MA}}}u_{\mathrm{NMA}}(n) \tag{18}$$

which is a rectangular pulse in fact.

The moving average is realized as the summing of N_{MA} samples and weighting them by its number before letting them into the Goertzel filter. The design of the Goertzel coefficients then has to include N_{MA} times decimation as a substitution

$$\frac{k}{N}\rightarrow\frac{N_{\mathrm{MA}}k}{N}=N_{\mathrm{MA}}\frac{f_{\mathrm{c}}}{f_{\mathrm{S}}}, \tag{19}$$

i.e. to give the desired frequency response, the k/N ratio of the Goertzel filter is N_{MA} times raised.

The same approach can be used to design the high-pass filter resulting in the impulse response

$$h_{\mathrm{HP}}(n)=\frac{1}{N_{\mathrm{HP}}}e^{j\pi n}=\frac{1}{N_{\mathrm{HP}}}(-1)^{n} \tag{20}$$

and magnitude frequency response

$$\left|H_{\mathrm{HP}}(f)\right|=H_{\mathrm{G}}(f)_{k=\frac{N}{2}}=\frac{\sin\left(\pi N_{\mathrm{HP}}\left(f-\frac{1}{2}\right)\right)}{N_{\mathrm{MA}}\sin\left(\pi\left(f-\frac{1}{2}\right)\right)}. \tag{21}$$

From the impulse response it can be derived that realization of the high-pass filter lies also in summing of N_{HP} samples, but with alternating signs.

The design of the Goertzel filter coefficients then has to include substitution according to formula

$$\frac{k}{N}\rightarrow N_{\mathrm{HP}}\frac{N-k}{N}=N_{\mathrm{HP}}\frac{f_{\mathrm{S}}-f_{\mathrm{c}}}{f_{\mathrm{S}}}. \tag{22}$$

This proposed solution helps to eliminate the quantization error of the coefficient C, however, very simple filters cause aliasing and distortion; therefore to reach the best performance, the system has to be optimized, or other better filters have to be used.

The compensation coefficients Q_{M1}, Q_{M2}, Q_{M3} do not affect the recursive part of the filter and therefore, their impact on the results is weak and can be easily predicted from their values, as seen in Fig. 6.

5. Efficiency Comparison

Finally, the proposed method is briefly compared to the FFT methods from an efficiency point of view. The efficiency or computation demands can be assessed from the theoretical number of multiplications and additions per frame or in our case per fixed number of samples.

The realization equations of the Goertzel algorithm are summarized as

$$d(n)=x(n)+Cd(n-1)-d(n-2) \tag{23}$$

and

$$\begin{aligned}Y_{\mathrm{k}}(n)=&Q_{\mathrm{M1}}d(n)^{2}+Q_{\mathrm{M2}}d(n-1)^{2}\\&-CQ_{\mathrm{M3}}d(n)d(n-1).\end{aligned} \tag{24}$$

From the equations the number of real multiplications per frame is $N+6$ and the number of real additions is $2(N+1)$. Assuming the log-scale distribution of base 2, as in Fig. 3, results in the number of multiplications

$$M_{\mathrm{G}}=2^{K}\left(K2^{p-1}+6\right)-6 \tag{25}$$

and additions

$$A_{\mathrm{G}}=2^{K}\left(K2^{p}+4\right)-4 \tag{26}$$

for K log-scale bins. The frame size of each bin is different, so the number of samples, to which the results are compared, is the length of the longest frame. The longest frame also defines the length of the FFT algorithm used for the same frequency resolution within other spectral analysis algorithms. The value

$$p=\log_{2}N_{\mathrm{K}} \tag{27}$$

defines the bandwidth of the bin tuned to the highest frequency, i.e. with the widest bandwidth N_{K}.

The fast Fourier transform radix-2 algorithm has the following number of multiplications

$$M_{\text{FFT}} = N\log_2 N + 2N \tag{28}$$

and additions

$$A_{\text{FFT}} = 2N\log_2 N + N\,, \tag{29}$$

when used for power spectrum computation.

In Tab. 1, there are a few values of the Goertzel algorithm computational demands in comparison with its equivalent FFT algorithm for $p = 3$. For example for the row for eight log-scale bins at relative frequencies $\left[\frac{2}{8},\frac{2}{16},\frac{2}{32},\frac{2}{64},\frac{2}{128},\frac{2}{256},\frac{2}{512},\frac{2}{1024}\right]_{p=3}$ the relative deviation is -20.9% and -19.1% for multiplications and additions respectively compared at 1024 samples of the input signal. The comparison was provided as a relative deviation of the computational demands against the reference FFT radix-2 power spectrum algorithm. In all cases, there appears the minus sign which means an improvement. Notice also that for very large number of bins K, equation (25) and (26) converge to the FFT demands (28) and (29), because in such a case $K = \log_2 N$.

K	M_G	M_{FFT}	δ_{rM}	A_G	A_{FFT}	δ_{rA}
4	346	512	-32.4%	572	832	-31.3%
8	9722	12288	-20.9%	17404	21504	-19,1%
16	4587514	5242880	-12.5%	8650748	9699328	-10.8%
32	5.76E+11	6.18E+11	-6.9%	1.12E+12	1.19E+12	-5.8%
64	4.83E+21	5.02E+21	-3.7%	9.52E+21	9.81E+21	-3.0%

Tab. 1. An example of computational demands.

6. Conclusion

The Goertzel algorithm has been introduced in a role of an efficient bank of filters for an arbitrary frequency scale spectral analysis. The algorithm implicitly uses a rectangular weighting window and this causes strong spectral leakage, which is a price paid for its simplicity. The Goertzel algorithm is able to be tuned to whatever real frequency within the sampled range, but has stronger leakage caused by the mirrored central frequency. The novel method of effective mirror image elimination has been introduced. The elimination lies in compensation provided on the output power side with 3 additional, mostly non-critical, coefficients which are much more efficient compared to window weighting.

Implementation issues are a very important part of the system design and the Goertzel filter has a very critical issue in the quantization error of the tuning coefficient, which affects the recursive part. The proposed solution helps to significantly reduce the tuning error, but the price paid is the aliasing of leaking signals from the stop-band.

The efficiency of the algorithm was assessed by comparing the FFT radix-2 algorithm of equivalent length. The number of multiplication and addition operations per a fixed number of samples was compared. The result showed that the algorithm is slightly less demanding, but considering that the FFT spectral evaluation is only a part not giving the desired arbitrary frequency scale power spectrum, the proposed algorithm is even more efficient.

The algorithm has already been implemented within an audio amplifier to visualize an input signal spectrum in the logarithmic frequency scale. The display resolution is 16 bins of 4-bit depth. In such cases the spectral leakage of the rectangular window is not disturbing; moreover the speed of the algorithm is excellent.

Acknowledgments

The presented research was financed by the Czech Ministry of Education in frame of the National Sustainability Program, the grant LO1401 INWITE. For the research, infrastructure of the SIX Center was used.

References

[1] ZHENG, F., ZHANG, G., SONG, Z. Comparison of different implementations of MFCC. *Journal of Computer Science and Technology*, 2001, vol. 16, no. 6, p. 582–589. DOI: 10.1007/BF02943243

[2] DAVIS, S. B., MERMELSTEIN, P. Comparison of parametric representations for monosyllabic word recognition in continuously spoken sentences. *IEEE Transactions on Acoustics, Speech and Signal Processing*, 1980, vol. 28, no. 4, p. 357–366. DOI: 10.1109/TASSP.1980.1163420

[3] BROWN, J. C. Calculation of constant Q spectral transform. *Journal of the Acoustical Society of America*, 1991, vol. 89, no. 1, p. 425–434. DOI: 10.1121/1.400476

[4] GANCHEV, T., FAKOTAKIS, N., KOKKINAKIS, G. Comparative evaluation of various MFCC implementations on the speaker verification task. In *Proceedings of the 10th International Conference on Speech and Computer SPECOM 2005*. Patras (Greece), October 2005, vol. 1, p. 191–194.

[5] MOLAU, S., PITZ, M., SCHLÜTER, R., NEY, H. Computing MEL-frequency cepstral coefficients on the power spectrum. In *IEEE International Conference on Acoustics, Speech and Signal Processing*. Salt Lake City (USA), May 2001, vol. 1, p. 73–76. DOI: 10.1109/ICASSP.2001.940770

[6] BROWN, J. C., PUCKETTE, M. S. An efficient algorithm for the calculation of a constant Q transform. *Journal of the Acoustical Society of America*, Nov. 1992, vol. 92, no. 5, p. 2698–2701. DOI: 10.1121/1.404385

[7] DOS SANTOS, C. N., NETTO, S. L., BISCAINHO, L. W. P., GRAZIOSI, D. B. A modified constant Q-transform for audio signals. In *IEEE International Conference on Acoustics, Speech, and Signal Processing ICASSP2004*. Montreal (Canada), 2004, vol. 2, p. 469–472. DOI: 10.1109/ICASSP.2004.1326296

[8] TRÖBS, M., HEINZEL, G. Improved spectrum estimation from digitized time series on a logarithmic frequency axis. *Measurement*, 2006, vol. 39, no. 2, p. 120–129. DOI: 10.1016/j.measurement.2005.10.010

[9] ZAPLATA, F., KASAL, M. Using the Goertzel algorithm as a filter. In *Proceedings of the 24th International Conference Radioelektronika 2014*. Bratislava (Slovakia), 2014, p. 1–3. DOI: 10.1109/Radioelek.2014.6828441

[10] ZAPLATA, F., KASAL, M. Software defined DCF77 receiver. *Radioengineering*, 2013, vol. 22, no. 4, p. 1211–1217. ISSN: 1210-2512.

[11] DE JESUS, M. A., TEIXEIRA, M., VICENTE, L., RODRÍGUEZ, Y. Nonuniform discrete short-time Fourier transform: A Goertzel filter bank versus a FIR filtering approach. In *IEEE International Midwest Symposium on Circuits and Systems MWSCAS 2006*. San Juan (Puerto Rico), 2006, vol. 2, p. 188–192. DOI: 10.1109/MWSCAS.2006.382241

[12] ZAPLATA, F., KASAL, M. SDR implementation for DCF77. In *Proceedings of the 23rd International Conference Radioelektronika 2013*. Pardubice (Czech Republic), 2013, p. 340–345. DOI: 10.1109/RadioElek.2013.6530943

About the Authors ...

Filip ZAPLATA was born in Chrudim, Czech Republic in 1987. He received the M.Sc. degree in Electronics and Communications from the Brno University of Technology (BUT) in 2011. From 2011, he was a combined Ph.D. student and simultaneously working as an R&D Engineer in Vesla s.r.o. company. Since 2012 he has been a full-time Ph.D. student at the Department of Radio Electronics, BUT. His research interests include digital signal processing algorithms for software defined radio and digital receiver technology.

Miroslav KASAL was born in Litomysl, Czech Republic. He graduated in Communication Engineering from the Faculty of Electrical Engineering, Brno University of Technology, in 1970. In 1984 he obtained his Ph.D. degree in Metering Engineering. He was the head of the NMR Department and Electronics Laboratory of the Institute of Scientific Instruments, Academy of Science of the Czech Republic (1991-2002). Since 2002 he has been with the Department of Radio Engineering, Faculty of Electrical Engineering and Communication, Brno University of Technology as a professor. He is a senior member of the IEEE. In 2007, prof. Kasal received the Prize for Research of the Minister of Education of the Czech Republic.

32

Design of a Wideband Inductively Coupled Loop Feed Patch Antenna for UHF RFID Tag

Mohd Saiful Riza BASHRI, Muhammad Ibn IBRAHIMY, S. M. A MOTAKABBER

Dept. of Electrical and Computer Engineering, International Islamic University Malaysia, 53100 Kuala Lumpur, Malaysia

mohdsaifulriza@yahoo.com, ibrahimy@iium.edu.my, amotakabber@iium.edu.my

Abstract. *A planar wideband patch antenna for ultra-high frequency (UHF) radio frequency identification (RFID) tag for metallic applications is presented in this research work. Three different shape patches are inductively coupled to a triangle loop to form wide impedance bandwidth for universal application UHF (860–960 MHz) RFID. The structure of the proposed antenna exhibits planar profile to provide ease of fabrication for cost reduction well suited for mass production. The simulation of the antenna was carried out using Finite Element Method (FEM) based software, Ansoft HFSS v13. The simulated and measured impedance bandwidth of 113 MHz and 117 MHz (Return Loss ≥ 6 dB) were achieved to cover the entire UHF RFID operating frequency band worldwide. The simulated and measured radiation patterns at the operating frequency of 915 MHz are in good agreement. Moreover the simulated maximum antenna gain at the bore sight direction in free space and when mounted on 200 × 200 mm² metal plate are -5.5 dBi and -9 dBi respectively which is enough to provide reasonable read range over the entire UHF RFID system operating band.*

Keywords

Complex impedance matching, patch antenna, radio frequency identification (RFID), metallic object, ultra high frequency (UHF)

1. Introduction

Recently, Radio Frequency Identification (RFID) technology is gaining traction in various sectors due to its numerous advantages such as it does not require line of sight, high read distance, fast date rate and large storage capacity as compared to conventional barcode technology [1]. Some of the sectors utilizing RFID are supply chain management, logistics, access control and real time location service (RTLS) etc. RFID in its basic form consists of tag, attached to an object to be tracked and reader whose function is to read the information contained inside the tag memory. Generally, RFID can be categorized into several types based on their operating frequencies, power source and protocols that govern its communication. Low frequency (LF) and high frequency (HF) systems are operated based on near-field communication thus having limited read range up to only 1 meter. As for ultra-high frequency (UHF) and microwave systems, the interaction between tag and reader is accomplished via propagating electromagnetic wave hence able to provide longer read range. As such, UHF based RFID technology is rapidly becoming the preferred solution.

Tag is made up of a microchip and an antenna connected together. To operate the microchip, ample power is needed. Due to cost factors, most systems employ passive tag where there is no on board power source such as battery to provide the power to the microchip. To circumvent this matter, tag antenna extracts the energy from the incident radio wave emitted by the reader to be delivered to the microchip. In addition, in the absence of transmitter on the tag, a special modulation technique is utilized in RFID called backscattering modulation [1]. In this method, the electromagnetic wave from the reader is modulated and reflected back to the reader. The modulation is performed by the tag microchip by changing its input impedance between two states which are matched and mismatched to the antenna input impedance to represent the binary code '1' and '0' of the information to be transmitted to the reader. The corresponding high (mismatched) and low (matched) power of the reflected wave received is then demodulated by the reader.

Antenna design is of great importance in passive UHF RFID system to ensure tag is able to operate properly [2], [3]. Although numerous works have been done in designing tag antenna, there are still many open issues that require further studies and research in order to truly exploit its potential. One of the issues is performance degradation of commonly used label typed dipole tag antenna [4–6] when placed on metal surface due to cancellation of tangential electric current at the boundary between the antenna and the metal surface [7], [8]. One of the many attempts to mitigate the problem is to separate the antenna and the metal surface by using a foam spacer to create constructive interference between the incoming and reflected signal. However, it results in thicker antenna structure which is unsuitable for RFID applications.

Another method that has been widely adopted is the use of microstrip patch antenna due to its grounded structure. When mounted on metal objects, the metal plane will act as an extension of its ground plane hence giving little effect to the antenna performance. Several microstrip antennas for UHF RFID tag have been proposed by [9–13]. However, they exhibit narrow bandwidth. To operate worldwide, the required impedance bandwidth should be able to cover the whole frequency range of UHF RFID band (860–960 MHz) [14]. List of operating frequency of several countries is shown in Tab. 1. Several solutions to improve the impedance bandwidth of patch antenna for RFID were presented by [15–19]. However, the structures of the presented antennas require multi or cross-layered configuration which will add significant manufacturing cost to the antenna fabrication due to additional process required. Moreover, the impedance bandwidth performances of the antenna were evaluated based on the half-power bandwidth (Return Loss ≥ 3 dB) that accounts for only half of the power accepted by the tag antenna to be actually delivered to the tag's microchip. Several complete planar patch antenna have been proposed by [20–23] although with limited bandwidth.

Region/country	Operating frequency, f (MHz)
North America	902–928
Europe	865–868
China	917–922
Japan	916–921 & 952–956
Australia	918–926
Hong Kong	865–868 & 920–925
Taiwan	922–928

Tab. 1. List of operating frequency for several major countries.

This letter proposes a planar wideband microstrip patch RFID tag antenna design for metallic applications. The wide impedance bandwidth is achieved by utilizing three radiating elements to excite three resonances close to each other. The complex impedance matching with the referenced microchip, Alien Higgs-3, with impedance value of $Z_c = 31 - j212\ \Omega$ and sensitivity, P_{th} of -18 dBm is realized by using inductively coupled triangle loop structure. The structure of the proposed antenna does not incorporate any via hole or shorting wall/plate which further simplify its fabrication process. The proposed antenna design concept and configuration will be explained in Sec. 2. Section 3 demonstrates the simulation and measurement results while conclusions are drawn in Sec. 4.

2. Antenna Design and Configuration

Several important factors for designing antenna for UHF RFID tag have been comprehensively presented by [2], [3]. The aim of this research is to design metal mountable tag antenna for use in metallic applications where typical label-type dipole antenna suffers performance degradation like shift in operating frequency that leads to impedance mismatch and distorted radiation pattern. Moreover, to realize a universal tag that is able to operate across the world, a wideband characteristic is required which is quite challenging for patch antenna due to its inherent narrow bandwidth. To begin with the antenna design, Alien Higgs-3 was selected as a referenced microchip [24]. The impedance of the microchip, Z_c, is $31 - j212\ \Omega$ at 915 MHz. Typically, most antenna is designed to match with 50 Ω characteristic impedance of feeding line such as coaxial cable. However, for tag antenna, its impedance must be conjugate matched with the impedance of the microchip which is connected to [25]. This is very crucial particularly for passive UHF RFID system where the tag itself does not possess its own power source to operate the microchip [2] and all the power needed is extracted from the electromagnetic signal emitted by the reader. To ensure sufficient power is delivered to the microchip, impedance matching is crucial. The evaluation of the matching efficiency can be evaluated based on return loss, *RL* as expressed in (1) below [26]

$$RL(\mathrm{dB}) = -20\log_{10}|\Gamma| \tag{1}$$

where Γ is the reflection coefficient at the antenna input terminal. Γ can be calculated as shown in (2) below [25]

$$\Gamma = \frac{Z_c - Z_{in}^*}{Z_c + Z_{in}^*} \tag{2}$$

where Z_{in} is the antenna input impedance.

There are several impedance matching techniques that have been proposed such as T-matching network, inductive coupled feed loop, nested loop, open end microstrip line shorted to ground, proximity-coupled feed and open stub feed [2], [12], [23], [27]. In this work, an inductively coupled loop structure in the form of triangle was used for complex impedance matching with the referenced microchip. The resulting input impedance seen at the input terminal of the antenna due to the triangle feed loop is given by (3) [27]

$$Z_{in} = Z_{loop} + \frac{(2\pi f M)^2}{Z_A} \tag{3}$$

where $Z_{loop} = j2\pi f L_{loop}$ is the input impedance of the feed loop and Z_A is the antenna impedance without the matching element. Based on (3), resistance and reactance at the input terminal of the antenna can be calculated based on (4) and (5)

$$R_{in}(f_0) = \frac{(2\pi f_0 M)^2}{R_A(f_0)}, \tag{4}$$

$$X_{in}(f_0) = 2\pi f_0 L_{loop}. \tag{5}$$

It can be seen from (4) and (5) that the input resistance depends on the mutual inductance between the feeding loop and the patches while the reactance value is solely contributed by the loop's inductance. The mutual coupling, M is then determined by the size of the loop and its distance from the patches. As for the loop inductance, it is mainly affected by its aspect ratio. The inductance of the triangle loop feed network, L_{loop} can be approximated using (6) [28]

$$L_{loop} \approx N^2 \frac{\mu_0 \mu_r}{2\pi} \left[\begin{array}{l} 2c\ln\left(\frac{2c}{0.5s}\right) + b\ln\left(\frac{2c}{0.5s}\right) \\ -2(b+c)\sinh^{-1}\left(\frac{b^2}{\sqrt{4b^2c^2 - b^4}}\right) \\ -2c\sinh^{-1}\left(\frac{2c^2 - b^2}{\sqrt{4b^2c^2 - b^4}}\right) - (2c+b) \end{array} \right] \quad (6)$$

where b, c, and s are the dimension of the triangle loop. N, μ_0, and μ_r are number of turns of the loop, permittivity of free space and effective permittivity of the substrate. The approximate geometrical dimension of the matching loop calculated based (5) and (6) is shown in Tab. 2.

The impedance bandwidth of the proposed antenna is enhanced by utilizing coplanar multi-resonator configuration. Three patches of different shapes were constructed to be the radiating elements. The patches resonate at three different frequencies closed to each other to form a wide impedance bandwidth to cover the entire frequency range of UHF RFID. The radiating elements of the antenna are composed of one narrow rectangular patch and two meandered patches. The physical length, L of the patches can be initially approximated using the closed form expression as shown in (7)[29]

$$L = \frac{1}{2 f_r \sqrt{\varepsilon_{\text{reff}}} \sqrt{\mu_0 \varepsilon_0}} - 2\Delta L \quad (7)$$

where f_r, $\varepsilon_{\text{reff}}$ and ΔL are the resonant frequency, effective dielectric constant and extension of patch length due to fringing field effect. The effective dielectric constant is calculated as in (8) [30]

$$\varepsilon_{\text{reff}} = \frac{(\varepsilon_r + 1)}{2} + \frac{(\varepsilon_r - 1)}{2}\left[1 + \frac{12h}{W}\right]^{-1/2} \quad (8)$$

where W is the width of the patch. In this design, the width of the patches is chosen to be less than the effective width to reduce the overall size of the antenna. Nevertheless, a good balance between gain performance and size is observed. Then, all three patches are fed by the triangle feed loop at each side as illustrated in Fig. 1. The antenna design was simulated using commercial electromagnetic simulator Ansoft HFSS v13 based on the approximated calculation of the antenna parameters. Afterwards, parametric refinement on the antenna parameter for the matching loop and the radiating patches geometry were carried out. The input resistance was matched by varying the distances, $d1$, $d2$ and $d3$ of the patches from the loop. As for the reactance part, the geometry of the triangle loop, b, c and s were varied to get $X_{\text{in}} = 212\text{j}\ \Omega$ required to cancel the capacitive value of the microchip. FR-4 epoxy glass substrate with a dielectric constant, ε_r of 4.4 and thickness of 1.6 mm was used due to its cheap cost [30]. It also lowers the Q-value of the antenna thus contributes to increased bandwidth. The optimal design parameter is tabulated in Tab. 2. The antenna was then fabricated using photolithography and etching technique. The final prototype of the antenna is shown in Fig. 2. A test fixture shown in Fig. 3 was used to probe the antenna. The input impedance of the proposed antenna was then measured using the two port differential probe technique proposed by [31], [32] due the balanced feed structure of the antenna. The impedance measurement setup is shown in Fig. 4. The input impedance of the antenna was then extracted from the measured S-parameters.

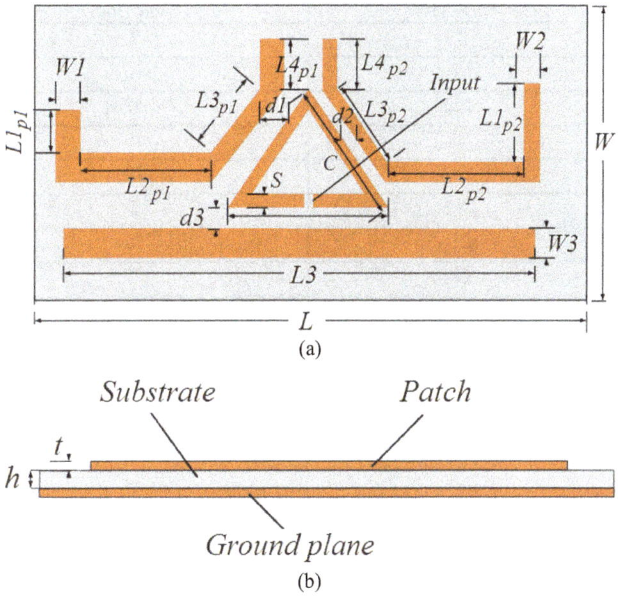

Fig. 1. Structure of the antenna. (a) Top view and (b) side view.

Fig. 2. Prototype of the antenna.

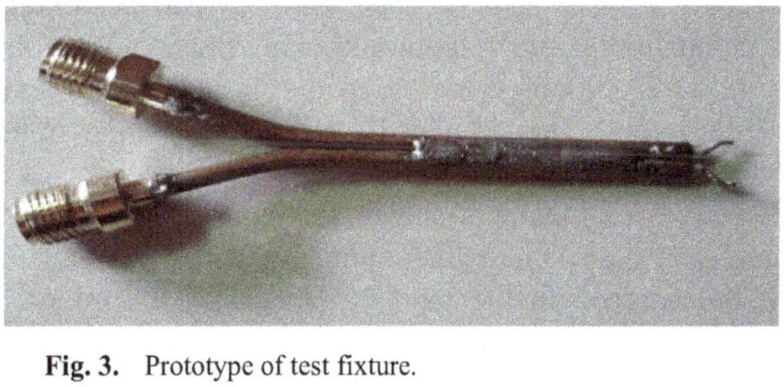

Fig. 3. Prototype of test fixture.

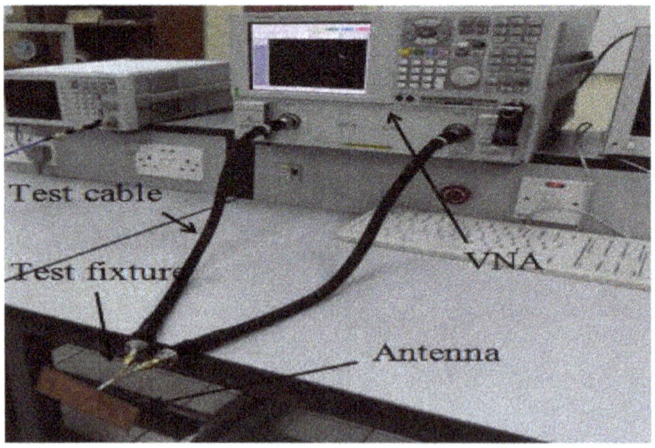

Fig. 4. Impedance measurement setup.

Parameter	Value (mm)
$W1$	7
$L1_{p1}$	10
$L2_{p1}$	37
$L3_{p1}$	10
$L4_{p1}$	12
$W2$	5
$L1_{p2}$	15
$L2_{p2}$	37
$L3_{p2}$	14
$L4_{p2}$	14
$W3$	5
$L3$	86
c	29.4
b	34
s	2
$d1$	2
$d2$	1
$d3$	2
t	0.0358
h	1.6
Ground plane and substrate	130 x 63

Tab. 2. Optimized design parameter of the proposed antenna.

3. Results and Discussion

The simulation and measurement results of the design are presented in this section to verify the initial deduction. The surface current density of the antenna illustrated in Fig. 5 to show the proposed antenna resonates at three different resonant frequencies 882 MHz, 908 MHz and 949 MHz. It can also be seen that the antenna exhibits linear horizontal polarization although slight cross polarization might be observed due to the meandering structure of the patches. The simulated and measured input impedance of the antenna against the conjugate impedance of the referenced microchip is shown in Fig. 6. By varying the parameter, d_1, d_2 and d_3 the input resistance of the antenna can be matched to the input resistance of the microchip over the UHF RFID frequency band while the size of the triangle loop feed, b, c and s can be adjusted to present the

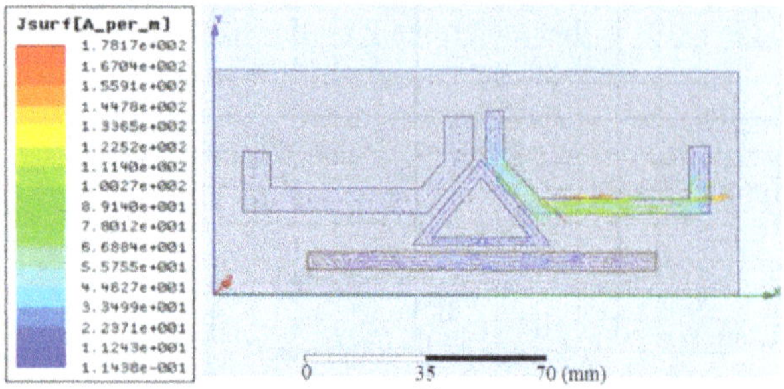

(a)

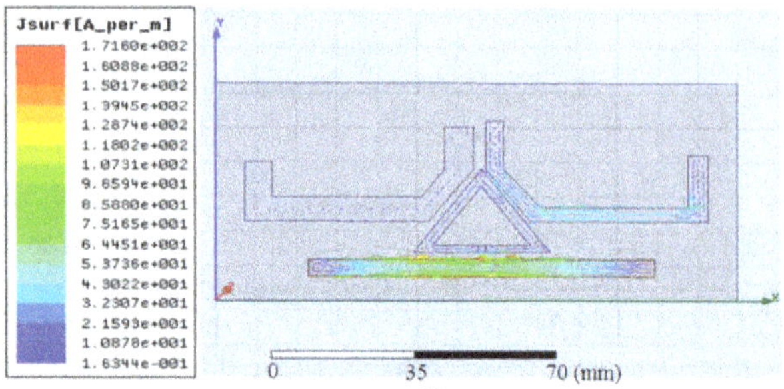

(b)

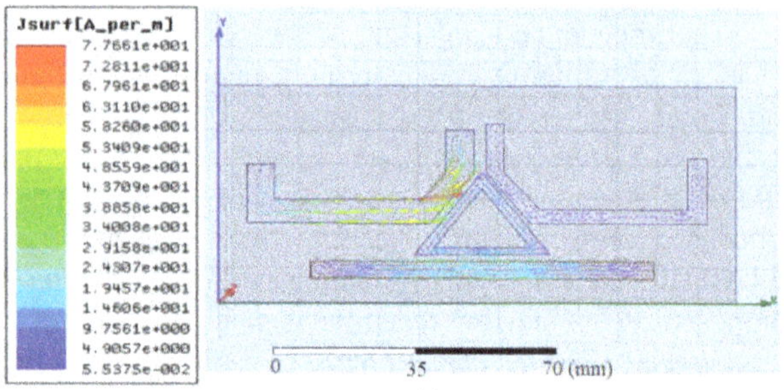

(c)

Fig. 5. Surface current distribution at three resonant frequencies. (a) 882.4 MHz, (b) 908.6 MHz and (c) 949.3 MHz.

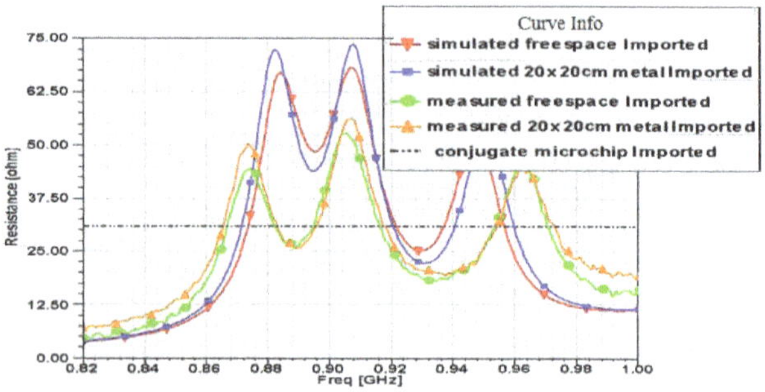

(a)

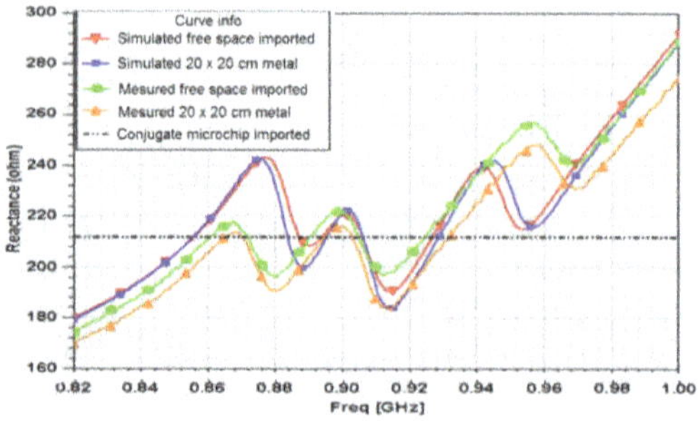

(b)

Fig. 6. Simulated and measured (a) resistance and (b) reactance value of the input impedance of the antenna against conjugate impedance of referenced microchip.

required reactance to cancel the capacitive value of the microchip. To observe the effect of metal surface to its performance, the antenna was simulated on top of metal plane with dimension of 200 × 200 mm^2 in addition to the free space scenario. For actual measurement, the antenna was then attached to a metal plate of the same size and the impedance of the antenna was measured experimentally.

The slight difference between the simulated and measured input impedance of the proposed antenna shown in Fig. 6 was due to the fabrication inaccuracy as well as the surrounding effects when the measurement was taken.

To evaluate the impedance bandwidth of the antenna, equation (1) is used to calculate the return loss of the antenna [12], [15], [16], [22], [33]. The simulated and measured return loss for both free space and mounted on metal plane is shown in Fig. 7. Despite the difference of input impedance between the simulation and measurement, both return loss performance for both cases are 113 MHz and 117 MHz, well over the required 100 MHz for the entire UHF RFID operating frequency band. As for comparison of the antenna performance on free space and when it is mounted on a metal plate, a slight shift on the resonant frequency is observed on the simulation result. However, based on the measurement, the antenna input impedances on free space and when it was attached on the metal surface is almost identical. Hence, it is evident that the proposed antenna works well when being mounted on the metallic surface. In order to further investigate the performance of the antenna, the peak gain of the antenna over the UHF RFID operating frequency band was simulated and the result is shown in Fig. 8 while the simulated antenna efficiency is shown in Fig. 9. It is seen that the efficiency of the antenna is quite low around 30% thus result in low gain. This is primarily due to the thin and lossy nature of the substrate. On the other hand, the use of the low permittivity substrate would result in increase of antenna size and cost. The peak gain of the antenna is more uniform across the operating frequency of 860–960 MHz when mounted on the metal plate as compared to free space condition. This is likely because of the reduced backside radiation due to reflection by the larger metallic surface as opposed to smaller size ground plane without the metal plate. Nevertheless, the low gain for tag antenna was not uncommon as reported in previous works since a read range of about 1 to 3 meters is enough for some of the RFID applications [17], [22], [34].

In addition, the far field radiation patterns of the antenna were measured in the anechoic chamber and compared to simulation results. The simulated and measured normalized E-field and H-field pattern at the operating frequency of 915 MHz are depicted in Fig. 10 respectively. It can be safely concluded the radiation pattern for both simulation and measurement results are in good agreement.

To further evaluate the performance of the antenna, the theoretical maximum read range of the antenna was calculated using Friis transmission equation (8) at three different operating frequencies as listed in Tab. 2 [3]. The

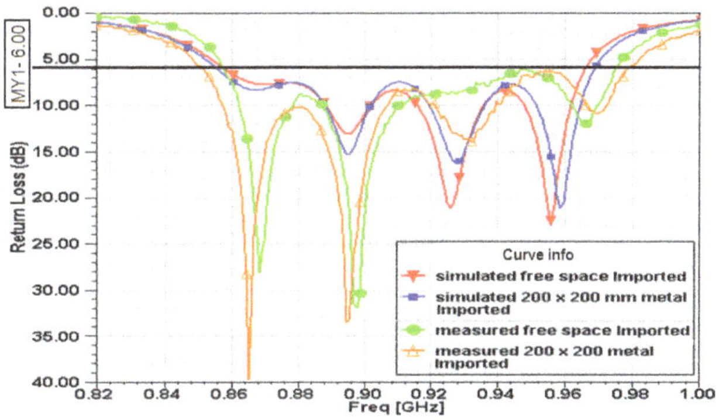

Fig. 7. Simulated and measured return loss of the antenna.

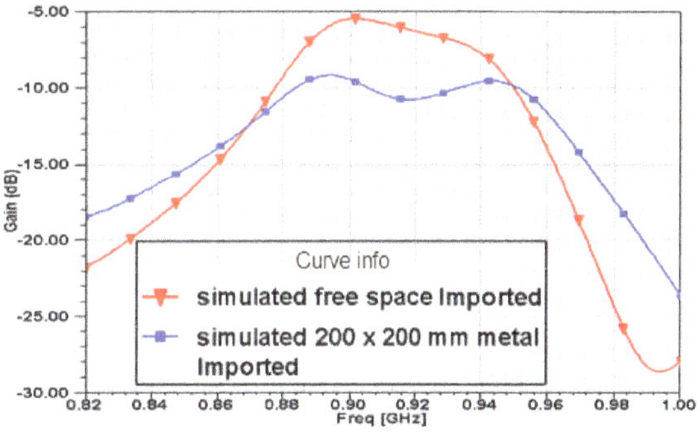

Fig. 8. Simulated peak gain of the antenna.

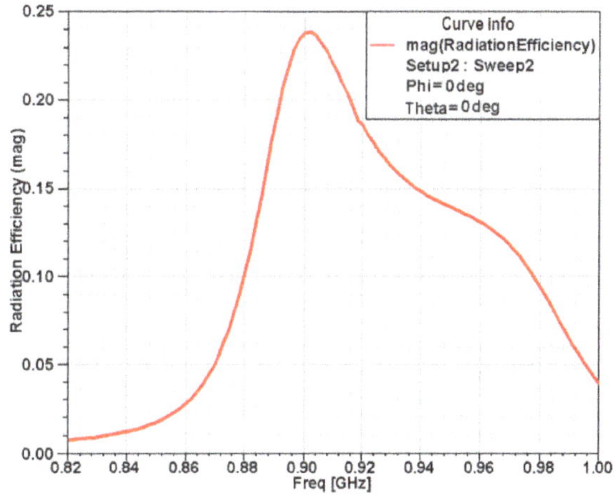

Fig. 9. Simulated antenna efficiency.

Country/ Region	Center freq., f_c (MHz)	EIRP (W)	Simulated peak gain (dBi)		Calculated read range (m)	
			Free space	200 × 200 mm^2 metal sheet	Free space	200 × 200 mm^2 metal sheet
Europe	886	3.3	-13.28	-11.50	2.68	3.33
North America	915	4	-6.05	-10.56	6.18	3.61
Japan	954	4	-11.54	-12.06	2.99	2.80

Tab. 3. Theoretical calculated read range of the antenna.

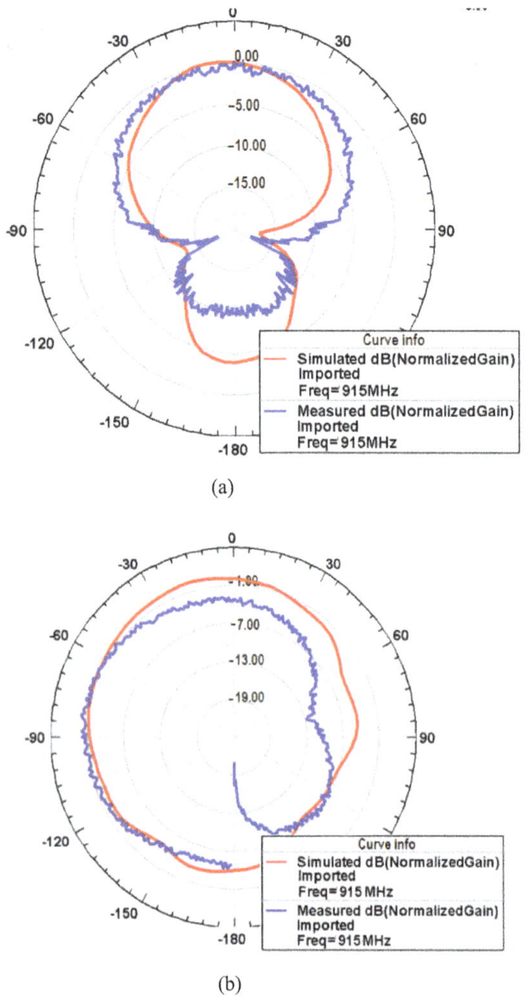

Fig. 10. Simulated and measured normalized radiation pattern at 915 MHz. (a) E-plane and (b) H-plane field pattern.

results show the antenna is able to give minimum read range of at least 2 meter worldwide as shown in Tab. 3.

$$r = \frac{\lambda}{4\pi}\sqrt{\frac{P_t G_t(\theta,\phi) G_r(\theta,\phi) p\tau}{P_{th}}}. \tag{9}$$

λ is the free-space wavelength of the operating frequency, P_t represents the reader's transmitted power, G_t is the gain of the reader's transmitting antenna, G_r is the gain of the receiving tag antenna, p accounts for the polarization mismatch between the antenna, P_{th} is the sensitivity of microchip and τ is the power transmission coefficient given by

$$\tau = 1 - |\Gamma|^2. \tag{10}$$

4. Conclusion

A new wideband microstrip antenna for tagging metallic objects is presented in this paper. The designed tag antenna demonstrates simulated and measured impedance bandwidth of 113 MHz and 117 MHz based on 6 dB return loss. Moreover, the antenna exhibits planar configuration without any multi or cross-layered configuration which significantly reduces the fabrication cost especially for mass production. For future work, the antenna will be integrated with the referenced microchip to experimentally measure the read range of the antenna and comparison will be made with the theoretical results.

Acknowledgements

This research has been supported by the Ministry of Higher Education of Malaysia through the Fundamental Research Grant Scheme FRGS13-027-0268.

References

[1] DOBKIN, D. M. *The RF in RFID : Passive UHF RFID in Practice*. Massachusetts: Elsevier Inc., 2008.

[2] MARROCCO, G. The art of UHF RFID antenna design: impedance-matching and size-reduction techniques. *IEEE Antennas and Propagation Magazine,* 2008, vol. 50, no. 1, p. 66 to 79. DOI: 10.1109/MAP.2008.4494504

[3] RAO, K. V. S., et al. Antenna design for UHF RFID tags: A review and a practical application. *IEEE Transactions on Antennas and Propagation*, 2005, vol. 53, no. 12, p. 3870–3876. DOI: 10.1109/TAP.2005.859919

[4] YANG, B., FENG, Q. A folded dipole antenna for RFID tag. In *Proceedings of International Conference on Microwave and Millimeter Wave Technology*. Nanjing (China), 2008, p. 1047–1049. DOI: 10.1109/ICMMT.2008.4540601

[5] CHOI, Y., et al. Design of modified folded dipole antenna for UHF RFID tag. *Electronics Letters*, 2009, vol. 45, p. 387–389. DOI: 10.1049/el.2009.0198

[6] MONTI, G., et al. Broad-band dipole for RFID applications. *Progress In Electromagnetics Research C*, 2010, vol. 12, p. 163 to 172. doi:10.2528/PIERC10012606

[7] PROTHRO, J. T., et al. The effects of a metal ground plane on RFID tag antennas. In *Proceedings of the IEEE Antennas and Propagation Society International Symposium*, 2006. DOI: 10.1109/APS.2006.1711302

[8] GHANNAY, N., et al. Effects of metal plate to RFID tag antenna parameters. In *Proceedings of the Mediterrannean Microwave Symposium 2009*. Tangiers (Morrocco), 2009, p. 1–3. DOI: 10.1109/MMS.2009.5409801

[9] TASHI, T., et al. A complete planner design of microstrip patch antenna for a passive UHF RFID tag. In *Proceedings of the 17th International Conference on Automation and ComputingICAC 2011*. Huddersfield (UK), 2011, p. 12–17.

[10] SON, H.-W., et al. Design of wideband RFID tag antenna for metallic surfaces. *Electronics Letters*, 2006, vol. 42, no. 5, p. 263 to 265.DOI: 10.1049/el:20064323

[11] CHOI, W., et al. An RFID tag using a planar inverted-F antenna capable of being stuck to metallic objects. *ETRI Journal*, 2006, vol. 20, p. 216–218. DOI: DOI: 10.4218/etrij.06.0205.0082

[12] SON, H.-W., JEONG, S.-H. Wideband RFID tag antenna for metallic surfaces using proximity-coupled feed. *IEEE Antennas and Wireless Propagation Letters*, 2011, vol. 10, p. 377–380. DOI: 10.1109/LAWP.2011.2148151

[13] MO, L., QIN, C. Tunable compact UHFRFID metal tag based on CPW open stub feed PIFA antenna. *International Journal of Antennas and Propagation*, 2012, 8 p. DOI: 10.1155/2012/167658

[14] Regulation, U. *Regulatory status for using RFID in the UHF spectrum*, Cited 28-09-2012. Available: http://www.gs1.org/docs /epcglobal/UHF_Regulations.pdf

[15] MO, L., et al. Broadband UHF RFID tag antenna with a pair of U slots mountable on metallic objects. *Electronics Letters*, 2008, vol. 44, p. 1173–1174. DOI: 10.1049/el:20089813

[16] HUANG, J. Z., et al. A compact broadband patch antenna for UHF RFID tags. In *Proceedings of the Asia Pacific Microwave Conference APMC2009*. Singapore, 2009, p. 1044–1047. DOI: 10.1109/APMC.2009.5384364

[17] LAI, M., et al. Low-profile broadband RFID tag antennas mountable on metallic objects. In *Proceedings of the IEEE Antennas and Propagation Society International Symposium*. Toronto (Canada), 2010, 4 p. DOI: 10.1109/APS.2010.5561167

[18] TAN, L. R., WU, R. X. Miniaturized broadband tag antenna for multi-standard UHF RFID applications. In *Proceedings of the IEEE International Conference on Microwave Technology and Computational Electromagnetics*. Beijing (China), 2011, p. 274 to 276. DOI: 10.1109/ICMTCE.2011.5915510

[19] LU, J.-H., ZHENG, G.-T. Planar broadband tag antenna mounted on the metallic material for UHF RFID system. *IEEE Antennas and Propagation Magazine*, 2011, vol. 10, p. 1405–1408. DOI: 10.1109/LAWP.2011.217899

[20] EUNNI, M. S. M., DEAVOURS, D. D. A novel planar microstrip antenna design for UHF RFID. *Journal of Systemics, Cybernetics and Informatics*, 2007, vol. 5, no. 1, p. 6–10.

[21] TASHI, et al. Design and simulation of UHF RFID tag antennas and performance evaluation in presence of a metallic surface. In *Proceedings of the 5th International Conference on Software, Knowledge Information, Industrial Management and Applications*. Benevento (Italy), 2011, 5 p. DOI: 10.1109/SKIMA.2011.6089974

[22] CHO, H.-G., et al. Design of an embedded-feed type microstrip patch antenna for UHF radio frequency identification tag on metallic objects. *IET Microwaves, Antennas & Propagation*, 2010, vol. 4, p. 1232–1239. DOI:10.1049/iet-map.2009.040

[23] MO, L., QIN, C. Planar UHF RFID tag antenna with open stub feed for metallic objects. *IEEE Transactions on Antennas and Propagation*, 2010, vol. 58, no. 9, p. 3037–3043. DOI: 10.1109/TAP.2010.2052570

[24] Alien. *Alien Higgs 3 EPC Class 1 Gen 2 RFID Tag IC*. Cited 20-10-2012. Available: http://www.alientechnology.com/docs/ products/Alien-Technology-Higgs-3-ALC-360.pdf

[25] LOO, C.-H., et al. Chip impedance matching for UHF RFID tag antenna design. *Progress In Electromagnetics Research*, 2008, vol. 81, p. 359–370. DOI:10.2528/PIER08011804

[26] BIRD, T. S. Definition and misuse of return loss. *IEEE Antennas and Propagation Magazine*, 2009, vol. 51, p. 166–167.

[27] SON, H.-W., PYO, C.-S. Design of RFID tag antennas using an inductively coupled feed. *Electronics Letters*, 2005, vol. 41, no. 18, p. 994–996. DOI: 10.1049/el:20051536

[28] GROVER, F. W. *Inductance Calculations: Working Formulas and Tables*. New York: D. Van Nostrand, 1946.

[29] Balanis, C. A. *Antenna Theory Analysis and Design*. 3rd ed. New Jersey: John Wiley & Sons, 2005.

[30] KUMAR, G., RAY, K. P. *Broadband Microstrip Antenna*. Noorwood, MA: Artech House, 2003.

[31] KUO, S.-K., et al. An accurate method for impedance measurement of RFID tag antenna. *Progress In Electromagnetics Research*, 2008, vol. 83, p. 93–106. DOI:10.2528/PIER08042104

[32] QING, X. et al. Impedance characterization of RFID tag antennas and application in tag co-design. *IEEE Transactions on Microwave Theory and Techniques*, 2009, vol. 57, p. 1268–1274. DOI: 10.1109/TMTT.2009.2017288

[33] XU, L., et al. UHF RFID tag antenna with broadband characteristic. *Electronics Letters*, 2008, vol. 44, no. 2, p. 79–80. DOI: 10.1049/el:20083009

[34] LU, J.-H., HUNG, K.-T. Planar inverted-E antenna for UHF RFID tag onmetallic objects with bandwidth enhancement. *Electronics Letters*, 2010, vol. 46, no. 17, p. 1182–1183. DOI: 10.1049/el.2010.0817

About the Authors ...

Mohd Saiful Riza BASHRI was born in Selangor, Malaysia, in May 1985. He received the B.Sc. degrees in Communication Engineering from International Islamic University Malaysia (IIUM) in 2009. He worked for Telekom Malaysia (TM) from 2009 to 2011 as an assistant manager which oversaw the planning of submarine cable network in the South East Asia region. Currently he is pursuing his master degree at IIUM. His research interests include antenna design, RFID and microwave devices.

Muhammad Ibn IBRAHIMY was born in Sirajgonj, Bangladesh, in August 1962. He received the B.Sc. and M.Sc. degrees in Applied Physics and Electronics from the University of Rajshahi, Bangladesh, in 1985 and 1986, respectively, and the Ph.D. degree in Biomedical Signal Processing from the National University of Malaysia, in 2001. From 2001 to 2003, he was a Postdoctoral Fellow in the Dept. of Electrical and Electronic Engineering at the Mie University of Japan. He is now Associate Professor in the Dept. of Electrical and Computer Engineering, IIUM. Dr. Ibrahimy is a senior member of the Inst. of Electrical and Electronics Engineers (IEEE), a member of the Bangladesh Computer Society (BCS) and the Bangladesh Electronics Society (BEC). His research interests are Analog and Digital Electronic System Design, Medical and Industrial Instrumentation, Biomedical Signal Processing, VLSI Design, RFID and Computer Networks in Telemedicine. He has published about 50 research articles in peer reviewed journals, conferences and 2 books.

S. M. A. MOTAKABBER was born in Naogaon, Bangladesh, in May 1966. He received the B.Sc. and M.Sc. degrees in Applied Physics and Electronics from the University of Rajshahi, Bangladesh, in 1986 and 1987, respectively, and the Ph.D. degree in Electrical, Electronic and Systems Engineering from the National University of Malaysia, in 2011. From 1993 to 2011, he was an Associate Professor in the Dept. of Applied Physics and Electronic Engineering, University of Rajshahi, Bangladesh. He is now an Assistant Professor in the Dept. of Electrical and Computer Engineering, IIUM. Dr. Motakabber is a member of the Inst. of Electrical and Electronics Engineers (IEEE), a life member of the Bangladesh Association for Advancement of Science (BAAS) and the Bangladesh Electronics Society (BEC). His research interests are Analog and Digital Electronic System Design, Medical and Industrial Instrumentation, VLSI Design, RFID, Robotics, Automation and Computer Control Systems. He has published about 20 research articles in peer reviewed journals, conferences and 1 book.

Permissions

All chapters in this book were first published in Radioengineering, by Společnost pro radioelektronické inženýrství; hereby published with permission under the Creative Commons Attribution License or equivalent. Every chapter published in this book has been scrutinized by our experts. Their significance has been extensively debated. The topics covered herein carry significant findings which will fuel the growth of the discipline. They may even be implemented as practical applications or may be referred to as a beginning point for another development.

The contributors of this book come from diverse backgrounds, making this book a truly international effort. This book will bring forth new frontiers with its revolutionizing research information and detailed analysis of the nascent developments around the world.

We would like to thank all the contributing authors for lending their expertise to make the book truly unique. They have played a crucial role in the development of this book. Without their invaluable contributions this book wouldn't have been possible. They have made vital efforts to compile up to date information on the varied aspects of this subject to make this book a valuable addition to the collection of many professionals and students.

This book was conceptualized with the vision of imparting up-to-date information and advanced data in this field. To ensure the same, a matchless editorial board was set up. Every individual on the board went through rigorous rounds of assessment to prove their worth. After which they invested a large part of their time researching and compiling the most relevant data for our readers.

The editorial board has been involved in producing this book since its inception. They have spent rigorous hours researching and exploring the diverse topics which have resulted in the successful publishing of this book. They have passed on their knowledge of decades through this book. To expedite this challenging task, the publisher supported the team at every step. A small team of assistant editors was also appointed to further simplify the editing procedure and attain best results for the readers.

Apart from the editorial board, the designing team has also invested a significant amount of their time in understanding the subject and creating the most relevant covers. They scrutinized every image to scout for the most suitable representation of the subject and create an appropriate cover for the book.

The publishing team has been an ardent support to the editorial, designing and production team. Their endless efforts to recruit the best for this project, has resulted in the accomplishment of this book. They are a veteran in the field of academics and their pool of knowledge is as vast as their experience in printing. Their expertise and guidance has proved useful at every step. Their uncompromising quality standards have made this book an exceptional effort. Their encouragement from time to time has been an inspiration for everyone.

The publisher and the editorial board hope that this book will prove to be a valuable piece of knowledge for researchers, students, practitioners and scholars across the globe.

List of Contributors

Stanislav Zvanovec and Petr Chvojka
Dept. of Electromagnetic Field, Czech Technical University in Prague, Technicka 2, 166 27 Prague, Czech Republic

Paul Anthony Haigh
Faculty of Engineering, University of Bristol, Bristol, BS8 1TR, UK

Zabih Ghassemlooy
Optical Communications Research Group, Faculty of Engineering and Environment, Northumbria University, Newcastle-upon-Tyne NE1 8ST, UK

Muhammad Zaka ur Rehman
Dept. of Electrical and Electronics Engineering, Universiti Teknologi Petronas, Tronoh, Perak, Malaysia
Dept. of Physics, COMSATS Institute of Information Technology, Park Road, Islamabad, Pakistan

Zuhairi Baharudin, Mohd Azman Zakariya, Mohd Haris Md. Khir and Muhammad Taha Jilani
Dept. of Electrical and Electronics Engineering, Universiti Teknologi Petronas, Tronoh, Perak, Malaysia

Paiboon Yoiyod and Monai Krairiksh
Faculty of Engineering, King Mongkut's Institute of Technology Ladkrabang, Bangkok 10520, Thailand

Sabeena Fatima, S. Sheikh Muhammad and A. D. Raza
Dept. of Electrical Engineering, National University of Computer and Emerging Sciences, Lahore, Pakistan

Mohsen Hayati, Ashkan Abdipour and Arash Abdipour
Electrical Engineering Dept., Faculty of Engineering, Razi University, Tagh-E-Bostan, Kermanshah-67149, Iran

Siming Peng, Zhigang Yuan, Yuehong Shen and Wei Jian
Dept. of Wireless Communications, PLA University of Science and Technology, 210014 Nanjing, China

Jun You
Dept. of Command Information System, PLA University of Science and Technology, 210014 Nanjing, China

Cong-Hui Qi and Zhi-Qin Zhao
School of Electronic Engineering, University of Electronic Science and Technology of China, Xiyuan Ave 2006, Chengdu, Sichuan, China

Hassan Elkamchouchi and Mohamed Hassan
Dept. of Electrical Engineering, University of Alexandria, Alhuria Street, Alexandria, Egypt

Ahmet Oncu
Dept. of Electrical and Electronics Engineering, Bogazici University, Istanbul, 34342, Turkey

Václav Paňko, Stanislav Banáš and Jan Divín
Dept. of Radio Engineering, Czech Technical University in Prague, Technická 2, 166 27 Praha 6, Czech Republic
ON Semiconductor, SCG Czech Design Center, 1. maje 2594, 75661 Roznov p. R., Czech Republic

Richard Burton
ON Semiconductor, 5005 East McDowell Road, Phoenix, AZ 85008, USA

Karel Ptáček
Dept. of Microelectronics, Brno University of Technology, Technicka 3058/10, 61600 Brno, Czech Republic
ON Semiconductor, SCG Czech Design Center, 1. maje 2594, 75661 Roznov p. R., Czech Republic

Josef Dobeš
Dept. of Radio Engineering, Czech Technical University in Prague, Technická 2, 166 27 Praha 6, Czech Republic

Xue Li
Inst. for Pattern Recognition and Artificial Intelligence, Huazhong University of Science and Technology, Wuhan 430074, China
Defense Forces Academy, Zhengzhou 450052, China

Jinwen Tian
Inst. for Pattern Recognition and Artificial Intelligence, Huazhong University of Science and Technology, Wuhan 430074, China

Josef Slezak and Tomas Gotthans
Dept. of Radio Electronics, Brno University of Technology, Technická 12, 616 00 Brno, Czech Republic

Gollakota Venkata Krishna Sharma
Dept. of ECE, GITAM Institute of Technology, GITAM University, India

Konduri Raja Rajeswari
Dept. of ECE, College of Engineering Autonomous, Andhra University, India

Branimir Jaksic, Mihajlo Stefanovic and Vladeta Milenkovic
Faculty of Electrical Engineering, University of Nis, Aleksandra Medvedeva 14, 18000 Nis, Serbia

Dusan Stefanovic
College of Applied Technical Sciences, Aleksandra Medvedeva 20, 18000 Nis, Serbia

Petar Spalevic
Faculty of Technical Sciences, University of Pristina, Knjaza Milosa 7, 38220 Kosovska Mitrovica, Serbia

Bojan Dimitrijevic, Bojana Nikolic, Slavoljub Aleksic and Nebojsa Raicevic
Faculty of Electronic Engineering, University of Nis, A. Medvedeva 14, 18 000 Nis, Serbia

M. Tahir Mushtaq and Otto Koudelka
Inst. of Communications Networks and Satellite Communications, TU Graz, Austria

Inayatullah Khan
Centers of Excellence in Science and Applied Technologies, Islamabad

M. S. Khan
Dept. of Electrical Engineering, University of Gujrat, Gujrat, Pakistan

Nafiseh Khajavi
Dept. of Electrical Engineering, Dezful Branch, Islamic Azad University, Dezful, Iran

Seyed Vahab Al-Din Makki
Dept. of Electrical Engineering, Razi University, Kermanshah, Iran

Sohrab Majidifar
Dept. of Electrical Engineering, Kermanshah University of Technology, Kermanshah, Iran

Asuman Savascihabes and Ozgur Ertug
Telecommunications and Signal Processing Laboratory, Dept. of Electrical and Electronics
Engineering, University of Gazi, Ankara, Turkey

Erdem Yazgan
Dept. of Electrical and Electronics Engineering, University of Hacettepe, Ankara, Turkey

Soobum Cho
Dept. of Electrical Engineering, Stanford University, Stanford, CA 94305, USA

Sang Kyu Park
Dept. of Electronics and Computer Engineering, Hanyang University, Seoul 133-791, South Korea

Mohamed Nasrun Osman, Mohamad Kamal A. Rahim, Mohamad Rijal Hamid, Mohd Fairus Mohd Yusoff and Huda A. Majid
Communication Engineering Dept., Faculty of Electrical Engineering, Universiti Teknologi Malaysia, 81310 Skudai, Johor, Malaysia

Peter Gardner
School of Electronic, Electrical and Computer Engineering, University of Birmingham, Edgbaston, Birmingham, B15 2TT, United Kingdom

Chenguang Shi, Fei Wang, Jianjiang Zhou and Jun Chen
Key Laboratory of Radar Imaging and Microwave Photonics, Ministry of Education, Nanjing University of Aeronautics and Astronautics, No. 29, Yudao Street, Qinhuai District, Nanjing city, Jiangsu Province, Nanjing 210016, P. R. China

Maciej Czyżak, Jacek Horiszny and Robert Smyk
Faculty of Electrical and Control Engineering, Gdansk University of Technology, G Narutowicza 11/12, 80-233

Tao Zhang, Weiwei Yang and Yueming Cai
College of Communications Engineering, PLA University of Science and Technology, Yudao Street, Nanjing City, Jiangsu Province, 210007, China

Kaleeswaran Rajeswari
Dept. of Electronics and Communication Engg., Thiagarajar College of Engineering, 625 015 Madurai, TamilNadu, India

S. Jayaraman Thiruvengadam
Dept. of Electronics and Communication Engg., Thiagarajar College of Engineering, 625 015 Madurai, TamilNadu, India
TIFAC CORE in Wireless Technologies, Thiagarajar College of Engineering, 625 015 Madurai, TamilNadu, India

Hui Chen, Qun Wan, Rong Fan and Fei Wen
School of Communication & Information Engineering, University of Electronic Science and Technology of China, Chengdu 611731, China

Ran Guo, Xing-Peng Mao, Shao-Bin Li and Xiu-Hong Wang
School of Electronic and Information Engineering, Harbin Institute of Technology, Harbin, Heilongjiang Province, China

Yi-Ming Wang
The First Institute of Oceanography, SOA, Qingdao, Shandong Province, China

Milan Kvicera and Pavel Pechac
Faculty of Electrical Engineering, Czech Technical University in Prague, Technicka 2, 166 27 Prague, Czech Republic

Aydin Kizilkaya, Adem Ukte and Mehmet Dogan Elbi
Dept. of Electrical and Electronics Engineering, University of Pamukkale, 20070 Denizli, Turkey

Milan R. Dincic and Zoran H. Peric
Faculty of Electronic Engineering, University of Nis, Aleksandra Medvedeva 14, 18000 Nis, Serbia

Manuel Cedillo-Hernandez, Antonio Cedillo-Hernandez and Francisco Garcia-Ugalde
Electric Engineering Division, Engineering Faculty, National Autonomous University of Mexico, Circuito Exterior, Ciudad Universitaria, Coyoacan 04510, Mexico City, Mexico

Mariko Nakano-Miyatake and Hector Perez-Meana
Postgraduate Section, Mechanical Electrical Engineering School, National Polytechnic Institute of Mexico, 1000 Santa Ana Avenue, San Francisco Culhuacan, Coyoacan 04430, Mexico City, Mexico

Filip Zaplata and Miroslav Kasal
Dept. of Radio Electronics, Brno University of Technology, Technická 12, 612 00 Brno, Czech Republic

Mohd Saiful Riza Bashri, Muhammad Ibn Ibrahimy and S. M. A Motakabber
Dept. of Electrical and Computer Engineering, International Islamic University Malaysia, 53100 Kuala Lumpur, Malaysia